Photovoltaics for Space

Photovoltaics for Space

KEY ISSUES, MISSIONS, AND ALTERNATIVE TECHNOLOGIES

Edited by

SHEILA G. BAILEY
NASA Glenn Research Center (Retired), Cleveland, OH, United States

ALOYSIUS F. HEPP
Nanotech Innovations LLC, Oberlin, OH, United States

DALE C. FERGUSON
Air Force Research Laboratory, Albuquerque, NM, United States

RYNE P. RAFFAELLE
Rochester Institute of Technology, Rochester, NY, United States

STEVEN M. DURBIN
Western Michigan University, Kalamazoo, MI, United States

ELSEVIER

Elsevier
Radarweg 29, PO Box 211, 1000 AE Amsterdam, Netherlands
The Boulevard, Langford Lane, Kidlington, Oxford OX5 1GB, United Kingdom
50 Hampshire Street, 5th Floor, Cambridge, MA 02139, United States

ISBN: 978-0-12-823300-9

For information on all Elsevier publications visit our website at
https://www.elsevier.com/books-and-journals

Publisher: Matthew Deans
Acquisitions Editor: Chiara Giglio
Editorial Project Manager: Tom Mearns
Production Project Manager: Kamesh Ramajogi
Cover Designer: Matthew Limbert

Typeset by TNQ Technologies

Dedication

This book is dedicated to our mentor, colleague, and friend, Dr. Dennis J. Flood (Fig. D1). The original motivation was to honor his 80th birthday, which he recently celebrated. However, due to pandemic-related delays, we are also approaching the 30th anniversary of the first World Congress on Photovoltaic Energy Conversion (WCPEC) in 1994 at Waikaloa, Hawai'i. Dennis was a key driving force behind this gathering and served as its general conference Chair. The eighth World Congress (WCPEC-8) is being held at the end of September 2022 in Milan, Italy. Since its inception, it has been the world's largest platform for photovoltaic (PV) research and development, enabling a unique platform of international exchange by combining the three largest and most prominent international PV conferences: the European PV Solar Energy Conference (EU PVSEC), the Institute of Electrical and Electronics Engineers Photovoltaic Specialists Conference (IEEE PVSC), and the International PV Science and Engineering Conference (PVSEC). In 2018, Dr. Flood was a recipient of the prestigious World Photovoltaic Energy Award at WCPEC-7. According to the conference citation, the award was made for "his outstanding contribution to the development of science and technology of high efficiency and radiation-tolerant space solar cells, arrays and systems, and for extraordinary contributions to international cooperation in PV technology, bringing our world PV community together with his vision in founding the WCPEC series." A native of Michigan, Dennis received a B.S. in physics from Ohio's Wittenberg University; he subsequently returned home to earn the M.S. and Ph.D. degrees in solid state physics from Michigan State University. He has more than 40 years of experience in developing solar cell and array technology for both space and terrestrial applications. Dr. Flood is an inventor or co-inventor on several patents and patent applications in PV and nanotechnology, and he has more than 100 peer-reviewed publications and presentations in solar energy, electron devices and materials science to his credit.

Figure D.1 Dr. Dennis J. Flood. *Courtesy: Dennis M. Flood*

At the NASA Glenn Research Center in Cleveland, Dr. Flood served for 15 years as Chief of the Photovoltaic and Space Environments Branch, where he led programs in advanced PV systems development. He received two agency awards for his pioneering work on advanced solar cells for space applications and for research that established the feasibility of powering a human outpost on the surface of Mars with solar energy. Dennis served as Chair of the IEEE Electron Device Society's Photovoltaic Device Technical Committee, served as a member of the Society's Education Committee, and participated in its Distinguished Lecturer Series. He presently serves on the International Advisory Committees of the European, the United States, the Japan/Asia, and the World Photovoltaic Conference organizing committees. Dr. Flood along with his son (and current General Manager), Dennis M. Flood, was a co-founder and member of the scientific advisory board of Nanotech Innovations LLC in 2008 in Oberlin, OH. CoreWater Technologies (CWT), a 2020 spin-off of Nanotech Innovations, is a privately held corporation registered in Ohio; it was formed to enter the water filtration market at multiple levels, both in the United States and globally. CWT has developed an adsorbent particle that removes polyfluorinated organics from drinking water. Throughout his long and productive career, however, what Dennis is perhaps best known for is his quiet passion for identifying and supporting the next generation of researchers. His genuine concern for the individual, enthusiasm for PV technology, and extraordinary leadership produced many new contributors to PV. Supporting the growth of the PV community, particularly the space PV sector, is without a doubt one of his greatest legacies.

Finally, we would like to acknowledge our colleague, friend, and originally co-editor of this book, Phillip P. Jenkins (Fig. D2) of the Naval Research Laboratory, who passed away suddenly in December 2019. Born in California, Phil spent his formative years in the suburbs of Cleveland. After graduating with a B.S. in physics from Denison University, he earned a master's degree in electrical engineering from Cleveland State University. He continued his work in space PV at the NASA Glenn Research Center. Phil spent nearly all of his engineering career researching new technologies and applications for aerospace solar cells, contributing experiments to several space flight missions, including NASA's Mars Pathfinder in 1997. Phillip's collaboration with physicists at the US Naval Research Laboratory took him to Washington, DC, in 2006, where he most recently served as the head of the laboratory's PV section. In that role, he applied his experience to projects making terrestrial power systems more efficient, less reliant on fossil fuels, and better adapted to harsh environments. Phil was a highly respected researcher and valued colleague: rigorous, creative, enthusiastic, and productive. We will also remember Phil as a caring parent, devoted husband, earnest public servant, avid cyclist, problem solver, and observer of the world. Phil's lively sense of humor, creative insights, friendship, and mentorship leave those who knew him grieving a life well lived, and cut entirely too short.

Figure D.2 Phillip P. Jenkins (1956–2019). *Source: Public domain*

Contents

PART 3 Near-Earth and deep-space missions

List of contributors

Tariq Rizvi Alam
Nuclear Environments and System Assessments, Applied Research Associates, Santa Barbara, CA, United States

S. Aranya
Department of Physics, University College, Trivandrum, Kerala, India; Department of Physics, Sree Narayana College, Trivandrum, Kerala, India

Carsten Baur
European Space Agency, Noordwijk, The Netherlands

Philip T. Chiu
Spectrolab, Sylmar, CA, United States

Donald L. Chubb
Retired from NASA Glenn Research Center, Cleveland, OH, United States

Kyle Crowley
NASA Glenn Research Center, Cleveland, OH, United States

Alejandro Datas
Instituto de Energía Solar – Universidad Politécnica de Madrid, Madrid, Spain

Brandon K. Durant
Homer L. Dodge Department of Physics and Astronomy, University of Oklahoma, Norman, OK, United States

D.P. Engelhart
Assurance Technology Corp., Carlisle, MA, United States

D.C. Ferguson
Air Force Research Laboratory, Kirtland Air Force Base, Albuquerque, NM, United States

T.E. Girish
Department of Physics, University College, Trivandrum, Kerala, India

José Ramón González
European Space Agency, Noordwijk, The Netherlands

G. Gopkumar
Department of Physics, University College, Trivandrum, Kerala, India

Aloysius F. Hepp
Nanotech Innovations LLC, Oberlin, OH, United States

R.C. Hoffmann
Air Force Research Laboratory, Kirtland Air Force Base, Albuquerque, NM, United States

Varnana M. Kumar
Department of Physics, University College, Trivandrum, Kerala, India

Geoffrey A. Landis
NASA John Glenn Research Center, Photovoltaic & Electrochemical Systems Branch, Cleveland, OH, United States

Emilio Fernández Lisbona
European Space Agency, Noordwijk, The Netherlands

Ina T. Martin
Department of Materials Science & Engineering, Case Western Reserve University, Cleveland, OH, United States

Lyndsey McMillon-Brown
NASA Glenn Research Center, Photovoltaic & Electrochemical Systems Branch, Cleveland, OH, United States

Carolyn R. Mercer
NASA Glenn Research Center, Cleveland, OH, United States

Naoya Miyashita
Research Center for Advanced Science and Technology (RCAST), The University of Tokyo, Tokyo, Japan; Department of Engineering Science, The University of Electro-Communications, Chofu, Japan

V.J. Murray
AFRL Kirtland AFB, Albuquerque, NM, United States

Mark J. O'Neill
Mark J. O'Neill, LLC, Fort Worth, TX, United States

Yoshitaka Okada
Research Center for Advanced Science and Technology (RCAST), The University of Tokyo, Tokyo, Japan

Timothy J. Peshek
NASA Glenn Research Center, Photovoltaic & Electrochemical Systems Branch, Cleveland, OH, United States

E.A. Plis
Assurance Technology Corp., Carlisle, MA, United States

Mark Antonio Prelas
Electrical Engineering, University of Missouri, Columbia, MO, United States

Ryne P. Raffaelle
Rochester Institute of Technology, Rochester, NY, United States

Bibhudutta Rout
Department of Physics, University of North Texas, Denton, TX, United States

Thara N. Sathyan
Department of Physics, University College, Trivandrum, Kerala, India

Ian R. Sellers
Homer L. Dodge Department of Physics and Astronomy, University of Oklahoma, Norman, OK, United States

Modeste Tchakoua Tchouaso
Department of Physics and Astronomomy, Howard University, Washington, DC, United States; Physics, North Carolina A&T State University, Greensboro, NC, United States

Biographies

Sheila G. Bailey

Dr. Sheila Bailey was a senior scientist in the Photovoltaics and Electrochemical Systems Branch at NASA Glenn Research Center, where she has worked from 1985 to 2018. She received her B.S. (Duke) and M.S. (University of North Carolina) degrees in Physics and her Ph.D. in Solid State Physics (University of Manchester, UK). She taught at Baldwin Wallace University for 27 years and has been an associate faculty member of the International Space University Dr. Bailey's photovoltaics work has centered around advanced etching, epitaxial lift off, and quantum dot solar cells. She is co-author of over 150 publications in various aspects of solar cell materials processing and applications for space exploration. She has also authored numerous chapters in books and co-edited several books on this topic. She serves on editorial boards of several PV journals as a space PV expert. Dr. Sheila Bailey received the 2022 Irving Weinberg Award at the 27th SPRAT (Space Photovoltaics R&T) Meeting. The award is presented at every SPRAT conference to an individual who has made significant contributions to the field of space photovoltaics.

Affiliation

Senior Physicist (Retired), NASA Glenn Research Center, Cleveland, OH, USA

Aloysius F. Hepp

Dr. Aloysius Hepp earned a B.S. in Chemistry from Carnegie Mellon University and a Ph.D. in Inorganic Photochemistry from M.I.T. He retired in December 2016 from the Photovoltaics and Electrochemical Systems Branch of the NASA Glenn Research Center. Dr. Hepp has co-authored nearly 200 publications (including six patents) focused on processing of thin film and nanomaterials for I-III-VI solar cells, Li-ion batteries, integrated power devices, and flight experiments, as well as novel precursors and spray pyrolysis methods for deposition of sulfides and carbon nanotubes. He has co-edited 12 books on advanced materials processing, energy conversion and electronics, biomimicry, and aerospace technologies: six for Elsevier over the past year. He is currently Editor-in-Chief (2010–15) Emeritus of *Materials Science in Semiconductor Processing* (MSSP) and Chair of the International Advisory Board of MSSP, as well as serving on the Editorial Advisory Boards of *Mater. Sci.* and *Engin. B* and *Heliyon*, all Elsevier journals.

Affiliation

Chief Technologist, Nanotech Innovations LLC, Oberlin, OH, USA

Dale C. Ferguson

Dr. Dale Ferguson received the Ph.D. degree from The University of Arizona, Tucson, in 1974. He is currently the Lead for Spacecraft Charging Science and Technology with the Space Vehicles Division, Air Force Research Laboratory (AFRL), Kirtland Air Force Base, Albuquerque, NM. For nearly 40 years, he has been addressing spacecraft charging problems, first with the NASA and now with AFRL. He is the author of more than 200 publications. He has been the Principal Investigator on numerous spaceflight experiments, including the Wheel Abrasion Experiment on the Mars Pathfinder Sojourner Rover. Dr. Ferguson has recently been awarded an AFRL Fellowship; this is the highest award that the AFRL confers on Air Force researchers.

Affiliation

Space Vehicles Division, Air Force Research Laboratory (AFRL), Kirtland Air Force Base, Albuquerque, NM, USA

Ryne P. Raffaelle

Prof. Ryne Raffaelle earned both a B.S. and M.S. in Physics from Southern Illinois University and a Ph.D. in Physics from Missouri University of Science and Technology. He is the Vice President for Research and Associate Provost at Rochester Institute of Technology (RIT). He is the former Director of the National Center for Photovoltaics at the National Renewable Energy Lab of the US Department of Energy. Prior to serving at NREL, he was the Academic Director for the Golisano Institute for Sustainability and Director of the NanoPower Research Laboratory at RIT. He has worked as a visiting scientist at the NASA Glenn Research Center, NASA Kennedy Research Center, and DOE's Oak Ridge National Laboratory. He is the author of over 200 refereed publications. He is on the Advisory Board of Elsevier's *Materials Science in Semiconductor Processing* and is the Managing Editor of *Progress in Photovoltaics*, published by Wiley Interscience. He is the co-editor of several books on photovoltaics and nanotechnologies.

Affiliation

Vice President and Associate Provost for Research, Rochester Institute of Technology, Rochester, NY, USA

Steven M. Durbin

Prof. Steven Durbin received the B.S., M.S., and Ph.D. degrees in Electrical Engineering from Purdue University Prior to joining Western Michigan University in 2013, he taught at the Florida State University, the University of Canterbury (New Zealand), and the University at Buffalo (SUNY). He is a senior member of the IEEE, and a member of the American Physical Society, the Materials Research Society, and the Royal Society of New Zealand. His interests include novel semiconductors, oxide and nitride compounds, molecular beam epitaxy, pulsed laser deposition, and Schottky contact-based devices and have resulted in over 150 publications. Prof. Durbin is Editor-in-Chief Emeritus (2016–21) of *Materials Science in Semiconductor Processing*.

Affiliation

Department of Electrical and Computer Engineering, Western Michigan University, Kalamazoo, MI, USA

Preface

The history of space photovoltaic(s) (PV) is in many ways the history of PV. Although the early development of the PV solar cell, or "solar battery" as it was called by the inventors at Bell Labs, was driven by visions of numerous potential terrestrial uses for the new source of electrical power back in 1954, it was, in fact, the development of space solar power systems that drove most of the early development of PV solar cells and arrays. As part of the response to the former Soviet Union's launching of the Sputnik satellites, the United States launched the first solar-powered satellite, Vanguard 1, on March 17, 1958. It was the fourth satellite to be successfully launched, after the Sputnik 1 and 2 (launched October 4th, 1957, and November 3rd, 1957, respectively) and the US Explorer 1 (January 31, 1958). Vanguard 1 was not only the first solar-powered satellite but also holds the distinction of being the oldest man-made object in orbit around Earth. The solar power industry today owes a tremendous debt of gratitude to the space power scientists and engineers, without whom much of the technology being used in terrestrial power systems might still not exist. The co-editors of *Space Photovoltaics: Materials, Missions and Alternative Technologies* have selected and organized 17 chapters authored by experts in the fields of space power, electronic materials, PV, and aerospace systems. The book is divided into three sections, each devoted to a different aspect of space PV: an introductory section with representative key technologies background, materials-focused content, and space exploration- and/or mission-oriented chapters. This book is dedicated to our colleague, mentor, and friend, Dr. Dennis J. Flood, distinguished and well-respected member of the space PV community for nearly 50 years. We also honor our collaborator and friend, Phillip P. Jenkins, originally a key member of the editorial team. Sadly, Phil, a highly regarded, enthusiastic researcher in his own right, passed away in late 2019 as this project commenced. The reader is directed to the Dedication at the front of the book.

The initial section of this book is an introduction to space PV with some historical background to the technologies, issues, and applications of space PV, and it includes six chapters. Chapter 1, contributed by co-editor Prof. Raffaelle of the Rochester Institute of Technology, begins with a short account of the early days of PV solar cell development. The chapter is organized chronologically, correlating with missions and cell (and array) technologies. After early satellite missions utilizing silicon (Si) solar cells, various applications and the challenges that arose during the Apollo and Skylab era of the 1960s and 1970s are discussed. This is followed by the 1980s and 1990s that saw great progress in the evolution of space solar power with the launch of the Space Shuttle, Hubble Space Telescope, operation of Mir and the International Space Station, and the introduction of gallium arsenide (III-V) technology. The chapter concludes with the description of some 21st century technologies and missions, such as the exploration of Mars and advanced cell technologies, such as concentrators and even thin-film III-V multijunction (MJ) technology. Further details on both the advanced cell designs and the use of solar power in space exploration missions are addressed in later chapters. Chapter 2, authored by a team led by a co-editor, Dr. Dale Ferguson, along with his colleagues from the US Air Force Research Laboratory (Kirtland Air Force Base, NM) and Assurance Technology Corp. (Carlisle, MA), explores the need for space solar arrays to survive in the hostile space environment. In this chapter, the authors discuss why solar arrays are often the spacecraft components most likely to arc and how this is related to electrical charging of the spacecraft. The basic charging equations are given, and the factors involved in charging and arcing are enumerated. This involves how charging is related to the space plasma environment and what are the different types of charging that are the associated mechanisms of arcing on solar arrays, such as transient arcs and sustained arcs. Standards related to charging and arcing are listed and described, and mitigation strategies (both passive and active) are surveyed. Because charging and arcing are driven by the space environment, models of the space plasma environment and charging models are listed and described. Primary and secondary calibration as well as electrical performance measurements of space solar cells are discussed in Chapter 3, contributed by Dr. Carsten Baur and his colleagues from the European Space Agency (Noordwijk, the Netherlands). The distinction between high-altitude and synthetic calibration methods for primary calibrations is explained. The synthetic calibration technique based on the differential spectral responsivity measurement clearly benefits from the update of the air mass zero (AM0) spectrum with respect to the relative spectral distribution. Special attention is given to the spectral responsivity measurement where measurement artifacts must be addressed, and the adjustment procedures for solar simulators when measuring MJ cells are introduced. Finally, results from recent round robin testing are presented that demonstrate the impressive capability of calibration laboratories to perform highly accurate measurements of MJ solar cells. In Chapter 4, space applications of III-V single-junction (SJ) and MJ solar cells are outlined by Dr. Philip Chiu of Spectrolab, Inc., of Sylmar, CA. The chapter primarily establishes a solid-state physics formalism necessary to model the performance of both SJ and MJ space solar cells both at beginning of life (BOL) and at end of life (EOL). The chapter then proceeds to utilize that model to predict the BOL and EOL performance of III-V-based technologies ranging from SJ GaAs cells, to lattice-matched triple-junction (3J) structures, all the way up to complex 5J lattice-matched structures. Each MJ cell technology is evaluated for its viability in space-based applications based upon two key metrics: the EOL efficiency and EOL $/W. Chapter 5 is the first of two chapters that focus on perovskite solar cells (PSCs) and provides a background and prospects for space power applications; it is co-authored by Drs. Lyndsey McMillon-Brown and Timothy J. Peshek of the NASA

Glenn Research Center (GRC), Photovoltaics & Electrochemical Systems Branch, Cleveland, Ohio 44135, USA. Traditionally space solar cells are high-cost, high-efficiency, high-fidelity III-V- or Si-based devices. In this chapter, the authors present an of overview of a variety of solar cells with potential to perform in niche aerospace applications at lower costs without sacrificing performance or power. A review of recent advances in the PV light harvesting efficiency of PSCs is presented. It is further demonstrated that PSCs present numerous unique characteristics that make them attractive for space applications. Finally, the authors outline opportunities for further advances in this materials technology by addressing current materials and device challenges, thereby enabling further potential aerospace applications. Chapter 6, the final chapter of the introductory section introduces the alternative technology of a local nuclear energy source (versus the Sun) to generate electricity. This is the first of three chapters co-authored by Prof. Mark Prelas (University of MO S&T) and his colleagues, Prof. T.R. Alam (ARA, Fort Belvoir, VA) and Dr. M.T. Tchouaso (Howard University), with a rotating primary authorship. This chapter provides a background on nuclear energy conversion and PV for space applications. Devices based on betavoltaics, which harvest the energy of beta emitters, alphavoltaics, which harvest the energy of alpha emitters, and photon intermediate direct energy conversion, which use photons produced through nuclear interactions with matter to power PV, are presented. These devices and other potential manifestations of direct energy conversion systems based on nuclear power in concert with *p-n* junction devices are discussed along with the key issues associated with their use for space power.

The five chapters in the second part of the book have a materials focus, addressing new technologies and advanced processing in the context of space applications. Chapter 7, co-authored by Prof. Ian Sellers and Dr. Brandon Durant of the University of Oklahoma's Physics Department and Prof. B. Rout of the University of North Texas, is the second chapter on the topic of PSCs, with an increased emphasis on materials science in the context of radiation damage resistance. Mixed organic-inorganic PSCs are not only leading contenders to disrupt terrestrial PV but also the proliferating space PV markets. This is due to their rapid improvement in conversion efficiencies, potential for lower cost manufacturing, and the potential for large-scale flexible architectures. The chapter presents the current status of research into the use of these devices for applications in space power, as well as their unique properties and defect tolerance, which should prove useful for the harsh conditions of space, especially in regard to the exposure of high-energy protons and electrons. The behavior of several different kinds of perovskites that are being assessed by groups around the world is summarized. Some of the unique properties outlined earlier suggest the potential use of perovskites in space, where the radiation environment is particularly harsh, especially if their stability can improve. This critical issue is discussed in terms of future avenues for advanced encapsulation and device architectures developed specifically for the rigorous requirements for space applications.

Chapter 8, co-authored by Prof. Alejandro Datas (Universidad Politécnica de Madrid, Spain) and Dr. Donald L. Chubb (NASA GRC, retired), explores thermophotovoltaic (TPV) energy conversion in space. For space applications, TPV (direct conversion of radiant energy from heat into electricity through using PV) has been primarily investigated for the development of radioisotope power systems that could be used in missions where the solar resource is too weak or intermittent, such as deep space or planetary settlements. The main advantages of TPV are its high efficiency, the lack of moving parts, and that there is the potential that they could be very long lived. The main drawbacks are unproven reliability and the requirement of low rejection temperatures, which necessitate the use of large heat radiators. This chapter reviews the state of the art of TPV, along with other technologies for converting thermal to electrical energy that are being developed within the framework of space power research programs. Chapter 9 addresses legacy (non-PSC or organic-based) thin-film materials (thin Si, CIGS, and CdTe) for space PV; it is co-authored by Dr. Ina Martin of Case Western Reserve University, Dr. Kyle Crowley of NASA GRC, and co-editor Dr. A.F. Hepp, retired from NASA GRC and Chief Technologist of Nanotech Innovations LLC. Studies of inorganic thin-film solar cell materials have yielded several potential space applications as well as numerous scientific and technologic insights over the past 50 years. A short discussion of the difference between space (air mass zero, AM0) and terrestrial (AM1.0 or AM1.5) solar radiation begins the chapter. Next, the unique stressors in the space environment, varying with location, encountered by PV devices are summarized. After contrasting crystalline and TFSCs, key components of TFSC devices of the main inorganic absorber materials are then considered, in context of overcoming environmental challenges. Issues such as mass specific power and impact of processing, as well as testing relevant to use of TFSCs for space missions are examined. A brief summary of past, present, and future applications for space exploration concludes the chapter. Finally, examples of technology transfer from studies of these material systems to other PV systems and devices are highlighted. Chapter 10 is authored by Profs. Yoshitaka Okada and Naoya Miyashita of the University of Tokyo. The chapter introduces recent results on the development of a dilute nitride alloy, gallium indium nitride arsenide antimonide (GaInNAsSb), as a 1.0 eV absorber material in III-V-based multijunction solar cells (MJSCs). Main topics addressed include improvement of the GaInNAsSb material quality and solar cell properties, attempts of a hybrid growth by a combination of metalorganic chemical vapor deposition and molecular beam epitaxy for developing MJSCs, and epitaxial lifted-off lattice-matched GaInP/GaAs/GaInNAsSb triple-junction solar cells. Relevant literature concerning use of III-V MJSC technology for space applications is briefly discussed throughout the chapter. Chapter 11, the last chapter of the materials section, continues the exploration of the utilization of nuclear energy sources to generate electricity via betavoltaics, presented from a materials perspective. This is the second chapter co-authored by Prof. Mark Prelas (University of MO S&T) and his colleagues,

Dr. T.R. Alam (ARA, Fort Belvoir, VA) and Dr. M.T. Tchouaso (Howard University), with Dr. Alam as the primary author. This chapter provides a background on nuclear energy conversion and PV for space applications. Devices based on betavoltaics, introduced in Chapter 6, show how the harvesting of the energy from beta emitters could be used in space. The essential elements, a beta emitting radioisotope and a semiconductor-based converter, are described. This device functions essentially as a battery whose design requires the concurrent optimization of the radioisotope selection and the semiconductor materials. The optimization of semiconductor materials parameters (i.e., as doping concentrations, minority carrier diffusion lengths, width of the depletion region, surface geometry, thickness, resistance, temperature coefficients, and effective surface area) is done in an analogous fashion to that of a PV used to convert light. This chapter provides a critical review of the literature, summarizes the key design and operational principles, and gives an original analysis of the end-to-end design of betavoltaic batteries including electron transport and semiconductor charge collection. Recent advancements in betavoltaic development for small-scale space applications is presented.

The third and final section consists of six chapters covering near-Earth and deep space missions: past, present, and future. In Chapter 12, Dr. Carolyn Mercer from NASA GRC provides an overview of solar power systems that have enabled planetary science missions to much of the Solar System. The chapter outlines the types of solar array structures that have been used on US-led robotic planetary science missions since the beginning of the space age, including several in development for future launch. The first solar-powered spacecraft employed body-mounted solar cells, but designs quickly moved to extended panels to generate more electrical power. Early "paddle" designs were replaced by deployable twin rectangular "wings," which are still common in science missions and nearly ubiquitous in commercial telecommunications satellites. The increasing complexity of science missions investigating the nature of the space environment drove the development of the solar-powered satellites from the very beginning, and unique mission requirements to explore ever more distant and harsher regions of our solar system continue to drive innovations in spacecraft design, including solar arrays. This general progression to ever larger solar arrays to power ever more sophisticated spacecraft in ever harsher environments, punctuated by missions that emphasize lower cost and complexity, is a pattern that began in the 1960s and continues to this day. This chapter describes the solar array configurations used on deep space missions, with an emphasis on NASA's planetary science missions, including specific design requirements that led to individual design selections. Future trends and missions are also described. Chapter 13 discusses lunar science based on Apollo solar cell measurements, authored by S. Aranya, G. Gopkumar, and T.E. Girish, Department of Physics, University College, Trivandrum, India. Apollo solar cell observations made 5 decades ago are still a valuable resource for understanding lunar science. Detailed output variations of three different Apollo 14 Si solar cells observed during the years 1971–76 have been studied. The amplitude of annual variations of solar radiation near the lunar equator is inferred to be >100 Wm^{-2}. The number of days of observations of the output of Apollo solar cells during a lunar day or lunation period is shown to vary between 12 and 15 days, which may indicate the variability of the length of the lunar daytime or the number of sunlit days per lunation period near the Moon's equator. The association between long-term degradation of Apollo 14 solar cells and occurrences of solar proton events is discussed. The observations of slight variations in the lunation-to-lunation change in the dawn voltage of the Apollo 14 solar cells are likely related to solar insulation and possibly dust. The aforementioned results will also be compared with the Chang'E 3 mission observations using triple-junction solar cells on the mid-latitude of the Moon during 2013−14. Chapter 14, authored by Dr. Geoffrey A. Landis of NASA GRC, explores the development of space PV for extreme high-temperature missions. This chapter highlights approaches to solar array design for near-Sun missions including thermal management at the systems level, to optimize efficiency at elevated temperature, or the use of novel device design to reduce the incident solar energy to limit operating temperature. Several of these have been successfully demonstrated to enable solar-powered spacecraft to explore the near-Sun planets such as Mercury and Venus as well as the Sun itself. Chapter 15 reviews space PV concentrators for outer planet and near-Sun missions using ultra-light Fresnel lenses, authored by Mark J. O'Neill. This chapter provides a brief history of space PV concentrator technology and discusses the basic optical and thermal constraints on the technology. The latest versions of these line-focus and point-focus concentrators are described. Both 4X line-focus concentrators requiring only single-axis sun-tracking (with radiator/solar cell articulation for large longitudinal incidence angles) and 25X point-focus concentrators requiring two-axis sun-tracking have been developed and tested at the fully functional prototype level. Finally, the performance, mass, and cost advantages that they offer for various space missions, including outer planet missions and near-Sun missions, is examined and quantified. Chapter 16, co-authored by Prof. T.E. Girish, G.M. Anupama, and G. Lakshmi from the Department of Physics, University College, Trivandrum, India, provides an overview of the relative abundance, spatial distribution, and technologic applications of high-silica deposits on Mars. As space PV power is a proven technology with 60 years of significant advancements compared with space nuclear power, which is yet to be proven for high-power applications, it is safe to assume that some harvesting of solar energy is indispensable to meet the power needs of future Mars exploration and habitation missions. After investigating the expected performance of Si solar cells on Mars, the in situ manufacturing possibilities of such solar cells on Mars will be then explained. The presence of widely distributed (extended up to the Southern Polar region) and significant silica deposits on Mars supports the extensive use of Si solar power on this planet in future. A major focus of Chapter 17 is radioisotope power systems (RPSs), which have been used for space missions since the 1960s. They are attractive because they offer high power densities and provide continuous, reliable, maintenance-free, and long-lasting power independent of the distance from the Sun and radiation environment. A RPS converts decay heat generated by radioisotopes into electricity via static or dynamic conversion technologies. The radioisotope thermal generator (RTG) is the most popular RPS and has been used to provide reliable power for several space missions. Several radioisotope systems that can achieve higher efficiencies than RTGs are discussed in the context of future space missions. The driver for developing new high-efficiency systems is to reduce the quantity of the already limited stockpile of Pu-239 and the size, cost, and safety concerns of future missions. In this chapter, the available radioactive sources for space applications, space nuclear power systems and their critical components, future space nuclear power systems, and safety and regulation regarding nuclear space power are discussed.

The success of this edited book project is the result of the full commitment of each contributing author. Without their availability in sharing their valuable knowledge and critical review of respective topics, this book could not be published at such a high standard. Our heartfelt appreciation also goes to Mariana Kuhl, our project manager during the development and editorial phases of the publication process. Carrie Bolger, our original acquisitions editor, is acknowledged for her constant support given throughout the editorial process. The production staff were very helpful, thorough, and professional throughout the final stages of publication. Finally, we are proud to honor the career and work of Dennis Flood and the memory of Phil Jenkins throughout and by the publication of this book.

Sheila G. Bailey — NASA Glenn Research Center (Retired)
Aloysius F. Hepp — Nanotech Innovations LLC
Ryne P. Raffaelle — Rochester Institute of Technology
Dale J. Ferguson — USAF Research Laboratory
Steven Durbin — Western Michigan University

PART ONE

Introduction: Technologies, issues, and applications

CHAPTER ONE

An introduction to space photovoltaics: Technologies, issues, and missions

Ryne P. Raffaelle
Rochester Institute of Technology, Rochester, NY, United States

1.1 Introduction to the photovoltaic effect and solar cell

The development of photovoltaic(s) (PV) technology and devices began with the discovery of the "photovoltaic effect" by Alexandre-Edmund Bequerel in 1839. Experimenting in his father Henri's laboratory at only 19 years of age, he observed the generation of a voltage and current in an illuminated electrochemical cell. He described this in *Comptes Rendus de l'Académie des Sciences* as "the production of an electric current when two plates of platinum or gold immersed in an acid, neutral, or alkaline solution are exposed in an uneven way to solar radiation" or in other words, when one plate was illuminated and the other was not [1]. After seeing similar results in other electrode-electrolyte systems, attention turned to all solid-state systems.

Willoughby Smith described the "effect of light on selenium during the passage of an electric current" in the February 20, 1873 issue of *Nature* [2]. This was followed by "The Action of Light on Selenium" published by William Grylls Adams, Professor at Kings College in England, and his student, Richard Evans Day, in the *Proceedings of the Royal Society* published in 1877 [3]. They described "that light had caused a flow of electricity through a solid material" when they placed a platinum electrode onto a piece of selenium (Se) and subjected it to illumination. The American inventor Charles Fritts used this effect to create what many believe to be the first solid-state solar cell by coating this same Se with a thin layer of gold [4]. In 1884, these devices were used to make the world's first rooftop solar array, which was installed in New York City.

1.1.1 Photovoltaics into the 20th century

The turn of the 20th century ushered in the quantum era with many of the fundamental discoveries leading to the development of quantum mechanics (i.e., Planck's description of blackbody radiation in 1900, Einstein's description of the photoelectric effect in 1905, Bohr's model for the atom in 1913, Compton's discovery of Compton scattering in 1923, etc.) [5]. Looking back, it is interesting to note how much of the early development of quantum mechanics was devoted to the understanding of the electrical, optical,

Photovoltaics for Space
ISBN 978-0-12-823300-9, https://doi.org/10.1016/B978-0-12-823300-9.00001-7

and thermomechanical properties of solids, and of course all those things that were instrumental to the understanding and development of PV.

During this period, scientists continued to work on the development of solid-state solar cells, studying not only Se, but other compound materials such as lead sulfide (PbS) and copper (I) oxide (Cu_2O). These were followed by a $Cu-Cu_2O$ "photoelectric generator;" these earlier cells are now known as Schottky barrier cells or metal-semiconductor junctions [6]. It was also at this stage that the future of solar power and more importantly space solar power was about to take a dramatic change from that of metal-semiconductor junction to semiconductor-semiconductor junctions.

The first necessary element for this quantum leap in solar cell development was that the "purity" of the semiconducting materials needed to improve, and the ability to control the dopants or added "impurities" would also have to improve. It was during this time that Russel Ohl of Bell Labs patented his new method for producing a semi-conducting ingot of silicon (Si) with "P and N zones" [7]. This paved the way in 1946 for one of the most important discoveries of 20th century, and certainly for the future of solar power, that of the semiconducting "*p-n* junction." This *p-n* junction was also the basis of the US patent Ohl received in 1946 for a "light sensitive device" (Fig. 1.1) [8].

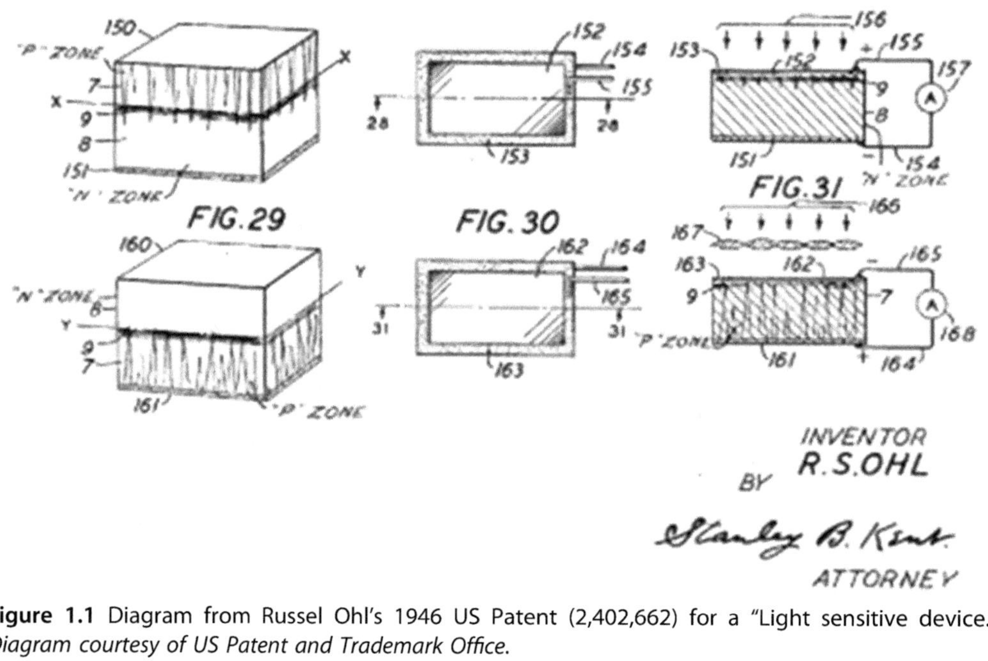

Figure 1.1 Diagram from Russel Ohl's 1946 US Patent (2,402,662) for a "Light sensitive device." *Diagram courtesy of US Patent and Trademark Office.*

1.1.2 The invention of the practical photovoltaic device

On April 25, 1954, Bell Labs held a press conference to introduce their new "solar battery," or what we would call today a PV solar cell. They used these devices to power a toy Ferris wheel and a radio transmitter solely from sunlight. These early cells had an unprecedented solar conversion efficiency for that time of 6% [9].

These devices were developed by Daryl Chapin who was working on the development of a new power source to replace batteries for telephone systems in remote and humid areas, and Calvin Fuller and Gerald Pearson, who were following in the footsteps of Ohl on doping and controlling the properties of semiconductors. Chapin, Fuller, and Pearson created a single-crystal Si *p-n* junction solar cell, and they chained together multiple single-crystal silicon *p-n* junction solar cells to form a solar panel; see Fig. 1.2 [10]. The material systems in these record results for the time were arsenic- and boron-doped silicon.

By 1958, small-area Si solar cells had reached an efficiency of 14% under terrestrial sunlight [11]. Although Bell Labs saw commercial potential in the terrestrial use of solar energy, it would be several decades before any real terrestrial deployment at scale of solar power systems would be realized. However, with the dawn of the "space race," a new driving force for the continued development in PV solar cells was soon to arise, that for space solar power systems. Throughout the remainder of the chapter, we will summarize

Figure 1.2 Daryl Chapin, Calvin Fuller, and Gerald Pearson of Bell Labs. *Image courtesy of AT&T Archives.*

key advances in space PV, including new materials, novel device technologies, practical applications and missions, and issues driving technologic breakthroughs. Previous reviews have focused on more specific aspects of developments in (space) PV technologies [12,13]. Whenever appropriate, relevant chapters from the remainder of this book will be recommended to the reader for further and more in-depth discussion.

1.2 First-generation space photovoltaics: Missions, technologies, and issues

As part of the response to the former Soviet Union's launching of the Sputnik satellites, the United States launched the first solar-powered satellite, Vanguard 1, on March 17, 1958; see Fig. 1.3 [14]. Vanguard 1 was the second US satellite placed in orbit around Earth and the first with solar power. It was the fourth satellite to be successfully launched, after the Sputnik 1 and 2 (launched October 4th, 1957 and November 3rd, 1957, respectively) and the US Explorer 1 (January 31, 1958). Vanguard 1 has six body-mounted solar panels (see Fig. 1.3b), each with 18 2 × 0.5 cm p-on-n Si solar cells, with a 0.16-cm-thick quartz cover glass. The cells were fabricated by Hoffman Electronics for the US Army Signal Research and Development Laboratory at Fort Monmouth New Jersey. The solar cells continued to power a radio signal until February 1965. Vanguard 1 continues as the oldest man-made object in orbit around Earth.

1.2.1 Explorer 6

In an effort to expand a satellite's ability to capture sunlight, Explorer 6, which was launched on August 7, 1959, incorporated solar cells mounted onto four 51-cm^2 external paddles (see Fig. 1.4) [15]. It was a small, spheroidal satellite designed to study

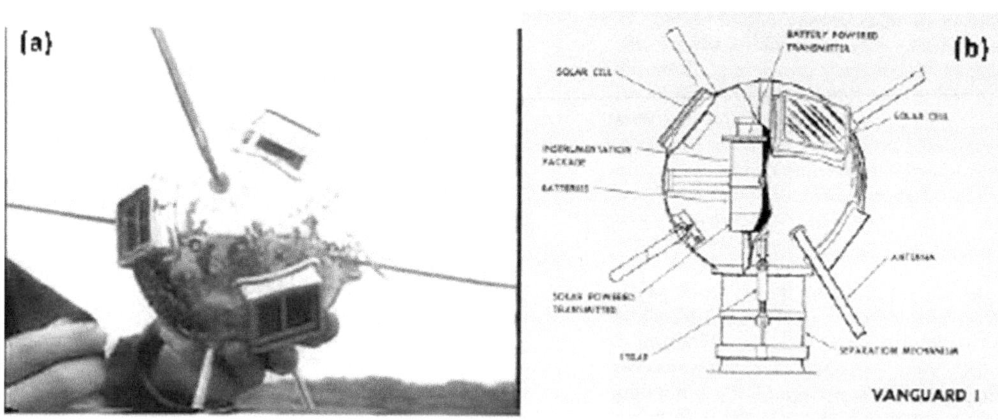

Figure 1.3 (a) Photograph and (b) schematic of Vanguard 1. *Images courtesy of US Naval Research Laboratory.*

Figure 1.4 US satellite Explorer 6, launched in 1959, with externally mounted solar arrays. *Image courtesy of NASA.*

trapped radiation in the atmosphere, galactic cosmic rays, geomagnetism, radio propagation in the upper atmosphere, and the flux of micrometeorites. Unfortunately, one of the panels failed to fully deploy. The solar cells quickly degraded due to radiation from the Van Allen belts; see discussion in Section 1.2.4. Explorer 6 lost communication in less than 2 months (September 11, 1959), and the last contact with the payload was made on October 6, 1959. However, it was the first satellite to take an image of Earth from space and did serve as a model for the use of space solar arrays for the spacecraft that were due to follow.

1.2.2 Dawn of telecommunications

Telstar is the name given to a series of over 20 communications satellites that began to be launched in the early 1960s and continue to the present day [16]. These satellites continued to grow in size, complexity, and power demands with each new generation. They have ranged from 77 to 79 kg for Telstar 1 and 2, respectively, to over 7000 kg for Telstar 19V launched in 2018 aboard a SpaceX Falcon 9 (Fig. 1.5). Telstar 19V became the heaviest commercial communications satellite ever launched. It can be argued that this series of satellites has been the greatest human achievement in communication since the telephone, or perhaps even the Guttenberg printing press. Ironically, according to a US Information Agency poll, Telstar was better known in Great Britain in 1962 than Sputnik had been in 1957 [17].

The first of the Telstar satellites had body-mounted panels, not unlike Vanguard 1. Telstar 1 was covered with a number of solar panels containing 3600 solar cells that were

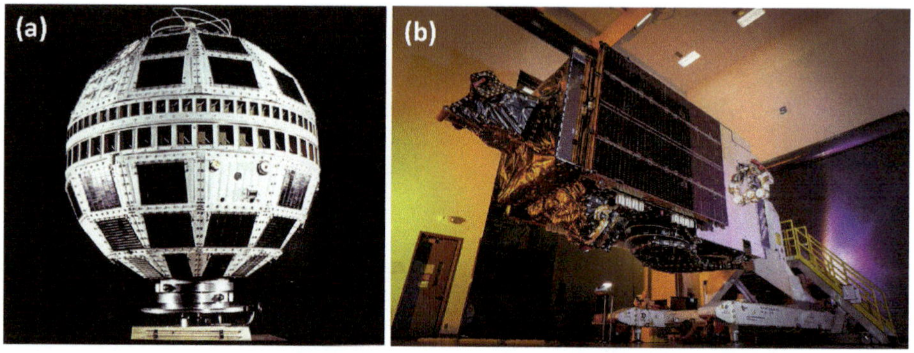

Figure 1.5 (a) Telstar 1 and (b) Telestar 19 Vantage (19V). *Images courtesy of Space Systems Loral.*

capable of producing 14 W of electrical power (see Fig. 1.5a). It was launched on July 10, 1962, using a Delta-DM19 rocket. It successfully relayed the first television pictures, telephone calls, and telegraph images via satellite and was responsible for the first live transatlantic television feed, which featured CBS's Walter Cronkite and NBC's Chet Huntley in New York and the BBC's Richard Dimbleby in Brussels [18]. The first pictures transmitted were of the Statue of Liberty in New York and the Eiffel Tower in Paris [19]. The first broadcast featured remarks by President John F. Kennedy, but the signal was acquired before the president was ready, so engineers filled the lead-in time with a short segment of a televised game between the Philadelphia Phillies and the Chicago Cubs at Wrigley Field [20].

Telstar 1 was part of a multinational agreement between AT&T (USA), Bell Telephone Laboratories (USA), NASA (USA), GPO (United Kingdom), and the National PTT (France) to establish satellite communications over the Atlantic Ocean. The almost identical, albeit slightly larger at 79 kg, Telstar 2 was launched less than a year later onboard a Delta B rocket. Although no longer functional, these satellites, like their predecessor Vanguard 1, continue to orbit Earth. There have been 21 Telstar satellites deployed between 1962 and 2019, with the most recent being Telstar 19V, launched on July 22, 2018, and built by Space Systems Loral (see Fig. 1.5b) [21]. Bell Labs was responsible for developing the first two, Hughes for the next three, Lockheed Martin for the next three, and Space Systems Loral the remaining 13.

1.2.3 Early space solar arrays

The first satellites operated on less than a few hundred watts, and since the cost of the solar arrays was a small fraction of the overall satellite cost, the premium was put on reliability, as it still is today. However, as the size and complexity of the satellites continued to increase, so did their need for increased solar power. With launch costs approaching $10,000 or more per kilogram, there was considerable attention paid to solar array specific power (W/kg), and therefore the efficiency of the solar cells. In addition, many satellites were still using body-mounted arrays, which severely limited the array size and therefore required the highest efficiency cells available.

The initial development of space solar power systems focused on silicon. They had been shown to be relatively efficient and reliable. Theoretical predictions suggested that practical Si solar cells could be produced with efficiencies approaching 20%, although its band gap was 0.4 eV below the optimum band gap (~1.5 eV) for terrestrial illumination [22]. Fundamental solar cell research was focused on understanding and mitigating the factors that limited cell efficiency (e.g., minority carrier lifetime, surface recombination velocity, series resistance, reflection of incident light, and nonideal diode behavior). However, new materials and cell designs were beginning to emerge, including use of III-V semiconductors or potential tandem or multijunction (MJ) cell designs. An optimized triple-junction (3J) cell was shown to have a theoretical efficiency of 37% [23].

1.2.4 Radiation damage

A new concern for space solar cell design emerged with the launching of Explorer I and the discovery of the Van Allen radiation belts, that of electron and proton irradiation damage. This was made painfully evident with the US high-altitude nuclear weapon test "Starfish." On July 9, 1962, the Starfish Prime device was detonated at an altitude of 250 miles (400 km). The 1.4 MT nuclear warhead detonated 13 min 41 s after liftoff of the Thor missile from Johnston Atoll [24]. Starfish Prime created an extremely large, and greater than expected, electromagnetic pulse. It drove much of the measurement instrumentation aboard 27 separately launched smaller rockets off scale, eliminating the ability to get accurate measurements of the radiation.

The result of all the radiation that was dumped into the Van Allen radiation belts was that a number of satellites ceased to function as their solar arrays failed due to the radiation damage. Even Telstar was negatively impacted, with a decreased solar power output following the Starfish detonation [25]. The lessons learned from Explorer I and Telstar prompted a surge of activity in radiation protection of space solar cells. The Naval Research Laboratories provided much needed guidance to spacecraft designers in how to account for natural cell degradation due to radiation in space [26]. Silicon solar cells migrated away from the conventional *p*-on-*n* design to an *n*-on-*p* structure to provide better radiation resistance [27]. The next chapter in this book by Ferguson et al. is suggested for a more in-depth treatment of space environmental effects [28].

1.3 The next era in space: New materials and missions in low-earth orbit

During the 1970s and 1980s, solar cell efficiencies rose dramatically [11,12]. While Si-based solar arrays still provided power for many missions, new III-V compound semiconductor materials were introduced [29]. These III-V semiconductors required new processing methods [30]. This section describes their introduction, new device structures, and early low-Earth orbit (LEO) missions.

1.3.1 Skylab

The first manned "space station," named Skylab, was operated by NASA in LEO from May 1973 until February 1974 (Fig. 1.6) [31]. Skylab was powered by the largest solar power system deployed up to that time, which contained two solar arrays, the Orbital Workshop array and the Apollo Telescope Mount array [32]. The Orbital Workshop array had two deployable wings, each with 73,920 (2 × 4 cm) n-on-p Si cells that provided over 6 kW of power. Unfortunately, one of these wings was lost during launch. The Apollo Telescope Mount array had four wings with 123,120 (2 × 4 cm) cells and 41,040 (2 × 6 cm) cells providing over 10 kW of power.

1.3.2 III-V semiconductor material systems and multijunction structures

Fig. 1.7 shows III-V semiconductor material devices and associated MJ cells. As cell efficiencies continued to rise, the gap between theoretical efficiencies and experimental efficiencies for Si, gallium arsenide (GaAs), and indium phosphide (InP) became almost nonexistent [33]. In addition, this period saw a host of other improvements to space solar cells such as the first use of shallow junction silicon cells for increased blue response and

Figure 1.6 The Skylab Orbital Workshop in LEO as photographed from the Skylab 4 Command and Service Module. *Image courtesy of NASA.*

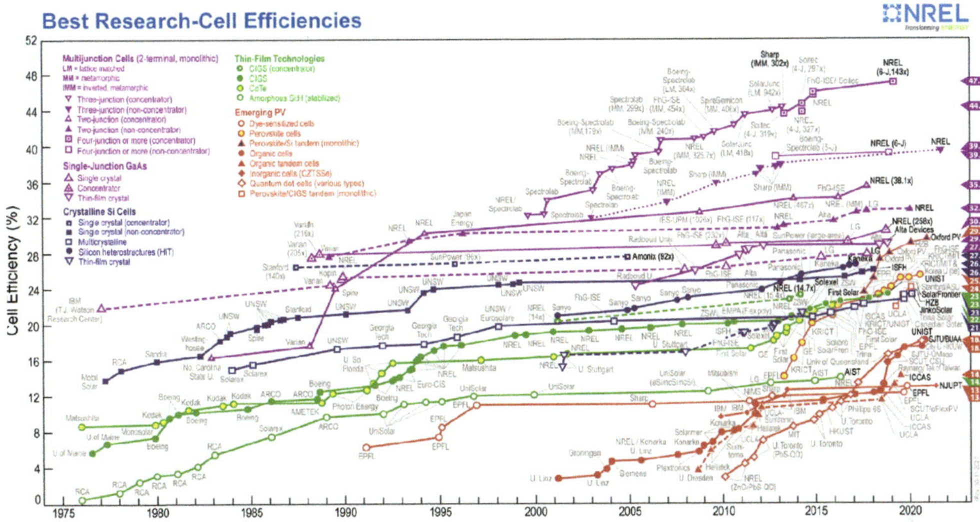

Figure 1.7 US Department of Energy (DOE) National renewable Energy Laboratory (NREL), Best Record Research-Cell Efficiencies Chart, https://www.nrel.gov/pv/cell-efficiency.html (accessed 11/16/2021). *Courtesy of US DOE/NREL.*

current output, the use of a back surface field, the low–high junction theory for increased silicon cell voltage output, and the development of wraparound contacts to enable automated array assembly and to reduce costs [34].

During this period MJ III-V solar cells began to displace silicon as the solar cell material system of choice for space power systems [12,33]. However, silicon would continue to see some use including some of the initial arrays on the largest space solar power system ever deployed, those of International Space Station (ISS) launched on November 20, 1998 (Fig.1.8).

1.3.3 The International Space Station

The ISS currently (2022) has eight retractable solar array wings (Fig. 1.9). The first pair was deployed in 1998, and the system was completed in March 2009 [35]. The arrays contain a total of 262,400 solar cells and cover an area of 2500 m^2. Altogether, the arrays can generate about 240 kW of electrical power at 160 V.

The ISS is in LEO, so the array power output has been continuously degrading from radiation, primarily due to low-energy electrons trapped in the aforementioned Van Allen radiation belts (in Section 1.2.4). NASA has plans to replace six of the eight - existing power channels of the space station with new solar arrays. Boeing, NASA's prime contractor for space station operations, will be providing the new arrays, which will be manufactured by Deployable Space Systems using Spectrolab cells. The new

Figure 1.8 The International Space Station, showing photovoltaic arrays. *Image courtesy of NASA.*

Figure 1.9 A spacewalker using the International Space Station robotic arm to help retract one of the solar arrays during STS-116. *Image courtesy of NASA.*

arrays will be a larger version of the Roll-Out Solar Array (ROSA) technology that was demonstrated aboard the space station in June 2017; see Fig. 1.10 [36].

These new arrays will be smaller, due to their high efficiency cells, and positioned in front of six of the current arrays; they will have Spectrolab NeXt Triple Junction (XTJ Prime) 3J III-V solar cells [29]. The arrays will be delivered by the SpaceX Dragon cargo

Figure 1.10 The ROSA (Roll-Out Solar Array) technology undergoes testing. *Image courtesy of Deployable Space Systems.*

spacecraft during three resupply missions starting in 2021. These new cells are estimated to provide up to a 30% increase in power for the ISS; see Fig. 1.11. Chapter 4 of this book provides an in-depth discussion of III-V single- and MJ solar cells and arrays for space applications [29].

1.3.4 Use of GaAs solar cells on Mir

Although it was launched before the ISS, the Mir modular space station that was assembled in LEO 1986 to 1996, and operated until 2001 by the Soviet Union and later

Figure 1.11 The SpaceX Dragon spacecraft arriving at the International Space Station. *Image courtesy of NASA.*

Figure 1.12 The MIR Space Station viewed from the Space Shuttle Endeavor during the SRS-89 rendezvous. *Image courtesy of NASA.*

by Russia, actually used GaAs solar cells; see Fig. 1.12 [37]. Unfortunately, the deployment of its arrays, which occurred over an 11-year period, was slower than planned, and the station continually suffered from a shortage of power as a result. Two of the final arrays were delivered by the Space Shuttle Atlantis during STS-74. Altogether, the solar arrays aboard Mir provided approximately 26 KW. Mir made history as the first continuously inhabited long-term research station. It provided a human presence in space for 3644 days, a record that was finally surpassed by the ISS on October 23, 2010 [38].

1.3.5 The Hubble Space Telescope

The Hubble Space Telescope (often referred to as HST or Hubble) was launched into LEO in 1990. Its original solar power system incorporated a novel "roll-out" solar array that was compared to a window shade; see Fig. 1.13 [39]. Unfortunately, problems with this array and the "jitter" it caused with the early images resulted in the need for the famous Hubble "servicing" mission in 1993 (STS-61) [40]. During this mission and its associated spacewalk, the roll-out array was disconnected and jettisoned to burn up in Earth's atmosphere and was replaced with more conventional flat-panel arrays.

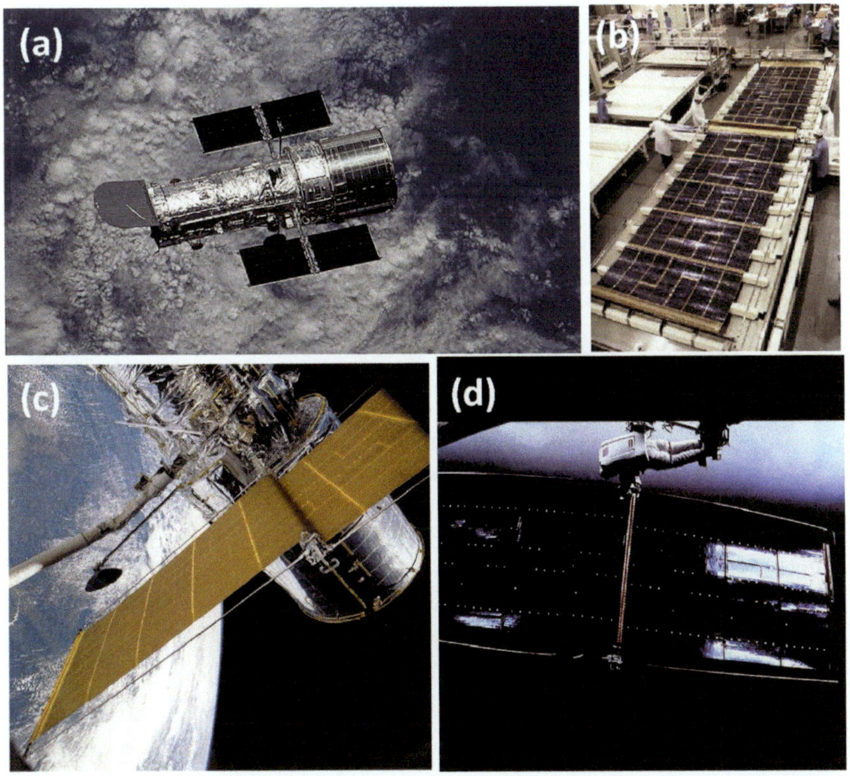

Figure 1.13 (a) The Hubble Space Telescope (HST); (b) preparation of the HST roll-out arrays; (c) close-up image of the HST roll-out array deployment system; and (d) a spacewalker servicing the HST arrays. *Images courtesy of NASA.*

Hubble, named after astronomer Edwin Hubble, was not the first space telescope, but it is arguably the most famous, and one of the most important, tools ever used in astronomy. After the servicing mission, which also included the installation of a new main camera and corrective optics package because of an aberration in the original mirror, the resulting improvement was remarkable (Fig. 1.14).

1.4 Into the twenty-first century: New device and advanced materials technologies

The transition from Si to III-V and III-V MJ solar cells occurred much more rapidly for most space power systems being developed than had occurred with the ISS. During the 1990s and 2000s, satellites continued to grow in both size and power requirements, and much more innovative structures were designed to deploy ever larger solar arrays. The mass and fuel penalty for attitude control of very large arrays continued

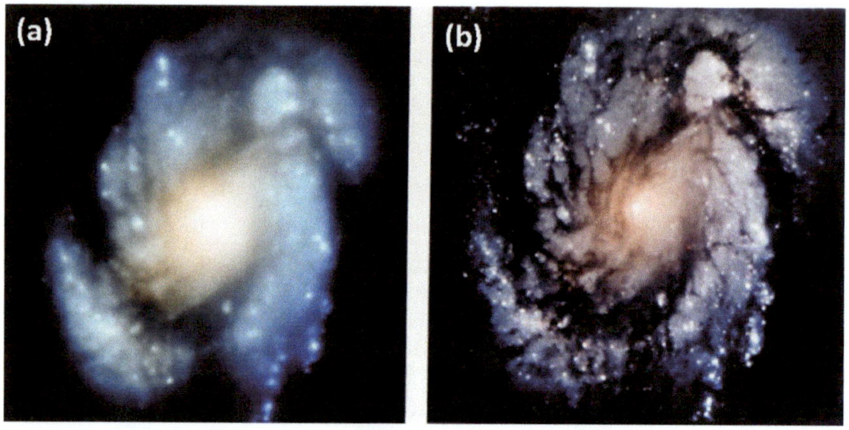

Figure 1.14 Images taken with the Hubble Space Telescope of galaxy M100 (a) before and (b) after its first servicing mission in December of 1993. *Images courtesy of NASA.*

to be a huge driver for ever more efficient cells. III-V 3J solar cells on Ge or GaAs substrates became the de facto standard during this time. Spacecraft developers were still willing to pay a premium for the best cells, with satellite power systems continuing to be in the $1000/W range, so there was steady improvement in the efficiency of these cells as seen in the NREL efficiency chart (Fig. 1.7) [13,33].

Another way to decrease array cost is by the use of concentrator arrays. Concentrator arrays use either refractive or reflective optics to direct concentrated sunlight onto a smaller active area of solar cells. Deep Space 1, launched by NASA in October 1998, was the first spacecraft to rely upon concentrator arrays; see Fig. 1.15 [41]. It had two such arrays with each capable of producing 2.5 kW at 100 V (DC). The SCARLET arrays were developed by AEC-ABLE Engineering through support provided by the US Ballistic Missile Defense Organization. These arrays performed flawlessly in this inaugural demonstration.

Unfortunately, not all concentrator arrays used in the 1990s fared as well. The original Boeing Space Systems (BSS-702) bus used for communication satellites exhibited faster than expected degradation [42]. The problem was isolated to the concentrator reflector surfaces, which degraded after becoming coated with contaminants outgassing from the array because of its increased temperature. Insurance underwriters paid out some $875 million in claims to the first 702-model owners—PanAmSat, now part of Intelsat; Thuraya; Telesat Canada; and XM Satellite Radio, now Sirius XM Radio—related to the solar array issue. A thorough, practical overview of Fresnel lenses instead of reflectors for concentrator technologies for space is provided by O'Neill in the final chapter of this book [43].

State-of-the-art (SOA) solar arrays today have an AM0 (space) efficiency of over 30% and are typically 3J III-V cells grown on Ge or GaAs. However, four-junction and even

Figure 1.15 (a) The Deep Space 1 satellite with it SCARLET solar concentrator arrays in the stowed position and (b) the SCARLET array deployed. *Images courtesy of NASA/AEC Able.*

five-junction cells do exist (see Fig. 1.7) [29]. Some of the more advanced techniques used to develop these stacks require them to be grown metamorphically, with buffer layers that allow a relaxation of the conventional lattice-matched epitaxial growth, or by growing the structure in an inverted fashion and requiring substrate removal. Details of these particular structures and the various techniques used to achieve them will be covered in later chapters [29,30].

Suffice it to say that the advance in the ability to grow new III-V materials epitaxially or metamorphically has given space cell and array manufacturers an unprecedented ability to tailor their systems to specific illumination and environmental conditions. Conventional arrays have reached panel-level specific power of over 100 W/kg, and pathways now exist for even higher specific power levels (150—500 W/kg), allowing manufacturers to meet the new demands for solar electric propulsion systems that can be used to explore the solar system and neighboring planets or systems that may have harsh or challenging environments (i.e., low solar intensity and low temperature, high solar high intensities, and high-radiation exposure) [44]. In the future, challenging missions may also be enabled via other advanced materials systems utilizing lightweight substrates [45—49] as detailed in several other chapters in this book [50—52].

1.5 Exploration of Mars and beyond: Notable solar-powered spacecraft

In addition to satellites, space telescopes, and space stations, space solar arrays have been utilized on many other types of spacecraft and space probes. A few of the more noteworthy examples include rovers, landers, and even aircraft designed to fly over other planets. These are exemplified in the following subsections. A more thorough discussion of solar arrays for science exploration missions is provided in Chapter 12 by Mercer [53].

1.5.1 Early Mars rover missions: Pathfinder, explorer rover, and science laboratory

So far, there have been five successfully deployed Mars rovers [54]. The first was Sojourner, which touched down in the Ares Vallis region of the Martian surface on July 4, 1997 [55]. It was part of the Mars Pathfinder mission and was the first wheeled vehicle to rove another planet; see Fig. 1.16a. It was in operation for a total of 92 sols or the equivalent of 95 days. It was 2 feet long and had a flat-panel array mounted on its top surface. This array was made up of 200 18% efficiency GaAs on Ge solar cells and was capable of producing 16 W of power at "noon" or when the Sun was directly overhead on Mars (Fig. 1.16b). The reader is reminded that since Mars is approximately 1.5 times farther from the Sun, the solar arrays will produce less than half the equivalent power they could produce if on Earth.

The Mars Pathfinder Sojourner rover has since been followed by Spirit and its twin Opportunity [56], and subsequently Curiosity [57] and its twin Perseverance [58]; see Fig. 1.17. All five rovers have been operated by the NASA Jet Propulsion Laboratory [54]. The Spirit/Opportunity and Curiosity/Perseverance twins were considerably larger than their forerunner Sojourner, at 5.2 feet and 10 ft long, respectively. Spirit and Opportunity used solar panels constructed with high efficiency III-V 3J solar cells [56]. Initially, these solar arrays were able to produce about 900 W-hours of energy per Martian day, or sol. Curiosity and Perseverance did not have solar arrays, although the helicopter Ingenuity launched by Perseverance does [59].

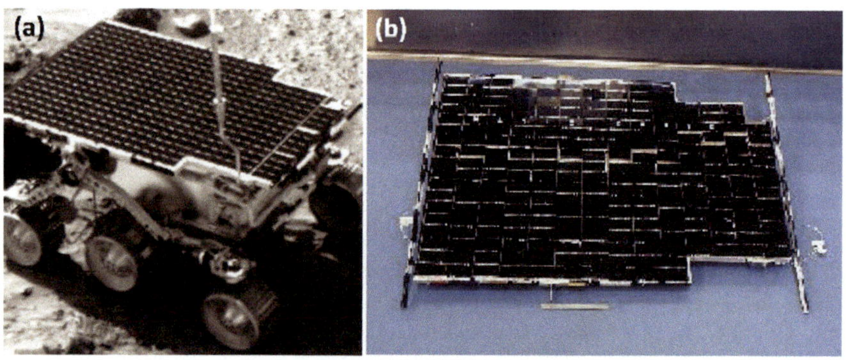

Figure 1.16 (a) Image of Sojourner rover on the surface of Mars; (b) GaAs on Ge solar array used on NASA's Mars Sojourner rover. *Images courtesy of NASA.*

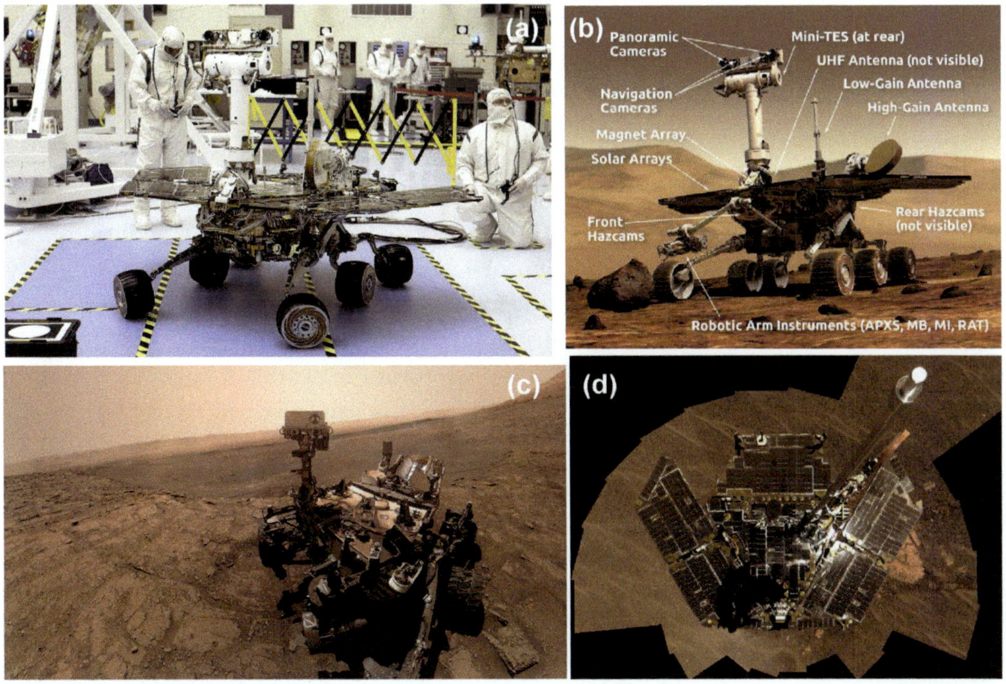

Figure 1.17 (a) Mars Exploration Rover Spirit is tested for mobility and maneuverability in the Payload Hazardous Servicing Facility at NASA Kennedy Space Center; the MER Mission consisted of two identical rovers designed to cover roughly 110 yards each Martian day; each rover carried five scientific instruments that to allow it to search for evidence of liquid water that may have been present in the planet's past; (b) Mars Exploration Rover instruments diagram; (c) self-portrait of NASA's MSL-Curiosity rover taken on Oct. 11, 2019 (2,553rd Martian day, or sol, of its mission); self-portraits are created using images taken by Curiosity's Mars Hand Lens Imager; (d) overhead view of 3J solar panels on the twin rovers Spirit and Opportunity. *Photographs courtesy of NASA (a, b, and d) or NASA/JPL-Caltech (c).*

There are a number of challenges that arise when generating power using solar cells on Mars, in addition to the previously mentioned lower solar intensity due to its greater distance from the Sun than Earth's. The solar spectrum is also quite different from the AM0 spectrum experienced in space, as a result of light absorption by the Martian atmosphere. The spectrum is comparatively depleted in shorter wavelength, higher energy photons. Of course, the designers of solar cells for use on Mars must tune their MJs to the Martian spectrum to optimize their performance. In addition, solar arrays used on Mars must confront the challenge posed by dust accumulation and the resulting obscuration due to the famous Martian dust storms. Although the accumulation of dust is a real issue that must be confronted (i.e., electrostatic dust shields, movable cover glass, piezoelectric or mechanical methods), occasionally the dust storms can work to their advantage [60–62]. It was observed that Spirit was able to complete its planned 90-sol mission with the aid of "cleaning events" or windstorms that blew its solar arrays clean. It actually went on to function effectively over 20 times longer than NASA planners expected, logging about 10 km of driving distance. Unfortunately, after over 5 years, Spirit became stuck in soft soil.

1.5.2 Philae lander

Rosetta was a space probe built by the European Space Agency and launched on March 2, 2004 (Fig. 1.18) [63]. Rosetta was equipped with a lander module named Philae. It was designed to perform a detailed study of comet 67P/Churyumov-Gerasimenko, which was due to pass near Earth. During its journey to the comet, the spacecraft performed flybys of Earth, Mars, and two asteroids. It was launched as the third cornerstone mission of the ESA's Horizon 2000 program after SOHO-Cluster and XMM-Newton.

On August 6, 2014, Rosetta reached the comet and performed a series of orbital maneuvers. It then deployed the lander module Philae, which achieved the first successful landing on a comet. Unfortunately, communication with Philae only lasted 2 days due to the loss of the lander's battery power when it landed in a shadowed region. Communications with Philae were briefly restored in June and July 2015, but due to diminishing solar power, Rosetta's communications module with the lander was turned off on July 27, 2016. On September 30, 2016, the Rosetta spacecraft ended its mission with a designed hard-landing on the comet.

1.5.3 Juno spacecraft

Juno is a NASA space probe designed to orbit the planet Jupiter; see Fig. 1.19 [64,65]. It was built by Lockheed Martin and is operated by NASA's Jet Propulsion Laboratory. The

Figure 1.18 (Left) ESA spacecraft Rosetta with Philae lander, before separation. Frame from the movie "Chasing A Comet – The Rosetta Mission" (Right) Depiction of the ESA lander Philae on Comet 67P/Churyumov–Gerasimenko. *Images courtesy of German Aerospace Center (DLR), Creative Commons Attribution 3.0 Germany (CC BY 3.0 DE) (Left) https://www.flickr.com/photos/dlr_de/11963777196 (Right) https://www.flickr.com/photos/dlr_de/15307538379/.*

Figure 1.19 Juno spacecraft undergoes weight and balance testing at the Astrotech payload processing facility, Titusville, Florida, June 16, 2011. *Image courtesy of NASA.*

spacecraft was launched from Cape Canaveral Air Force Station on August 5, 2011, as part of the New Frontiers program. Juno entered a polar orbit of Jupiter on July 5, 2016 UTC, to begin its scientific investigation of the planet. Its mission was to measure Jupiter's composition, gravitational field, magnetic field, and polar magnetosphere. After completing its mission, Juno was to be deorbited into Jupiter's atmosphere; this was originally scheduled to occur in July 2021. However, NASA has decided to extend the mission through September 2025 [66].

Juno is the second spacecraft to orbit Jupiter, after the nuclear-powered Galileo orbiter, which orbited from 1995 to 2003. Unlike all earlier spacecraft sent to the outer planets, Juno is powered by solar panels. Before Juno, all spacecraft used to investigate the outer planets had relied on radioisotope thermoelectric generators [67].

Juno has the three largest solar panel wings ever deployed on a planetary probe. The three solar panels are symmetrically arranged around the spacecraft. The three panels were deployed shortly after the spacecraft cleared Earth's atmosphere. Two of the panels have four hinged segments each, and the third panel has three segments and a magnetometer. Each panel is 2.7 by 8.9 m long. These panels not only provide power but also play an integral role in stabilizing the spacecraft.

The combined mass of the three panels is nearly 340 kg. Comparably sized panels designed to operate at Earth would produce something on the order of 12–14 kW of power. Of course, at the distance of Jupiter from the Sun, this is dramatically reduced. On Juno's arrival at Jupiter the arrays produced 486 W of power. Once deployed, the solar panels are sun-facing throughout Juno's mission, except for a few brief periods during the operation of the main engine or in eclipses by Jupiter when orbiting.

1.5.4 NASA's Mars 2020 mission: Perseverance and ingenuity

Ingenuity is a small robotic helicopter developed as part of NASA's Mars 2020 mission; see Fig. 1.20 [59]. On April 19, 2020, the 4-pound 19.3-inch-tall helicopter was deployed by the aforementioned Perseverance Mars rover and was successfully flown in the Martian atmosphere. Ingenuity completed the first powered controlled flight by an aircraft on another planet and had completed 12 successful flights as of August 16, 2021. Ingenuity was built by NASA's Jet Propulsion Laboratory (JPL). Other contributors include AeroVironment, NASA Ames Research Center, NASA Langley Research Center, SolAero Technologies, and Lockheed Martin Space.

Ingenuity carries a piece of fabric from the wing of the 1903 Wright brothers' airplane (i.e., the Wright Flyer) used in the first controlled, powered, heavier-than-air (i.e., nonballoon) flight on Earth. The initial take-off and landing area for Ingenuity is named Wright Brothers Field as a tribute to the Wright brothers' first powered flight in 1903. Before Ingenuity, the first flight of any kind on a planet beyond Earth was an unpowered balloon flight on Venus, by the Soviet Union's Vega 1 spacecraft in 1985.

The expected lateral range was exceeded in the third flight, and the expected flight duration was exceeded in the fourth flight. With those technical successes, Ingenuity had achieved all of its original objectives. On its fourth flight, April 30, 2021, Ingenuity became the first interplanetary spacecraft whose sound was recorded by the Perseverance rover, another interplanetary spacecraft.

Figure 1.20 (a) NASA's Ingenuity helicopter and (b) a close-up of Ingenuity's solar array. Martian dust can be seen contaminating its array. These images were captured by the Mastcam-Z imager aboard NASA's Perseverance Mars rover from which Ingenuity was deployed. *Images courtesy of NASA.*

Ingenuity is powered by SolAero Technologies' 33.0% efficient inverted metamorphic multijunction (IMM) solar cells. These IMM cells are approximately 40% lighter than typical space-grade solar cells with conventional substrates. As previously discussed (Section 1.4), growing the cell structure in an inverted fashion with the removal of the substrate can dramatically improve the specific power. The panels are approximately 425 × 165 mm and are used to charge six lithium-ion batteries with the helicopter. The solar panel for Ingenuity was manufactured in SolAero Technologies' SOA production facility in Albuquerque, NM.

1.5.5 Solar blankets

There are quite a number of PV "blankets" similar to the one developed for Ingenuity that are currently under development. Boeing has developed a solar panel that integrates the Spectrolab 33% IMM, which they call the Integrated Blanket/Interconnect System. There are also a number of flat-folding *flexible* arrays produced by Lockheed Martin, which have incorporated conventional high-efficiency 3J solar cells, with specific power exceeding 100 W/kg. Examples of their use include the Mars Phoenix Lander and Terra: the EOS-AM flagship. Japan Aerospace Exploration Agency (JAXA) has also developed what they refer to as a "Space Solar Sheet" [68]. This PV blanket incorporates thin-film 2J cells (InGaP/GaAs) with a one-sun AM0 efficiency of 25% into a flexible laminate of a transparent resin polymer sheet. It was designed for LEO applications or as a thin cover glass for Geosynchronous Earth Orbit (GEO) applications. The specific power of current sheets yield ~500 W/kg. In fact, there are designs utilizing IMMs under development that could potentially deliver even higher (i.e., 700 W/kg) mass specific power; this technology has been proposed for solar electric propulsion missions to the outer solar system [69].

1.6 Conclusions

In this chapter we have explored a short history of space PV from the early days of PV solar cell development to current space exploration missions that utilize PV. We have emphasized the impact that it continues to have on the exploration and development of space. Clearly, the history of space PV is in many ways the history of PV. In fact, the development of space solar power systems drove much of the development of early PV solar cells and arrays. The solar power industry today owes a tremendous debt of gratitude to the space power scientists and engineers without whom much of the technology being used in terrestrial power systems might still not exist. The remaining chapters in this book outline many more intriguing details. The reader is also encouraged to consult several omnibus sources for information related to space PV technologies for future science missions [70], space exploration [71,72], and environments, needs, and opportunities for space PV [73].

References

[1] E. Becquerel, On electron effects under the influence of solar radiation, Comptes Rendus, Academie des Sciences, Paris 9 (1839) 561−567.

[2] W. Smith, Effect of light on selenium during the passage of an electric current, Nature 7 (1873) 303, https://doi.org/10.1038/007303e0.

[3] W.G. Adams, R.E. Day, The action of light on selenium, Proc. Royal Soc. 25 (1877) 171−178.

[4] C.E. Fritts, On a new form of selenium photocell, Am. J. Sci. 26 (1883) 465−472, https://doi.org/10.2475/ajs.s3-26.156.465.

[5] K. Krane, Modern Physics, fourth ed., John Wiley & Sons, Hoboken, NJ, 2020.

[6] R.T. Tung, The physics and chemistry of the Schottky barrier height, Appl. Phys. Rev. 1 (2014) 011304, https://doi.org/10.1063/1.4858400.

[7] R.S. Ohl, Alternating current rectifier, U.S. Patent 2 (402) (June 25, 1946) 661.

[8] R.S. Ohl, Light-sensitive electric device, U.S. Patent 2 (402) (June 25, 1946) 662.

[9] D.M. Chapin, C.S. Fuller, G.L. Pearson, New silicon p-n junction photocells for conversion of solar radiation into electrical power, J. Appl. Phys. 25 (1954) 676−677.

[10] D.M. Chapin, C.S. Fuller, G.L. Pearson, Solar energy converting apparatus, U.S. Patent 2 (780) (1957) 765.

[11] A. Goetzberger, C. Hebling, H.-W. Schock, Photovoltaic materials, history, status and outlook, Mater. Sci. Eng. R Rep. 40 (1) (January 2003) 1−46.

[12] A.F. Hepp, S.G. Bailey, R.P. Raffaelle, Inorganic photovoltaic materials and devices: Past, present, and future, in: S.-S. Sun, N.S. Sariciftci (Eds.), Organic Photovoltaics: Mechanisms, Materials and Devices, CRC Press, Boca Raton, FL, USA, 2005, pp. 19−36.

[13] S. Bailey, R. Raffaelle, Chapter II-4-B - operation of solar cells in a space environment, in: S. Kalogirou (Ed.), McEvoy's Handbook of Photovoltaics (Third Edition) Fundamentals and Applications, Academic Press, 2017, pp. 987−1003.

[14] R. Easton, M. Votaw, I. Vanguard, Satellite IGY (1958 Beta), Rev. Sci. Instrum. 30 (2) (1959) 70−75.

[15] Explorer 6, https://nssdc.gsfc.nasa.gov/nmc/spacecraft/display.action?id=1959-004A, accessed November 26, 2021.

[16] D.F. Hoth, E.F. O'Neill, I. Welber, The Telstar satellite system, Bell Syst. Tech. J. 42 (4) (1963) 765−799.

[17] D.R. Glover, NASA experimental communications satellites, 1958−1995, in: A.J. Butrica (Ed.), Beyond the Ionosphere: Fifty Years of Satellite Communication SP-4217, Washington, D.C., 1997 available at: https://history.nasa.gov/SP-4217/ch6.htm.

[18] W. Cronkite, retrieved from: https://www.youtube.com/watch?v=FgplIWibv4Q, accessed November 12, 2021.

[19] G. Clary, retrieved from: https://lightyears.blogs.cnn.com/2012/07/13/50th-anniversary-of-satellite-celebrated/, accessed November 12, 2021.

[20] Philadelphia Phillies vs Chicago Cubs, retrieved from: https://www.baseball-almanac.com/box-scores/boxscore.php?boxid=196207230CHN, accessed on November 12, 2021.

[21] I. Telstar, Telstar Satellite Design, Construction, Ground Facilities, and Uses, vol. 1, NASA SP-32, NASA Goddard Space Flight Center, Greenbelt, MD USA, June 1963 available at: https://ntrs.nasa.gov/citations/19640000959. (Accessed 26 November 2021).

[22] J.J. Loeferski, Theoretical considerations governing the choice of the optimum semiconductor for photovoltaic solar energy conversion, J. Appl. Phys. 27 (7) (1956) 777−785.

[23] E.D. Jackson, Areas for improving of the semiconductor solar energy converter, in: Transactions of the Conference on the Use of Solar Energy, 1955 vol 5, University of Arizona Press, Tucson, AZ, USA, 1958, pp. 122−128.

[24] P. Dyal, W. Simmons, Air Force Weapons Laboratory Operation Dominic, Fish Bowl Series, Debris Expansion Experiment, Report ADA995428, December 1965 extracted version (September 1985), available at: https://apps.dtic.mil/sti/pdfs/ADA995428.pdf.

[25] H.S. Rauschenbach, Solar Cell Array Design Handbook, JPL SP43-38, vol 1, NASA CR-149364, 1976 available at: https://ntrs.nasa.gov/citations/19770007250.

[26] D.J. Curtin, R.L. Statler, Review of radiation damage to silicon solar cells, IEEE Trans. Aerospace Electr. Syst. AES-11 4 (July 1975) 499–513.
[27] R.L. Crabb, Solar cell radiation damage, Radiat. Phys. Chem. 43 (1–2) (January–February 1994) 93–103.
[28] D.C. Ferguson, D.P. Engelhart, R.C. Hoffmann, V.J. Murray, E.A. Plis, Space solar arrays and spacecraft charging, in: S.G. Bailey, A.F. Hepp, D.C. Ferguson, R.P. Raffaelle, S.M. Durbin (Eds.), Space Photovoltaics: Materials, Missions and Alternative Technologies, Elsevier, Cambridge, MA USA, 2022, pp. 29–50.
[29] P.T. Chiu, Space applications of III-V single- and multi-junction solar cells, in: S.G. Bailey, A.F. Hepp, D.C. Ferguson, R.P. Raffaelle, S.M. Durbin (Eds.), Space Photovoltaics: Materials, Missions and Alternative Technologies, Elsevier, Cambridge, MA USA, 2022, pp. 79–127.
[30] Y. Okada, N. Miyashita, Inverted lattice-matched GaInP/GaAs/GaInNAsSb triple junction solar cells: Epitaxial lift-off thin film devices and potential space applications, in: S.G. Bailey, A.F. Hepp, D.C. Ferguson, R.P. Raffaelle, S.M. Durbin (Eds.), Space Photovoltaics: Materials, Missions and Alternative Technologies, Elsevier, Cambridge, MA USA, 2022, pp. 265–291.
[31] L.F. Belew (Ed.), Skylab, Our First Space Station, NASA SP-400, Washington D.C. USA, 1977 available at: https://history.nasa.gov/SP-400/sp400.htm.
[32] N. North, D. Baker, in: Proc. 9th IEEE Photovoltaic Specialist Conf, 1972, pp. 263–270.
[33] S.G. Bailey, R. Raffaelle, K. Emery, Space and terrestrial photovoltaics: Synergy and diversity, Prog. Photovoltaics Res. Appl. 10 (6) (September 2002) 399–406.
[34] AIAA Solar Cells and Solar Panels Committee on Standards, Qualification and Quality Requirements for Space Solar Cells, AIAA S-111A-2014 (Revision of AIAA S-111-2005), American Institute of Aeronautics and Astronautics, Reston, VA USA, June 2014, p. 25, https://doi.org/10.2514/4.102806.
[35] L.M. Hague, K.J. Metcalf, G.M. Shannon, R.C. Hill, C.-Y. Lu, Performance of International Space Station electric power system during station assembly, in: Proc. 31st Intersociety Energy Conversion Engineering Conf. Part 1 (Aug. 11–16, 1996) 154–159, doi: 10.1109/IECEC.1996.552863.
[36] New solar arrays to power NASA's International Space Station research, retrieved from: https://www.nasa.gov/feature/new-solar-arrays-to-power-nasa-s-international-space-station-research, accessed on November 12, 2021.
[37] P. Bouisset, V.D. Nguyen, Y.A. Akatov, M. Siegrist, N. Parmentier, V.V. Archangelsky, A.S. Vorojtsov, V.M. Petrov, E.E. Kovalev, Quality factor and dose equivalent investigations aboard the soviet space station MIR, Adv. Space Res. 12 (2–3) (1992) 363–367.
[38] M.G. Smith, M. Kelley, M. Basner, A brief history of spaceflight from 1961 to 2020: An analysis of missions and astronaut demographics, Acta Astronaut. 175 (2020) 290–299.
[39] Hubble Space Telescope, electrical power: https://www.nasa.gov/content/goddard/hubble-space-telescope-electrical-power-system, accessed on November 24, 2021.
[40] NASA STS-61, https://www.nasa.gov/mission_pages/shuttle/shuttlemissions/archives/sts-61.html, accessed on November 24, 2021.
[41] M.J. O'Neill, M.F. McDanal, M.F. Piszczor, D.L. Edwards, M.I. Eskenazi, H.W. Brandhorst, Recent technology advances for the stretched lens array (SLA), a space solar array offering state of the art performance at low cost and ultra-light mass, in: Conference Record of the Thirty-First IEEE Photovoltaic Specialists Conference, 2005, pp. 810–813, https://doi.org/10.1109/PVSC.2005.1488256.
[42] G.A. Landis, Tabulation of Power-Related Satellite Failure Causes, AIAA 2013-3736, 11th International Energy Conversion Engineering Conference, July 14–17, 2013, San Jose, CA, USA, https://doi.org/10.2514/6.2013-3736.
[43] M.J. O'Neill, Space PV concentrators for outer planet and near-Sun missions using ultra-light Fresnel lenses, in: S.G. Bailey, A.F. Hepp, D.C. Ferguson, R.P. Raffaelle, S.M. Durbin (Eds.), Space Photovoltaics: Materials, Missions and Alternative Technologies, Elsevier, Cambridge, MA USA, 2022, pp. 411–432.

[44] S. Bailey, R. Raffaelle, Space solar cells and arrays, in: A. Luque, S. Hegedus (Eds.), Handbook of Photovoltaics Science and Engineering, John Wiley & Sons, 2011, pp. 365–401. Available online at: https://doi.org/10.1002/9780470974704.ch9.

[45] A.F. Hepp, J.S. McNatt, S.G. Bailey, R.P. Raffaelle, B.J. Landi, S.-S. Sun, C.E. Bonner, K.K. Banger, D. Rauh, Ultra-lightweight space power from hybrid thin-film solar cells, IEEE aerospace and elec, Systems 23 (9) (2008) 31–41.

[46] F. Lang, N.H. Nickel, J. Bundesmann, S. Seidel, A. Denker, S. Albrecht, V.V. Brus, J. Rappich, B. Rech, G. Landi, H.C. Neitzert, Radiation hardness and self-healing of perovskite solar cells, Adv. Mater. 28 (2016) 8726–8731.

[47] J. Ramanujam, D.M. Bishop, T.K. Todorov, O. Gunawan, J. Rath, R. Nekovei, E. Artegiani, A. Romeo, Flexible CIGS, CdTe and a-Si:H based thin film solar cells: A review, Prog. Mater. Sci. 110 (May 2020). Article number 100619.

[48] F. Lang, M. Jošt, K. Frohna, E. Köhnen, A. Al-Ashouri, A.R. Bowman, T. Bertram, et al., Proton radiation hardness of perovskite tandem photovoltaics, Joule 4 (5) (2020) 1054–1069.

[49] B.K. Durant, H. Afshari, S. Sourabh, V. Yeddu, M.T. Bamidele, S. Singh, B. Rout, G.E. Eperon, D.Y. Kim, I.R. Sellers, Radiation stability of mixed tin–lead halide perovskites: Implications for space applications, Sol. Energy Mater. Sol. Cell. 230 (2021) 111232.

[50] L. McMillon-Brown, T.J. Peshek, in: S.G. Bailey, A.F. Hepp, D.C. Ferguson, R.P. Raffaelle, S.M. Durbin (Eds.), Space Photovoltaics: Materials, Missions and Alternative Technologies, Elsevier, Cambridge, MA USA, 2022, pp. 129–156.

[51] B.K. Durant, I.R. Sellers, B. Rout, Perovskite solar cells on the horizon for space power systems, in: S.G. Bailey, A.F. Hepp, D.C. Ferguson, R.P. Raffaelle, S.M. Durbin (Eds.), Space Photovoltaics: Materials, Missions and Alternative Technologies, Elsevier, Cambridge, MA USA, 2022, pp. 175–195.

[52] I. Martin, K. Crowley, A.F. Hepp, Thin film solar cells and arrays for space power, in: S.G. Bailey, A.F. Hepp, D.C. Ferguson, R.P. Raffaelle, S.M. Durbin (Eds.), Space Photovoltaics: Materials, Missions and Alternative Technologies, Elsevier, Cambridge, MA USA, 2022, pp. 215–263.

[53] C.R. Mercer, Solar array designs for Deep space science missions, in: S.G. Bailey, A.F. Hepp, D.C. Ferguson, R.P. Raffaelle, S.M. Durbin (Eds.), Space Photovoltaics: Materials, Missions and Alternative Technologies, Elsevier, Cambridge, MA USA, 2022, pp. 349–378.

[54] https://www.jpl.nasa.gov/missions?mission_target=Mars, accessed Nov. 16, 2021.

[55] https://mars.nasa.gov/MPF/index1.html, accessed Nov. 16, 2021.

[56] https://www.nasa.gov/mission_pages/mer/index.html, accessed Nov. 16, 2021.

[57] https://www.jpl.nasa.gov/missions/mars-science-laboratory-curiosity-rover-msl/, accessed Nov. 16, 2021.

[58] https://mars.nasa.gov/mars2020/, accessed Nov. 16, 2021.

[59] https://mars.nasa.gov/technology/helicopter/, accessed Nov. 16, 2021.

[60] Scientists Developing Ways to Mitigate Dust Problem for Explorers, retrieved from: https://www.nasa.gov/content/scientists-developing-ways-to-mitigate-dust-problem-for-explorers, accessed November 12, 2021.

[61] NASA Mars Rover Churns Up Questions With Sulfur-Rich Soil, retrieved from: https://www.jpl.nasa.gov/news/nasa-mars-rover-churns-up-questions-with-sulfur-rich-soil, accessed November 12, 2021.

[62] P. Cohen, Mystery of Mars Rover's 'Carwash' Rolls on, New Scientist, Dec. 23, 2004. https://www.newscientist.com/article/dn6824-mystery-of-mars-rovers-carwash-rolls-on/. (Accessed 12 November 2021).

[63] M. Ashman, M. Barthélémy, L. O'Rourke, M. Almeida, N. Altobelli, M. Costa Sitjà, J.J. García Beteta, et al., Rosetta science operations in support of the Philae mission, Acta Astronaut. 125 (August–September 2016) 41–64.

[64] S.J. Bolton, J. Lunine, D. Stevenson, J.E.P. Connerney, S. Levin, T.C. Owen, F. Bagenal, et al., The Juno mission, Space Sci. Rev. 213 (2017) 5–37.

[65] S.J. Bolton, Juno celebrates a year at Jupiter, Nat. Astron. 1 (2017) 0178. Available online at: https://doi.org/10.1038/s41550-017-0178.

[66] https://www.nasa.gov/mission_pages/juno/main/index.html, accessed Nov. 16, 2021.
[67] G.L. Bennett, Mission interplanetary: using radioisotope power to explore the solar system, Energy Convers. Manag. 49 (3) (March 2008) 382–392.
[68] H. Yamaguchi, R. Ijichi, Y. Suzuki, S. Ooka, K. Shimada, N. Takahashi, H. Washio, et al., Development of space solar sheet with inverted triple-junction cells, in: 2015 IEEE 42nd Photovoltaic Specialist Conference, 2015, pp. 1–5, https://doi.org/10.1109/PVSC.2015.7356138.
[69] Y. Takao, O. Mori, M. Matsushita, A.K. Sugihara, Solar electric propulsion by a solar power sail for small spacecraft missions to the outer solar system, Acta Astronaut. 181 (April 2021) 362–376.
[70] P.M. Beauchamp, J.A. Cutts, Solar Power Technologies for Future Planetary Science Missions, Report JPL D-101316, JPL-Caltech, Pasadena, CA USA, December 2017 report available at: https://solarsystem.nasa.gov/resources/548/solar-power-technologies-for-future-planetary-science-missions/. (Accessed 16 November 2021).
[71] A.A. Siddiqi, Beyond Earth: A Chronicle of Deep Space Exploration, 1958–2016, Second edition, Washington, DC : National Aeronautics and Space Administration, Office of Communications, NASA History Division, 2018, NASA SP2018-4041, The NASA history series, report available at: https://www.nasa.gov/connect/ebooks/beyond_earth_detail.html, accessed November 16, 2021.
[72] https://www.nasa.gov/content/solar-missions-list, Accessed November 16, 2021.
[73] A. Bermudez-Garcia, P. Voarino, Olivier Raccurt, Environments, needs and opportunities for future space photovoltaic power generation: A review, Appl. Energy 290 (2021) 116757.

CHAPTER TWO

Space solar arrays and spacecraft charging

D.C. Ferguson[1], D.P. Engelhart[2], R.C. Hoffmann[1], V.J. Murray[3] and E.A. Plis[2]
[1]Air Force Research Laboratory, Kirtland Air Force Base, Albuquerque, NM, United States
[2]Assurance Technology Corp., Carlisle, MA, United States
[3]AFRL Kirtland AFB, Albuquerque, NM, United States

2.1 Introduction to spacecraft charging

Most of space near Earth is filled with a dilute plasma of electrons and positive ions from gas ionized by ultraviolet light from the Sun and trapped in Earth's magnetic field. Satellites orbit through this plasma, which is for the most part too tenuous to cause significant drag. At geosynchronous altitudes, where most communications satellites orbit, the plasma consists almost entirely of hydrogen ions (protons) and electrons in equal parts by number. However, the plasma impinges on satellite surfaces, and the flux of electrons is much greater than that of the ions. At a given plasma temperature, the ratio of the average electron velocity to that of the ions is the inverse square root of the mass ratio, which for electrons and protons is about 43. Therefore, the electron flux on a surface will be about 43 times that of the protons. The surface will charge negatively until it repels enough of the incoming electrons so that those remaining balance the incoming ions (the so-called "charge balance" condition) [1]. In practice, all other things being equal, this means that surfaces can and will charge up to a potential equal to the electron temperature (in eV). This is called spacecraft charging.

2.1.1 Issues presented by solar array space utilization

If all spacecraft surfaces charged equally, charging would not be a concern for designers. However, surface charging is modified by the photoelectric effect and secondary electron emission, both of which are inherent properties of a material. Moreover, the rate of charging of different surfaces depends on their capacitance with respect to spacecraft "ground." Consequently, different surfaces will charge to different potentials. This is known as differential charging. If the potential difference between adjacent surfaces is sufficiently great, an electric arc will occur, eliminating the potential difference [1]. Arcing is most likely to occur at "triple junctions," where an insulator, a conductor, and the space plasma meet (Fig. 2.1) [2].

Photovoltaics for Space
ISBN 978-0-12-823300-9, https://doi.org/10.1016/B978-0-12-823300-9.00002-9

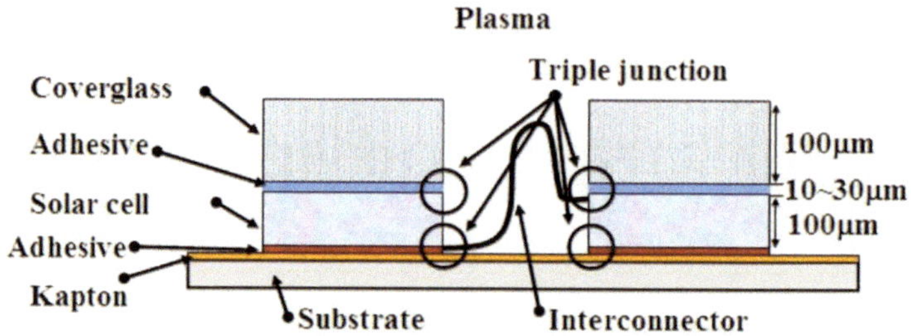

Figure 2.1 Plasma triple junctions in a solar array. *Courtesy: NASA.*

The risk of arcing can be significantly reduced by covering all satellite surfaces with conductive materials that are grounded to the spacecraft's chassis. However, this technique cannot usually be used on solar arrays, the source of satellite power. There are many triple junctions between the solar cell material (conductor), the solar cell interconnects, and the solar cell cover glasses. Cover glasses are employed to protect the solar cells from radiation. They are usually good insulators and often have an antireflection coating that is also prone to charging. Unfortunately, there are very few conducting materials that are sufficiently transparent to be used as cover glass coatings [6]. If arcs occur on the solar arrays, the entire cover glass capacitance can be discharged in a matter of microseconds in a localized arc with peak currents >10 A [3]. This arc current can couple to the power system creating a power transient [4], can spread contaminants onto the cover glass surfaces [4], and create radio frequency interference [5]. In the worst case, the arc can transition into a sustained discharge between two solar cell strings [4], powered by the solar array itself, until the discharge destroys the solar cell circuit [4]. This must be avoided at all costs.

2.1.2 The current balance equation

The equation governing spacecraft surface charging is called the current balance equation (Eq. 2.1). It can be used to calculate the potential of an entire spacecraft or an individual surface [1]:

$$F_e(V) - F_i(V) - F_{cond}(V) - F_{see} - F_{pe} = 0. \tag{2.1}$$

Here, V is the potential of the surface, F_e is the electron flux onto the surface, F_i is the ion flux, F_{cond} is the conduction flux through an imperfect insulator (for an entire spacecraft, omit this term), F_{see} is the secondary electron flux (may contain electron and ion secondary electron emission, and backscattered electrons), and F_{pe} is the photoelectron flux from the surface. The term V must include any voltage impressed by the power system. The complicating factor here is that the surface potential (V) depends on the previous charging history. The equation can be solved for known or measured environments by using simplifying assumptions, such as $Q = CV$, etc., but in many cases the geometry of the spacecraft determines the fluxes that strike surfaces. Therefore, spacecraft charging codes that incorporate 3D geometries are often used to calculate all surface potentials on a spacecraft [7]. Surface potentials can also be calculated using the capacitance of the spacecraft with respect to space (a very small quantity). To calculate charging of individual dielectric surfaces with respect to spacecraft "ground" (the spacecraft chassis), the capacitance of each surface is needed (usually a significantly larger number). The time it takes to charge is therefore very short (seconds) for the entire spacecraft, but it may be much longer (even hours) for individual insulating surfaces. A few specific items of nomenclature are as follows:

1. plasma potential: the potential a surface in space would reach if electron and ion fluxes were equal;
2. absolute charging: the potential of the entire spacecraft relative to the plasma potential;
3. differential charging: the potential of surfaces relative to each other.

Another complicating factor for predicting spacecraft charging is the secondary electron yield of its constituent materials. A typical secondary electron emission yield curve is shown in Fig. 2.2 and is defined by the ratio of electrons escaping the surface to electrons impinging upon it. The secondary electron emission yield δ from a surface can have a dependence on impinging electron energy. For some impinging electron energies, more than one electron is emitted per electron that hits the surface (yield >1), resulting in a more positive surface potential (Fig. 2.2) [1]. This effect can discharge a negative surface or increase the charge on a positive surface. The energy at which the yield first becomes greater than 1 is called the first crossover point, and the energy at which the yield drops below 1 again is the second crossover point [8]. The energy at

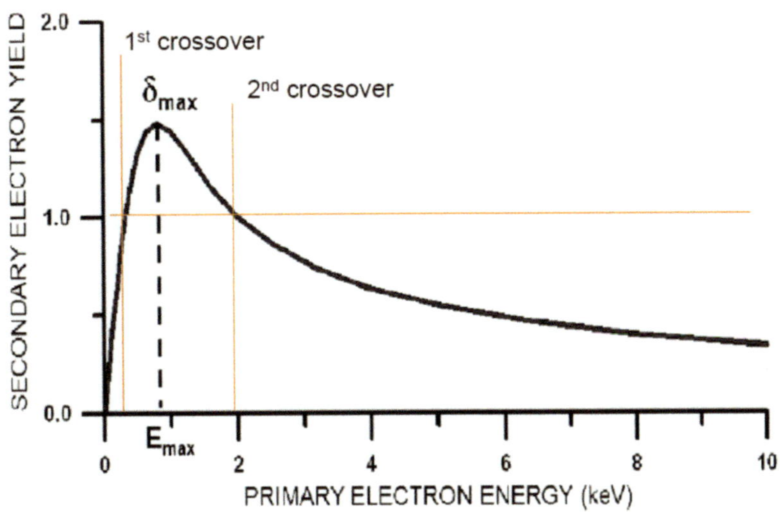

Figure 2.2 A hypothetical secondary electron emission curve. *Courtesy: NASA.*

which δ is a maximum is E_{max}. When the secondary yield is less than 1, the surface will charge more negatively. Some materials, such as graphite, have no impingement energy at which they have a yield greater than 1 [9]. The secondary electron emission is strictly a surface phenomenon, so for a coated surface, it is representative of the coating, not the underlying substrate.

2.1.3 Natural plasma fluxes

The equatorial low-Earth orbit (LEO) plasma is a dense and relatively cool plasma that tends to discharge surfaces rather than charge them [10]. To build up differential potentials that can cause arcing on surfaces, it is necessary for the power system to provide the voltages. Thus, the simple solution of limiting the solar array string voltage can be used to suppress charging and mitigate arcing. For reasons beyond the scope of this chapter, solar array string voltages of less than about 50 V are almost never prone to arcing in LEO, and even voltages as high as 160 V (International Space Station, ISS) can be used in LEO. By contrast, geosynchronous Earth orbit (GEO), where most communications satellites reside, features a tenuous and hot plasma that can lead to severe charging [1]. For example, a plasma "temperature" of 0.1–0.25 eV (1100–2750 K) is typical for LEO (outside of the polar caps) [10], whereas in GEO, temperatures of 7000–25,000 eV (75 million to 200 million K) abound [1]. Under these conditions, satellites can charge negatively to thousands of volts, and differential charging levels can reach arcing thresholds in a short period of time.

There are several peculiarities associated with charging in GEO [11]. Due to the low electron densities, the fluxes on surfaces are often low enough that the sunlit side of the spacecraft surfaces are almost completely discharged by photoemission. However, the dark side may be negatively charged to a potential of many thousands of volts. Consequently, it can be particularly dangerous when a satellite emerges from eclipse because highly charged surfaces can be quickly discharged by photoemission, leading to large differential potentials. Thus, geostationary satellites are more vulnerable to arcing during eclipse seasons (near the equinox dates) than at other times [12]. Additionally, electrons present in GEO plasma may have energies above 9 or 10 keV, well above the second crossover point of secondary electron emission curve for most materials, resulting in more negative surface potentials. When the Earth's magnetosphere is hit by solar storm plasmas, the electron fluxes may dramatically increase in a period of minutes, further enhancing the risk of differential charging [13]. Electron fluxes vary daily, as the satellite comes into or goes out of regions where the electron density or temperature are higher than at surrounding areas. Electron fluxes also vary seasonally, as the Earth's magnetic polar caps point toward or away from the sun, changing the ease with which solar storm particles can enter the magnetosphere and how many of these particles surround the satellite in its orbit [14].

The harsh GEO environment readily leads to spacecraft charging. During a strong charging event, such as a solar storm, the solar array string voltage becomes relatively unimportant, as charging voltages overwhelm them. Therefore, differential charging between surfaces is a bigger concern for arcing than interstring potentials. Clever selection of spacecraft construction materials can be used to mitigate differential charging. The most important electrical properties to consider when mitigating differential charging are the materials' bulk and surface conductivity, secondary electron emission, and photoemission work function.

2.1.4 Arcing

2.1.4.1 Arcing voltages versus electric fields

Our discussion thus far has concerned surface potentials, but arcing is the result of large electric fields (the spatial derivative of voltages) [15] developing to the point of exceeding the dielectric breakdown strength of the native material. Arcing is an inherently local phenomenon (Fig. 2.3) [3,4,16]; however, often the microscopic conditions (surface roughness, proximity of conductors to insulators, exact potentials at the arc site) that would allow us to determine the electric field thresholds for arcing are unknown. For this reason, scientists and engineers refer to the differential voltage threshold for arcing on solar cells and arrays [15]. It is a truism that lower voltages mean weaker electric fields. If the threshold voltage for arcing is poorly defined for a tested sample, it means that local conditions modify the fields appreciably, leading to a wide, environmentally dependent variability in threshold voltage.

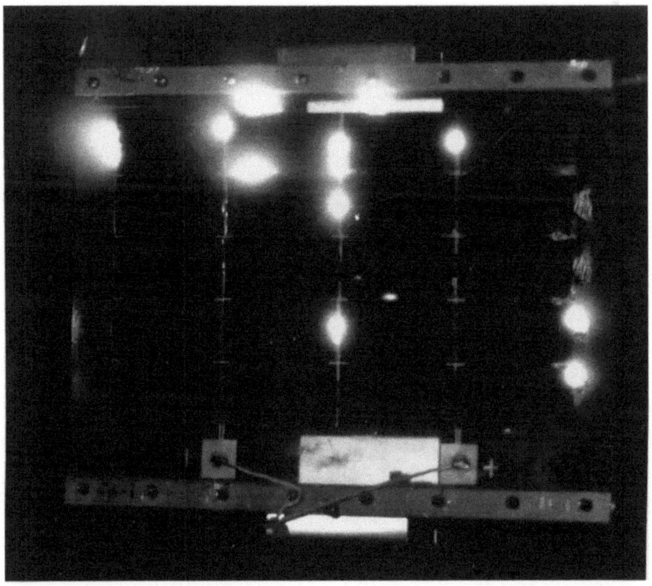

Figure 2.3 Electrostatic discharge (ESD) on a space solar array in low-Earth orbit conditions. *Courtesy: NASA.*

2.1.4.2 Primary versus secondary arcs

The initial discharge in an arc, sometimes called the electrostatic discharge (ESD), lasts only a short time (microseconds) and is often impulsive. Such an arc is said to be a primary arc. The arc current in such a discharge originates at a microscopic region (the arc site, which is typically a triple point), and intense transient currents vaporize a small amount of the surrounding material, creating an expanding plasma cloud. On a solar array, the initial current spike is between the cell conductor or interconnect and the cover glass surface. Primary arcs have a finite duration determined by the time it takes for the local plasma generated by the ESD to sweep across and discharge nearby surfaces, such as cover glasses [18]. For large arcs and large arrays, the primary arcs may last 100 μs or longer and are extinguished when the capacitive energy of the array is exhausted. Since the plasma generated by a primary arc is conductive, as it sweeps out over the array, it can cause an interstring "short" to occur, know as a secondary arc. An arc that extends for a few times this duration is called a secondary arc, or a temporary sustained arc. Usually, it is considered that an arc that extends longer than 1 ms in duration (a permanent sustained arc) may last until the power supply is turned off or the string burns up, whichever comes first [17]. A secondary arc can draw charge from other strings adjacent to the arc site [17]. It is these permanent sustained arcs that can destroy one or two solar array strings at a time [12].

2.1.4.3 Primary arc versus secondary arc thresholds

Primary arcs on solar arrays are due to the electric field between the cell or interconnect and the cover glass surface. This is a result of differential charging caused by the ambient environment. Primary arc voltage thresholds in GEO can vary from 500 to 2000 V. Secondary arcs arise from the electric field between adjacent solar array strings, and the voltages are usually not greater than the string voltage [19]. Modern spacecraft may have string voltages of 100 or 120 V. The secondary arc threshold therefore must be within this range, or the arcs will not occur, but it is usually much lower, 60 V or so for arrays where secondary arc mitigation has not been attempted [20]. In the presence of mitigation techniques, such as grouting cell edges [21], this arcing threshold may be raised above the string voltage and make the array (at least the pristine array) immune from secondary arcing [21].

2.1.4.4 Arcing transients versus sustained arcs

The rise time for primary arc currents is very short (usually less than a microsecond) and can lead to power system transients on the same timescale [22]. In addition, primary arcs usually finally cut off in nanoseconds, leading to even shorter transients. If a secondary arc is started and extends for more than about 1 ms, we say the arc has become sustained, and it has transitioned into an arc between, and powered by, the solar array strings themselves (Fig. 2.4) [23]. These sustained arcs sometimes continue until the string's circuit is destroyed. The possibility of sustained arcs has led most space solar array designers to go to extraordinary lengths to prevent their occurrence, for even one event may disable a satellite [12]. ISS has seen sustained arcing on its solar arrays, where the arc initiation was from a micrometeoroid or debris impact, not a primary plasma arc (Fig. 2.5) [24].

2.2 Arcing: Effects, standards, and mitigation

Arcing due to spacecraft charging is the major process that must be mitigated or prevented. In this section, we review its effects and standards that have been produced to help prevent or mitigate it, as well as best practices.

2.2.1 Effects of arcing

2.2.1.1 Primary arcing: Loss of cells, cell-junction shorting

2.2.1.1.1 Optical flashes and radio frequency interference (RFI)

Visible light produced by solar array arcs was previously used to detect arcing in laboratory testing [25]. Low-level arcs can sometimes be seen with the eye or camera even without measurable arc currents [26]. Some arcs produce broadband RFI [27], allowing simple antennas in vacuum chambers to be used to precisely determine when arcs occur and allow oscilloscopes and other equipment to be triggered to record high time resolution data [28]. Broadband arcing RFI from arcs on orbiting satellites has been detected and measured with radio telescopes on the ground [29].

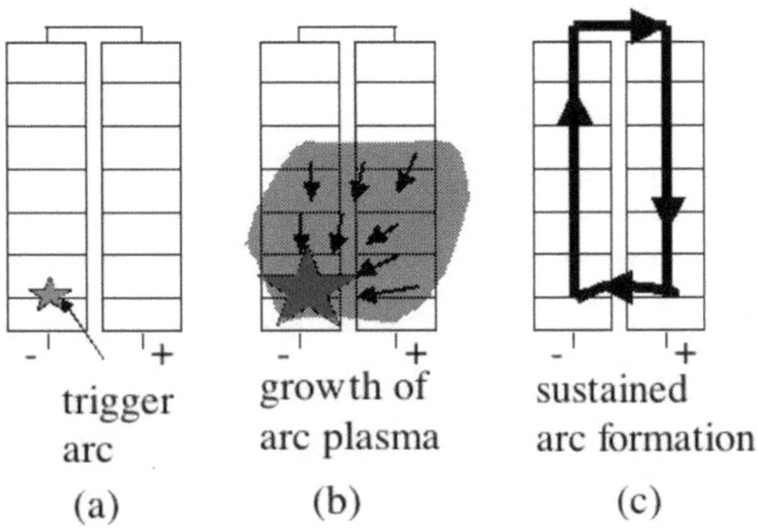

Figure 2.4 Sustained arc development from a primary or trigger arc. *Courtesy: NASA.*

Figure 2.5 ISS solar array sustained arcing damage from micrometeoroid or orbital debris (MMOD) impact. *Courtesy: NASA.*

2.2.1.1.2 Contamination and arc tracking

Contamination from individual arcs has been known to short out junctions in multi-junction solar cells ([30,31], Fig. 2.6). Over long periods, cover glasses may become so contaminated by material blown off in arcs that power production may be degraded (see Fig. 2.7) [31]; the colors of cover glasses may be affected [32,33]. In rare cases, individual arcs may be large enough that they affect optical measurements of spacecraft from the ground [34]. Another possible effect is "arc tracking" where charred Kapton, from the extremely high temperatures at the arc site, makes a conductive path between two supposedly isolated conductors (Fig. 2.5) [35].

2.2.1.1.3 Loss of sensitive circuits

Arc-caused transients in spacecraft power and conducted currents within spacecraft can damage or destroy sensitive circuits, especially digital electronic switches [13]. On the Galaxy 15 spacecraft, an inadvertent bit flip in an electronic switch after an arc caused a loss of command control for over 10 months [13].

2.2.1.2 Sustained arcing: Loss of solar array strings, tethers, etc.

It is possible for initial primary arcs to transition into sustained arcs, powered by the solar array itself. This happens when an arc between two adjacent cells that are at different

Figure 2.6 Arc contamination on Mir solar array on-orbit. Contaminated areas are cloudy semicircles near cell edges, as in the circled area. *Courtesy: United States Air Force.*

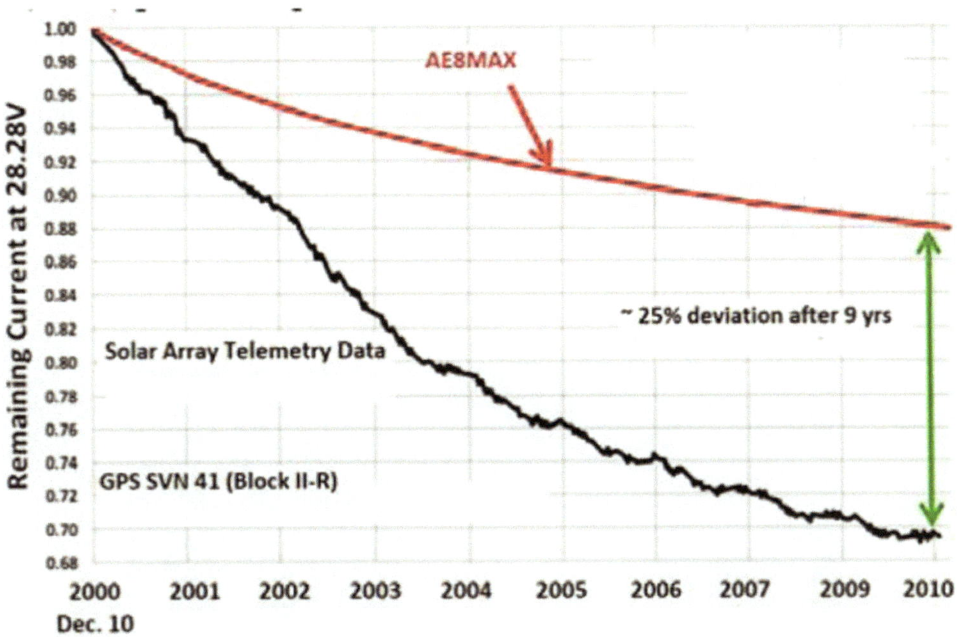

Figure 2.7 Power loss on a GPS satellite over 10 years in excess of radiation loss. *Courtesy: NASA.*

voltages starts delivering current between the cells. Usually, the cells are at different voltages because they are on different strings or on opposite ends of a single string. When a sustained arc develops, it can continue as long as the strings are active or until a short occurs that disrupts the arc circuit, as illustrated in Fig. 2.8. This may permanently damage one or two strings of the array. In Fig. 2.9, a sustained arc has resulted in arc tracking that has destroyed at least two strings. The voltage threshold for sustained arcs between adjacent cells is often less than the voltage threshold for the initial primary arc.

An unusual sustained arc occurred on the Space Shuttle TSS-1R tether. In this case, the arc started between the tether and a reel enclosure because of gas leaking from a tether insulation puncture (a Paschen breakdown of neutral gas) when the tether conductor was a few thousand volts negative. It transitioned into a plasma discharge when the gas leak site exited the enclosure, burned the 20-km-long tether into two parts, and continued arcing for a few minutes after the tethered satellite was flying off into space, only extinguishing when the shuttle went into eclipse, and the surrounding plasma density became so low that the arc current could no longer be sustained; see Fig. 2.10 for a photo of the severed tether. Obviously, sustained arcs are to be avoided at all costs.

2.2.2 Standards

There are many national and international standards relevant to solar array charging and arcing. While details may differ, the main requirements are generally agreed upon. Here is a partial listing of relevant documents: NASA TP-2361 [1]; NASA-HDBK-4002A [35]; NASA-STD-4005A and NASA-HDBK-4006A [4]; BSR/ANSI/AIAA S-115 [36]; ECSS-E-ST-20-06C Rev. 1 [37]; JERG-2-211A [38]; ISO 11221 [39]; ISO 19923:2017 [40]. A brief overview of each document constitutes the remainder of this subsection.

Figure 2.8 Frame from laboratory video of sustained arcing on solar array. *Source: Authors.*

Figure 2.9 Result of sustained arc on Eureca-A. *Courtesy: NASA.*

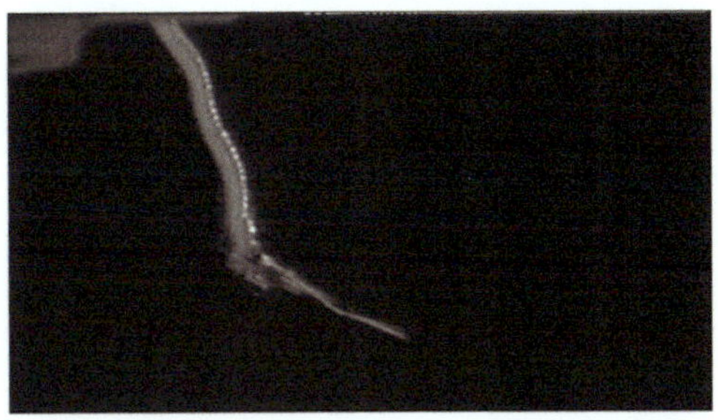

Figure 2.10 Result of sustained arc on TSS-1R. *Courtesy NASA.*

2.2.2.1 NASA TP-2361

The "Design Guidelines for Assessing and Controlling Spacecraft Charging Effects" [1] were issued in 1984. They only apply to spacecraft in GEO orbits, but they contain most of the guidance necessary for understanding and controlling GEO spacecraft charging. They are the best place to start to learn about charging.

2.2.2.2 NASA-HDBK- 4002A

"Mitigating In-Space Charging Effects—A Guideline," [35] issued in 2011 by JPL, is the first revised version. NASA-HDBK- 4002-B is due out in 2021. This is an attempt to incorporate all that has been learned about GEO charging since 1984.

2.2.2.3 NASA-STD-4005A and NASA-HDBK-4006A

"Low Earth Orbit Spacecraft Charging Design Standard" and " .... Handbook," (2018) [4] are only applicable to LEO equatorial orbits (−50 <inclination <50 degrees) and power system voltages of about 55 V or greater. There are, however, sections on sustained arcing that apply to any spacecraft. The charging environment in equatorial LEO is unique and requires different design considerations than GEO standards. Of primary importance here is the solar array string voltage.

2.2.2.4 BSR/ANSI/AIAA S-115

The "Low Earth Orbit Spacecraft Charging Design Standard Requirement and Associated Handbook" [36] is an American National Standard. It was last affirmed by vote in 2017. This standard incorporates all the material in NASA-STD-4005 and NASA-HDBK-4006 in one place. Like them, it is an equatorial LEO orbit standard only.

2.2.2.5 ECSS-E-ST-20-06C Rev.1

"Space Engineering, Spacecraft Charging," 2019 [37] is a European standard. Although it claims to be for all spacecraft, it is noticeably deficient in LEO-specific design requirements. The sections on grounding are well written and pertinent.

2.2.2.6 JERG-2-211A

"Design Standard – Spacecraft Charging and Discharging," 2012 [38] is the Japanese national standard. It is a comprehensive GEO standard patterned after the European standard. It is especially concerned with active charge control.

2.2.2.7 ISO 11,221

"Space Systems – Space Solar Panels - Spacecraft Charging Induced – Electrostatic Discharge Test Methods," 2011 [39], provides acceptance and qualification standards for space solar array hardware against plasma arcing. A truly international standard, it was authored by experts from the United States, Europe, and Asia.

2.2.2.8 ISO 19,923:2017

"Space Environment (Natural and Artificial) – Plasma Environments for Generation of Worst Case Electrical Potential Differences for Spacecraft," 2017 [40], gives natural worst-case charging environments to be used with spacecraft charging codes.

2.2.3 Mitigation strategies

While there are many mitigation strategies against solar array arcing, what has been chosen for a given program depends on engineering trades of cost versus benefit. There are two main categories of arc mitigation, passive and active. Passive approaches use design decisions of geometry, surface materials, and grounding to control differential charging. Active solutions use devices that control the surrounding plasma to eliminate differential charging. Some of the design solutions are discussed subsequently.

2.2.3.1 *Passive methods*

2.2.3.1.1 Grouting

Grouting refers to covering cell edges with a pliable dielectric material to prevent the space plasma from contacting the cell and arc plasmas from reaching other cells and strings. Although mainly used to prevent sustained arcs [41], it can also be used to prevent primary arcs. The main drawback to this practice is the aging of the grout in the space environment, which can cause cracking and delamination.

2.2.3.1.2 String layout

Another ploy to prevent sustained arcing involves serpentine or leapfrog string layouts to prevent adjacent cells in high voltage strings from having high differential potentials [42]. This technique may be used in conjunction with other methods.

2.2.3.1.3 Diodes

Sometimes it is possible to prevent current from communicating between strings during an arc by limiting forward-bias current below the sustained arcing current limit, and preventing back-bias current flow, using opposing diodes. Usually, this means adding extra mass to the array, and it may mean removing cells from each string to accommodate flat-pack diodes [43].

2.2.3.1.4 Conductive cover glass coatings

The number of transparent conductive coatings for solar cell cover glasses is quite limited. Traditionally, ITO (indium tin oxide) has been used [44]. However, ITO is hard to bond to grounding wires (a necessary requirement, or else one is left with an ungrounded conductor), and there have been reports of grounding tabs falling off during vibe testing. Also, most space solar arrays use magnesium fluoride (MgF_2) as an antireflective coating, but MgF_2 is a dielectric and incompatible with ITO use. These complications could explain why measurements at the Arecibo radio telescope have detected arcing on solar arrays of satellites that are supposedly arc-proof due to ITO coatings.

2.2.3.1.5 Bulk-conductive cover glass(es)

First suggested after the SS/L Tempo-2 and PAS-6 disasters [12], this remedy requires use of somewhat conductive solar cell cover materials (usually glass) [45]. Solar cell cover glasses are often ceria-doped borosilicate glass. Three typical solar cell cover glass types are called CMG, CMX, and CMO (Excelitas Technologies, Billerica, MA). Of the three, CMX has been shown to have a low enough bulk resistivity (in ohm-cm) to prevent cover glass surface charging from exceeding the arc voltage threshold [46]. However, the CMX light transmission cutoff is at a longer wavelength than is optimum for modern multijunction solar cells [45].

2.2.3.1.6 Conductive paints

Instead of focusing on solar cell charging, grounded conductive paints seek to keep differential charging to a minimum by reducing surface charging of the entire spacecraft or the backs of the solar arrays [47]. Paints, however, may deteriorate in the GEO environment, which can affect material properties, and in the worst case can lead to loss of flexibility and flaking.

2.2.3.1.7 Tailoring photoemission and secondary electron emission

Surface material properties tend to evolve toward those of carbon (very low secondary electron emission) after some time in space [48]. This unintentional change in surface properties leads to the idea of manipulating the design of solar cell cover glasses to change the secondary electron emission properties (and/or photoemission properties) to mitigate spacecraft charging [49]. Shu Lai has been a proponent of shiny, reflective surfaces to suppress photoemission [50].

2.2.3.1.8 Passive electron emitters: Spindt cathodes or ELF, electron-emitting paints, and films

Spacecraft charging and arcing can be prevented by using materials that emit electrons when they are charged negatively. These materials may be made of electron-emitting point cathodes (Spindt cathodes) [51], films constructed so that there is a breakdown and discharge at very low negative voltages (ELF, or electron-emitting film, for one example [52]), or paints and films with very low electron work function energies [53]. None of these are currently used on operating satellites because they are novel, prone to contamination or burnout, or are just now in development. Because of these difficulties, active electron emitters (detailed subsequently) have been used on some spacecraft.

2.2.3.2 Active methods

2.2.3.2.1 Plasma contactors or electron guns

Active charge control has a long and somewhat successful history. Of these techniques, the only one that does not require gas emission and control systems is the electron gun. Active charge control got its start with SCATHA (spacecraft charging at high altitudes [54]), which employed an electron gun and a xenon ion emitter to investigate charge control. The electron gun proved to be somewhat effective at controlling negative charging, although a barrier potential is believed to have been set up that prevented the original −6 kV charging from being discharged to beyond −1 kV. The ion gun was found to be severely space-charge limited. These limitations have been seen in other flight experiments since that time. This led to the advent of plasma contactors. In a plasma contactor, a high-density plasma of electrons and ions is emitted, so there are ostensibly no space-charge or potential barrier limitations, and electrons (or ions) may be freely emitted to control charge buildup [55,56]. A notable success for their use is the

ISS, where they have been used from the beginning to prevent arcing from astronaut suits during spacewalks.

2.2.3.2.2 Cold gas thrusters

Spacecraft attitude control is often done by emitting cold gas through nozzles. Cold gas thrusters may function as charge control devices if the spacecraft potential is more negative than the gas ionization potential, in which case the emitted gas may be ionized by Paschen discharge, and electrons may be emitted. This technique has been tested out on numerous rocket flights [57]. Since attitude control mainly uses cold gas thrusters, charge control may be an extra added benefit.

2.2.3.2.3 Ion and Hall thrusters

Ion and Hall thrusters achieve space-charge neutrality by the use of beam neutralizers, usually hollow cathodes arranged near the exhaust. If the thruster's cathode is "grounded," the thruster plume may emit electrons or ions to also neutralize the spacecraft frame potential. For this technique, it is imperative that the neutralizer electrons are able to efficiently enter the main thruster beam, or charge neutralization will only be partial.

2.3 Models of environment and charging

Models are very important in design of spacecraft. To accurately model spacecraft charging, three types of models are employed. Environment models are used to predict what electron and ion densities, temperatures, and fluxes your satellite will encounter. Most environmental models are empirical, as our physical knowledge of spacecraft environments is incomplete. Surface charging is predicted using geometrical spacecraft models that calculate how different surfaces will charge when placed in a certain environment. These are usually 3D, physics-based, time-dependent models. Finally, models of the interior charging of a satellite or spacecraft are needed. Many high-energy electrons can pass right through the outer layers of a spacecraft to imbed themselves in interior dielectrics, a phenomenon referred to as deep-dielectric charging [58].

2.3.1 Environment models

Standard models of the space environment extending out to GEO are detailed in the remainder of this subsection. Our discussion includes their heritage and details of output(s).

2.3.1.1 IRENE (AP9/AE9/SPM)

A direct descendant of the AP8/AE8 (Aerospace Proton and Electron) model, the IRENE (International Radiation Environment Near Earth) empirical model (formerly called AP9/AE9/SPM [Space Plasma Model]) is based on many years of satellite measurements of the local environment for a wide range of altitudes, longitudes, and latitudes [59]. Interpolation is used to give estimates for gaps in the coverage. It calculates electron and proton fluxes, as well as their statistical probabilities (percentiles) for a wide range of energies, orbits, and durations.

2.3.1.2 AF MSM/MSFM

MSM/MSFM is an acronym for Magnetospheric Specification Forecast Model. The code originated at Rice University, and it is the US Air Force (and Space Force) Standard [60]. The MSFM specifies electron and ion fluxes of energies between 100 eV and 100 keV in the distance range of 2–10 Earth radii on a two-dimensional grid. A semiempirical model, AF MSM/MSFM uses physics to predict and interpolate fluxes for positions where no data exist.

2.3.1.3 International Reference Ionosphere

The International Reference Ionosphere (IRI) is an international project sponsored by the Committee on Space Research (COSPAR) and the International Union of Radio Science (URSI). It is an empirical standard model of the ionosphere, based on all available data sources [61]. Several steadily improved editions of the model have been released. For given location, time, and date, IRI provides monthly averages of the electron density, electron temperature, ion temperature, and ion composition in the ionospheric altitude range.

2.3.2 Surface charging models

The standard models for surface charging are Nascap-2K, SPIS, MUSCAT, Coulomb 2, and the Spacecraft Charging Index. Because Nascap-2K is EAR (Export Administration Regulations) restricted, and may not be exported abroad, other countries than the United States have developed their own 3D spacecraft charging codes. The codes listed next are widely used and give answers similar enough that any can be used to calculate the charging on an entire spacecraft or any spacecraft surface, such as solar arrays.

2.3.2.1 Nascap-2K

The de facto US standard spacecraft charging code, Nascap-2K (originally an acronym for NASA Air Force Spacecraft Charging Analysis Program) has been extensively verified against actual spacecraft potential and environment measurements [62]. It can calculate fluxes and potentials on different surfaces by full-PIC (particle in cell), hybrid PIC (analytical electron fluxes plus PIC for ions), or fully analytical means, and it has approximate algorithms for sheath solutions. Because of its analytical solutions, it is very fast, and it can give full surface potential histories for a spacecraft with 2000 surface nodes on a desktop computer in a matter of minutes. It has full graphical user interface, and materials properties may be specified or taken from a menu of common spacecraft materials. It can do calculations for LEO, LEO polar (auroral), Medium Earth Orbit (MEO), and GEO environments, and it allows quick trade studies for different designs to be made.

2.3.2.2 SPIS

SPIS (Spacecraft Plasma Interaction Software) is an ESA (European Space Agency) and CNES (French Space Agency) code that has many of the same capabilities as Nascap-2K [63], and it is part of the SPENVIS (Space Environment Visualization) suite of codes. However, SPIS is a full-PIC code, which makes it slower than Nascap. While Nascap is a standalone code, SPIS may be accessed online. Like most 3D codes, SPIS sets up a grid for its calculations, but the Nascap and SPIS gridding schemes are very different. SPIS is widely used in Europe and in the United States by users willing to divulge models of their spacecraft designs. The US military does not use SPIS because it is not considered secure.

2.3.2.3 MUSCAT

MUSCAT (Multi-Utility Spacecraft Charging Analysis Tool), developed by JAXA (the Japanese Space Agency) and Kyushu Institute of Technology, is the Japanese equivalent of Nascap or SPIS. Comparative results of these three charging codes for a typical GEO spacecraft model are given in ISO 19923:2017 [64]. It is a full-PIC code, and so is slower than Nascap, even though the calculations are parallelized.

2.3.2.4 Coulomb-2

The Russian spacecraft charging code, Coulomb-2 [64], gives results comparable to the other three aforementioned codes [65]. It gives possible reasons for the divergence when the codes do not agree.

2.3.2.5 Spacecraft Charging Index

Taking a different tack from detailed 3D, physics-based spacecraft models, the Spacecraft Charging Index [11] seeks to find a single factor that is closely related to spacecraft charging (both absolute and differential) for typical spacecraft. Assuming that Nascap-2K calculates the surface charge correctly, different spacecraft models were subjected to various sunlit environments, and the results were compared with possible charging factors. The charging index mostly closely correlated with spacecraft charging was the total electron flux above a threshold energy of about 9 keV. This can be justified by noting that the ion fluxes in all GEO environments are much less than the electron fluxes, and for electrons between the first and second crossover points in energy, incoming electron fluxes discharge, rather than charge, spacecraft surfaces, due to the secondary electron yield being greater than one. Comparisons with real spacecraft charging values for cases where electron fluxes were measured validated the approach.

2.3.3 Deep-dielectric charging models

Although not discussed in this chapter, deep-dielectric charging is an important topic. Notable deep-dielectric charging models include NUMIT (Numerical Integration [66]) and derivative codes NUMIT-3D, NUMIT2 and AF-NUMIT2, MCICT (Monte-Carlo Internal Charging Code [67]), and GEANT4 [68].

2.4 Conclusions: TRhe necessity for testing

While models of the environment and spacecraft interactions with the environment are helpful in determining potentials and differential potentials, our knowledge of the physics of arc thresholds is not sufficient to determine whether a given solar array geometry on a given spacecraft will arc. For that, we need laboratory measurements of arcing voltage thresholds, both for primary arcs and for sustained arcs [69]. Laboratories for these types of measurements exist both within governments and in private industries. Rules of thumb are not sufficient. Even supposedly identical arrays or cells may have different thresholds and arc rates. For an introduction to such qualification test measurements, Ref. [39] is a good resource. Reference [70] provides a good introduction to and overview of spacecraft charging issues.

References

[1] C.K. Purvis, H.B. Garrett, A.C. Whittlesey, N.J. Stevens, Design Guidelines for Assessing and Controlling Spacecraft Charging Effects, NASA-TP-2361, National Aeronautics and Space Administration, 1984. 47 pp., available at: https://ntrs.nasa.gov/api/citations/19840025381/downloads/19840025381.pdf.

[2] B. Vayner, Arcing on solar array surface in space and laboratory, in: 3rd International Workshop on Mechanisms of Vacuum Arcs, Albuquerque, New Mexico, USA, October 1–4, 2012.

[3] D.C. Ferguson, B.V. Vayner, J.T. Galofaro and G. Barry Hillard, Arcing in LEO - does the whole array discharge? in: Proceedings of the 9th Spacecraft Charging Technology Conference, (JAXA-SP-05-001e), Tsukuba, Japan, April 4–8, 2005.

[4] Low Earth Orbit Spacecraft Charging Design Handbook, NASA-HDBK-4006a, National Aeronautics and Space Administration, 2018. https://standards.nasa.gov/standard/nasa/nasa-std-i-4005. (Accessed 10 May 2022).

[5] D. Ferguson, S. White, R. Rast, E. Holeman, The case for Global Positioning System arcing and high satellite arc rates, IEEE Trans. Plasma Sci. 47 (8) (August 2019) 3834–3841, https://doi.org/10.1109/TPS.2019.2922556.

[6] Z.S. Levin, D.M. Wilt, R. Hoffman, D. Ferguson, Conductive space solar cell cover-glass replacement technology, SPIE OPTO, in: 2014, San Francisco, Proceedings Volume 8981, Physics, Simulation, and Photonic Engineering of Photovoltaic Devices III, 2014, p. 89810W, https://doi.org/10.1117/12.2045450.

[7] V.A. Davis, B. Gardner, M.J. Mandell, NASCAP-2K Version 4.3 User's Manual, AFRL-RV-PS TR-2017-0003, Leidos, Inc., 10260 Campus Point Drive, Mailstop C4, San Diego, CA 92121 USA, December 31, 2016. Available at: https://apps.dtic.mil/dtic/tr/fulltext/u2/1034384.pdf.

[8] A. Chvyreva, A. Pemen, Experimental investigation of electron emission from dielectric surfaces due to primary electron beam: A review, IEEE Trans. Dielectr. Electr. Insul. 21 (2014) 2274–2282.

[9] K. Ueno, T. Kumihashi, K. Saiki, A. Koma, Characteristic secondary electron emission from graphite and glassy carbon surfaces, Jpn. J. Appl. Phys. 27 (5A) (1988) L759, https://doi.org/10.1143/JJAP.27.L759.

[10] D.C. Ferguson, G.B. Hillard, Low Earth Orbit Spacecraft Charging Design Guidelines, NASA/TP—2003-212287, 2003.
[11] D.C. Ferguson, R.V. Hilmer, V.A. Davis, The best GEO daytime spacecraft charging index, J. Spacecraft Rockets 52 (2) (2015) 526–543.
[12] D.C. Ferguson, R.C. Hoffmann, R.J. Cooper, J.A. Hughes, 1997–2002 Solar array string failures revisited, J. Spacecraft Rockets (54) (2017) 542–553, https://doi.org/10.2514/1.A33637.
[13] D.C. Ferguson, W.F. Denig, J.V. Rodriguez, Plasma conditions during the Galaxy 15 anomaly and the possibility of ESD from subsurface charging, in: 49th AIAA Aerospace Sciences Meeting, January 4–7, 2011, Orlando, FL, USA, AIAA 2011-1061.
[14] D. Vassiliadis, A.J. Klimas, S.G. Kanekal, D.N. Baker, R.S. Weigel, Long-term-average, solar cycle, and seasonal response of magnetospheric energetic electrons to the solar wind speed, J. Geophys. Res. Space Phys. 107 (A11) (November 2002). SMP 22-1-SMP 22-18.
[15] A.R. Frederickson, J.E. Nanevicz, J.S. Thayer, C.L. Enloe, E.G. Mullen, D.B. Parkinson, Leaky insulating paint for preventing discharge anomalies on circuit boards, IEEE Trans. Nucl. Sci. 36 (6) (December 1989) 1405–1410, https://doi.org/10.1109/23.45455.
[16] A.C. Tribble, Professional Development Short Course on: The Space Environment Implications for Spacecraft Design, Applied Technology Institute, for further details, see: https://aticourses.arlo.co/w/courses/6-space-environment-implications-for-spacecraft-design
[17] M. Bodeau, Updated current and voltage thresholds for sustained arcs in power systems, IEEE Trans. Plasma Sci. 42 (7) (July 2014) 1917–1921, https://doi.org/10.1109/TPS.2014.2324491.
[18] D. Ferguson, R.C. Hoffmann, A.T. Wheelock, A U.S. round-robin experiment on characteristics of arc plasma expansion, in: 4th AIAA Atmospheric and Space Environments Conference, June 25–28, 2012, New Orleans, LA, USA, AIAA 2012-2940, https://doi.org/10.2514/6.2012-2940.
[19] M. Cho, Modeling of secondary arc conditions on satellite solar array, in: International Symposium on Discharges and Electrical Insulation in Vacuum, Matsue, Japan, 2006, pp. 726–729, https://doi.org/10.1109/DEIV.2006.357405.
[20] H. Masui, K. Toyoda, M. Cho, Secondary arc threshold measurement on solar arrays for Japanese Spacecraft Charging Guideline, in: 47th AIAA Aerospace Sciences Meeting including The New Horizons Forum and Aerospace Exposition, AIAA 2009-114, January 5–8, 2009, Orlando, FL USA, Available at: https://doi.org/10.2514/6.2009-114.
[21] B. Hoang, F.K. Wong, R.L. Corey, G. Gardiner, V.V. Funderburk, R.L. Gahart, K.H. Wright, T.A. Schneider, J.A. Vaughn, Combined space environmental exposure tests of multi-junction GaAs/Ge solar array coupons, IEEE Trans. Plasma Sci. 40 (2) (February 2012) 324–333, https://doi.org/10.1109/TPS.2011.2174161.
[22] D.C. Ferguson, R.C. Hoffmann, E.A. Plis, D.P. Engelhart, Atypical normal potential gradient arcing on solar arrays, J. Spacecraft Rockets 55 (3) (May–June 2018) 698–711.
[23] C.F. Hoeber, E.A. Robertson, I. Katz, V.A. Davis, D.B. Snyder, Solar Array Augmented Electrostatic Discharge in GEO, AIM-Paper 98-1401, Yokohama, Japan, Feb. 1998.
[24] J.L. Hyde, E.L. Christiansen, D.M. Lear, Observations of MMOD impact damage to the ISS, in: First International Orbital Debris Conference, December 9–12, 2019, Houston, TX, USA.
[25] D.C. Ferguson, J. Murray-Krezan, D.A. Barton, J.R. Dennison, S.A. Gregory, Feasibility of detecting spacecraft charging and arcing by remote sensing, J. Spacecraft Rockets 51 (6) (November–December 2014) 1907–1913.
[26] G. Murphy, D. Croley, M. Ratliff, P. Leung, The role of external circuit impedance in dielectric breakdown, in: 30th Aerospace Sciences Meeting and Exhibit, January 6–9, 1992, Reno, NV USA (AIAA Paper 92-0821).
[27] D. Ferguson, B. Vayner, Radiofrequency emissions from satellite solar array arcing, in: Proceedings of the 15th Spacecraft Charging Technology Conference, Kobe, Japan, June 25–29, 2018.
[28] D. Ferguson, R. Hoffmann, J. Patton, A.T. Wheelock, J.A. Young, M.W. Crofton, J.L. Prebola, et al., AFRL round-robin test results on plasma propagation velocity, in: 13th Spacecraft Charging Technology Conference, Pasadena, CA, June 23-27, 2014.
[29] D. Ferguson, S. White, R. Rast, D. Thompson, D. Suszcynsky, K. Balasubramaniam, E. Holeman, Ground-based surveillance campaign to detect Global Positioning System arcing: First preliminary

results, J. Spacecraft Rockets 54 (3) (May-June 2017) 566–571, https://doi.org/10.2514/1.A33724. Available at:.

[30] T. Okumura, M. Imaizumi, K. Toyoda, M. Cho, Characterization of solar cell degradation due to electrostatic discharge on multi-junction solar cell, in: 35th IEEE Photovoltaic Specialists Conference, 2010, pp. 2599–2603, https://doi.org/10.1109/PVSC.2010.5614144.

[31] D. Ferguson, P. Crabtree, S. White, B. Vayner, Anomalous global positioning system power degradation from arc-induced contamination, J. Spacecraft Rockets (April 15, 2016), https://doi.org/10.2514/1.A33438.

[32] D.L. Liu, S.H. Liu, C.J. Panetta, K.R. Olson, S.M. Hong, D.R. Alann, C.J. Mann, K.T. Luey, Effects of contamination on solar cell coverglasses, in: 35th IEEE Photovoltaic Specialists Conference, 2010, pp. 2563–2568, https://doi.org/10.1109/PVSC.2010.5614601.

[33] D.P. Engelhart, E.A. Plis, D. Ferguson, K. Artyushkova, D. Wellems, R. Cooper, R. Hoffmann, XPS investigation of the source of GPS arc contamination, IEEE Trans. Plasma Sci. 47 (8) (August 2019) 3848–3851, https://doi.org/10.1109/TPS.2019.2921937.

[34] D. Hall, AMOS Galaxy 15 satellite observations and analysis, in: Advanced Maui Optical and Space Surveillance Technologies Conference, September 2011, Maui, HI, USA, Available at: https://amostech.com/TechnicalPapers/2011/NROC/HALL.pdf.

[35] NASA-HDBK- 4002A, Mitigating In-Space Charging Effects —A Guideline, National Aeronautics and Space Administration, 2011. Available at: https://www.document-center.com/standards/show/NASA-HDBK-4002.

[36] BSR/ANSI/AIAA S-115, Low Earth Orbit Spacecraft Charging Design Standard Requirement and Associated Handbook, American National Standards Institute (ANSI), 2017.

[37] ECSS-E-ST-20-06C Rev. 1, Space Engineering, Spacecraft Charging, European Cooperation for Space Standardization (ECSS), 2019.

[38] JERG-2-211A, Design Standard – Spacecraft Charging and Discharging, Japanese Aerospace Exploration Agency (JAXA), 2012, https://sma.jaxa.jp/en/TechDoc/Docs/E_JAXA-JERG-2-211A.pdf.

[39] ISO 11221, Space Systems – Space Solar Panels - Spacecraft Charging Induced – Electrostatic Discharge Test Methods, International Standards Organization, 2011. Available at: https://www.iso.org/home.html.

[40] ISO 19923, Space Environment (Natural and Artificial – Plasma Environments for Generation of Worst Case Electrical Potential Differences for Spacecraft, International Standards Organization, 2017. Available at: https://www.iso.org/home.html.

[41] I. Katz, V.A. Davis, D.B. Snyder, Solar array augmented electrostatic discharge in GEO, in: Proc. 17th ICSSC, Apr. 1998, pp. 1401–1417, https://doi.org/10.2514/6.1998-1401.

[42] I. Katz, V.A. Davis, D.B. Snyder, E.A., Robertson, ESD triggered solar array failure mechanism, in: Proceedings of the 6th Spacecraft Charging Conference, November 2–6, 1998, AFRL Science Center, Hanscom AFB, MA, USA., pp. 39–42.

[43] G. Bonin, N. Orr, R.E. Zee, J. Cain, Solar array arcing mitigation for low-Earth orbit spacecraft, in: 24th Annual AIAA/USU Conference on Small Satellites, 2010. SSC10-X-7.

[44] T. Stern, D. Krumweide, E. Gaddy, I. Katz, Development of an electrostatically clean solar array panel, in: Conference Record of the Twenty-Eighth IEEE Photovoltaic Specialists Conference - 2000 (Cat. No.00CH37036), Anchorage, Alaska, USA, 15–22 September, 2000, 10.1109/PVSC.2000.916141.

[45] I. Katz, V.A. Davis, D.B. Snyder, Mitigation techniques for spacecraft charging induced arcing on solar arrays, in: 37th Aerospace Sciences Meeting, AIAA-99-0215, Reno, NV USA, January 11–14, 1999, https://doi.org/10.2514/6.1999-215.

[46] D.C. Ferguson, E. Plis, R. Hoffmann, D. Engelhart, A proven method to prevent solar array arcing in geosynchronous orbit – bulk-conductive coverglasses, in: AIAA SciTech Forum, 6–10 January 2020, Orlando, FL, USA, https://doi.org/10.2514/6.2020-2013.

[47] D.C Ferguson, M.S. Rhee, Controlling charging and arcing on a solar powered auroral orbiting spacecraft, in: 33rd IEEE Photovoltaic Specialists Conference, San Diego, CA, May 11–16, 2008.

[48] W. Chang, J.R. Dennison, J. Kite, R. Davies, Effects of evolving surface contamination on spacecraft charging, in: 38th Aerospace Sciences Meeting and Exhibit, January 2000. AIAA-2000-0861.

[49] D.C. Ferguson, Secondary electron emission from solar cell cover-glasses and its effect on absolute vehicle charging, in: 9th European Space Power Conference, Saint-Raphaël, France, 5–10 June, 2011.

[50] S.T. Lai, Charging of mirror surfaces in space, J. Geophys. Res. 110 (2005) A01204, https://doi.org/10.1029/2002JA009447.

[51] V.M. Aguero, R.C. Adamo, Space applications of spindt cathode field emission arrays, in: Proceedings of the 6th Spacecraft Charging Conference, November 2–6, 1998, AFRL Science Center, Hanscom AFB, MA, USA., pp. 347–352.

[52] M. Iwata, A.R. Khan, H. Igawa, K. Toyoda, M. Cho, T. Fujita, Development of electron-emitting film for spacecraft charging mitigation, J. Spacecraft Rockets 49 (3) (May–June 2012) 546–552, https://doi.org/10.2514/1.50735.

[53] J.D. Williams, J.R. Sanmartin, L.P. Rand, Low work-function coating for an entirely propellantless bare electrodynamic tether, IEEE Trans. Plasma Sci. 40 (5) (May 2012) 1441–1445.

[54] P.R. Stannard, I. Katz, M.J. Mandell, J.J. Cassidy, D.E. Parks, M. Rotenberg, P.G. Steen, Analysis of the Charging of the SCATHA (P78-2) Satellite, NASA CR-165348 (SSS-R-81-4798), December 1980.

[55] V.A. Davis, I. Katz, M.J. Mandell, D.E. Parks, Hollow cathodes as electron emitting plasma contactors - theory and computer modeling, J. Spacecraft Rockets 25 (2) (March-April 1988) 175–179, https://doi.org/10.2514/3.25967.

[56] M.J. Patterson, J.A. Hamley, C.J. Sarmiento, D.H. Manzella, T. Sarver-Verhey, G.C. Soulas, A. Nelson, Plasma Contactor Development for Space Station, NASA Technical Memorandum 106425 (IEPC-93-246), 1993.

[57] W.J. Raitt, Charging mitigation experiments on sounding rockets, in: Proceedings of the 6th Spacecraft Charging Conference, November 2-6, 1998, AFRL Science Center, Hanscom AFB, MA, USA, 2000, pp. 315–321. AFRL-VS-TR-20001578.

[58] A.R. Frederickson, Spacecraft Charging Phenomena and Spacecraft Design Guidelines, JPL/NASA, 2003. https://trs.jpl.nasa.gov/bitstream/handle/2014/39316/05-0815.pdf?sequence=1.

[59] B. Johnston, P. O'Brien, G. Ginet, S. Huston, T. Guild, C. Roth, Y.-J. Su, et al., IRENE: AE9/AP9/SPM Radiation Environment Model User's Guide, Version 1.50.001, 2017. https://www.vdl.afrl.af.mil/programs/ae9ap9/files/package/Ae9_Ap9_SPM_Users_Guide.pdf.

[60] R.L. Lambour, Calibration of the Rice Magnetospheric Specification and Forecast Model for the Inner Magnetosphere, Dissertation, Rice University, Houston, TX, USA, 1994. Available at: https://hdl.handle.net/1911/16749.

[61] International Reference Ionosphere, http://irimodel.org/. (Accessed May 10, 2022).

[62] M.J. Mandell, V.A. Davis, B.M. Gardner, I.G. Mikellides, D.L. Cooke, J. Minor, NASCAP-2K: An overview, in: Proceedings of the 8th Spacecraft Charging Technology Conference, 2004. https://ntrs.nasa.gov/api/citations/20040111035/downloads/20040111035.pdf.

[63] B. Thiébault, et al., SPIS 5.1: An innovative approach for spacecraft plasma modeling, IEEE Trans. Plasma Sci. 43 (9) (September 2015) 2782–2788, https://doi.org/10.1109/TPS.2015.2425300.

[64] ISO 19923:2017, Space Environment (Natural and Artificial) — Plasma Environments for Generation of Worst Case Electrical Potential Differences for Spacecraft, Technical Committee: ISO/TC 20/SC 14 Space systems and operations, International Standards Organization, 2017, p. 14.

[65] L.S. Novikov, A.A. Makletsov, V.V. Sinolits, Comparison of Coulomb-2, NASCAP-2K, MUSCAT and SPIS codes for geosynchronous spacecraft charging, Adv. Space Res. 57 (2) (January 2016) 671–680.

[66] R. Frederickson, S. Woolf, J.C. Garth, Model for space charge evolution and dose in irradiated insulators at high electric fields, IEEE Trans. Nucl. Sci. 40 (6) (Dec. 1993) 1393–1401.

[67] F. Lei, D. Rodgers, P. Truscott, MCICT - Monte-Carlo internal charging Tool, in: 14th Spacecraft Charging Technology Conference, ESA/ESTEC, Noordwijk, Netherlands, April 4-8, 2016.

[68] S. Hurtado, M. Garcia-León, R. Garcia-Tenorio, GEANT4 code for simulation of a germanium gamma-ray detector and its application to efficiency calibration, Nucl. Instrum. Methods Phys. Res. Sect. A Accel. Spectrom. Detect. Assoc. Equip. 518 (3) (February 2004) 764–774.

[69] D. Ferguson, A. Wheelock, Necessity of testing for affordable survivable spacecraft, in: 14th Spacecraft Charging Technology Conference, Noordwijk, Netherlands, April 4-8, 2016, available at: http://d54z.ru/Sctc2016/351/1000_Ferguson.pdf.

[70] J.I. Minow, L.N. Parker, Spacecraft charging: Anomaly and failure mechanisms, in: Spacecraft Anomalies and Failures Workshop July 24, 2014, Chantilly, VA, USA, Presentation available at: https://ntrs.nasa.gov/citations/20140012586.

CHAPTER THREE

Air mass zero (AM0) studies and solar cell calibration

Carsten Baur, Emilio Fernández Lisbona and José Ramón González
European Space Agency, Noordwijk, The Netherlands

3.1 Introduction

The calibration of solar cells potentially refers to different types of measurements and characterization methods. Thus, as a starting point, it might be helpful to clarify what is meant by "calibration." Typically, solar cell calibration can be subdivided into two or three categories: primary calibration, secondary calibration, and measurement (which is in the end, also a secondary calibration, although not directly from the primary standard). All these different calibration categories involve different experimental techniques; these will subsequently be extensively discussed in this chapter.

Calibrations all serve one purpose, to provide the most accurate performance evaluation of a solar cell technology under specific conditions. In the case of space photovoltaics, this means to measure the power output of a solar cell under an outer space solar spectrum that a solar array will see during operation. The electrical performance figure and its related uncertainties have an impact not only on the economic benefit of a solar cell manufacturer but also serve a solar array manufacturer as an input when sizing the solar generator for a particular space mission.

Solar cell calibration for space applications is on the one hand very similar to calibration for terrestrial applications. This is particularly true for secondary calibrations. However, there are also fundamental differences. While on Earth there is no well-defined solar spectrum under which the solar cells are operating, this is completely different in space. In space, the solar spectrum is the so-called air mass zero (AM0) spectrum, which today is known quite accurately by the measurements performed by a number of spaceborne instruments. The term "air mass" is an indication of the length of the light path through the atmosphere before reaching a certain location; the "0" in "AM0" then means that no atmosphere is passed. Therefore, the AM0 spectrum is the pure spectrum coming from the sun not disturbed or filtered by Earth's atmosphere. On the ground, obviously, the light is filtered by the atmosphere. The amount of atmosphere the light has to travel through depends mainly on the location on Earth and the air pressure (under clear sky conditions). The air mass for any location on ground is defined (Eq. 3.1) as follows:

Photovoltaics for Space
ISBN 978-0-12-823300-9, https://doi.org/10.1016/B978-0-12-823300-9.00008-X

$$\text{AM } X \text{ where } X = \frac{p}{p_0}\frac{1}{\cos\alpha}, \tag{3.1}$$

with α being the angle under which the light is entering the earth's atmosphere with respect to the zenith and with p being the local air pressure, while p_0 indicates the standard atmospheric pressure of 1013.25 hPa. AM1 would be a spectrum at normal incidence at the equator at standard atmospheric pressure.

For ground applications and for the sake of comparing the performance of different solar cells, as a reference the "AM1.5 global" spectrum is used to represent an average spectrum. According to Eq. (3.1) and assuming standard atmospheric pressure, AM1.5 conditions occur when the light enters the atmosphere at an angle of 48 degrees against the zenith. A comparison of both spectra that are applicable today is shown in Fig. 3.1. In contrast to space where the solar cells in fact always see the AM0 spectrum, it is quite obvious that AM1.5 conditions are nominal and rarely occur in reality.

3.2 Primary calibration

In the context of photovoltaics, a primary calibration is the method by which a so-called primary reference solar cell is created. The primary reference value (the measurement quantity) of such a primary reference solar cell is its short circuit current under standard testing conditions (STC), and it involves the reference spectrum and the temperature. In the space applications case, these are the AM0 spectrum and a temperature of 25°C. One of the main important requirements for a solar cell for getting the "primary reference" label is traceability. In the terminology of metrology, this means that

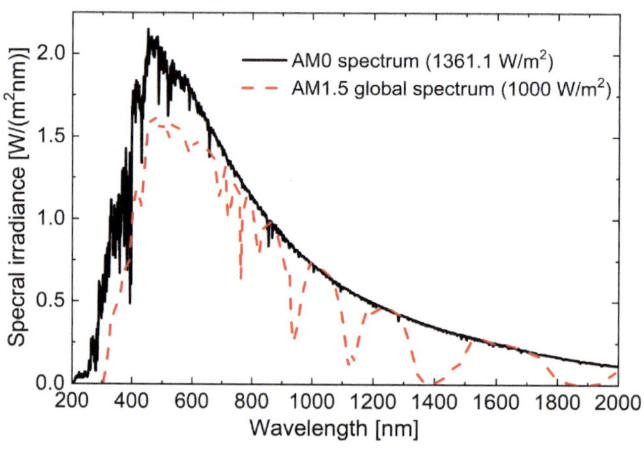

Figure 3.1 Solar reference spectra outside Earth's atmosphere (AM0 spectrum) and the "average spectrum" used for ground-based applications (AM1.5 global spectrum). *Adapted from Refs. [1,2] by authors.*

such a reference device is traceable by an unbroken calibration chain back to a primary standard that is represented by SI units (International System of Units). The calibration value of a primary reference solar cell establishes a relation between a well-defined irradiance source (the solar spectrum) and the current of the solar cell. Therefore, a primary standard could, for example, be a standard lamp, a cryogenic trap detector, or a group of cavity radiometers constituting the World Standard Group of the World Radiometric Reference. In other words, it is a device that allows defining an irradiance. There exists also another possibility for a primary reference standard, which is relevant in this context: that is to choose an artifact by convention [3].

Primary calibration methods for space applications are defined in the ISO standard 15387 [4]. There, a distinction is made between so-called "extraterrestrial" (space) or "high-altitude" calibration procedures and "synthetic calibration" procedures. The major difference between the two is that for the high-altitude (this label will be used in the updated version) calibration procedures, the reference standard is in fact defined as an artifact as mentioned earlier. This means that primary reference cells, following these procedures, are directly measured under "standard" conditions.

3.2.1 High-altitude calibration techniques

Calibration techniques that have the "high altitude" label are balloon flights or aircraft, which all can perform measurements at an altitude above 15 km where the residual atmosphere is on the order of 1% or less than that at sea level. Although for those methods some minor corrections still have to be applied, e.g., correcting for the residual atmosphere and correcting to the average Earth—Sun distance, those methods were historically considered to provide the most accurate results.

A round-robin experiment was performed from 2001 to 2003 where the three high-altitude calibration providers (NASA, JPL, and CNES (Center National d'Etudes Spatiales = French Space Agency)) each calibrated the same solar cells with their methods. The solar cells involved were two triple-junction (3J) solar cells and two "component cells" per each of the subcells of the triple-junction cell. A component cell is a specially manufactured solar cell, with an almost identical layer setup as the corresponding multijunction (MJ) solar cell but with only one active junction. The main important feature of a component cell is that it has identical or very close optical properties to the corresponding subcell in a MJ solar cell. While NASA used a Lear Jet, CNES and JPL had a balloon platform for calibrating the devices. As shown in Fig. 3.2, the deviation of all entities from the mean value for all cells is less then $\pm 1\%$ and in most of the cases less than 0.5%. This deviation is even better if only NASA and CNES results were considered. Overall, the outcome of this round-robin demonstrated quite good agreement, and therefore, the majority of the space community mainly used primary reference cells generated by these high-altitude methods. Calibration values provided by CNES are typically reported with an uncertainty of 0.7%.

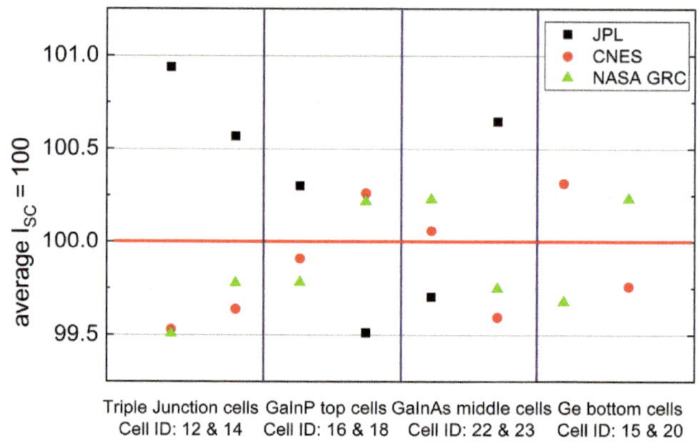

Figure 3.2 Deviation of average short circuit current of different solar cells when calibrated by three different entities. *Adapted from Ref. [5] by authors.*

3.2.2 Synthetic calibration techniques

So-called "synthetic calibration methods," on the other hand, are all other calibration techniques that do not directly use the actual AM0 spectrum but either use light sources that artificially simulate AM0 conditions, like sun simulators or terrestrial sunlight, or use other ground-based methods like the differential spectral responsivity (DSR) measurement. There are a number of these methods still in the current issue of ISO 15387 taken from the corresponding IEC standard 60904-4 that describe primary calibration techniques for ground-based applications. However, among those methods and at the present time, the DSR measurement technique [6,7] is applied for space-related primary calibrations. The short circuit current that is identical to the calibration value of a primary reference solar cell is here derived from the absolute measurement of the spectral responsivity (SR) of the specimen. Fig. 3.3 shows an example of such a measurement. There, the absolute SR is shown for three component cells of a 3G28 solar cell structure from AZUR SPACE Solar Power. The measurements have been performed by PTB (Physikalisch-Technische Bundesanstalt, German national metrological institute).

The SR of a solar cell provides the current output at a given wavelength and a given input power. Therefore, the short circuit current (I_{sc}) of the solar cell under any spectrum can be calculated according to Eq. (3.2):

$$I_{sc} = \int SR(\lambda) \cdot E(\lambda) \cdot d\lambda \tag{3.2}$$

with $SR(\lambda)$ being the absolute SR and $E(\lambda)$ being the spectrum. For space applications $E(\lambda)$ would be the AM0 spectrum. This method is a very elegant way to establish calibration values since with one measurement one would basically get the calibration value for any kind of reference spectrum. This could be the AM0 spectrum but also, e.g.,

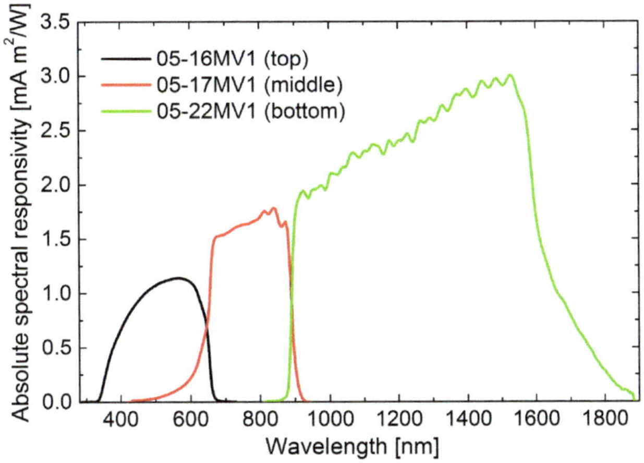

Figure 3.3 Absolute spectral responsivity data of the reference cell sets calibrated at PTB. Three component cells from 3G28 structure. *Reproduced with permission from Ref. [8] by IEEE.*

the AM1.5 spectrum. However, in contrast to the high-altitude balloon methods that could claim to measure under the "real" spectrum, the DSR method requires a very good knowledge of the solar spectrum as well. Metrological institutes such as PTB could improve the DSR technique, so they can claim today an uncertainty in their DSR method of 0.5% on the calibration values. This is taking into account no uncertainties in the spectrum. Thus, any uncertainty in the spectrum would immediately translate into an additional uncertainty in the calibration value. Therefore, it is very important that the AM0 spectrum is well known together with its uncertainties. This is, by the way, different from ground applications. There, the spectrum is artificially defined. It is not an actual measured spectrum but calculated by software (NREL's Simple Model of the Atmospheric Radiative Transfer of Sunshine [SMARTS2]) and then tabulated to be used as a reference. As such, it comes without any uncertainties. As already mentioned before, for ground-based applications the reference spectrum mainly serves the purpose of having a reference for comparison. The energy harvest of a solar cell technology depends then very much on the location and weather conditions under which it is operating and is predictable only to a certain extent. In space, the situation is completely different. The spectral input is practically unchanging. Therefore, the performance of a solar generator in space is quite predictable. And this is also very important since after the spacecraft is launched, there is usually no way to add more solar panels to it if the power sizing was not done properly. As mentioned before, that is why it is so important to measure the performance of a solar cell that shall operate in space as accurately as possible. Of course, one could decide to oversize the solar generator to be on the safe side, but any extra weight launched into space not only costs a lot of money but also takes away mass and volume that could otherwise be used for the satellite payload, its main purpose.

Coming back now to the AM0 spectrum, as a matter of fact, the AM0 spectrum has undergone a number of changes historically. By that, it is not meant that the solar spectrum itself changed, but the development of better instruments to measure both the spectral distribution of the solar spectrum and its irradiance value have improved significantly over the years. This has resulted in continuous reevaluation of the AM0 spectrum and the total solar irradiance value. More details can be found in the literature [9,10].

3.2.3 Total solar irradiance and spectral distribution of the AM0 spectrum

The total solar irradiance (TSI), also often called the "solar constant," is defined as the incoming solar electromagnetic radiation or the power per unit area that would be incident on a plane perpendicular to the Sun's rays at a distance of 1 astronomical unit (AU), i.e., the mean distance between Sun and Earth. Note that this solar irradiance is measured just outside the earth's atmosphere. Thus, the solar constant is actually equivalent to the integrated value of the AM0 spectrum or vice versa. As a matter of fact, the solar constant is not exactly a constant but varies slightly over the 11-year solar cycle.

Only very recently has the TSI value been updated or corrected from a value of around 1367 W/m^2 (defined in 1982 by the World Meteorological Organisation [WMO] and adopted also in ISO 15387 [4]), or 1366.1 W/m^2 as defined in ASTM E−490, to a new value of around 1361 W/m^2. This correction goes back to findings by Kopp et al. and is obviously based on more accurate measurements [10]. In the solar cell community, the most commonly used values even today have been 1367 W/m^2 or 1366.1 W/m^2, and several solar cell manufacturers still like to use the old value of 1353 W/m^2, which goes back to Thekaekara in 1973 [11]. This is mainly because the lower TSI value of the Thekaekara results in efficiencies being about 1% (relative) higher than if one would use the WMO TSI value of 1367 W/m^2. However, in the updated version of the ISO 15387, it has already been agreed by the members of the working group that the new TSI value will be 1361.1 W/m^2 as suggested by C. Gueymard [12].

For the reader with interest in understanding in detail the evolution of the TSI value over time, it is recommended to access another publication by Gueymard [9] where an overview of different TSI values reported in the last 80 years is provided. It is quite interesting to see that TSI values ranged from 1322 to 1430 W/m^2. But those two extreme values stem from a time when measurements still had to be conducted within Earth's atmosphere, which generally produced spectra with limited detail in the UV as well as limited spectral resolution. Unsurprisingly, large fluctuations occurred during this period in the quantification of both the TSI and the UV spectrum, in particular.

Also regarding the spectral distribution of the AM0 spectrum, important updates have been suggested only very recently [1]. For many years (probably since the mid-80s until the present day), many solar cell manufacturers either used a spectrum that goes back to "Wehrli" and the World Radiometric Center [13], which was normalized to the WMO TSI value of 1367 W/m^2, or the Thekaekara spectrum, with a TSI value of 1353 W/m^2 (the latter often being replaced, since approximately 2000, by the ASTM spectrum with a TSI value of 1366.1 W/m^2). Coming back to the initial discussion on the uncertainty of the DSR method for primary calibration, all the mentioned different spectra would obviously lead to different calibration values, and what is more, it would be difficult to tell which one is more accurate. Fortunately, with the introduction of the latest spectrum suggested by C. Gueymard together with a TSI value of 1361.1 W/m^2, both of them provided with their respective uncertainties, it appears that this controversy can be overcome. A total of 31 sources of spectral irradiance have been considered from observations and solar models to arrive at the spectrum by C. Gueymard [1] allowing for each listed wavelength (wavelength steps of 0.5, 1, and 5 nm in the relevant wavelength region between 250 and 2000 nm) to assign a dedicated uncertainty, raising the credibility of this spectrum over the ones used in the past.

Fig. 3.4 shows the AM0 solar spectra that were mainly used in the last years ([4,14]) in comparison to the two more recent spectra suggested by C. Gueymard. The Gueymard spectrum from 2004 was still normalized to yield an integrated value of 1366.1 W/m^2. It is clear that the graph itself does not really reveal the differences in the spectra. This can be better derived from Fig. 3.5 where the deviations for different wavelength bins with respect to the latest Gueymard spectrum (from 2018) are provided. Deviations per 100-nm wavelength bin can get up to almost 4%. The first bin that ranges from 280 to 2000 nm, the complete wavelength range of interest for solar cells, shows a deviation of about 0.5%, which can be to some extent referred back to the difference in TSI value of the different spectra. In the following, it will be shown how the different spectra will alter the calibration value when using the DSR method.

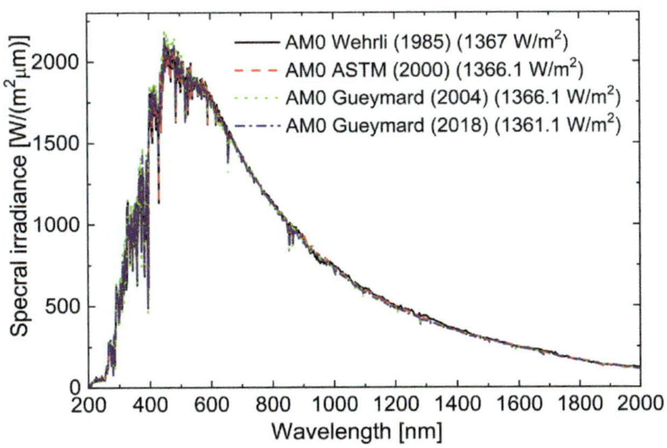

Figure 3.4 Different AM0 spectra used historically and today. *Adapted from Refs. [1,4,14,15] by authors.*

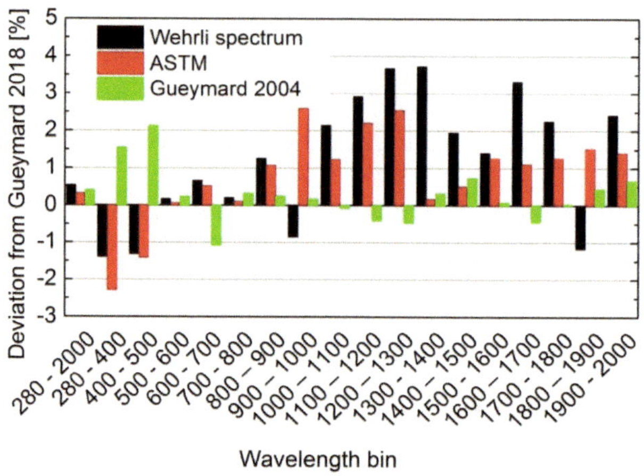

Figure 3.5 Deviations in terms of integrated power for different wavelength bins of historical spectra with respect to today's most credible spectrum "Gueymard 2018" with a TSI value of 1361.1 W/m^2. *Adapted from Ref. [8] by authors.*

3.2.4 High-altitude versus synthetic calibration methods

Due to the fact that high-altitude calibration methods are quite costly and test campaigns did not happen very frequently (they were not even available for quite some time), both CNES and JPL stopped their service in the early 2000s. In fact, the last CNES calibration flight took place in 2005. Only recently, CNES performed a qualification flight with a new platform in 2017 aiming to reestablish their service on a biannual basis. However, for quite a long time the possibility to calibrate primary reference cells using high-altitude methods was quite limited. This created considerable problems due to the fact that quite a number of new solar cell products were put on the market after 2005. Therefore, already in 2010 an effort was made by ESA (European Space Agency) to verify if the DSR method was a credible alternative calibration method [16] for single-junction solar cells. At that time, and mainly to avoid harming the "gold standards" that were flown on the CNES balloon in 2005, a direct comparison between the CNES balloon and the synthetic DSR method was not possible. However, the gold standards calibrated by the CNES balloon were compared to a twin set (manufactured as a backup and with equivalent short circuit currents and external quantum efficiencies) calibrated in 2010 by PTB using the DSR method. At that time, without having the most recent understanding of the AM0 spectrum and TSI value, good agreement in calibration values was found at least for the top and the middle component cells (derived from a 3G28 triple-junction solar cell) ($\leq$1%) [16]. Only for the bottom component cell was there a higher deviation, on the order of 2.5%. However, at that time, the Wehrli spectrum was used to arrive at the calibration values for the twin set that was calibrated by PTB.

Meanwhile, a direct comparison of calibration values established by high-altitude and synthetic methods is possible since a number of different reference solar cells are available. These cells have been calibrated by both methods, the DSR and the balloon calibration method. However, when CNES was reestablishing their service in 2017, a direct comparison was performed between the CNES balloon and the PTB DSR method using the identical list of primary reference cells given in Table 3.1 in both calibrations. On this occasion, the most recent solar cell technology, component cells from AZUR SPACEs 4G32C solar cell, was used.

Table 3.1 lists the four component cells derived from a 4G32 upright metamorphic cell structure obtained from AZUR SPACE Solar Power that have been included in the intercomparison between CNES and PTB. This set of component cells was calibrated with the DSR method in January 2017, after that mounted on the CNES balloon in April 2017, and then again calibrated by PTB using the DSR method in January and February 2018.

The absolute spectral responsivities of the component cells listed in Table 3.1 are displayed in Fig. 3.6. The graph shows, for each cell, the two measurements before and after the CNES balloon flight. From the graph, it is possible to see the good match between the two measurements, which demonstrates the excellent reproducibility of the DSR method. However, junction 3 (J3) shows a drop in absolute SR in the measurement after the flight. This can be traced back to an incident occurring during the balloon flight that resulted in some cell breakages. One of the affected cells was J3. Even so, J3 could still be measured on the balloon before breakage, so a comparison can still be performed with the PTB measurement carried out before the flight.

The results of this comparison are provided in Table 3.2. There, under the row "CNES measurement," the calibration values obtained from the CNES balloon flight for each cell are given. In addition, the calibration values calculated for the two historical, as well as the more recently suggested reference spectra, are provided in the table. Deviations are given with respect to the CNES values used as reference. Two main lessons can be drawn from this table:

Table 3.1 List of primary reference cells used during the 2017/18 intercomparison between the CNES balloon flight and the PTB DSR calibration method. PTB measured the reference cells before flight in 2017 and after the flight in 2018 [8].

Cell reference	Cell type	CNES	PTB
2016_25_MV1	J1	2017	2017/18
2016_27_MV1	J2	2017	2017/18
2016_29_MV1	J3	2017	2017/18
2016_31_MV1	J4	2017	2017/18

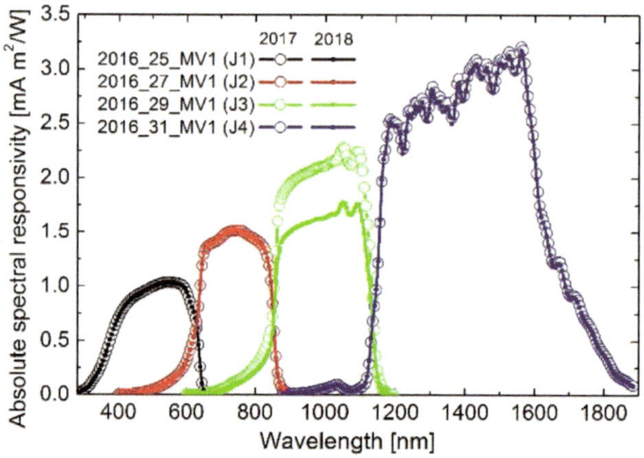

Figure 3.6 Absolute spectral responsivity data of the 4G32 reference cell set calibrated at PTB before (2017) and after CNES balloon flight (2018). *Reproduced with permission from Ref. [8] by IEEE.*

Table 3.2 Comparison of the calibration values established for the 4G32 component cell when calibrated with the CNES balloon flight or the PTB DSR (values after flight are given here except for J3, which was damaged during flight). Deviations are given with respect to the CNES values. Calibration values have been calculated for different reference spectra. Data obtained from the literature [8].

Cell type	J1	J2	J3	J4
CNES measurement				
I_{sc} [mA]	445.8	461.3	468.0	509.4
PTB measurement				
I_{sc} (Wehrli, 1367 W/m^2) [mA]	440.5	459.3	471.5	524.3
Deviation [%]	−1.2	−0.4	+0.7	+2.9
I_{sc} (ASTM, 1366.1 W/m^2) [mA]	439.3	458.2	474.6	515.9
Deviation [%]	−1.4	−0.7	+1.4	+1.4
I_{sc} (Gueymard 2004, 1366.1 W/m^2) [mA]	445.9	455.7	466.6	508.9
Deviation [%]	0.0	−1.2	−0.3	−0.1
I_{sc} (Gueymard 2018, 1361.1 W/m^2) [mA]	441.9	457.6	466.4	508.9
Deviation [%]	**−0.9**	**−0.8**	**−0.3**	**−0.1**

1. The calibration values in the DSR method are strongly dependent on the reference spectrum used. Comparing only all of the calibration values for the different spectra for each cell, one finds differences up to 3% (for J4). Other differences are greater than 1% for almost all cells.

2. When comparing the calibration values for the DSR method to the CNES reference, the deviation for all cells is below 1% only for the Gueymard 2018 spectrum. It should also be mentioned that the other spectra are slightly positively biased due to the fact that they are using a higher TSI value (which today is understood to be incorrect). Thus, if these values were calculated normalizing the spectra to the new TSI value of 1361.1 W/m^2, each of the deviations would have to be lowered by another 0.3%–0.4%, which would further reduce the agreement with the CNES values.

Although, this should not be viewed as rock-solid proof, the good agreement between DSR calibration values for the Gueymard 2018 and the CNES calibration values gives one some confidence that with this new reference spectrum, the DSR method can be used equivalently to high-altitude methods. As explained earlier, it is important to also have synthetic methods that are less expensive and more facile than balloon flights and high-altitude aircraft.

In summary, applying the most recent Gueymard 2018 AM0 spectrum for calculating the DSR method reference values, the agreement between the DSR method and the CNES balloon is better than 1%. Thus, the values are in excellent agreement within the quoted uncertainties for both methods. CNES gives a standard uncertainty of about 0.7%. However, in the 2017 qualification flight, CNES reported a potential systematic error due to stray light effects that led to a systematic positive bias of the CNES calibration values. Since CNES has always flown some cells on each balloon flight, it was possible to assess this bias to a few tenths of a percent [8]. This might be an explanation for the fact that for most of the cells included in the intercomparison, the reference values reported for the DSR method are lower than the CNES reference, especially in the shorter wavelength range that is covered by J1 and J2.

PTB typically quotes an uncertainty of 0.5% without taking into account any uncertainties in the spectrum. The Gueymard 2018 spectrum and the TSI value, however, are both reported with uncertainties. If those uncertainties are propagated on the calibration value, one arrives to uncertainties on the order of 0.8%–1.1% (the latter value represents a worst case when all uncertainties in the spectrum would be fully correlated). In any case, the main conclusion here is that the new TSI value along with the new spectral distribution of the AM0 spectrum should provide enough credibility that the DSR method could be acceptable as a valid alternative to the high-altitude calibration methods.

3.3 Secondary calibration and measurements of multi-junction solar cells

The primary reference solar cell is a prerequisite for any subsequent calibration or solar cell measurement (often those terms "calibration" and "measurement" are used synonymously). Any (secondary) calibration has to be traceable back to a primary reference solar cell. In contrast to primary calibrations, (secondary) calibrations for space applications are exclusively performed in laboratories under solar simulators. The adjustment of the solar simulator is typically performed with reference solar cells. Those reference solar cells can be primary or secondary reference solar cells or most often so-called "working standards." In case of a secondary calibration in its truest sense, the reference cell that is used is a primary reference solar cell.

Calibrating a solar cell means actually to understand its performance under operating conditions or previously defined STCs (AM0 spectrum, test temperature of 25°C). A temperature of 25°C is actually not a value where the solar generators will operate in space (except when coming out of eclipse); typical operational temperatures in space are higher and, depending on the mission, more on the order of 60−100°C. However, 25°C is chosen by convention since it is a temperature easily adjustable and controllable in laboratories. The much larger difficulty to be reproduced in a laboratory environment is the solar spectrum. As a matter of fact, there is no ideal solar simulator that can exactly reproduce the solar spectrum. Fortunately, this is also not necessary. The most important goal that needs to be achieved for an accurate calibration is that the photocurrent that is created in the test specimen under the solar simulator $I_{photo,test}^{sim}$ is equal to the photocurrent that would be generated under the reference conditions $I_{photo,test}^{AM0}$, i.e., under the AM0 spectrum in the case of space solar cells (Eq. 3.3):

$$I_{photo,test}^{sim} = I_{photo,test}^{AM0} \tag{3.3}$$

Apart from the solar simulator, as mentioned before, a reference cell is needed that allows the correct adjustment of the simulator. Since the calibration value of the primary reference solar cell is the short circuit current (which can be considered identical to the photocurrent), one could think to set the solar simulator so the reference cell would deliver its calibration value and then measure the test cell. However, this is in almost all cases not correct. There are two reasons: there is no "perfect" solar simulator in that sense that it perfectly reproduces the solar spectrum, and there is also no "perfectly" matched reference cell that has exactly the same SR as the test cell. This ultimately leads to the fact that the reference cell "weighs" a spectrum differently compared to the test cell. An example of a solar simulator spectrum is shown in Fig. 3.7 in comparison with the AM0 spectrum. Furthermore, the external quantum efficiencies (EQE) of two GaInAs middle cells of a triple-junction cell are shown. Note that EQE and SR are equivalent quantities. However, EQE is the more intuitive quantity in a graphic representation. It gives the percentage of photons that are absorbed and create electron-hole pairs that are extracted to the terminals of the cell. Therefore, it is used preferably from here on in figures, while for the calculation, the respective SR is used.

While one curve belongs to a component single-junction (SJ) cell (dashed blue), the other one is from the subcell in the full device. It is obvious that both EQEs are quite

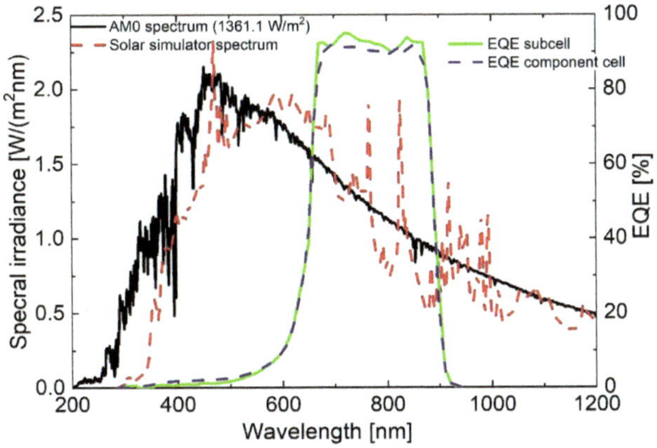

Figure 3.7 AM0 spectrum and the example of a spectrum of a solar simulator. Furthermore, the external quantum efficiency of two GaInAs middle cells of a triple-junction cell is shown. While the dashed blue line is a component single-junction cell, the solid green curve is the subcell in the full device. *Source: Authors.*

similar but, of course, not identical. This example represents a very typical situation in a lab where the component cell is used as the reference cell. The solar simulator spectrum in the figure has been set so the photocurrent from the component cell under the simulator spectrum is exactly matching the photocurrent under the AM0 spectrum (applying Eq. 3.2). Now the photocurrent is calculated for the subcell under both spectra values of 17.43 mA/cm^2 (AM0) and 16.93 mA/cm^2 (simulator spectrum). This is a difference of about 3%, which is not negligible anymore. So, the adjustment procedure of the solar simulator has to take into account that subcell and reference component cell will weigh the reference and simulator spectra differently.

Using Eq. (3.2) we can obtain the photocurrents of the test and reference cell under the simulator and AM0 spectrum, respectively, using Eqs. (3.4)—(3.7):

$$I_{photo,test}^{sim} = \int sr_{test}(\lambda) \cdot e(\lambda) \cdot d\lambda \tag{3.4}$$

$$I_{photo,test}^{AM0} = \int sr_{test}(\lambda) \cdot E_{AM0}(\lambda) \cdot d\lambda \tag{3.5}$$

$$I_{photo,ref}^{sim} = \int SR_{ref}(\lambda) \cdot e_{sim}(\lambda) \cdot d\lambda \tag{3.6}$$

$$I_{photo,ref}^{AM0} = \int SR_{ref}(\lambda) \cdot E_{AM0}(\lambda) \cdot d\lambda, \tag{3.7}$$

with the solar simulator spectrum $e_{sim}(\lambda)$, the AM0 spectrum $E_{AM0}(\lambda)$, and the spectral responsivities of test and reference cell $sr_{test}(\lambda)$ and $SR_{ref}(\lambda)$. Small characters in the quantities mean that the quantity is only known in relative terms. This typically applies to the solar simulator spectrum and the SR of the test cell.

All these quantities have to be known at least in relative terms. In the end, the knowledge of the primary calibration value (Eq. 3.7) will allow for the correct setting of the solar simulator, as will be explained in the following. According to Eq. (3.3), we have to fulfill the condition that sets Eq. (3.4) = Eq. (3.5). This can be expressed as follows by Eq. (3.8):

$$\int sr_{test}(\lambda)\cdot E_{AM0}(\lambda)\cdot d\lambda = A\cdot \int sr_{test}(\lambda)\cdot e_{sim}(\lambda)\cdot d\lambda, \tag{3.8}$$

with a parameter A that needs to be identified. In practice, A can be considered to influence the light intensity of the solar simulator, which is typically done by adjusting the current of the solar simulator lamp. From Eq. (3.8) we get A simply expressed by Eq. (3.9):

$$A = \frac{\int SR_{test}(\lambda)\cdot E_{AM0}(\lambda)\cdot d\lambda}{\int SR_{test}(\lambda)\cdot e_{sim}(\lambda)\cdot d\lambda} \tag{3.9}$$

If all quantities in Eq. (3.9) were known in absolute terms, one could set the simulator using the test cell only. However, as mentioned earlier, in practice this is usually not the case. Measuring the absolute SR and also the absolute spectrum is quite difficult. Therefore, the reference cell is used for the calibration value, which is known in absolute terms (Eq. 3.7). The simulator spectrum has to be modified with "A," then we get the simulator photocurrent of the reference cell as determined by Eq. (3.10):

$$I^{sim}_{photo,ref} = A\cdot \int SR_{ref}(\lambda)\cdot e_{sim}(\lambda)\cdot d\lambda = \frac{\int sr_{test}(\lambda)\cdot E_{AM0}(\lambda)\cdot d\lambda}{\int sr_{test}(\lambda)\cdot e_{sim}(\lambda)\cdot d\lambda}\cdot \int SR_{ref}(\lambda)\cdot e_{sim}(\lambda)\cdot d\lambda \tag{3.10}$$

Often Eq. (3.10) is rewritten by using the calibration value of the reference cell under the AM0 spectrum using Eq. (3.7) to produce Eq. (3.11):

$$\begin{aligned} I^{sim}_{photo,ref} &= A\cdot \int SR_{ref}(\lambda)\cdot e_{sim}(\lambda)\cdot d\lambda \\ &= \frac{\int sr_{test}(\lambda)\cdot E_{AM0}(\lambda)\cdot d\lambda\cdot \int SR_{ref}(\lambda)\cdot e_{sim}(\lambda)\cdot d\lambda}{\int sr_{test}(\lambda)\cdot e_{sim}(\lambda)\cdot d\lambda\cdot \int SR_{ref}(\lambda)\cdot E_{AM0}(\lambda)\cdot d\lambda}\cdot I^{AM0}_{photo,ref} \end{aligned} \tag{3.11}$$

By introducing the so-called mismatch factor M [17] (Eq. 3.12),

$$M = \frac{\int sr_{test}(\lambda) \cdot e_{sim}(\lambda) \cdot d\lambda \cdot \int SR_{ref}(\lambda) \cdot E_{AM0}(\lambda) \cdot d\lambda}{\int sr_{test}(\lambda) \cdot E_{AM0}(\lambda) \cdot d\lambda \cdot \int SR_{ref}(\lambda) \cdot e_{sim}(\lambda) \cdot d\lambda} \tag{3.12}$$

Eq. (3.11) is transformed into Eq. (3.13):

$$I_{photo,ref}^{sim} = \frac{I_{photo,ref}^{AM0}}{M} \tag{3.13}$$

Thereby, the adjustment of the solar simulator is clearly traceable back to the calibration value of the reference cell. It has to be noted that whenever the intensity of a light source is modified in line with "*A*," the relative distribution of the spectrum might change as well. This would then require a remeasurement of the solar simulator spectrum and a recalculation of the mismatch factor. Only if "*A*" is close to "1," or in other words the light intensity needs to be hardly changed with respect to the conditions under which the solar simulator spectrum has been measured, can one assume that the adjustment procedure is finished.

The calculation and application of the mismatch factor are key elements for an accurate calibration of a measurement of the test device. From the definition of the mismatch factor, it is furthermore clear that both the spectral responses of test and reference cell as well as the spectra of the solar simulator and the sun are needed. The latter is tabulated in the respective standard. All other quantities need to be measured. The measurement of the solar simulator spectrum is typically performed using spectroradiometers. The SR, on the other hand, is measured using dedicated setups. Most commonly used are facilities based on a filter or a grating monochromator that produces tuneable monochromatic light. Fig. 3.8 shows a schematic of the setup used at the ESA.

Although, the setup might be different in other laboratories, the main components will probably be identical. One requires a white light source (xenon lamp in this case). The chopper wheel is for creating a sinusoidal signal that can be detected by the two lock-in amplifiers. The filter wheel suppresses higher orders of diffraction when the light is traveling through the monochromator. In the ESA setup, mainly optical fibers are used to guide both the monochromatic light and the bias light onto the test sample. Monitor cells are used to account for lamp instabilities between the system calibration using a reference cell and the subsequent measurement.

The SR is the current output per incident monochromatic light power input with the cell being operated at its short circuit. Therefore, the current output of the cell is measured when illuminated with monochromatic light of different wavelengths. To understand the input power per wavelength, the setup needs to be calibrated using a reference solar cell or similar device of which the SR is known. Once the setup is calibrated the test cell is then measured. For single-junction solar cells, this is quite

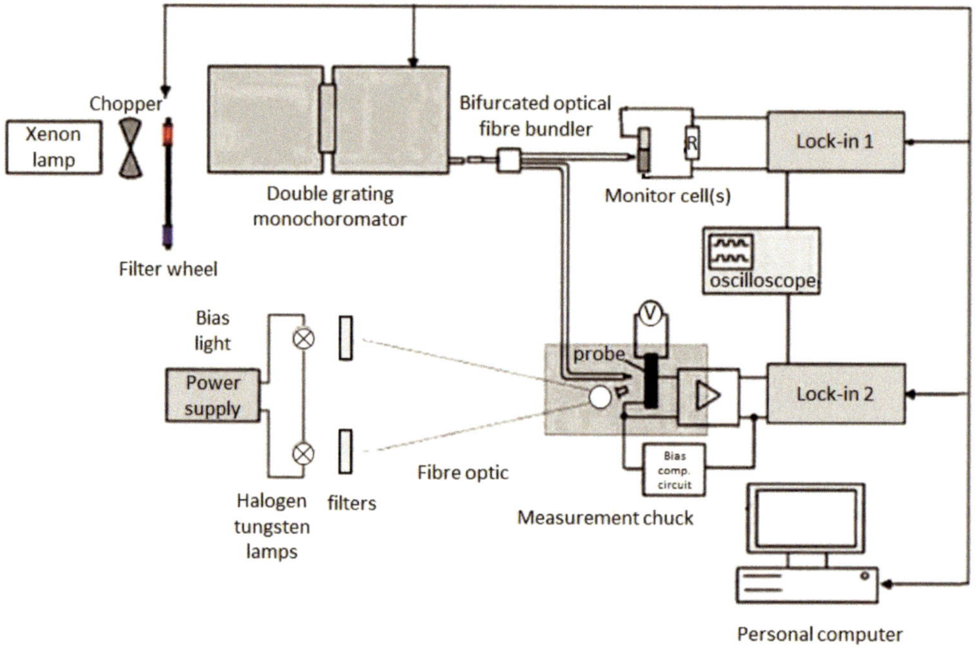

Figure 3.8 Schematic of the spectral responsivity measurement setup used at ESA.

straightforward. A bias light is applied that brings the solar cell into a dedicated operating point. Then, one has to make sure that the cell is kept at 0 V by a power supply.

Procedures for measuring SR and the I-V characteristic of single-junction solar cells are all well documented in standards [18,19]. However, nowadays in space, MJ solar cells are used almost exclusively. In those type of cells a number of different semiconductor layers are used to more efficiently use the sunlight. In most cases, three or more different solar cells are connected in series on top of each other by tunnel diodes. These cells are two-terminal devices, and due to the internal series connection, the current that can be provided by such a device is limited by the subcell that produces the smallest current.

Fig. 3.9 shows the principal layout of a triple-junction solar cell. The combination of GaInP/GaInAs/Ge is still widely used. This configuration has the big advantage that the single subcells can be grown lattice-matched on top of each other. However, the spectral harvesting is not optimum since the Ge subcell as shown on the right-hand side produces almost double the photocurrent of the other two subcells. The main message of the graph is, however, to show that the 3J cell current is limited by the subcell that produces the lowest current. On the other hand, at each current, the voltages of all subcells add up to the total voltage of the 3J cell. When it comes to the calibration of those cells, the series connection and the fact that the layers are grown on top of each other places some additional requirements on the measurement facilities and the measurement procedure

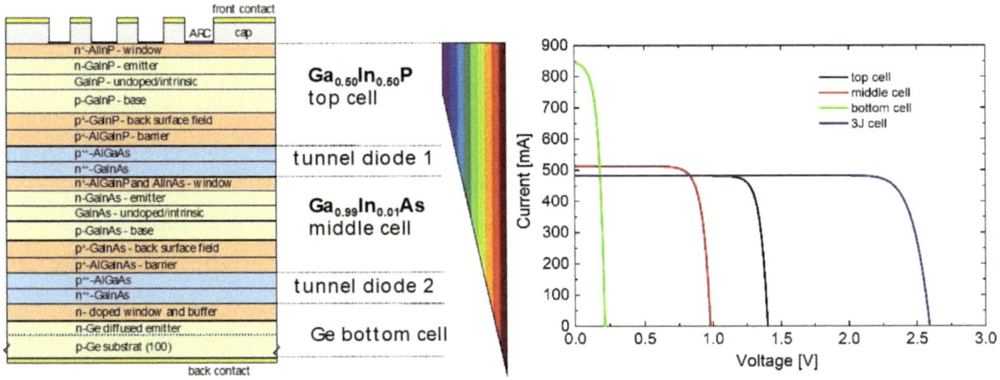

Figure 3.9 Left: Principal layout of a state-of-the-art triple-junction (3J) cell, reprinted with permission from Ref. [20], copyright Elsevier, 2018. Right: I-V characteristics of the three subcells in 3J cell and of the full device. *Source: Authors.*

that differ from the single-junction case. Those requirements will be addressed in detail in the following subsection.

3.3.1 Spectral responsivity of MJ solar cells

The basic principle of measuring the SR of MJ solar cells was first reported by Burdick and Gladfelter [21]. The SR of an MJ cell cannot be measured in one go. Each subcell has to be measured separately by using dedicated conditions. As was mentioned before, the subcell that generates the lowest current is the one that limits the current of the complete MJ. This principle of current limitation is used in the SR measurement. Thereby, bias light is applied to the MJ cell in such a way that one of the subcells becomes the current-limiting one. With this setting the SR of one subcell can be measured. Subsequently, the other subcells are measured by modifying the bias light conditions, so the other subcells become current limiting one by one.

As mentioned earlier, the SR should be measured when the cell is in the short circuit condition. However, in an illuminated MJ cell the limiting subcell is negatively biased when the full device is in short circuit. To compensate for this negative bias, a voltage bias also has to be applied to shift the operation point of the subcell to be measured close to 0 V (or any other desired voltage, in case the voltage dependence of the SR is of interest). Once light and voltage bias have been applied, the measurement of the SR of the subcell can be performed in accordance with the single-junction procedure that is also outlined in standards (i.e., [19]).

3.3.1.1 Bias light

The bias light in SR measurements on MJs has to be carefully adapted for each subcell. It is not sufficient that the subcell under test is only the limiting one under bias light conditions. In addition, the bias light conditions have to be so the subcell limitation does also not change when monochromatic light is added. Fig. 3.10 shows a typical (simplified) sketch of a setup in which bias light and bias voltage are applied to (in this case) a triple-junction cell. The bias light can be realized, for example, by filtered halogen tungsten lamps, light emitting diodes (LEDs), unfocused lasers, or combinations of different light sources.

Fig. 3.11 shows an example of what a bias light distribution could look like for the measurement of a bottom subcell in a triple-junction cell. The bias light (black curve) has a spectral irradiance with a high contribution only between 400 and 800 nm, which only overlaps with the spectral responses of the top and middle subcell. Then, a photocurrent is generated only in the top and middle subcells, thus making the bottom subcell current limiting. In the ideal case, an additional (small) contribution from the monochromatic light will then result in a signal change of the full MJ solar cell only if this light has a wavelength in the sensitivity range of the bottom subcell. This would be the "ideal case," since a signal can sometimes also be measured when the monochromatic light is illuminating the nonlimiting subcells, as discussed later.

3.3.1.2 *Bias voltage*

As explained before, the current limiting operates in reverse as soon as the full MJ device is illuminated and the cell is set to short circuit. Fig. 3.12 shows an example of an ideal situation in a triple-junction cell kept at 0 V when a bias light is applied to make the bottom subcell current limiting. The current of all subcells is dictated by the bottom cell, which makes the top and the middle subcells operate close to their open-circuit voltages. Since the sum of all voltages has to amount to 0 V (because the MJ cell is set to short circuit current conditions), the bottom subcell is operating in reverse. Therefore, a positive bias voltage has to be applied to the full MJ cell to move the operating point of

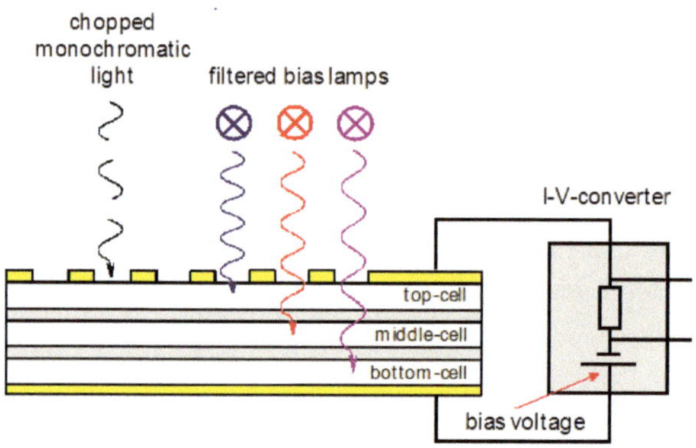

Figure 3.10 Simplified schematic of the application of bias light and bias voltage when measuring a triple-junction solar cell. *Source: Authors.*

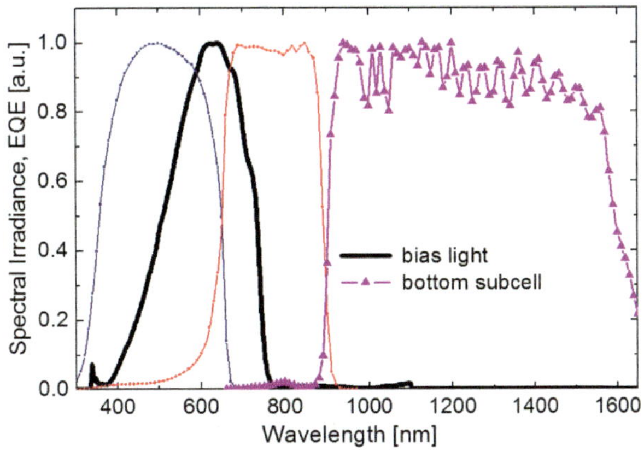

Figure 3.11 External quantum efficiency of a triple-junction cell and example of a bias light distribution for measuring the bottom subcell. In this case, the bias light will (almost) only generate photocurrent in top and middle subcells (*blue and red curves*), making the bottom subcell the current-limiting one. *Figure adapted from Ref. [22] by authors.*

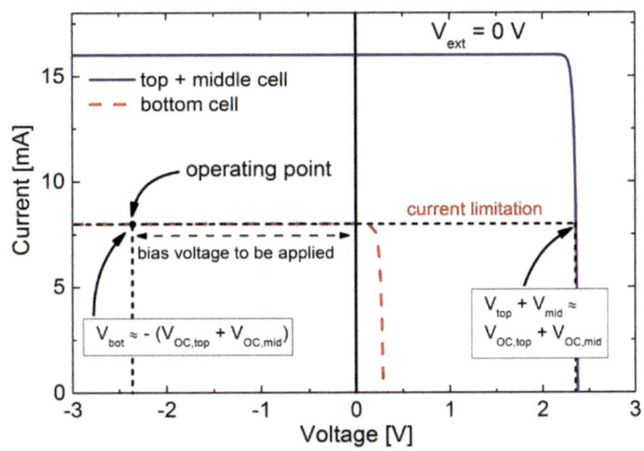

Figure 3.12 Modeled I-V curves of top + middle cell (represented by the blue curve) and the current-limiting bottom subcell (in red). If the full triple-junction cell is at 0 V (short circuit), the bottom cell is operating in reverse. *Figure adapted from Ref. [23] by authors.*

the bottom subcell to (or close to) 0 V. Normally, the bias voltage to be applied is the sum of the open-circuit voltages of the nonlimiting subcells (under the applied bias light conditions). It must be pointed out here that in practice an exact adjustment of the limiting cell to 0 V is not possible since the voltages of the nonlimiting cells under the bias light conditions are not easily accessible. Nevertheless, it is still possible to move the operating point of the current-limiting cell close to its I_{sc}.

3.3.1.3 Measurement artifacts and their correction

A type of feature only present when measuring MJ solar cells are so-called measurement artifacts, which can appear if certain conditions during the measurement and cell properties are met. They are called that since they are related to the specific way the measurement is performed, which differs from the actual use of the cell in application. Therefore, they can be considered not necessarily "real" effects. Measurement artifacts can have two completely independent origins. One is due to nonideal I-V characteristics of one or more subcells in the MJ solar cell, i.e., a low parallel resistance. The other one is due to optical coupling between solar cells. Both phenomena have been excessively discussed in the literature (e.g., Refs. [22–25]). The two artifacts and their correction will be explained in more detail in the following subsections (see also [26]).

3.3.1.3.1 Artifacts related to low shunt resistances

The objective of an SR measurement is to measure the extra current produced by monochromatic light in the (sub)cell under test. Due to the internal series connection in a MJ solar cell, a change in current in one subcell also affects the operating point in the other subcells. To be more clear, a change in current in the current-limiting cell will lead to a small change in voltage in the nonlimiting cell due to the external voltage that is kept constant, and also to a small change in the subcell under test. As long as the measurement is performed in a regime of the I-V curve where the curve is "flat," this does not lead to any artifact. If, however, the I-V curve is not flat, e.g., because the subcell under test has a low shunt resistance (or is operated in a breakdown regime), this change in operating voltage will lead to the additional current produced in the cell to be incorrectly measured. In the wavelength regime where the cell does "see" the monochromatic light, this leads to a lower signal, and in a wavelength regime where the cell is "blind" toward the monochromatic light, a signal is measured. Fig. 3.13 shows an example of a subcell with a measurement artifact. The graph shows the subcell SRs in a triple-junction cell

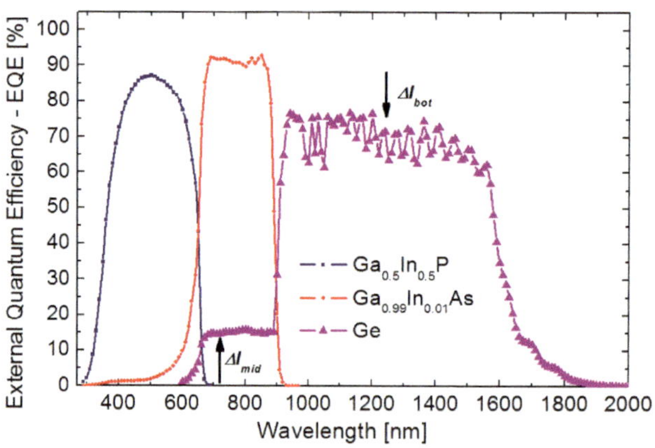

Figure 3.13 External quantum efficiency of a triple-junction cell. The germanium bottom cell shows a measurement artifact. *Figure adapted from Ref. [22] by authors.*

consisting of a GaInP top cell, GaInAs middle cell, and a Ge bottom cell. The Ge bottom cell has a low shunt resistance, and this leads to the shown nonzero signal in the wavelength regime in which the middle cell has already absorbed all the light and at the same time a lower signal in the wavelength regime where the Ge cell is sensitive.

There are ways to reduce the measurement artifact by applying appropriate light and voltage bias. However, when a cell behaves nonideally, it is not always possible to suppress the artifact completely. Most important, to suppress the measurement artifact related to nonideal I-V behavior, is the application of a bias light, where the absolute difference in photocurrents generated in the nonlimiting cells and the current at which the limiting cell operates are as high as possible [22,24,25]. Also, applying a forward bias voltage helps to reduce the measurement artifact since the current in the limiting cell will be reduced.

3.3.1.3.2 Artifacts related to luminescence coupling

When solar cells operate close to their open circuit, the electron-hole pairs created by the absorbed light will not be extracted into an external circuit. Thus, after their lifetime, these carriers will recombine. In high-quality direct semiconductors (as are present in many MJ cells used for space), the dominant recombination mechanism is radiative recombination. This means that the solar cell is emitting light. In SR measurements the nonlimiting cells obviously operate close to their open circuit. Therefore, in these situations the spectral response measurements of solar cells that are beneath subcells with high luminescence signal will show signals in the wavelength regime where they are not sensitive, and the signal is also going to be reduced in the wavelength regime where they are sensitive. The latter is due to the fact that whenever monochromatic light is put on the subcell under test, more current can also be extracted from the nonlimiting cell that emits radiative recombination. In these conditions, fewer carriers remain that can recombine radiatively. Thus, the measured signal is reduced by a certain contribution with respect to the situation when no monochromatic light is added. Consequently, the measured signal is reduced by a certain amount.

As for the shunt-related artifact, the luminescence artifact is "artificially" produced by the measurement conditions. In real operation the difference in photocurrents of the limiting and nonlimiting subcells will typically be very small, so luminescence effects are consequently also small. In addition, space solar cells are usually limited by the top cell at beginning of life, and the limiting cell does not emit any light since all the carriers are extracted into the external circuit. In operation due to particle irradiation in space, the cell limitation will change over time. However, once this happens, the defects created in the solar cell have suppressed radiative recombination, so then this effect can also be neglected.

While the artifact related to shunts is suppressed by increasing the bias light level in the nonlimiting cell, the contrary is true for the artifact related to luminescence coupling. In MJ solar cells where both effects are present, an optimum can be found [25]. If in doubt, it is recommended in any case to adjust the bias light conditions in a way to suppress potential shunt-related artifacts. Such conditions are recommended anyway to limit unwanted effects related to subcell capacitances, as shown in Ref. [27].

3.3.1.3.3 Correction procedure

Although the occurrence of measurement artifacts is in many cases unavoidable, there are ways to correct for them. To do that, it is required that the subcell that shows the artifact is at least measured covering a wavelength range of the subcell on top of it. For both artifacts the correction procedure is identical, at least for the first-order correction (which is typically sufficient). Here, it is assumed that for the shunt-related artifact correction, the artifact signal is predominantly found in the junction above the affected cell (e.g., the middle cell in case the bottom cell is measured and has a low shunt resistance). Furthermore, for the luminescence coupling-related artifact, only the first-order correction is considered, i.e., luminescence coupling is only coming from the cell directly above the cell under test.

Let J_n be the subcell under test showing the measurement artifact and J_{n-1} the cell on top of J_n . Then the step-by-step correction procedure is described as follows:

1. Identify the wavelength range $[\lambda_1, \lambda_2]$ in which the actual EQE of J_{n-1} is above 10% and the measured EQE of J_n is larger than 1%.
2. Within this wavelength range, calculate the ratios (Eq. 3.14):

$$\frac{EQE_{J_n}^{meas}(\lambda)}{EQE_{J_{n-1}}^{act}(\lambda)} \quad \lambda\varepsilon[\lambda_1, \lambda_2] \tag{3.14}$$

with $EQE_{J_n}^{meas}(\lambda)$ being the EQE of J_n as measured (including measurement artifacts), and $EQE_{J_{n-1}}^{act}(\lambda)$ is the actual (potentially corrected) EQE of J_{n-1}.

3. The minimum value of this ratio is the correction factor β as described by Eq. (3.15):

$$\beta = Min\left(\frac{EQE_{J_n}^{meas}(\lambda)}{EQE_{J_{n-1}}^{act}(\lambda)}\right) \quad \lambda\varepsilon[\lambda_1, \lambda_2] \tag{3.15}$$

4. The corrected (actual) EQE of J_{n-1} is then given by Eq. (3.16):

$$EQE_{J_n}^{act}(\lambda) = \frac{EQE_{J_n}^{meas}(\lambda)}{1-\beta} - \frac{\beta}{1-\beta} \cdot EQE_{J_{n-1}}^{act}(\lambda) \tag{3.16}$$

For further details the reader is referred to the following literature: [22,24,26].

The shunt-related artifact is predominantly seen in the wavelength range of the nonlimiting subcell producing the lowest photocurrent. This does not necessarily have to be the subcell on top of the one to be measured. It could even be one of the cells below the subcell under test. The correction procedure as described applies, however, in the same way (only J_{n-1} has then to be replaced with the respective subcell).

3.3.2 Calibration of the electrical performance of MJ cells under simulated sunlight

The electrical performance measurement or calibration of a solar cell is defined as the measurement of the I-V curve with determination of its maximum power output under well-defined conditions (STCs). While solar cells for terrestrial applications are also measured using natural sunlight, solar cells for space are actually exclusively measured in a lab environment using solar simulators. Measuring the I-V curve itself is usually quite straightforward, although capacitive effects can also require special attention, e.g., when the measurement is performed very fast (within 1 ms) when flash simulators are used.

However, the main task in the measurement process is the correct adjustment of the solar simulator. Since the typical solar cell for space is a MJ cell, this process is more complicated than for single-junction solar cells. In addition, while in the case of single-junction solar cells a correction to STC is basically also possible after the measurement, for MJ solar cells, this is almost impossible and should be avoided. The adjustment procedure as it was explained for SJ solar cells before must now be extended to MJ solar cells. This means that Eq. (3.3) then has to be fulfilled for all subcells (Eq. 3.17):

$$I_{photo,test,i}^{sim} = I_{photo,test,i}^{AM0} \quad i = 1, 2, 3, \ldots \tag{3.17}$$

In practice, this requires a simulator that is spectrally adjustable. This could be done by means of filters or by using solar simulators with a number of independent light sources that cover different wavelength ranges. The procedure for the adjustment of the simulator has a number of possibilities, and all can be referred back to the case described for SJ cells. One possibility is that for each subcell in the test device a dedicated reference cell is used. Then, the mismatch factor would be calculated for every subcell and corresponding reference cell. If the solar simulator consisted of individual light sources where each light source would only illuminate one subcell (which could be done, e.g., with an LED-based simulator), the procedure would be exactly identical to the single-junction cell case (Eq. 3.13), and one could set the simulator using Eq. (3.18):

$$I_{photo,ref,i}^{sim} = \frac{I_{photo,ref,i}^{AM0}}{M_i} \quad i = 1, 2, 3, \ldots \tag{3.18}$$

where the index i denotes different subcells and corresponding reference cells. In practice, LED-based solar simulators are still rare. And a simulator that consisted only of quite narrow-band light sources would also be very poor in terms of spectral match to AM0. Thus, in practice, individual light sources of the solar simulator would potentially contribute to more than only one subcell, or in other words, one subcell would see light from more than one light source. In this case, the procedures become more complex. Usually then this leads to solving a linear equation system. Let $e_{sim,j}(\lambda)$ be the spectral distribution of light source j in the solar simulator, so Eq. (3.8) is transformed to Eq. (3.19):

$$\int SR_{test,i}(\lambda)\cdot E_{AM0}(\lambda)\cdot d\lambda = \sum_j A_j\cdot \int SR_{test,i}(\lambda)\cdot e_{sim,j}(\lambda)\cdot d\lambda \quad i = 1,2,3,\ldots \tag{3.19}$$

with $SR_{test,i}(\lambda)$ being the SR of subcell i in the device under test. The unknowns in this equation system are the coefficients A_j that indicate the factor by which the light intensity of light source j has to be adjusted.

Once the A_j are derived from Eq. (3.19) the solar simulator can be adjusted using one or more reference cells. Assuming that k reference cells are used, we can calculate (Eq. 3.20) the photocurrent of each reference cell under light source j as follows:

$$I^{sim,j}_{photo,ref,k} = A_j\cdot \int SR_{ref,k}(\lambda)\cdot e_{sim,j}(\lambda)\cdot d\lambda. \tag{3.20}$$

Note that here the A_j coefficients cannot be linked to the mismatch factors as for single-junction solar cells. There are variants to this procedure where mismatch factors are also introduced in Eq. (3.19), but in the end, this represents simply a normalization to a calibration value and does not change the result. The underlying assumption here is that all reference cells have to be linear, i.e., they must satisfy Eq. (3.21),

$$I_{photo}(c\cdot E) = c\cdot I_{photo}(E), \tag{3.21}$$

since now the photocurrents of the reference cells calculated according to Eq. (3.20) can differ strongly from their calibration values. In practice, it is obvious, that in this case each light source is adjusted with a reference cell that provides the highest photocurrent.

Many companies that regularly do measurements of MJ solar cells are not equipped with a solar simulator with as many individual adjustable light sources as there are subcells in the MJ device. This will lead to the possibility that Eq. (3.17) might not be satisfied for all subcells at the same time. It is then important to make sure that during the measurement the limiting junction is identical to the one under the reference and Eq. (3.17) is fulfilled for that cell. However, deviations in photocurrent of the nonlimiting cells can also, of course, influence the measurement and lead to uncertainties. This has, for example, been discussed in Ref. [28]. Finally, whatever is done to meet Eq. (3.17), an important part of the measurement or calibration of an MJ solar cell is to perform a detailed measurement uncertainty analysis, and it is good practice to report measurement values together with the respective uncertainty evaluation. To assess the comparability and quality of electrical performance measurements of MJ cells on an international scale, the space community carried out a measurement round-robin, since MJ solar cells have now become the major product for space applications [29].

A total of eight laboratories participated in this round-robin. The solar cell under test was a triple-junction solar cell manufactured by Emcore (now SolAero). In the first round, each laboratory was asked to use its in-house measurement standards, while in a second round, matched reference cells that had to be used were included. Fig. 3.14 shows the results. The left graph shows the results where all labs were using their in-house standard, and the right side shows the results of the second round where all labs were using the matched reference cells. The values of each photovoltaic parameter were normalized to the average of all labs. This might be slightly misleading, especially in the first case where two outliers clearly would move the average toward higher values. The graph on the left side also shows the results of the primary calibration of this cell that was on the two balloons (JPL and CNES) and also on the Lear Jet. Those results actually indicate that the "real" average should be lower. In any case, the graph on the left already shows a reasonably good agreement between all labs. Only two clear outliers can be noticed, from AFRL and NRL, for both I_{sc} and the maximum power point P_{MPP} show deviations of up to 6%. It is quite likely that this is the result of a potentially degraded in-house reference. If a reference cell degrades, i.e., it produces less current at the same setting, this means that to arrive at the calibration value the light source would have to be increased to a higher level. Therefore, the measured I_{sc} and P_{MPP} are correspondingly overestimated. If the two outliers were ignored, the agreement between all labs would be ~2%.

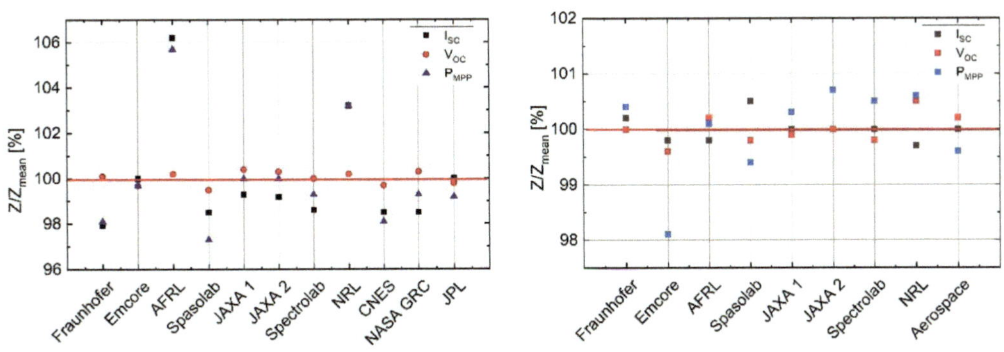

Figure 3.14 Photovoltaic parameters of a 3J cell normalized to the average values of all participating test entities. Left: Results where labs were using their in-house reference standards. Right: Results of the same cell, but this time each lab used matched reference cells that were shipped around with the 3J cell to be measured. *Figure adapted from Ref. [29] by authors.*

When all labs use the same reference cell, the agreement is again excellent within around 2% with the exception of one outlier. Interestingly, the outlier here is the P_{MPP} measured by Emcore, while the corresponding I_{sc} is very close to the average value. This may indicate that here, potentially due to limitations in the solar simulator, a higher discrepancy in the photocurrent of a nonlimiting subcell was accepted, which then led to a slightly larger offset with respect to the other labs. Nevertheless, given the variety in solar simulators and measurement procedures, the agreement demonstrated during this round-robin could be considered remarkably good, and it shows the competence of these labs to calibrate MJ cells.

3.4 Conclusions

The calibration of MJ solar cells for space application was discussed. When considering calibration, a distinction has to be made between primary and secondary calibration. Primary calibrations are still preferably performed using high-altitude methods where solar cells can be measured directly under the "real" spectrum. However, due to recent updates of the TSI value and a revision of the AM0 spectrum, other methods such as the DSR method are potentially a credible alternative. Primary reference cells are the initial source in a calibration chain that goes via secondary standards and working standards down to the calibration of space solar cells. A calibration in its truest sense is a best effort measurement of a solar cell under STCs. This includes the use of reference cells (primary, secondary of working standards) that are back-traceable to a primary standard. Furthermore, it requires the control of the STC temperature and the application of a spectrum with which the condition can be met that the photocurrent produced in each subcell under the solar simulator is equal to the subcell photocurrent under AM0. Since no perfect solar simulator exists, the spectrum of the solar simulator as well as the spectral responsivities of the reference cells and the subcells of the test cell need to be known, to apply appropriate corrections. The difficulty in measuring the SR of MJ solar cells has been explained, especially if accompanied by measurement artifacts. Related correction procedures for such artifacts have been provided. Even if all correction procedures are applied, it is good practice to perform a dedicated measurement uncertainty analysis. Finally, the results of an international measurement round-robin test show the generally quite strong capability of many labs to perform valid calibrations of MJ solar cells.

References

[1] C.A. Gueymard, Revised composite extraterrestrial spectrum based on recent solar irradiance observations, Sol. Energy. 169 (2018) 434–440, https://doi.org/10.1016/j.solener.2018.04.067.

[2] IEC 60904-3, Photovoltaic Devices - Part 3: Measurement Principles for Terrestrial Photovoltaic (PV) Solar Devices with Reference Spectral Irradiance Data, International Electrotechnical Commission, 2016, pp. 115.

[3] JCGM 200:2012, International Vocabulary of Metrology - Basic and General Concepts and Associated Terms (VIM) third edition, Joint Committee for Guides in Metrology, 2012, pp. 108.

[4] ISO 15387, Space Systems — Single-Junction Solar Cells — Measurement and Calibration Procedures, International Organization for Standardization, 2005, pp. 83.

[5] P. Jenkins, Preliminary results from the MJ calibration & measurement round robin, in: Presented at the 10th Workshop on Space Solar Cell Calibration and Measurement Techniques, 2005.

[6] J. Metzdorf, Calibration of solar cells. 1: the differential spectral responsivity method, In Appl. Opt. 26 (1987) 1701–1708.

[7] S. Winter, T. Fey, I. Kröger, D. Friedrich, K. Ladner, B. Ortel, S. Pendsa, F. Witt, Design, realization and uncertainty analysis of a laser-based primary calibration facility for solar cells at PTB, Measurement 51 (2014) 457–463, https://doi.org/10.1016/j.measurement.2013.12.001.

[8] C. Baur, I. Kroger, S. Winter, F. Witt, V. Pichetto, E. Rapp, Comparison of primary high altitude and synthetic calibration methods of solar cells for space applications in view of an updated AM0 spectrum, in: Proc 2018 IEEE 7th World Conf. Photovolt. Energy Convers. WCPEC Jt. Conf. 45th IEEE PVSC 28th PVSEC 34th EU PVSEC, IEEE, Waikoloa Village, HI, 2018, pp. 1570–1575, https://doi.org/10.1109/PVSC.2018.8548245.

[9] C.A. Gueymard, Reference solar spectra: their evolution, standardization issues, and comparison to recent measurements, Adv. Space Res. 37 (2006) 323–340.

[10] G. Kopp, J.L. Lean, A new, lower value of total solar irradiance: evidence and climate significance, Geophys. Res. Lett. 38 (2011) 1–7.

[11] M.P. Thekaekara, Extraterrestrial solar spectral irradiance, in: A.J. Drummond, M.P. Thekaekara (Eds.), Extraterr. Sol. Spectr, Institute of Environmental Sciences, Mount Prospect - Ill, 1973, pp. 71–133.

[12] C.A. Gueymard, A reevaluation of the solar constant based on a 42-year total solar irradiance time series and a reconciliation of spaceborne observations, Sol. Energy 168 (2018) 2–9, https://doi.org/10.1016/j.solener.2018.04.001.

[13] C. Wehrli, Extraterrestrial Solar Spectrum, World Radiometric Center, Davos, 1985. See: https://www.nrel.gov/grid/solar-resource/spectra-wehrli.html (accessed Sept. 18, 2022).

[14] ASTM E490-00a, Standard Solar Constant and Zero Air Mass Solar Spectral Irradiance Tables, American Society for Testing and Materials, 2014.

[15] C.A. Gueymard, The sun's total and spectral irradiance for solar energy applications and solar radiation models, Sol. Energy. 76 (2004) 423–453.

[16] C. Baur, G. Siefer, R. Kern, S. Winter, Primary solar cell standards - comparison of extraterrestrial and synthetic calibration, in: Proc. 9th Eur. Space Power Conf., Saint-Raphaël, France, 2011.

[17] IEC 60904-7, Photovoltaic Devices - Part 7: Computation of the Spectral Mismatch Correction for Measurements of Photovoltaic Devices, International Electrotechnical Commission, 2008.

[18] IEC 60904-1, Photovoltaic Devices - Part 1: Measurement of Photovoltaic Current-Voltage Characteristics, International Electrotechnical Commission, 2006, pp. 72.

[19] IEC 60904-8, Photovoltaic Devices - Part 8: Measurement of Spectral Responsivity of a Photovoltaic (PV) Device, International Electrotechnical Commission, 2014, pp. 48.

[20] S.P. Philipps, F. Dimroth, A.W. Bett, High-efficiency III–V multijunction solar cells, in: McEvoys Handb. Photovolt, Elsevier, 2018, pp. 439–472, https://doi.org/10.1016/B978-0-12-809921-6.00012-4.

[21] J. Burdick, T. Glatfelter, Spectral response and I-V measurements of tandem amorphous-silicon alloy solar cells, Sol. Cells. 18 (1986) 301–314.

[22] G. Siefer, C. Baur, A.W. Bett, External quantum efficiency measurements of germanium bottom subcells: measurement artifacts and correction procedures, in: Proc. 35th IEEE Photovolt. Spec. Conf., Honolulu, HI, USA, 2010, https://doi.org/10.1109/PVSC.2010.5616919.

[23] M. Meusel, C. Baur, G. Létay, A.W. Bett, W. Warta, E. Fernandez, Spectral response measurements of monolithic GaInP/Ga(In)As/Ge triple-junction solar cells: measurement artifacts and their explanation, Prog. Photovolt. Res. Appl. 11 (2003) 499–514.

[24] M.A. Steiner, J.F. Geisz, T.E. Moriarty, R.M. France, W.E. McMahon, J.M. Olson, S.R. Kurtz, D.J. Friedman, Measuring IV curves and subcell photocurrents in the presence of luminescent coupling, IEEE J. Photovolt. 3 (2013) 879–887.

[25] J.-J. Li, S.H. Lim, C.R. Allen, D. Ding, Y.-H. Zhang, Combined effects of shunt and luminescence coupling on external quantum efficiency measurements of multijunction solar cells, IEEE J. Photovolt. 1 (2011) 225–230.

[26] IEC 60904-8-1, Photovoltaic Devices - Part 8-1: Measurement of Spectral Responsivity of Multi-Junction Photovoltaic (PV) Devices, International Electrotechnical Commission, 2017, pp. 30.

[27] M.A. Steiner, S.R. Kurtz, J.F. Geisz, W.E. McMahon, J.M. Olson, Using phase effects to understand measurements of the quantum efficiency and related luminescent coupling in a multijunction solar cell, IEEE J. Photovolt. 2 (2012) 424–433.

[28] E. Fernández Lisbona, C. Baur, N. Klein, A.M. Grás, J.M. Fernandez Marin, Uncertainty analysis of three junction space solar cells electrical performance measurements with regard to the variation of the illumination on the germanium sub cell, in: Proc. 22nd Eur. Photovolt. Sol. Energy Conf., Milano, Italy, 2007, pp. 118–121.

[29] P. Jenkins, D. Scheiman, C. Goodbody, C. Baur, P. Sharps, M. Imaizumi, H. Yoo, T. Sahlstrom, R. Walters, J. Lorentzen, J. Nocerino, O. Khan, R. Cravens, J. Valles, C. Toporow, T. Gómez, L. Pasos Bazán, S. Bailey, Results from an international measurement round robin of III-V triple-junction solar cells under air mass zero, in: Proc. 4th World Conf. Photovolt. Energy Convers., Waikoloa, Hawaii, USA, 2006, pp. 1968–1970, https://doi.org/10.1109/WCPEC.2006.279885.

CHAPTER FOUR

Space applications of III-V single- and multijunction solar cells

Philip T. Chiu
Spectrolab, Sylmar, CA, United States

4.1 Space solar cells: Applications and challenges

On March 17, 1958, the first solar-powered satellite, Vanguard I, was launched into space. Vanguard I carried a 5 mW beacon transmitter powered by silicon (Si) solar cells producing approximately a total power of 1 W with 10% cell efficiency at beginning of life (BOL) [1]. The solar-powered transmitter lasted for 6 years, setting a record for durability at the time. The combination of low weight and reliability of photovoltaics in space exhibited by the Vanguard I mission catalyzed the use of space solar arrays to power almost all subsequent communication satellites, military satellites, and scientific space probes. Nevertheless, as the years have passed, the satellites launched into space have grown by multiple orders of magnitude in both complexity and power requirements. Three noteworthy recent examples are the ViaSat 2 satellite, the International Space Station, and the Juno exploration satellite. As shown in Fig. 4.1A, the ViaSat 2 satellite has a fully deployed wingspan of 158 feet (48.2 m) that outputs 18 kW of power [2]. In Fig. 4.1B, The ISS has over 3000 m^2 of solar cells (Fig. 4.1B) with a total average output power of 110 kW [3]; this is used to supply battery charging, life support, and power for research experiments. Finally, the Juno satellite array (shown in Fig. 4.1C) has a wingspan of 20 m [4]. If optimized to operate at earth, the arrays would produce between 12 and 14 kW of power [2,4].

The challenge of meeting the continuously increasing power demands from space satellites is compounded by the harsh characteristics of the space environment. One such challenging aspect is the considerable range of temperatures and solar intensity that can be encountered in space. A typical low-Earth orbit (LEO) satellite has an operating temperature of 60°C−80°C, dropping to −80°C in the eclipse cycle, with cycles occurring approximately 16 times per day. Temperature extrema for geosynchronous Earth orbit (GEO) satellites are even larger, with temperatures dropping to less than −150°C in eclipse. The wide temperature extremes place significant amounts of thermal stress on interconnection joints and adhesives used to bond solar cells to panel substrates. In addition, the solar cells must function properly at elevated temperatures.

Photovoltaics for Space
ISBN 978-0-12-823300-9, https://doi.org/10.1016/B978-0-12-823300-9.00004-2

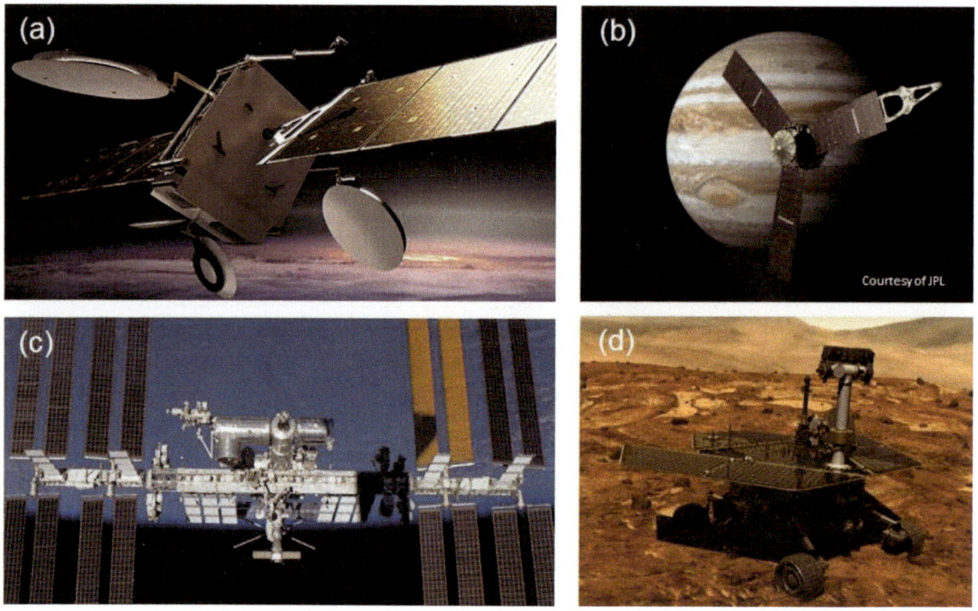

Figure 4.1 Examples of application III-V MJ solar cells in space. (A) Typical GEO synchronous space satellite, (B) Juno explorer probe, (C) International Space Station, (D) Mars Spirit rover. *(A) Image courtesy of Boeing Spectrolab, (B) Image courtesy of JPL, (C) Image courtesy of Boeing Spectrolab, (D) Image courtesy of JPL.*

Higher temperatures always result in a degradation of solar power that varies according to the type of solar cell. Challenges with respect to space environment temperature extrema, though certainly relevant, are not the focus of this chapter.

The presence of ionizing radiation that damages solar cells is the second harsh space environment characteristic that is a significant area of focus of this chapter. The space environment is actually composed of a spectrum of high-energy protons and electrons between 0.1 and 1 MeV that varies dramatically depending on the space orbit of interest. For simplicity of analysis, the complex charged particle spectrum is typically reduced to a single equivalent fluence of 1 MeV electrons using the JPL method [5]. LEO missions residing in orbits <1000 km from Earth vary between 10^{13} to 10^{14} e^-/cm^2 of total 1 MeV electron fluence. GEO missions at 35,780 km from Earth vary between 5×10^{14} to 2×10^{15} e^-/cm^2 of 1 MeV electron fluence. Finally, medium Earth orbit (MEO) missions that reside in the Van Allen radiation belts 2000–12,000 km from Earth see heavy doses of radiation (>1×10^{16} e^-/cm^2 1 MeV electron fluence) [6]. Deep space missions to Jupiter can also experience heavy radiation doses, with 1 MeV electron equivalent fluences from 1×10^{15} to 1×10^{16} e^-/cm^2, depending on the specific orbital path around the planet.

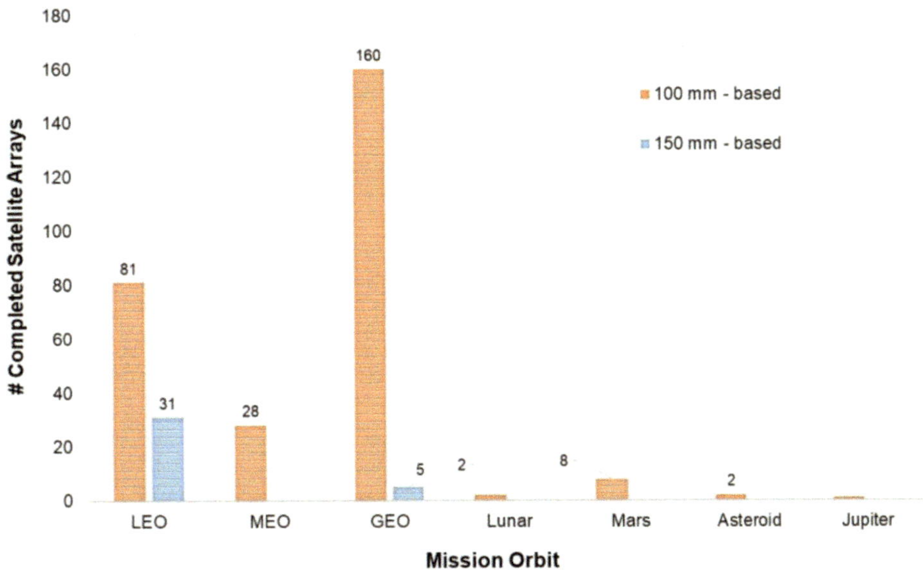

Figure 4.2 Bar graph of completed satellite arrays by mission orbit. *Orange* and *blue bars* are cells fabricated from 100 to 150 mm Ge wafers respectively. *Source: author.*

Of the potential space missions enumerated, the most prevalent over the past 60 years has been in GEO orbits. Fig. 4.2 shows a number of III-V-based satellite arrays completed by Spectrolab. GEO-based satellites comprise more than half of all completed solar arrays. Consequently, 1 MeV electron equivalent fluence associated with the GEO orbit of 1×10^{15} e^-/cm^2 is the benchmark fluence to express the radiation hardness of any space solar cell. Therefore, all subsequent solar cell efficiencies at end of life (EOL) or fraction of remaining power after radiation exposure (NP_{mp}) will be reported in terms of efficiency after 1×10^{15} e^-/cm^2 1 MeV electron fluence.

A multitude of solar cell structures with varying numbers of junctions have been introduced since Vanguard I. The first cell to be introduced was a single-junction (SJ) silicon cell. Since then, cell structures have evolved first to SJ cells based on III-V materials, to III-V-based multijunction (MJ) cells with increasing numbers of junctions. As indicated in Fig. 4.3A and B, the BOL and EOL efficiency has increased steadily with increasing complexity and number of junctions. For instance, the silicon SJ cell is the starting point for Fig. 4.3, with a BOL efficiency of 16%. The final cell structure has five junctions, with EOL efficiencies around 35.8%. This is a more than twofold increase in BOL efficiency achieved over the course of 20–30 years of cell development.

The purpose of this chapter is twofold. The first purpose is to provide the foundation in solid-state physics necessary to model BOL and EOL efficiency of complex MJ structures. The second is to use that foundation along with experimental data to evaluate

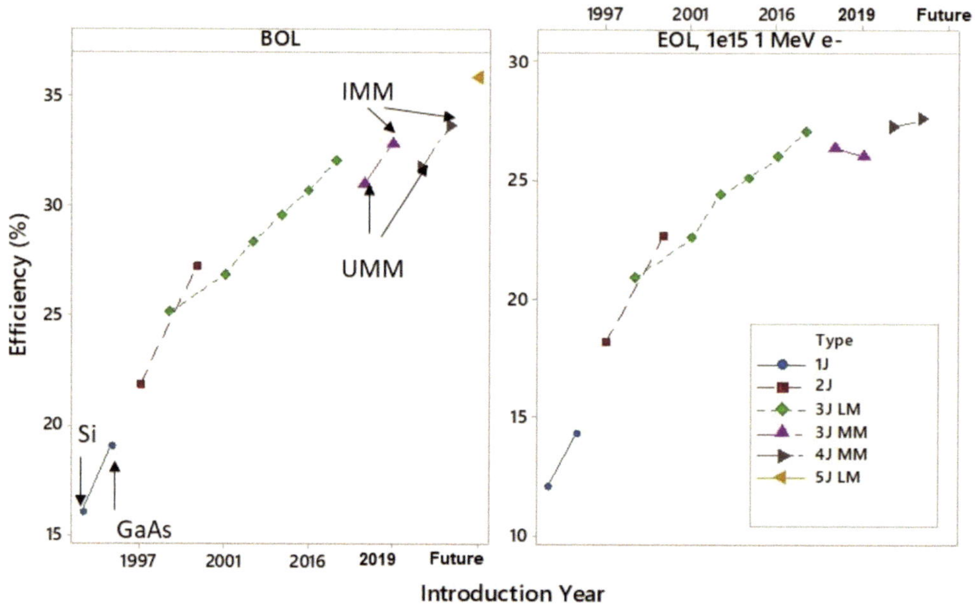

Figure 4.3 Graph of III-V-based cell technologies as a function of time. Left chart is the BOL efficiency. Right chart is EOL efficiency after exposure to equivalent fluence of 1×10^{15}, 1 MeV e^{-}/cm^{2} electrons. BOL and EOL efficiencies of Si are also provided for reference. *Source: author.*

all structures shown in Fig. 4.3 in terms of their absolute BOL/EOL efficiency. Moreover, each cell structure will be evaluated in terms of its viability for market adoption. The adoption of the various technologies shown in Fig. 4.3 has been primarily based upon two equally weighted metrics: EOL efficiency and EOL cost per watt ($/W). To meet success in the commercial marketplace for space solar power, new space technologies must demonstrate an increase in EOL efficiency relative to the incumbent technology. Moreover, the increase in EOL efficiency must also be achieved with consideration to the cost of manufacturing. If the relative cost in $/W significantly outweighs the relative increase in EOL efficiency, adoption of that technology is likely limited. Conversely, cell technologies that significantly lower cost and concurrently decrease efficiency are also unlikely to see widespread adoption.

4.2 Physics of single-junction solar cells

The purpose of this section is to summarize the analytical expressions necessary to model the light illuminated voltage (LIV) characteristics of a SJ cell at BOL and EOL. As shown in Fig. 4.4A, the equivalent circuit for an SJ solar cell can be expressed in terms of three parallel current paths: an ideal current source from photogeneration, a diode with

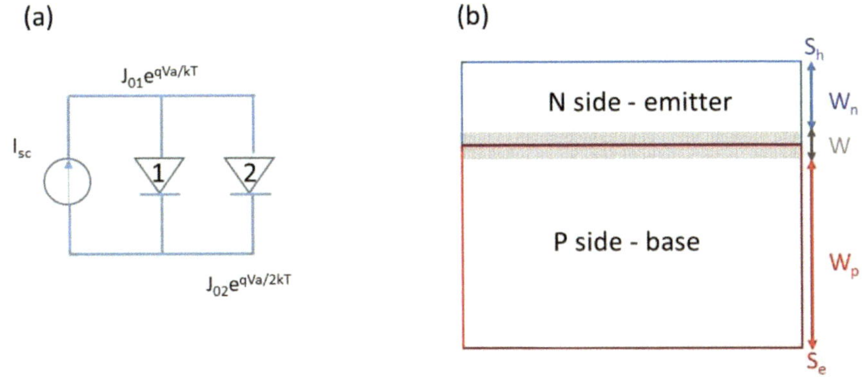

Figure 4.4 (A) Equivalent circuit diagram for solar cell. (B) Schematic for n-on-p solar cell. Doping on the n-side and p-side base is constant to form an idealized step junction profile. *Source: author.*

an ideality factor $n = 1$, and another diode with $n = 2$. It should be noted that the direction of the current for the current source is positive, and for diodes, it is negative. The corresponding expression for current (I) as a function of applied voltage (V) is given by Eq. (4.1):

$$I(V) = I_{sc} - I_{01}\left(e^{\frac{qV}{kT}} - 1\right) - I_{02}\left(e^{\frac{qV}{2kT}} - 1\right) \tag{4.1}$$

where each term corresponds to a path in the equivalent circuit diagram. The first term (I_{sc}) is the current source with a current equal to the short circuit current. The second and third terms represent the two diodes with idealities of 1 and 2, and dark saturation currents I_{01} *and* I_{02}, respectively. The terms I_{sc}, I_{01}, and I_{02} in Eq. (4.1) are complex expressions that depend on the configuration of the solar cell. Here we will consider the most relevant configuration of a finite thickness solar cell with an n-side emitter and p-side base (see Fig. 4.4B). This configuration has three main regions: the quasineutral portion of the emitter with thickness W_n, the depletion region with thickness W, and the quasineutral portion of the base with thickness W_p. On the front and rear surface of the solar cell, there is a surface recombination velocity for minority carriers (S_h and S_e, respectively).

For the specific solar cell configuration in Fig. 4.4B, I_{sc}, I_{01}, and I_{02} can be derived through a complete solution of the minority carrier diffusion equation [6,7]. We begin with the equations associated with the dark current for the $n = 1$ diode, I_{01}. I_{01} is equal to the sum of the dark current in the quasineutral regions on the n side ($I_{01,Wn}$) and p side ($I_{01,Wp}$) of the junction:

$$I_{01} = I_{01,Wn} + I_{01,Wp} \tag{4.2}$$

where

$$I_{01,Wn} = \frac{An_i^2}{N_d} \times \left\{ \frac{qD_h}{L_h} \frac{S_h \cosh(W_n/L_h) + D_h/L_h \sinh(W_n/L_h)}{D_h/L_h \cosh(W_n/L_h) + S_h \sinh(W_n/L_h)} \right\} \tag{4.3}$$

and

$$I_{01,Wp} = \frac{An_i^2}{N_a} \times \left\{ \frac{qD_e}{L_e} \frac{S_e \cosh(W_p/L_e) + D_e/L_e \sinh(W_p/L_e)}{D_e/L_e \cosh(W_p/L_e) + S_e \sinh(W_p/L_e)} \right\} \tag{4.4}$$

Eq. (4.3) relates $I_{01,Wn}$ to variables including intrinsic carrier concentration n_i, n-type doping concentration N_d, diffusivity D_h of minority holes, hole minority carrier diffusion length L_h, surface recombination velocity S_h, and thickness of the quasineutral region W_n. The first term in Eq. (4.3), n_i^2/N_d, is the concentration of minority carrier holes within W_n and is proportional to the negative exponential of the bandgap. The second term in Eq. (4.3) is related to material quality including hole surface recombination velocity (S_h) and diffusion length (L_h). Significantly, the ratio of $W_n/L_h << 1$ is indicative of high-quality material and a low $I_{01,Wn}$. A similar analysis is valid for the dark current term associated with the quasineutral region on the p-side $I_{01,Wp}$ shown in Eq. (4.4). In Eq. (4.4), a ratio of $W_p/L_e << 1$ is indicative of high-quality material and a low $I_{01,Wp}$.

Next, we present the equations associated with the dark current for the $n = 2$ diode, I_{02}. I_{02} is an often neglected dark current term associated with depletion region in the solar cell, but nonetheless it is critical to accurate modeling of EOL performance. It can be expressed as

$$I_{02} = qA\left(\frac{Wn_i}{\sqrt{\tau_e \tau_h}}\right) \tag{4.5}$$

where

$$W = \left[\frac{2K_s\varepsilon_0}{q}\left(\frac{N_a + N_d}{N_a N_d}\right)V_{bi}\right]^{1/2} \tag{4.6}$$

It is evident that the $n = 2$ dark current term is proportional to the depletion width W and the intrinsic carrier concentration n_i, while it is inversely proportional to the minority carrier lifetime of holes and electrons (τ_e and τ_h). As seen in Eq. (4.6), W is largely of function of the doping on the n side (N_d) and p side (N_a) of the junction higher doping on both the n side and p side reduces the depletion width. Therefore, higher doping and long minority carrier lifetime (high-quality material) of solar cells minimize the dark current in the depletion width.

The final set of equations related to the minority carrier diffusion equation relates to the short circuit current, I_{sc}. I_{sc} can be expressed in terms of the number of incident photons $F(\lambda)$.

$$I_{sc} = qA \int F(\lambda)\{(1-s)(1-R(\lambda))(1-A(\lambda))\}IQE(\lambda) \tag{4.7}$$

For space applications, $F(\lambda)$ is given by the well-established AM0 spectrum [8]. The incident photon flux is decremented by the term in curly braces that describes the photon losses prior to reaching the solar cell. These losses include grid shadowing $(1-s)$, front surface reflection $(1-R(\lambda))$, and absorption from layers above the solar cell $(1-A(\lambda))$. After factoring for external losses, we must also account for internal losses in the solar cell using the internal quantum efficiency $IQE(\lambda)$. IQE describes the fraction of photons that are converted in photogenerated carriers. The product of the terms within the integral specifies the current generated a single wavelength, λ. To calculate the total current over the entire AM0 spectrum, an integral over all wavelengths is required.

Of the many terms in Eq. (4.7), the analytical form of $IQE(\lambda)$ is the most difficult to determine. It has three components corresponding to three main regions of the cell:

$$IQE = IQE_{Wn} + IQE_{Wp} + IQE_{W} \tag{4.8}$$

where IQE_{Wn} is the IQE in the n-side quasineutral region, IQE_{Wp} is the IQE in the p-side quasineutral region, and IQE_{W} is the IQE in the depletion width:

$$IQE_{Wn} = \frac{1}{1-\alpha^{-2}L_h^{-2}} \left\{\frac{S_h/\alpha D_h + 1 - e^{-\alpha W_n}[(S_h/\alpha D_h)\cosh(W_n/L_h) + (1/\alpha D_h)\sinh(W_n/L_h)]}{(S_h L_h/D_h)\sinh(W_n/L_h) + \cosh(W_n/L_h)} - e^{-\alpha W_n}\right\} \tag{4.9}$$

$$IQE_{Wp} = \frac{e^{-\alpha(W_n+W)}}{1-\alpha^{-2}L_e^{-2}} \left\{1 - \frac{1}{\alpha L_e}\left[\frac{(S_e L_e/D_e)\cosh\left(W_p/L_e\right) - e^{-\alpha W_p} + \sinh\left(W_p/L_e\right) + \alpha L_e e^{-\alpha W_p}}{(S_e L_e/D_e)\sinh\left(W_p/L_e\right) + \cosh\left(W_p/L_e\right)}\right]\right\} \tag{4.10}$$

$$IQE_W = e^{-\alpha W_n}\left(1 - e^{-\alpha W}\right) \tag{4.11}$$

The equations for IQE_{Wn} and IQE_{Wp} are dependent upon the absorption coefficient of the cell material, α, and similar to the equation for dark saturation current, and terms associated with the quality of the materials such as L and S of their respective minority carriers. In contrast, the equation for IQE_W does not depend on the quality of material and is only related to the fraction of light absorbed in the depletion width. All photogenerated carriers are accelerated out of the depletion region due to the high electric field in this region, leading to 100% IQE_W regardless of the material quality.

By determining the terms I_{sc}, I_{01}, and I_{02} using Eqs. (4.2)−(4.11), one may now use Eq. (4.1) to model the BOL LIV curves for a solar cell and thereby calculate key figures of merit for that cell. An example of a modeled IV curve of a SJ cell is shown in Fig. 4.5. I_{sc} is the current when the applied voltage is zero. Consistent with Eq. (4.1), when the applied voltage increases, the positive diode diffusion current counteracts the negative photocurrent, and the solar cell current decreases rapidly. A second point of interest on the IV curve is where the power produced is at a maximum. This occurs in the largest area rectangle under the IV curve (see red box in Fig. 4.5).

The current and voltage associated with the largest rectangle in Fig. 4.5 are called I_{mp} and V_{mp}, respectively, and. the maximum power is $P_{mp} = V_{mp} \times I_{mp}$. The importance of P_{mp} is that the efficiency Eff is the ratio of P_{mp} to input power P_{in},

$$Eff = \frac{P_{mp}}{P_{in}} = \frac{V_{mp}I_{mp}}{P_{in}} = \frac{FFV_{oc}I_{sc}}{P_{in}} \tag{4.12}$$

The third point of interest in the graph is where $I = 0$, which corresponds to the open-circuit voltage or V_{oc}. For the two-diode model, V_{oc} may be expressed by

$$\begin{aligned} V_{oc} &= \frac{2kT}{q}\ln\left(\frac{-I_{02} + \sqrt{I_{02}^2 + 4I_{01}(I_{01} + I_{02} + I_{sc})}}{2I_{01}}\right) \\ &\approx \frac{2kT}{q}\ln\left(\frac{-I_{02} + \sqrt{I_{02}^2 + 4I_{01}I_{sc})}}{2I_{01}}\right) \end{aligned} \tag{4.13}$$

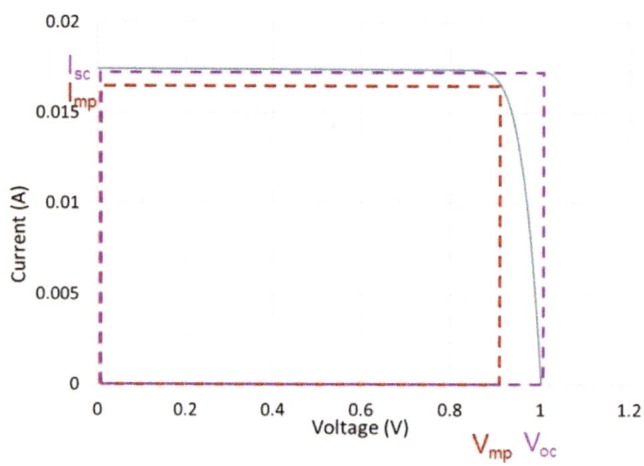

Figure 4.5 LIV curve of a typical single junction solar cell with key figures of merit labeled on the graph. *Source: author.*

The importance of the V_{oc} is that it is the LIV parameter that is most sensitive to the intrinsic quality of the material in the solar cell. Finally, the fill factor (FF) is defined as the ratio of the area of the red box over that of the purple box ($V_{oc} \times I_{sc}$) in Fig. 4.5, or

$$FF = \frac{V_{oc} - \frac{kT}{q}\ln\left(\frac{qV_{oc}}{kT} + 0.72\right)}{V_{oc} + kT/q} \tag{4.14}$$

FF is a measure of the squareness of the I-V and increases monotonically with increasing V_{oc} through Eq. (4.14) [9].

While Eq. (4.13) does provide an analytical means to calculate V_{oc} of the device, it does not provide much intuition regarding the factors that most affect V_{oc}. To provide some intuition, we model V_{oc} as a function of the cell bandgap, E_g. As seen in the blue solid line of Fig. 4.6, V_{oc} for an ideal material increases linearly with E_g. In fact, V_{oc} can alternatively be expressed by Eq. (4.15), where V_{oc} increases linearly with bandgap decremented by a factor called the voltage bandgap offset, W_{oc}. As evident from the dotted blue line in Fig. 4.6, W_{oc} is only weakly dependent on bandgap, ranging from 0.36 ± 0.05 V over the entire bandgap range for an ideal material. To a reasonable approximation, W_{oc} can be thought of as a constant with respect to E_g:

$$V_{oc}(E_g) = \frac{E_g}{q} - W_{oc}(E_g) \approx \frac{E_g}{q} - W_{oc} \tag{4.15}$$

The significance of W_{oc} is it describes the quality of the solar cell material in a single constant value, regardless of the bandgap of the cell. For near ideal materials, W_{oc} is ~0.36 V. The red dotted line in Fig. 4.6 shows a poor material, with poor minority carrier lifetimes/diffusion lengths. The resulting W_{oc} is near 0.6 V, regardless of bandgap. Therefore, a typical range of W_{oc} values goes from 0.36 to 0.6 V for ideal to poor quality

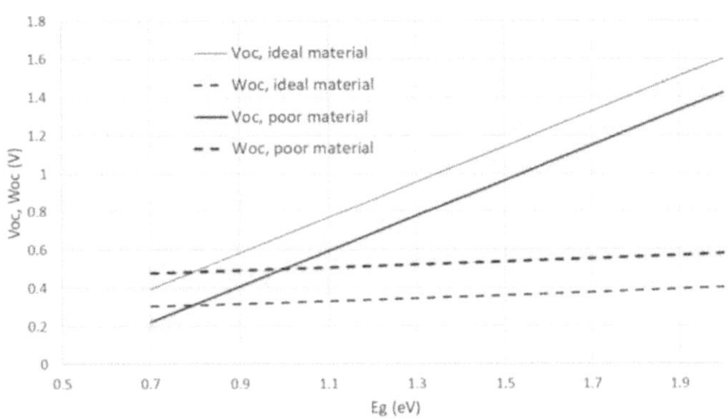

Figure 4.6 V_{oc} (*solid*) and W_{oc} (*dotted*) as a function of bandgap for ideal (*blue*) and poor (*red*) materials. *Source: author.*

cells, respectively. In subsequent sections of the chapter, we will repeatedly refer to the W_{oc} as a convenient gauge of the cell quality.

Calculation of the EOL LIV characteristics of a solar cell requires some understanding of the effect of radiation on solar cell. Space radiation is composed of protons and electrons, some of which have sufficient energy to displace atoms out of the lattice and thus create defects. Such defects can include antisites, vacancies, and interstitials [10], and they typically form deep levels within the forbidden bandgap of semiconductors. A review of Shockley Read Hall (SRH) recombination [7] indicates that deep level defects are an efficient means of nonradiative recombination that significantly reduce both the minority carrier diffusion length and minority carrier lifetime. The reduction in minority carrier diffusion length and lifetime are generally expressed as

$$\frac{1}{L^2} = \frac{1}{L_0^2} + K_L\Phi \tag{4.16}$$

$$\frac{1}{\tau} = \frac{1}{\tau_0} + K_\tau\Phi \tag{4.17}$$

In Eq. 4.16, L_0 and L are the minority carrier diffusion length at BOL and EOL, respectively, K_L is the diffusion length damage coefficient, and Φ is the fluence of 1 MeV electrons. We should note that Φ is most often expressed in terms of 1 MeV electrons because radiation in most GEO missions is dominated by this particle and energy, as previously mentioned.

Similarly, in Eq. (4.17), the loss in minority carrier lifetime $(\Delta 1/\tau)$ is linearly related in this case to the lifetime damage coefficient (K_τ) and again to the 1 MeV electron fluence, Φ. Although the damage coefficients are expressed in two different equations, it should be noted that modeling can be used to relate the two values to one another. Depending on the material and doping level, K_L is generally 10—100 times greater than K_τ.

For reference, Table 4.1 shows a list of K_L values for some group IV and III-V materials relevant to this chapter. Several material trends are evident from Table 4.1. First, higher doping increases the damage coefficient. Second, *p*-type materials generally have lower damage coefficients than *n*-type materials. Third, the damage coefficients of the III-V materials are generally higher than that of group IV materials. Within the class of III-V materials, phosphide-based materials have lower damage K_L values than arsenide-based materials. Such trends will become important as we discuss the radiation hardness of these materials and its effect on EOL cell performance.

Using the formalism developed in Eqs. (4.1)—(4.17), we can now calculate the full LIV curves and associated performance metrics of a solar cell *at EOL*. First, we use damage coefficient values from Table 4.1, determine values of L and τ at BOL, and then

Table 4.1 Diffusion length damage coefficients (K_L) for various relevant group IV and III-V materials.

Material	Polarity	Diffusion length damage coefficient K_L (e^{-1})		Reference
		Low doped	**High doped**	
Silicon	n	5e-11	4e-9	[53]
	p	8e-11	5e-10	
Germanium	n	1e-11	2e-10	
	p	7e-13	6e-11	
GaAs	p	5e-8	1e-7	[43]
InGaAs	p	>5e-8	>1e-7	
AlGaAs	p	<5e-8	<1e-7	
InGaP	p	1.5e-8	3e-8	

Adapted from sources indicated in the right column of the table.

calculate L and τ at EOL with Eqs. (4.16) and (4.17), respectively. We then substitute the EOL L and τ values into Eqs. (4.3)–(4.5) to determine the EOL dark saturation currents, I_{01} and I_{02} values. Similarly substituting EOL L values into Eqs. (4.8) and (4.9), we can calculate I_{sc} at EOL. With these three key current parameters, the EOL LIV curve can be fully determined, again using Eq. (4.1). We will use this exact methodology to determine EOL cell performance in upcoming sections for both SJ and MJ cells. Once the EOL efficiency is determined the radiation retention factor, or NP_{mp}, can be calculated, using the ratio of EOL and BOL efficiencies:

$$NP_{mp} = \frac{EOL\ Eff}{BOL\ Eff} \tag{4.18}$$

4.3 Silicon and gallium arsenide-based single-junction solar cells

For over 30 years, beginning in the late 1950s, silicon solar cells dominated solar arrays for communication satellites. The dominance of silicon over this time period was due primarily to its unmatched efficiency at both BOL and EOL and its relatively low cost of source materials and fabrication. The n-on-p silicon back surface field (BSF) solar cells most widely adopted for space applications increased their AM0 BOL efficiency from 10% to 16% [6]. In contrast, some early arrays based upon thin-film CdS and CdTe were attempted, but with efficiencies only reaching 7% [6]. The first serious challenge to silicon solar cells for space applications occurred in 1992 with the introduction of the GaAs solar cell. This cell was achieved using a relatively new growth technique, organometallic vapor phase epitaxy (OMVPE) [11], using an *n*-type Ge substrate as an epitaxial template. This III-V-based solar cell had a 19% BOL efficiency, exceeding the performance of the state-of-the-art silicon cell by 3% absolute and 19% relative. The reason for the significant increase in BOL efficiency for GaAs over silicon is attributed to a more optimized bandgap and relatively higher material quality. This chapter section will explore both these reasons in detail.

The rationale for higher efficiency of GaAs cells over silicon counterparts due to its bandgap can be explained with an intuitive argument based upon irradiance plots shown in Fig. 4.7. The area under the irradiance curve in Fig. 4.7 is equal to the full incident power for AM0 spectrum (136.7 mW/cm^2). The power generated by high (1.9 eV) and low (1.4 eV) bandgap solar cells is given by the blue and green areas in Fig. 4.7A and B, respectively. The efficiency of a solar cell is given by the ratio of the power areas over the area under the irradiance curve. For an ideal solar cell, incomplete filling under the AM0 irradiance curve under the y-axis is due to two voltage loss mechanisms. The first voltage loss mechanism is cell recombination losses. Recombination mechanisms including radiative and Auger occur even in defect-free cells, and they result in a nonzero W_{oc}. The recombination loss is represented by the gray area in Fig. 4.7. The area between the gray and the irradiance curve is attributed to thermalization loss, which occurs because photogenerated carriers excited with $E > E_g$ rapidly relax to the band extrema before they can be collected. A comparison of areas above the gray curves shows that the higher E_g cell in Fig. 4.7A has a lower thermalization loss than the lower E_g cell in Fig. 4.7B. Therefore, V_{oc} increases with increasing E_g values due to lower thermalization losses. However, increasing E_g to reduce thermalization losses comes with a tradeoff. Photons with $E < E_g$ transmit through the cell, and they do not contribute to I_{sc}. These transparency losses that reduce I_{sc} are represented by the unfilled areas along the x-axis. A comparison of the plots in Fig. 4.7A and B quickly demonstrates that higher E_g cells have higher transparency losses and therefore lower I_{sc} than lower E_g cells.

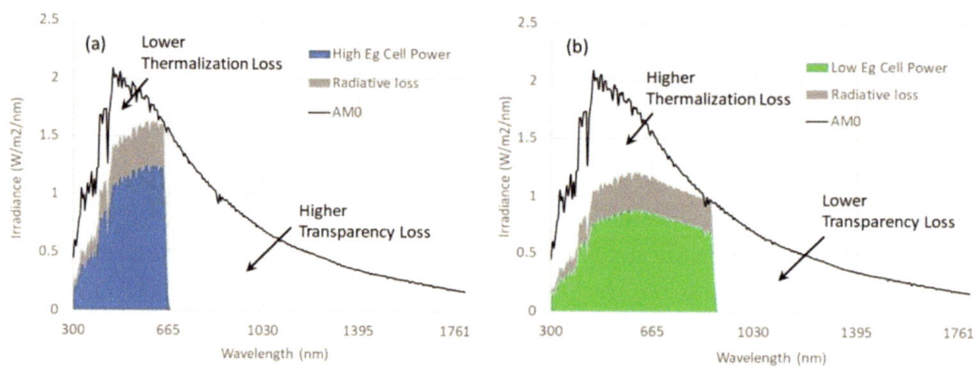

Figure 4.7 Comparison of power collected by ideal (A) high E_g cell and (B) low E_g cell in comparison with AM0 irradiance curve. Plots assume ideal cells where $EQE = 1$ for $E > E_g$ and $W_{oc} = 0.36$ V. Thermalization and transparency losses are indicated by the arrows. *Source: author.*

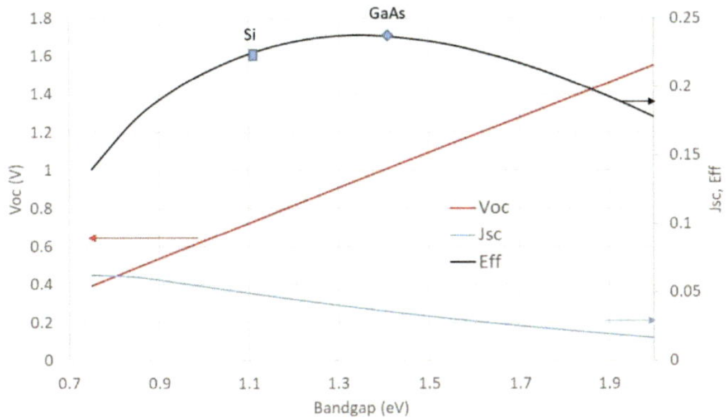

Figure 4.8 V_{oc} *(red curve)*, J_{sc} *(blue curve)*, and *Eff* *(black curve)* as a function of the cell bandgap. *Source: author.*

From the preceding irradiance/power analysis, it is not surprising that V_{oc} increases with E_g and I_{sc} decreases with E_g. This is confirmed with full modeling of cell performance parameters as a function of E_g, shown in Fig. 4.8. Since efficiency is related to the product of V_{oc} and I_{sc} in Eq. (4.12), it is evident that the efficiency would have a maximum at a specific bandgap. As shown in Fig. 4.8, the maximum in efficiency for the AM0 spectrum occurs at 1.36 eV. The actual bandgaps for silicon (1.1 eV) and for gallium arsenide (GaAs) (1.4 eV) are also indicated by the blue square and diamond symbols on the same plot, respectively. Since GaAs is far closer to the optimum bandgap than silicon, the corresponding solar cell will have a higher efficiency even if the quality of the materials is equal. According to the efficiency curve in Fig. 4.8, there is a 1.5% absolute difference in efficiency from GaAs and Si, which accounts for about half of the 3% absolute difference in reported BOL AM0 efficiencies.

The second reason for the higher efficiency of GaAs solar cells over that of silicon is due to the higher quality of GaAs cells. The superior quality is immediately evident in a direct comparison of the W_{oc} values for GaAs (<0.4V) and silicon (0.48 V). GaAs solar cells were able to achieve W_{oc} values close to the ideal level of 0.36 V, due in part to the unique growth process robustness of GaAs grown by OMVPE. High-quality and stoichiometric GaAs can be achieved over a wide range of OMVPE growth conditions including V-III ratio, growth rate, and growth temperature. Moreover, GaAs has low solubilities to a particularly harmful contaminant in OMVPE (oxygen), which is problematic because it forms deep-level defects in GaAs that can reduce the minority carrier lifetime of the material. This is concerning because relatively early OMVPE reactors and sources in the late 1980s were prone to leaks and contamination. Fortunately, even in the presence of large reactor leaks, GaAs maintains remarkably low levels of oxygen at $\sim 1 \times 10^{15}$ cm^{-3} that do not significantly affect minority carrier lifetime.

The high quality of GaAs cells can also be attributed to its low lattice mismatch, f, with respect to the n-type Ge substrate. The lattice mismatch is defined by

$$f = \frac{a_s - a_f}{a_s} \tag{4.19}$$

where a_s and a_f are the lattice constants of the substrate and film, respectively. GaAs has a slightly smaller lattice constant than Ge, leading to an f of 0.1%. At such a small lattice mismatch, even up to typical cell thicknesses of ~3 μm, GaAs on Ge epitaxy is pseudomorphic and generally free of defects that result from relaxation of lattice strain.

While it is clear that GaAs has a superior BOL efficiency over silicon, EOL efficiency is the key metric for space array applications. Using the formulas outlined in Section 4.2, we calculate that GaAs and silicon have a similar radiation retention factor (NP_{mp}) of ~0.75. This matches the experimental NP_{mp} values for both solar cells at the time of the GaAs space cell release in 1992. We note that a radiation retention of 0.75 is poor relative to other materials that will be discussed in forthcoming sections of this chapter. Given the BOL efficiency and retention factors, the EOL efficiencies of GaAs and silicon are 14% and 12% respectively, with GaAs maintaining the original 19% relative increase in BOL efficiency at EOL. This large relative increase was responsible for the adoption of the first III-V solar cell for space applications in 1992 despite the higher cost of manufacturing GaAs over silicon cells [12].

To gain insight into the poor but relatively equivalent radiation retention of GaAs and silicon space solar cells, we compare their respective W_p/L_e ratios at BOL and EOL (Table 4.2). The performance of the solar cell is a strong function of the W_p/L_e ratio. A high-quality solar cell requires a diffusion length of minority electrons (L_e) to be three to 10 times longer than the quasineutral thickness of the base (W_p). This results in W_p/L_e ratios between 0.1 and 0.3. At W_p/L_e ratios >1, a significant fraction of minority carrier electrons will recombine prior to being collected at the junction. The result of this recombination is a significant red wavelength response loss of IQE (and thus I_{sc}),

Table 4.2 Base quasineutral thickness, W_p, and ratio of base quasineutral thickness and hole diffusion length, W_p/L_e, for key materials. W_p/L_e is calculated at both BOL and EOL after exposure to 1×10^{15} fluence of 1 MeV electrons.

Material	W_p (μm)	W_p/L_e	
		BOL	EOL (1e15, 1 MeV e-)
Silicon	~140	0.33	2.33
GaAs	3	0.04	0.67-0.83
InGaP	0.3-0.8	0.01-0.03	0.07-0.11
Ge	~130	0.06-0.25	0.33-0.5

Source: author.

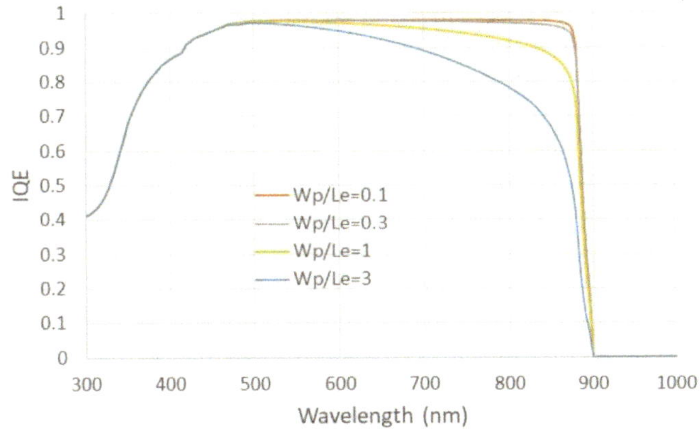

Figure 4.9 IQE curves for differing values of W_p/L_e ratios. Near 100% IQE of current occurs for W_p/L_e values between 0.1 and 0.3. Substantial losses of IQE occur for W_p/L_e ratios >1. *Source: author.*

illustrated in Fig. 4.9. The increased recombination also increases dark current, resulting in a reduction in V_{oc}. Of course, the same analysis can be performed on holes in the emitter (W_n/L_h ratio). However, for an *n*-on-*p* cell configuration of interest in space solar cells, the base is substantially thicker than the emitter, making it far more susceptible to damage. Therefore, it is sufficient to study the effect of radiation on the base or the W_p/L_e ratio to understand the effect of radiation on the overall solar cell.

We start with an analysis of the W_p/L_e ratio at BOL and EOL for GaAs space solar cells. At BOL, GaAs has a near ideal W_p/L_e ratio that ranges from 0.01 to 0.1 depending on the doping of the base. Due to the excellent material quality of GaAs that is nearly limited by radiative recombination, L_e values in the material can range between 20 and 300 μm. These diffusion lengths are far in excess of the base thickness required to fully absorb light up to the bandgap. As a direct gap semiconductor, GaAs is a highly efficient absorber of light [13], so only 3 μm of GaAs is required to absorb 99% of incident photons having energy greater than or equal to its bandgap of 1.42 eV. At EOL, radiation damage reduces the minority carrier diffusion length, L_e, by an amount proportional to the radiation damage coefficient K_L (see Eq. 4.16). Unfortunately, GaAs has the second highest K_L value in Table 4.1. The high K_L value significantly decreases the L_e by a factor ~30 for a 10^{15} e^-/cm^2 fluence of 1 MeV electrons. This increases W_p/L_e from a very healthy value less than 0.01 at BOL to a poor value of ~1 at EOL. Due to the strong decrease in I_{sc} and V_{oc} associated with a W_p/L_e ratio ≥1, GaAs is considered a radiation-soft material. The radiation softness is due largely to the high K_L value that decreases L_e at EOL.

By contrast, a similar W_p/L_e analysis of silicon space solar cells indicates that cell thickness is the cause of its poor radiation retention. At BOL, silicon has a borderline W_p/L_e ratio of 0.33 (Table 4.2). This relatively high W_p/L_e ratio is due not to a poor minority carrier diffusion length or material quality, but rather to the very high thickness value for W_p of 140 μm typically used for heritage silicon space BSF cells. As an indirect bandgap semiconductor, silicon is an inefficient absorber of light [13]. In fact, thickness values of 140 and 300 μm are needed to absorb 95% and 99% of incident photons having energy above its bandgap, respectively. As such, silicon requires ~100 times the thickness to absorb a comparable percentage of photons as GaAs. Fortunately, it has the second lowest K_L value of the relevant materials listed in Table 4.1. More specifically its K_L value of 10^{-10} is about 500–1000 times less than that of GaAs. The relatively low K_L causes a smaller reduction of L_e by about a factor of seven. However, even this relatively modest reduction in diffusion length pushes an already borderline W_p/L_e ratio at BOL to a ratio that exceeds two at EOL. As the EOL W_p/L_e exceeds unity, silicon is classified as a radiation-soft material, due to the high cell thickness required to sufficiently absorb light above its bandgap.

Over the past 2 decades, significant progress has been made in the efficiency and manufacturing cost of silicon solar cells, sparking a renewed interest in silicon technologies for space applications. Silicon cell costs in terms of $/W at BOL have decreased by nearly two orders of magnitude over the past 2 decades. Moreover, newer cell designs including heterojunction [14] and back contact [15] solar cells have achieved AM0 efficiencies above 21% [6]. However, while the changes associated with the newer cell designs optimize BOL performance, they are also detrimental to radiation resistance and EOL performance. For instance, many newer silicon solar cell designs increase thickness from 140 to 300 μm to increase BOL I_{sc}. The increase in thickness further increases the EOL W_p/L_e ratio by a factor of two. Furthermore the back contact and rear heterojunction designs use a p-on-n technology, where the base now becomes an n-type material. As seen in Table 4.1, the radiation damage coefficients for n-type Si are two to eight times higher than p-type. This again increases the EOL W_p/L_e ratio. In fact, Yamaguchi et al. show that the NP_{mp} values are so compromised by the increase in thickness and higher K_L values that the EOL efficiency of the newer silicon solar cell designs are actually less than that of the heritage Si BSF space solar cell [16]. Therefore, EOL efficiency remains a hurdle for silicon-based solar technology despite recent improvements in BOL efficiency and reductions in manufacturing cost.

4.4 Physics of multijunction solar cells

The purpose of this section is to first establish the motivation for MJ cells, and then to build the foundation necessary to fully model the LIV characteristics of MJ cells at both BOL and EOL. In the previous section, we demonstrated the tradeoff between thermalization and transparency losses for SJ cells. While higher bandgap junctions minimize thermalization losses, they also have higher transparency losses and vice versa. MJ cells offer a way to simultaneously lower both thermalization and transparency losses,

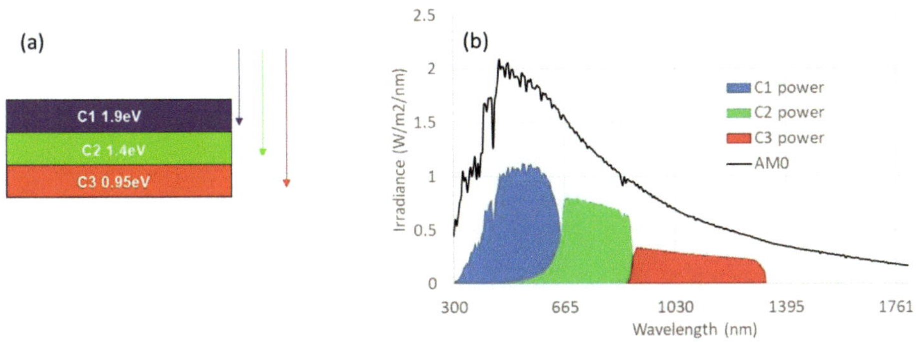

Figure 4.10 (A) Schematic of a 3J MJ cell with bandgaps of 1.9/1.4/0.95 eV. Arrows indicate light of differing bandgaps. (B) Irradiance curve and power curves for the same MJ cell shown in part (A) of the same figure. *Source: author.*

and therefore increase cell efficiency. As shown in Fig. 4.10A, an MJ cell consists of a stack of cells ordered from the first junction with the highest bandgap junction closest to the incident light, followed by the second junction with the next highest bandgap junction, and so forth. In this way, the highest energy light is absorbed by the most closely matched bandgap junction to minimize thermalization losses. The light unabsorbed by the first junction is not wasted to transparency losses, but rather is absorbed by the second junction. The advantage of the MJ cell is shown in Fig. 4.10B, which plots the power generation of the 3J cell relative to the AM0 irradiance curve. Comparing the irradiance/power plots of a 3J cell in Fig. 4.10B relative to those of SJs shown in Fig. 4.7, it is evident that the 3J cell better fills in the area under the AM0 irradiance curve. Therefore, the MJ cell would have a higher efficiency than its SJ counterparts. The MJ cell achieves the higher efficiency by tailoring the bandgap of each subcell to the energy of the incident light, thereby reducing the thermalization and transparency losses. Addition of more junctions would continue to reduce such losses (at increased cost), until ultimately the losses would be confined to unavoidable radiative recombination losses.

Although unspecified in Fig. 4.10, there are in fact multiple ways to implement an MJ cell. The two most common implementations are known as the four-terminal and two-terminal configurations. The four-terminal configuration connects a power lead to the n side and p side of each of the two subcells of the MJ stack. In the four-terminal approach, cells can be mechanically stacked one on top of the other or use optics to separate the light to the subcells (dichroic mirror approach). The advantage of this approach is that it sets no constraints on current or voltage. However, the four-terminal approach has $2n$ terminals for an MJ cell with n junctions. This results in highly complex cell processing and panel assembly due to the large number of terminals that require connections for power extraction. In contrast, the two-terminal or series-connected configuration is the

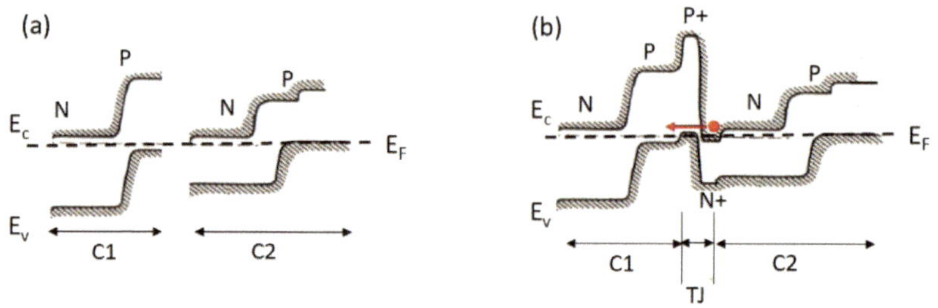

Figure 4.11 Band diagrams of two n-on-p junctions. C1 and C2 are high- and low-bandgap cells respectively. Windows and BSF layers are omitted for simplicity. (A) Case where C1 and C2 are physically separated. (B) Case where C1 and C2 are connected in series with a p^+/n^+ tunnel junction. *Red arrow* indicates tunneling of electrons from C1 to C2. *Adapted from Ref. [9].*

most widely adopted because it allows an MJ cell to be fabricated and integrated into panels with the same level of simplicity and cost as an SJ cell. Given its dominance in the space market, the two-terminal configuration will be the focus of all following sections and analysis.

Despite its simplicity, the two-terminal MJ cell comes with a few additional challenges relative to the four-terminal cell. If multiple junctions are simply series-connected, a pnp junction would form between the base of C1 and emitter/base of C2, as illustrated in Fig. 4.11A. The pnp junction and the associated potential barriers in the conduction and valence bands would completely block the flow of either electrons from flowing from C2 to C1 or holes from C1 to C2. This challenge has been overcome by placing a p^+/n^+ tunnel junction between C1 and C2 [7]. If doping in both the p and n sides of the junction are sufficient to drive the Fermi level into the valence and conduction bands (degenerate doping), the depletion width of the tunnel junction can be reduced to the order of ~ 10 Å in thickness. At this length scale, the electrons would not need to diffuse over a conduction band potential barrier, but rather could efficiently tunnel from C2 to C1, as shown in Fig. 4.11B. The second challenge is that two-terminal devices are subject to Kirchhoff's rules associated with any series-connected devices. In particular, the current entering a subcell must equal the current exiting the subcell. This rule places a constraint on the overall current of a two-terminal MJ cell that will discussed in the upcoming analysis.

With a known two-terminal configuration for the MJ cell, we can now present the analytical formalism used to determine BOL and EOL LIV characteristics for MJ cells. The formalism detailed in this section will build upon that previously established for SJ cells in Section 4.2. The first step in modeling an MJ cell is to iteratively determine the short circuit current for all n subcells in the MJ cell. The I_{sc} of the i-th subcell or $I_{sc,i}$ is given by:

$$I_{sc,i} = qA\int F(\lambda)\{(1-s)(1-R(\lambda))(1-A_{i-1}(\lambda))\}IQE_i(\lambda) \tag{4.20}$$

Eq. (4.20) is similar to Eq. (4.7), except one must account for the absorption loss of the $i-1$ junction, or the absorption loss of the junction above the ith junction. In addition, the $IQE_i(\lambda)$ of the i-th junction can be determined using Eqs. (4.8)−(4.11), without any alterations. Once $I_{sc,i}$ is calculated for all n subcells, I_{sc} of the overall MJ cell ($I_{sc,MJ}$) is given by

$$I_{sc,MJ} = \min\left(I_{sc,i}\right) \tag{4.21}$$

Eq. (4.21) indicates that $I_{sc,MJ}$ can only be as high as the lowest subcell current. Otherwise, the MJ cell would be in violation of Kirchhoff's law. The consequence of Eq. (4.21) is best illustrated with an example. Suppose we had two subcells with a large current mismatch, where $I_{sc,1} = 10$ mA and $I_{sc,2} = 20$ mA. The overall current of the two-junction cell would be only 10 mA; the additional 10 mA from subcell two would be wasted in the form of heat. On the other hand, if the bandgap of subcell one could be decreased or the thickness of subcell one could be increased so $I_{sc,1} = I_{sc,2} = 15$ mA, $I_{sc,MJ}$ would increase by 50% even though the total current collected by the subcells is equal in the two cases. Clearly given Eq. (4.21), the highest $I_{sc,MJ}$ occurs in any two-terminal MJ cell where all subcell I_{sc} values are equal to one another. This is known as the *current matched condition*. However, achieving the current matched condition is simpler in theory than in practice, generally requiring precise bandgaps and thicknesses for each subcell.

Similar to the $I_{sc,MJ}$, calculation of the overall MJ cell V_{oc} or $V_{oc,MJ}$ also requires an iterative calculation of the voltages of all n subcells. The V_{oc} of the ith subcell or $V_{oc,i}$ is still given by Eq. (4.13). The dark and illuminated currents, $I_{01,i}$, $I_{02,i}$, and $I_{sc,i}$, must be calculated for each of the ith subcells using Eqs. (4.2), (4.5) and (4.20). As shown in Eq. 4.22, $V_{oc,MJ}$ is the sum over $V_{oc,i}$ for i from 1 to n:

$$V_{OC,MJ} = \sum_{i}^{n} V_{oc,i} \tag{4.22}$$

The summation of subcell voltages in this manner is the quintessential benefit of an MJ cell, with a greater number of subcell voltages increasing the overall voltage and efficiency of that cell.

Finally, to determine the other MJ cell LIV parameters including P_{mp}, efficiency, and FF, we need to calculate the overall LIV curve for the MJ cell. The MJ cell LIV curve again comes directly from subcell LIV curves. Since $I_{01,i}$, $I_{02,i}$, and $I_{sc,i}$ are known from the calculation of $V_{oc,i}$, the subcell LIV curve is given directly by

$$I(V_i) = I_{sc,i} + I_{01,i}\left(e^{qV_i/kT} - 1\right) + I_{02,i}\left(e^{qV_i/2kT} - 1\right) \tag{4.23}$$

However, in the case of an MJ cell, it is easier to derive the MJ LIV curve from subcell LIV curves when Eq. (4.23) is inverted from an $I(V_i)$ function to a $V_i(I)$ function:

$$V_i(I) = \frac{2kT}{q}\ln\left(\frac{-I_{02,i} + \sqrt{I_{02,i}^2 + 4I_{01,i}(I_{sc,i} - I)}}{2I_{01,i}}\right) \tag{4.24}$$

Then $V_{MJ}(I)$ is the sum of $V_i(I)$ for $I < I_{sc,MJ}$. The voltage of an MJ cell at any forward bias current, I, is equal to the sum of the subcell voltages at that current:

$$V_{MJ}(I) = \sum_i^n V_i(I) \tag{4.25}$$

The constraint applying Eq. (4.25) for $I < I_{sc,MJ}$ enforces the fact that the current for any subcell does not exceed $I_{sc,MJ}$ as required by Eq. (4.21). We note that formalism and Eqs. (4.23)–(4.25) for MJ cell modeling are applicable at both BOL and EOL. The only difference between BOL and EOL is that dark and illuminated currents will change due to a decrease in minority carrier lifetime and diffusion length, discussed in Section 4.2.

Using the now fully developed MJ cell model for two terminal devices, we derive the maximum BOL AM0 efficiency and associated subcell bandgaps as a function of number of junctions, n. The major assumptions used to calculate the maximum efficiency are that the W_{oc} is ideal (~0.36 V) regardless of the subcell bandgap, and that each cell is sufficiently thick to fully absorb photons having energy greater than its bandgap. While we will prove in later sections that there is typically strong tradeoff between W_{oc} value and bandgap of a subcell, the results from the model are nonetheless instructive and serve as a reference point for future discussion. The results from the MJ cell LIV modeling are

(a)

# junctions	E_g1	E_g2	E_g3	E_g4	E_g5	Max AM0 Eff.
1	1.36					24%
2	1.83	1.22				32%
3	1.83	1.22	0.79			38%
4	2.07	1.52	1.14	0.84		41%
5	2.12	1.59	1.22	0.92	0.67	43%

(b)

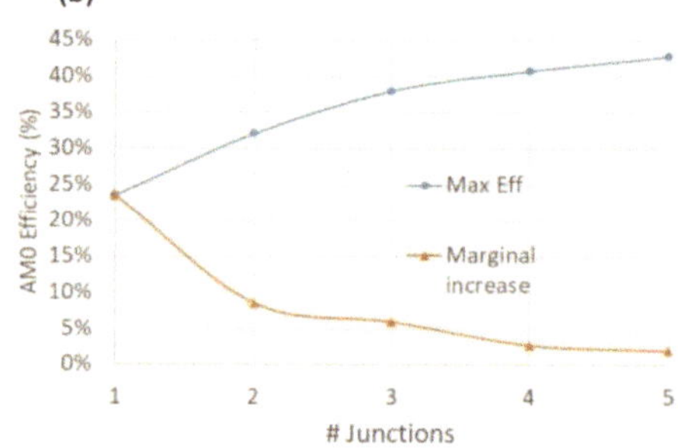

Figure 4.12 (A) Optimum bandgaps and maximum AM0 efficiency from 1J to 5J cells. (B) Corresponding graph of AM0 efficiency and marginal increase in efficiency as a function of number of junctions. *Source: author.*

summarized in Fig. 4.12A and B, illustrating the dependence of maximum AM0 efficiency versus n. Between $n = 1$ and 5, there is a clear increase in maximum efficiency from 24% to 43%, respectively. However, as indicated in Fig. 4.12B, the marginal increase in efficiency from addition of each junction decreases from $n = 1$ to 5. From $n = 1$ to $n = 2$, the maximum efficiency increases by 8% absolute. In contrast, from $n = 4$ to $n = 5$, the maximum efficiency increases only by 2% absolute. The diminishing return in efficiency with the increasing n will become important as we discuss the viability of more complex MJ structures in subsequent sections. In addition to the maximum efficiency, Fig. 4.12A lists the optimized bandgaps corresponding to the maximum efficiency as a function of number of junctions. For all MJ cells where $n > 1$, the maximum efficiency corresponds to a set of subcell bandgaps where the subcell currents are equal. This eliminates the loss in efficiency associated with current mismatch. These bandgaps will serve as important reference points as we discuss the bandgaps that are actually employed in MJ cells.

4.5 Indium gallium phosphide/gallium arsenide-based dual-junction solar cells

In this section we will discuss the implementation and advantages of the first MJ cell produced for use in space applications: the indium gallium phosphide/gallium arsenide (InGaP/GaAs) dual-junction (2J) cell grown on n-Ge substrates. Most early attempts of 2J cells involved using GaAs cell as the lower subcell. As discussed in Section 1.3, GaAs is lattice matched to two convenient substrates (GaAs and Ge), leading to epitaxial growth that is free of strain-related defects. Moreover, GaAs is a binary material, where high quality can be achieved over a wide range of growth conditions by metal organic vapor phase epitaxy (MOVPE). Finally in the case of SJs, GaAs has a bandgap of 1.42 eV, which nearly matches the optimum bandgap of 1.36 eV for AM0 cell efficiency.

In contrast to the SJ case, the bandgap of GaAs of is not well optimized for C2 of a 2J cell, leading to additional complications and considerations for its use in a 2J cell. As shown in Fig. 4.12A, the optimum bandgaps for a 2J cell occur at 1.83 and 1.22 eV for C1 and C2, respectively. The consequence of using a GaAs cell with a bandgap over 200 meV above the optimum is illustrated in Fig. 4.13A. If both C1 and C2 are grown with fully absorbing thickness, any bandgap that is chosen for C1 would lead to a significant and at least 8% absolute reduction in efficiency relative to the maximum for a 2J cell. The reason for this reduction is that with the C2 bandgap far higher than optimum, C1 has far more current than C2, leading to large current mismatch losses in efficiency. Assuming a fixed C2 bandgap at 1.42 eV, the current mismatch loss can be mitigated by an increase in the C1 bandgap and/or a decrease in C1 thickness. Both methods have the effect of allowing more photons through C1 into C2 to help balance the current between the subcells. The effect of these two methods to balance subcell currents is

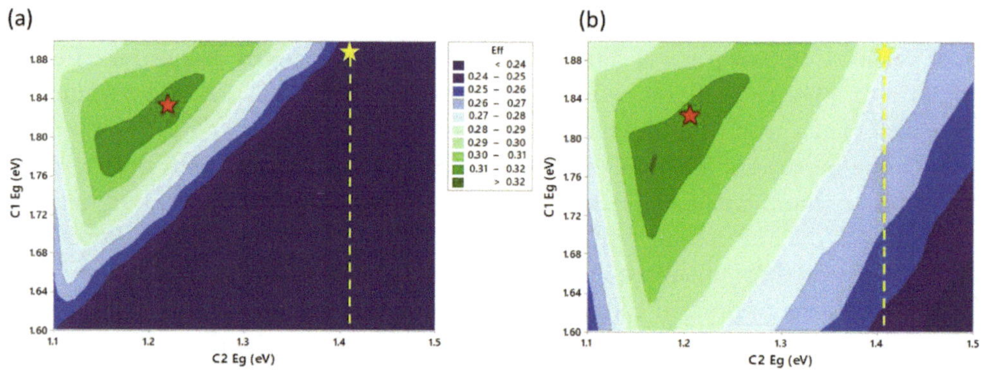

Figure 4.13 Contour maps for 2J cells where (A) C1 is full thickness and (B) C1 is thinned to current balance with C2. *Red star* indicates the optimum bandgap combination for a 2J cell. Line in yellow a fixed bandgap for C2 at the GaAs bandgap of 1.42 eV. *Source: author.*

illustrated in the contour map of Fig. 4.13B. In the case where C1 thinning is allowed to achieve current matching, the efficiency losses due to the use of a 1.42 eV C2 are significantly reduced. The use of higher C1 bandgaps up to the maximum explored in the contour plot of Fig. 4.13B further mitigate the efficiency losses. In fact, at 1.9 eV (yellow star in Fig. 4.13B)—the maximum C1 bandgap explored in the contour map—the efficiency loss is reduced to 3.5% absolute relative to the maximum efficiency.

Two of the early materials explored for use for C1 in a 2J cell were $Al_{0.4}Ga_{0.6}As$ and $In_{0.5}Ga_{0.5}P$ ternary alloys. For future reference, specific compositions of mixed group III ternaries will indicate the concentration of the first group III element (x), but the concentration for the second group III element ($1 - x$) will be omitted because it is readily determined implicitly. As indicated by the bandgap versus lattice constant chart in Fig. 4.14, $Al_{0.4}GaAs$ and $In_{0.5}GaP$ each have a bandgap close to the 1.9 eV value needed improve current matching with a GaAs C2 subcell. Moreover, the two ternary alloys are lattice matched to GaAs and Ge substrates, allowing them to be grown on top of the GaAs cell free of strain-related defects. The challenge with the $Al_{0.4}GaAs$ C1 material is not due to lattice mismatch—related defects but rather impurity-based defects. The aluminum in AlGaAs ternary alloys is extremely reactive, forming stable bonds to oxygen and silicon. For example, AlGaAs is highly sensitive to trace levels of oxygen in growth systems and source material [11]. In fact ~1 ppm of oxygen in a reactor can lead to $>10^{19}$ cm^{-3} oxygen incorporation in the alloy. Even in leak-tight systems, oxygen incorporation is typically in the 10^{17} cm^{-3} concentration level. This is in sharp contrast to the 10^{15} cm^{-3} levels for GaAs; oxygen impurities form high concentrations of deep level SRH defects. *n*-type AlGaAs is often doped with Si. Unfortunately, Al—Si bonds form a DX center complex that is also a deep-level SRH defect [17]. Due to the high concentration of SRH defects, some of the earliest results from 1982 indicated a poor W_{oc}

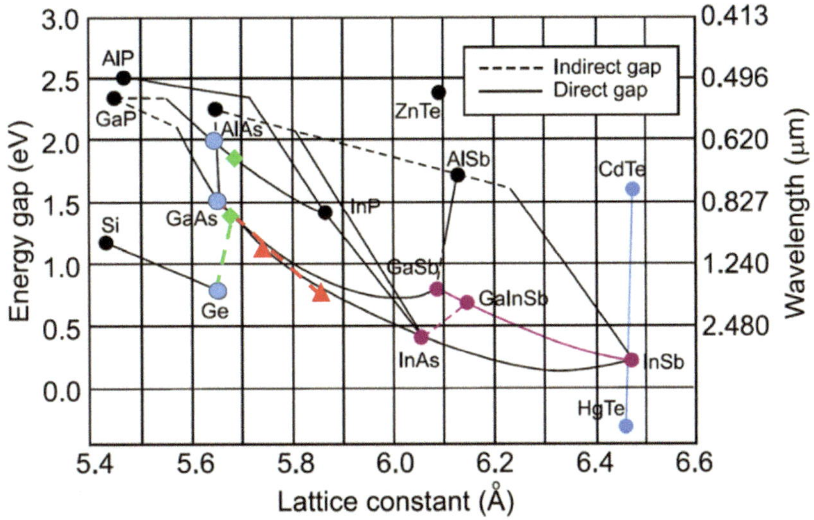

Figure 4.14 Bandgap versus lattice constant. *Blue circles* correspond to materials used in lattice-matched 2J and 3J cells. *Green diamonds* correspond to top two junctions of UMM 3J cell. *Red triangles* correspond to bottom junctions of the IMM 3J and IMM 4J cells. *Source: adapted from [54].*

of ~0.7 V for an $Al_{0.3}GaAs$ subcell [18]. Later work around 1989 for $Al_{0.4}GaAs$ cells showed relatively improved material quality but still poor W_{oc} values of ~0.53 V. The corresponding $Al_{0.4}GaAs$/GaAs 2J cell achieved an efficiency of 27.6% in an AM1.5G spectrum [19].

The second ternary alloy explored for use in 2J cells is InGaP, which is also a challenging alloy to grow by MOVPE. In contrast to GaAs, high-quality InGaP can only be achieved over a narrow range of growth temperatures, growth rates, and V-III ratios. To achieve lattice matching with respect to the GaAs or Ge substrates, the In-to-Ga ratio in the ternary alloy must be controlled to within ±0.5%. Otherwise, growth of a fully absorbing InGaP cell could exceed the critical thickness for relaxation and result in strain-related defects. Fortunately, modern MOVPE reactors have the instrumentation necessary to control the InGaP growth process within the required narrow process range. Another challenge of the InGaP material system is realization of the target bandgap of 1.9 eV. This is due to that fact that bandgap of InGaP is dependent upon the ordering of the In and Ga atoms on the group III sublattice [20]. Ordered InGaP has a CuPt structure with alternating planes of {111} planes of GaP and InP and a bandgap near 1.78 eV. To achieve a 1.9 eV bandgap, the group III sublattice must be close to fully disordered. This can be achieved by a combination using high growth temperatures, high substrate miscut [6], or the use of Sb as a surfactant [21,22].

Despite the growth challenges, InGaP has a significant advantage over the AlGaAs material system. Due to the absence of aluminum, InGaP is relatively insensitive to oxygen contamination in either the reactor or source materials, with oxygen incorporation levels typically in the mid 10^{15} cm^{-3} range. In addition, Si is again conveniently used as a dopant in *n*-type InGaP. The lack of aluminum in InGaP precludes the possibility of DX centers in the *n*-type emitter portion of the cell. The lower levels of SRH defects in InGaP results in improved material quality of InGaP bulk material over AlGaAs. Relatively early attempts at InGaP cells reached W_{oc} values of 0.47 V, which is over 60 mV better than that of AlGaAs. Due in large part to the improved W_{oc} values in C1, 2J cells using InGaP/GaAs subcells on a GaAs substrate achieved rapid progress, with a record efficiency of 29.5% in an AM1.5G spectrum in 1994 [23,24].

The rapid success of the InGaP/GaAs 2J cell led a successful adaptation for space applications. In 1997, Spectrolab introduced the first 2J cell with a 21.8% BOL efficiency in AM0 spectrum [12]. The 2J cell BOL efficiency is 15% relative higher than the GaAs SJ cell. The 2J space cell achieves this efficiency using a layer structure and equivalent circuit diagram shown in Fig. 4.15A and B, respectively. The cell is grown on *n*-Ge rather than *n*-GaAs due to the improved mechanical robustness of Ge over GaAs. An n-on-p GaAs subcell 2 is grown on top of the Ge substrate, followed by a p^+/n^+ tunnel junction. The earliest tunnel junctions utilized GaAs for both 3J layers due to that fact that GaAs is easy to degenerately dope *n*-type and *p*-type. However, it should be noted that GaAs tunnel layers also parasitically absorb light intended for subcell 2, and therefore

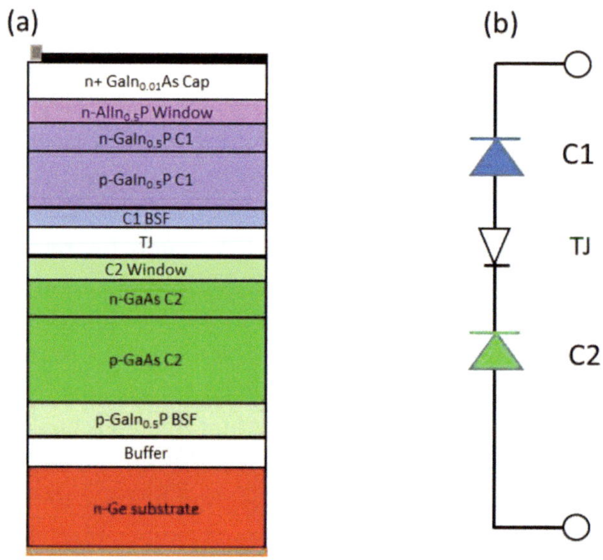

Figure 4.15 (A) Simplified layer structure and (B) circuit diagram for a lattice-matched 2J cell. *Source: author.*

reduce the efficiency of the 2J cell. Subcell 1 is a 1.85 eV InGaP cell. The bandgap does not reach the desired 1.9 eV level because the growth conditions and substrate miscut resulted in an only partially disordered group III sublattice. Instead, current matching between C1 and C2 was realized by thinning the C1 to the appropriate thickness (~0.3–0.8 μm).

The compelling BOL efficiency advantage of the InGaP/GaAs 2J over its SJ counterparts is further enhanced by a relative improvement in radiation retention. While GaAs and silicon SJ cells are both radiation-soft materials with NP_{mp} values of 0.75, the InGaP/GaAs space cell has far higher NP_{mp} of 0.84. This 12% relative increase in NP_{mp} is solely attributed to the radiation hardness of the InGaP subcell. Fig. 4.16A shows the modeled MJ and individual subcell LIV curves for the InGaP/GaAs 2J cell at both BOL and EOL. Using the subcell LIV curves at BOL and EOL, the radiation retention factors for each subcell are calculated and given in Fig. 4.16B. The NP_{mp} of the MJ cell is the average of the two subcell retention factors weighted by their fractional power contributions. From Fig. 4.16B, the GaAs C2 has an NP_{mp} of 0.73 that is consistent with the radiation retention of the GaAs SJ cell. In sharp contrast, the InGaP C1 has a far higher NP_{mp} of 0.93 relative to GaAs, which is indicative of the radiation hardness of the InGaP subcell.

The high radiation tolerance of InGaP can again be explained with an analysis of its W_p/L_e ratio at BOL and EOL. Similar to GaAs, at BOL, InGaP has a near ideal W_p/L_e ratio that nears 0.02. While the material quality of InGaP is not quite as good as GaAs, L_e values in the material are still in the range of 20 μm. This diffusion length is far greater than the base thickness required to fully absorb light up to the bandgap. InGaP is a direct gap semiconductor with an even higher absorption coefficient than GaAs, so only 1.2 μm of InGaP is required to absorb 99% of incident photons having energy greater

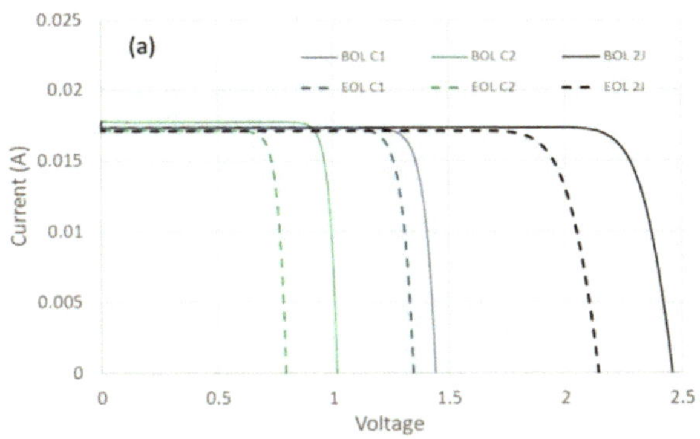

(b)

Subcell	Material	NP_{mp}
1	InGaP	0.93
2	$In_{0.01}GaAs$	0.73
DJ	-	0.84

Figure 4.16 (A) Modeled MJ LIV curve for 2J cell shown in Fig. 4.15; *solid lines* and *dotted lines* are LIV curves at BOL and EOL, respectively. MJ, subcell 1, and subcell 2 curves are given in *black, blue*, and *green* respectively. (B) Summary of power retention factors for subcells and MJ cell. *Source: author.*

than its bandgap. In InGaP/GaAs 2J cells for space, the thickness of the InGaP is further reduced to 0.3−0.8 μm to optimize current matching. Due to the low thickness (W_p) and reasonable diffusion length (L_e), the BOL W_p/L_e ratio of InGaP is comparable to that of GaAs. Moreover, InGaP also has a K_L value that is over three times lower than that of GaAs, as shown in Table 4.1. The lower K_L value results in small change in W_p/L_e ratio following radiation exposure. The EOL W_p/L_e ratio for InGaP of ~0.1 is 10 times better than GaAs and about 20 times better than Si and indicative of the radiation hardness of InGaP.

To provide some context regarding the significance of the first 2J space cell, we compare the EOL efficiency and EOL $/W of this technology relative to its SJ counterparts. Post 10^{15} e^-/cm^2 1 MeV exposure, the 2J cell has an EOL efficiency of 18.1%. This EOL efficiency is 32% and 51% higher than that of GaAs and silicon SJ cells, respectively. In particular, the compelling increase in relative EOL efficiency of 32% over the GaAs SJ cell was more than sufficient to overcome the increase in cost associated with the 2J cell. While there is a ~10% increase in growth cost for the tunnel junction and InGaP subcell, substrate and fabrication costs are nearly constant with respect to the GaAs SJ cell. Overall, then, the increase in total cost of the DJ relative to SJ GaAs is relatively minor when compared to the increase in EOL efficiency. From an EOL $/W and EOL efficiency perspective, the InGaP/GaAs 2J was therefore a compelling technologic upgrade over the GaAs SJ cell. This catalyzed rapid adoption of 2J cell in 1997, and from that point onward, it established III-V-based MJ cells as the dominant technology for space power applications.

4.6 InGaP/GaAs/Ge-based triple-junction solar cells

The lattice-matched (LM) 3J cell has dominated the space market for the past 20 years. Here, we explore the structure and the evolution of the BOL and EOL performance of the LM 3J cell in this time period. The LM 3J cell is a natural evolution of the 2J cell described in Section 4.5. As depicted in Fig. 4.17, the LM 3J cell is comprised of an InGaP C1 (E_g~1.86 eV), $In_{0.01}GaAs$ C2 ($E_g = 1.41$ eV), and Ge C3 ($E_g = 0.67$ eV). It is strikingly similar to the layer structure of the 2J cell in Fig. 4.15, with only three exceptions. First, the *n*-type Ge substrate used in the 2J cell is replaced with a *p*-type Ge substrate. Second, a key III-V-based nucleation layer is grown directly on top of the p-Ge substrate. The nucleation layer serves as a transparent front surface passivation layer or window. In addition, the group V material from the nucleation readily diffuses and counter-dopes the p-Ge to form an *n*-type emitter. The third change is the addition of a tunnel junction between the Ge and GaAs subcells, required to monolithically connect two subcells. In this case, a p + GaAs/n + GaAs similar to that used between the C1 and C2 can be used, and without the detraction of parasitic absorption losses for the Ge

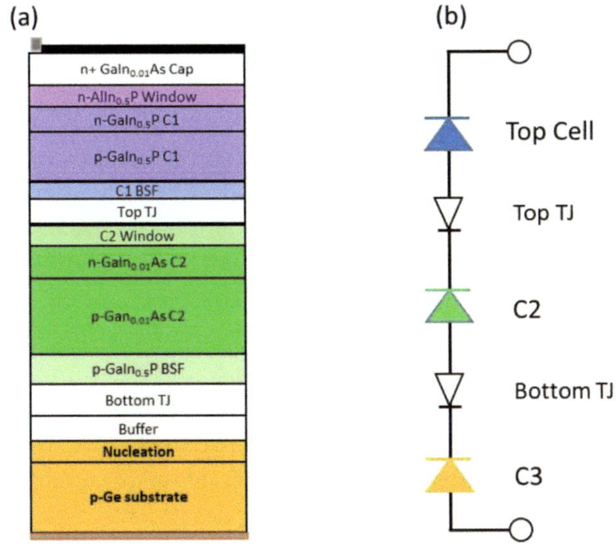

Figure 4.17 (A) Simplified layer structure and (B) circuit diagram for a lattice-matched 3J cell. *Source: author.*

subcell. Together the three changes: the p-Ge substrate, thin nucleation, and additional tunnel junction, are sufficient to form a functional Ge subcell in an LM 3J structure.

The first 3J cell using the structure shown in Fig. 4.17 was introduced in 1999 as a 25.1% BOL efficiency cell [12]. This is a 15% relative increase in BOL efficiency of the first 2J cell product. The increase in efficiency largely attributed to the addition of the Ge subcell, which increases the V_{oc} by ~ 0.2 V over the 2J cell. This increase in voltage comes with no tradeoff in cell current. Rather, the use of 1% InGaAs for subcell 2 slightly increases the current for the top two subcells relative to the 2J cell described in the previous section. While the Ge subcell does increase the BOL efficiency of the LM 3J cell, the increase is hindered by the use of Ge material, which has a nonoptimized bandgap for C3. According to Fig. 4.12, the optimum bandgap for C3 of 3J cell is 0.79 eV. This of course assumes relatively lower bandgaps for C1 (1.83 eV) and C2 (1.22 eV). For higher bandgaps used for C1 (1.86 eV) and C2 (1.41 eV) in the LM 3J, the optimum bandgap for C3 is even higher and closer to 1 eV. The consequence of using a low-bandgap material such as Ge is that C3 in the LM 3J cell has subcell current ~1.75 times greater than the average of subcells 1 and 2. The excess current in C3 is lost in the form of heat and does not increase MJ cell efficiency in either overall cell I_{sc} or V_{oc}. On a theoretical basis, the current mismatch loss associated with C3 reduces the maximum practical efficiency of the 3J cell 4% absolute, from 38% to 34%.

Despite the use of nonoptimized 1.86/1.41/0.67 eV bandgaps in the LM 3J cell structure, the maximum efficiency of 34% for these bandgaps is still far greater than the efficiency achieved for the first 3J cell at 25.1% (74% of maximum). Consequently, even after the realization of the first LM 3J cell in 1999, there was still ample opportunity to improve the efficiency by improving the quality of each subcell without changing the bandgaps of the 3J structure. The advantage of this approach is that improvements in subcell quality are often realized without additional epitaxial growth costs. For the past 20 years, a development focus on improving the subcell quality in LM 3J cells has proven to be an effective and fruitful strategy. It should be emphasized that the improvements in the LM 3J cell performance can be partially attributed to the absence of aluminum in the structure and the LM epitaxy; this eliminates two major sources of SRH defects that can limit cell efficiency. Fig. 4.18 shows the released product values for J_{sc}, V_{oc}, efficiency, and percentage of maximum efficiency specifically for LM 3J cells from 1999 to 2019. Steady but unmistakable gains have been made in every cell parameter. The J_{sc} and V_{oc} have increased 15% and 9% in the past 20 years. The cell efficiency has risen 28% from 25.1% in 1999% to 32.1% in 2019. The most recent LM 3J cell with a qualification efficiency of 32.1% is particularly impressive, exceeding 94% of the theoretical maximum for this cell structure [25]. This histogram in Fig. 4.19 demonstrates that the performance of the latest LM 3J cell structure is repeatable in a manufacturing environment. The production average of 32.3% actually exceeds the qualification average, for over 23k cells [26].

In addition to the gains in BOL efficiency, the LM 3J cell also offers a slight improvement in the power retention of NP_{mp} following exposure to radiation. The results from the full EOL/BOL LIV model for the LM 3J cell are given in Fig. 4.20. As

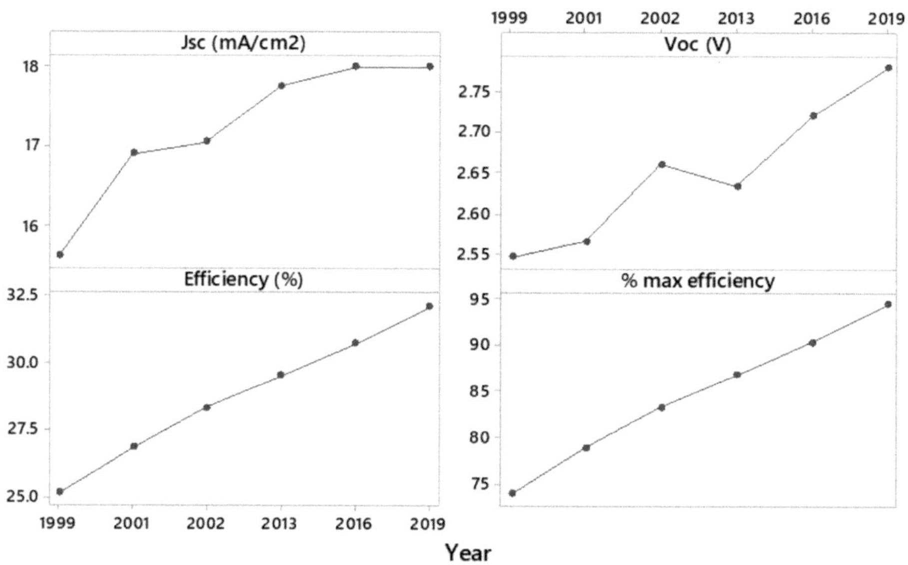

Figure 4.18 Scatterplot of J_{sc}, V_{oc}, Efficiency (%) and percent maximum possible efficiency as a function of year for the lattice-matched 3J cell. *Source: author.*

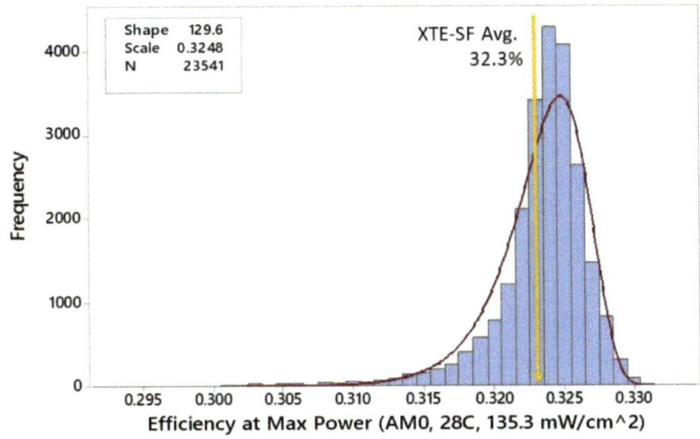

Figure 4.19 Efficiency histogram of the latest 3J cell introduced by Spectrolab in 2019. Average of the distribution occurs at 32.3% for >23.5 k cells. *Source: author.*

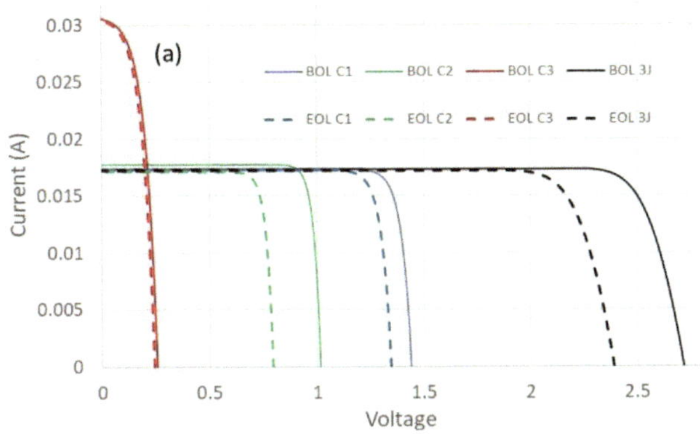

(b)

Subcell	Material	NP_{mp}
1	InGaP	0.93
2	$In_{0.01}GaAs$	0.73
3	Ge	0.94
3J	-	0.85

Figure 4.20 (A) Modeled MJ LIV curve for 3J cell shown in Fig. 4.17. Solid lines and dotted lines are LIV curves at BOL and EOL, respectively. (B) Summary of power retention factors for subcells and MJ cell. *Source: author.*

shown in Fig. 4.20, the power retention for subcells 1 and 2 are identical to that shown for the 2J cell at 0.93 and 0.73, respectively. This is predictable due to the fact that the material and thicknesses for the 2J are nearly identical to that of the LM 3J cell case. The added subcell 3 has an excellent power retention of 0.94 that is attributable to the low radiation damage coefficient of Ge. From Table 4.1, it is evident that Ge has by far the lowest K_L value for all relevant materials typically used in space solar cells. By comparison, Ge has a KL value approximately two orders of magnitude lower than silicon with the next lowest KL value. On the basis of its low K_L value, the EOL W_p/L_e ratio is reasonable at 0.4. The low EOL W_p/L_e is achieved despite the high thickness of Ge ($\sim$130 μm) needed to fully absorb light transmitted to the indirect bandgap layer.

However, it should be noted that despite the high subcell 3 power retention, the overall MJ cell NP_{mp} only increases by ~1%. This is due to the fact that overall MJ cell NP_{mp} is again a weighted average with respect to BOL subcell power. As the Ge cell contributes only 10% relative of the MJ cell BOL power, its effect on the overall MJ cell NP_{mp} is relatively small at 1%.

When we combine the BOL efficiency and power retention to obtain the EOL efficiency, unsurprisingly the EOL efficiency of the LM 3J cell compares favorably with respect to the 2J cell. The first 3J cell had an EOL efficiency of 22.5% that was 23% higher than the 2J cell. The 23% relative improvement in EOL efficiency is also significant because only a minor increase of ~1% in total thickness from the growth of the nucleation and additional tunnel junction was required to realize this gain. Therefore, a substantial decrease in the key EOL $/W metric of close to 23% was realized with the introduction of the first LM 3J cell. The compelling advantage of the LM 3J in both EOL efficiency and EOL $/W resulted in almost immediate adoption of this technology for use in space applications. Moreover, the near continuous improvements in the LM 3J cell have allowed its dominance in space power to persist for 20 years after the first 3J cell product. As of 2019, the EOL efficiency of the 3J cell has increased to 27.1%. This is a 75% and 52% increase in EOL efficiency relative to the 2J cell and first LM 3J cell, respectively. These cumulative gains in EOL efficiency have come largely as a result in changes in subcell quality and have only marginally increased the price of the LM 3J cell. Consequently, the gains in EOL efficiency have gone nearly directly toward a simultaneous and comparable relative decrease in EOL $/W.

Since the introduction of the 2J cell and all iterations of the LM 3J cell, MJ solar cells have achieved steady increases in EOL efficiency, and similar decreases in EOL $/W. Given the history of MJ cell product introductions, the space power industry has come to expect a continuation of this favorable trend. Unfortunately, the continuous improvement in LM 3J cell efficiency, without tradeoffs in cell cost, faces near-term formidable challenges. As previously stated, the maximum BOL cell efficiency for the LM 3J cell is nearly 34%. Therefore, even in the best-case scenario, one might expect at most a 6% relative improvement in cell efficiency over the existing record of 32.1%, if *all three* subcells in the LM 3J further approach the Shockley-Queisser limit for cell efficiency. Technologies focused on alternative bandgap 3J cells and/or MJ cells with >3 subcells will likely be required to further increase cell efficiency beyond that achievable for the LM 3J cell.

4.7 Lattice-mismatched triple-junction solar cells

To overcome the current mismatch losses in efficiency resulting from the non-optimized bandgaps utilized in the LM 3J cell, two metamorphic (MM) 3J cells have been developed as next-generation alternatives. As we recall from the previous section,

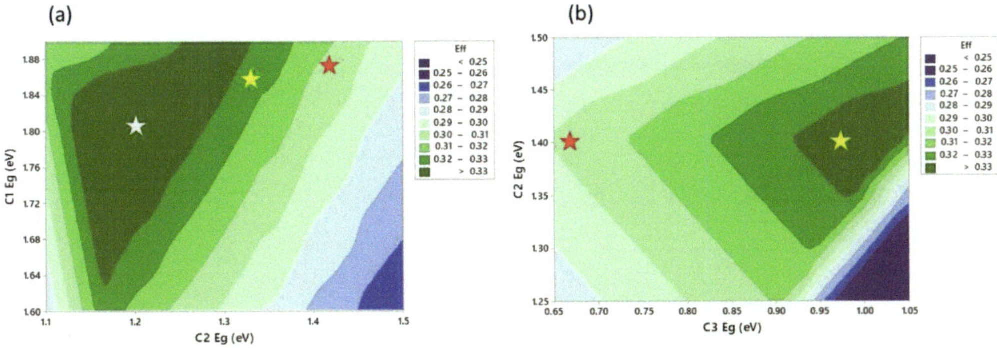

Figure 4.21 Modeled contour plots of MM cell efficiency. (A) Assumes a fixed C3, E_g and variable C1 E_g (*y*-axis) and C2 E_g (*x*-axis). (B) Assumes a fixed C1, E_g and variable C2 E_g (*y*-axis) and C3 E_g (*x*-axis). *Source: author.*

the bandgap of subcell 3 in the LM 3J cell in particular is too low at 0.67 eV, resulting in a subcell current 1.75 times greater than that of the top two subcells. The two 3J MM cells discussed in this section achieve improved current matching with subcell 3 and theoretically higher cell efficiency through two distinct cell structures. The first of the two 3J MM cell structures is known as the upright metamorphic (UMM) 3J. The UMM 3J achieves improved current matching by retaining the Ge subcell of the LM 3J (with a subcell 3 E_g at 0.67 eV) but then lowering the bandgaps of the top two junctions. Fig. 4.21A shows a contour plot of cell efficiency as a function of C1 and C2 E_g, assuming a fixed C3 E_g of 0.67 eV. The red star in Fig. 4.21A corresponding to the bandgaps of the C1 and C2 for the LM 3J cell resides in the 30%–31% BOL efficiency contour. On the other hand, the white star corresponds to a C1 $E_g = 1.8$ and C2 $E_g = 1.2$ eV, where current mismatch is minimized and the efficiency is maximized. By lowering the C1 and C2 E_g, the theoretical efficiency increases by 3% absolute. The more typical UMM 3J bandgaps have C1 and C2 at 1.8 and 1.31 eV, respectively, for practical implementation reasons. Even with the difference in bandgaps from the optimum, for the UMM 3J structure, as seen in the white star in Fig. 4.21A, the theoretical efficiency is 2% absolute higher than LM 3J cell.

The UMM 3J achieves a 1.8/1.31/0.67 eV bandgap combination using a cell structure similar to that depicted in Fig. 4.22A. The UMM 3J structure is identical to the LM 3J up to the bottom tunnel junction. However, the UMM 3J structure achieves the top two target bandgaps by increasing the lattice constant from that of Ge 5.65Å to 5.685 Å. More specifically, at a lattice constant of 5.685 Å, $In_{0.56}GaP$ and $In_{0.08}GaAs$ alloys meet the target of 1.8 and 1.31 eV bandgaps respectively. Unfortunately, a lattice constant of 5.685 Å has a 0.7% lattice mismatch with respect to that of the Ge substrate. Direct growth of thick layers >3 μm with even relatively small lattice mismatch of 0.7% will lead to islanding and catastrophic levels of defects. These issues can be largely

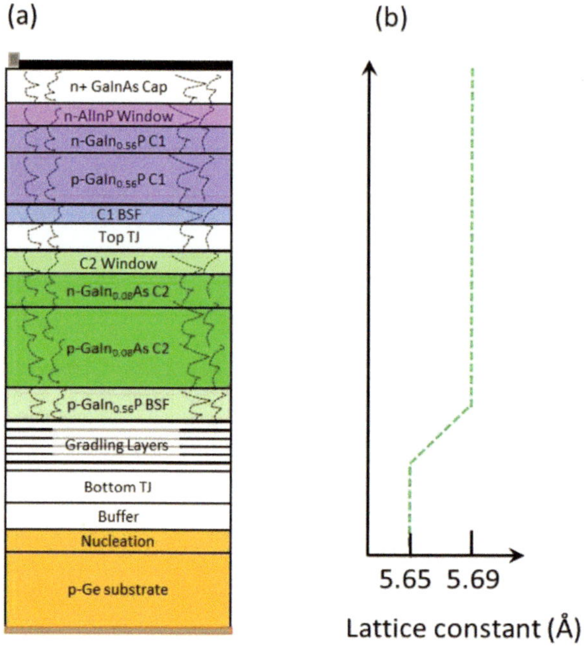

Figure 4.22 (A) Layer schematic and (B) lattice constant change as function of thickness for a UMM 3J cell. *Source: author.*

mitigated with the growth of compositional graded buffer layers (CGB). In the case of UMM 3J, the CGB are a series of layers with incrementally increasing lattice constant grown on top of the bottom tunnel junction. The changes in lattice constant between each CGB layer are insufficient to cause catastrophic islanding but do not prevent nucleation of threading dislocations in each layer. A large fraction of the threading dislocations generated in one constituent CGB layer are pinned and isolated at the interface to the next CGB layer. The $In_{0.08}GaAs$ and $In_{0.56}GaP$ are then grown LM to the final layer of the CGB to complete the UMM 3J structure. However, the remaining fraction of threading dislocations not pinned by the CGB propagate only in the direction of growth and penetrate through the top two junctions, as depicted in Fig. 4.22A. The threading dislocation density (TDD) in the top two junctions of the UMM 3J cell is highly dependent on the quality and design of the CGB layers.

Unfortunately, there is a deleterious effect of an increased TDD in the LIV performance of the top two junctions and overall UMM 3J cell. Fig. 4.23 compares the electron beam induced current (EBIC) and cathodoluminescence (CL) images of LM $In_{0.01}GaAs$ grown directly on Ge and lattice-mismatched $In_{0.08}GaAs$ grown on optimized CGB layers [27]. The dark spots in the EBIC and CL images correspond to a single threading dislocation. In a comparison of Fig. 4.23C and D, there is an increased

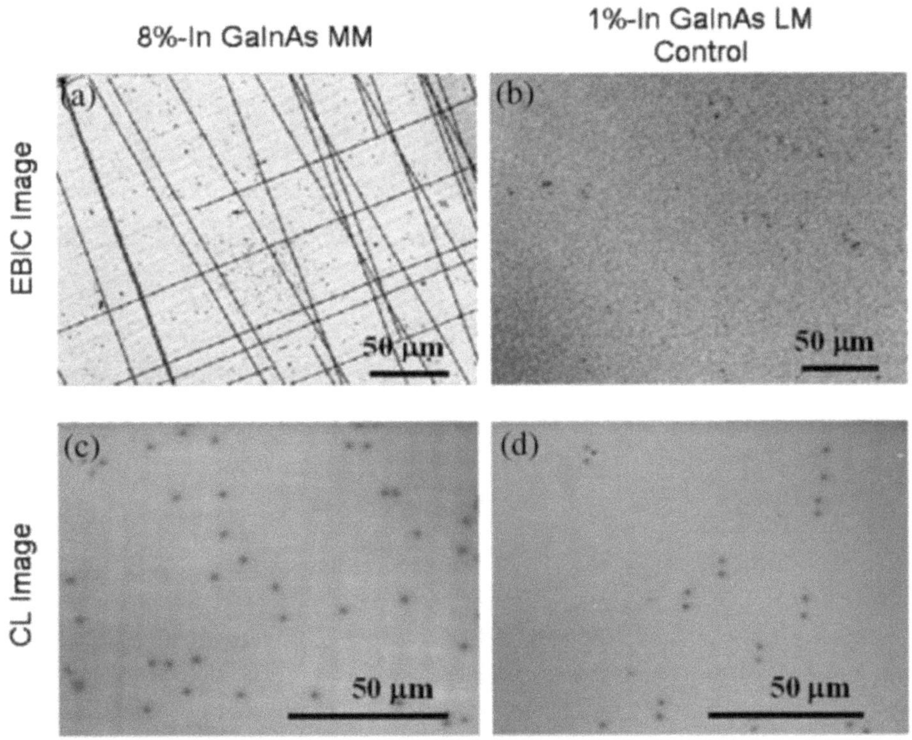

Figure 4.23 EBIC images for (A) 8% InGaAs and (B) 1% InGaAs cells. CL images for (C) 8% InGaAs and (D) 1% InGaAs cells. *Adapted from [27].*

TDD from 1×10^5 cm^{-2} to 2×10^5 cm^{-2} from 1% to 8% InGaAs cases. We note that the increase in TDD can vary depending on the quality and design of the CBG layers. As is the case with many defects, threading dislocations are also SRH defects. As such, even a small increase in TDD can reduce cell quality. This point is evident with the graphic comparison of W_{oc} values for LM (red triangles) versus upright lattice-mismatched materials (blue circles) in Fig. 4.24 [28]. In particular, the lattice-mismatched 8% InGaAs with a bandgap of 1.3 eV has a W_{oc} 20–40 mV higher than that of LM InGaAs. A similar observation occurs for LM InGaP relative to lattice-mismatched InGaP. The increase in TDD/SRH defects in the UMM 3J structure not only reduces V_{oc}, but it also reduces subcell FF via Eq. (4.14). Therefore, the increased TDD reduces subcell quality of the top two junctions in the form of reduced V_{oc} and FF. Moreover, the subcell quality is reduced in the top two junctions that produce 90% of the power UMM 3J cell. This in turn prevents the UMM 3J cell from achieving the modeled 2% absolute gain in efficiency relative to the LM 3J cell, as evident from the reported results for UMM 3J cells in Table 4.3. In 2017, Spectrolab and Azur reported 31.3% and 30.7% efficient UMM 3J cells, respectively [29,30]. These results are comparable to LM 3J cells produced by

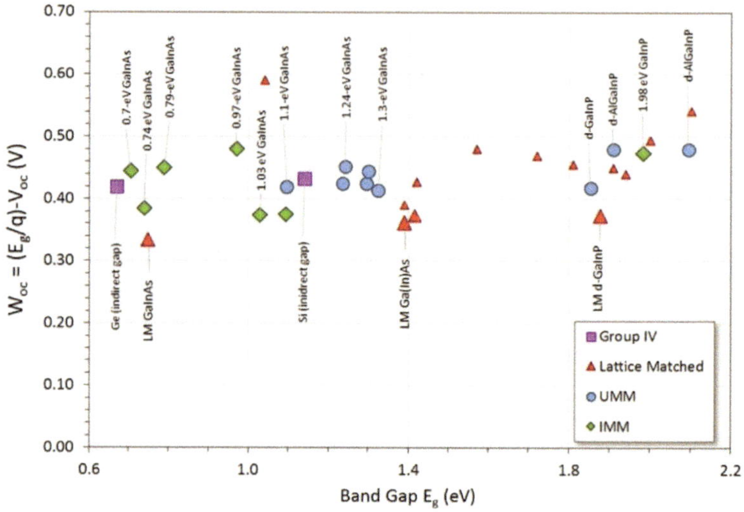

Figure 4.24 Graph of W_{oc} versus E_g of various materials. *Purple squares* are for group IV materials. *Red triangles* for LM materials. *Blue circles* for UMM materials. *Green diamonds* for IMM materials. *Adapted from [28].*

Table 4.3 Summary from literature sources of the BOL and EOL efficiency of MJ structures where the number of junctions is more than two.

Structure	Year	Bandgaps	Maker	V_{oc}	BOL eff	EOL eff	Status	Reference
LM 3J	2019	1.87/1.4/0.67	Spectrolab	2.78	32.1	27.1	Product	[25]
	2017	1.87/1.4/0.67		2.73	31.0	26.2	Product	[29]
UMM 3J	2017	1.83/1.3/0.67	Spectrolab	2.62	31.3	26.4	Research	[29]
	2017	1.83/1.34/0.67	Azur	2.605	30.7	—	Research	[30]
IMM 3J	2010	1.9/1.4/1.0	Spectrolab	2.98	32.8	26.2	Research	[37]
	2009	1.9/1.4/1.0	Solaero	3.05	32.0	—	Research	[36]
UMM 4J	2019	—	Azur	3.45	31.8	27.7	Product	[42]
	2018	1.94/1.44/1.1/0.67	Spectrolab	3.37	30.5	—	Research	[41]
IMM 4J	2010	1.9/1.42/1.02/0.7	Solaero	3.25	33.6	27.5	Research	[45]
	2019	1.9/1.45/1.0/0.7	Spectrolab	3.42	33.3	27.3	Research	[46]
SBT 5J	2016	2.1/1.9/1.4/1/0.7	Spectrolab	4.76	35.8	—	Research	[49]

Spectrolab in the same year [29], and they trail the 2019 3J cells by 3% in efficiency [25]. Therefore, the potential advantages in the UMM 3J BOL cell performance due to more optimized bandgap combination are mitigated by the loss in performance in C1 and C2 from an increase in TDD defects.

Although there have been no published reports of radiation retention of UMM 3J cells, EOL modeling indicates that retention factors for the UMM 3J cell structure are similar to that of the LM 3J case. This is due to the fact that the constituent materials of the UMM 3J cell are relatively small perturbations of the LM case, with a slight increase in indium for C1 and C2. Consequently, the NP_{mp} is still predicted to be close to 0.85 at a 10^{15} e^-/cm^2 fluence of 1 MeV electrons. This results in an EOL efficiency of approximately 26.4% that is comparable to the 2017 LM 3J case and less than the 2019 LM 3J cell. The UMM 3J cell upright growth structure requires no additional processing costs relative to the LM 3J cell. However, the UMM 3J cell is more costly to grow relative to the LM 3J cell. The UMM 3J cell requires the growth of CGB with thicknesses approaching that of C2. Generally, thicker and therefore costlier Ge substrates are also required for UMM 3J growth to prevent wafer warping during the growth of compressively strained layers. Warped wafers disrupt epitaxial growth processes and complicate photolithography processing steps in fabrication. The absence of a realized EOL efficiency advantage and the increase in cell cost has therefore prevented the adoption of the UMM 3J cell over the LM 3J.

The second of the two 3J MM cell structures is the inverted metamorphic 3J or IMM 3J cell. The IMM 3J achieves improvement by fixing the bandgap of the top two junctions at bandgaps close to LM 3J (C1 = 1.9 eV and C2 = 1.42 eV). The current mismatch with C3 is addressed by increasing its bandgap to 0.97 eV. In this way, power lost through current mismatch can be recovered in the form of increased voltage in C3. The effect of the increase in bandgap of C3 on efficiency is depicted by the contour map in Fig. 4.21B, where increasing the bandgap of the C3 from 0.67 eV (red star) to 0.97 eV (yellow star) increases the efficiency of the 3J cell by 3% absolute. While a 0.97 eV C3 material is ideal from a current matching perspective, an inspection of the bandgap versus lattice constant chart in Fig. 4.14 indicates few high-quality LM materials to Ge or GaAs with a bandgap near 1 eV.

The IMM 3J cell design addresses the absence of high-quality LM 1 eV materials with a novel inverted epitaxial growth design, with the highest bandgap junction grown first and lowest bandgap junction grown last. The inverted design allows the use of lattice-mismatched $In_{0.3}GaAs$ for the target 0.97 eV C3 material without affecting the top two junctions. As shown in Fig. 4.25B, the IMM 3J starts with 1.9 eV InGaP C1, followed by a 1.42 eV GaAs C2, both grown lattice matched to the GaAs substrate. In this way, the quality of the top two junctions is equal to that of the top two junctions of the LM 3J case. Then CGB layers are grown to transition to minimize the effect of the 2.2% lattice mismatch between the GaAs substrate (5.65 Å) and $In_{0.3}GaAs$ (5.78 Å). In this way, the threading dislocations that only propagate in the direction of growth affect the quality of the C3, but not the higher-power-producing top junctions.

Nevertheless, the inverted growth design renders the IMM 3J structure nonfunctional without additional processing. As grown, C3 would be left nearest to the incident photons, absorbing all photons, including those intended for C1 and C2. To orient the junctions in the highest bandgap C1 closest to the incident photons, at least two additional processing steps are required. As indicated in Fig. 4.25C, the as-grown wafer must

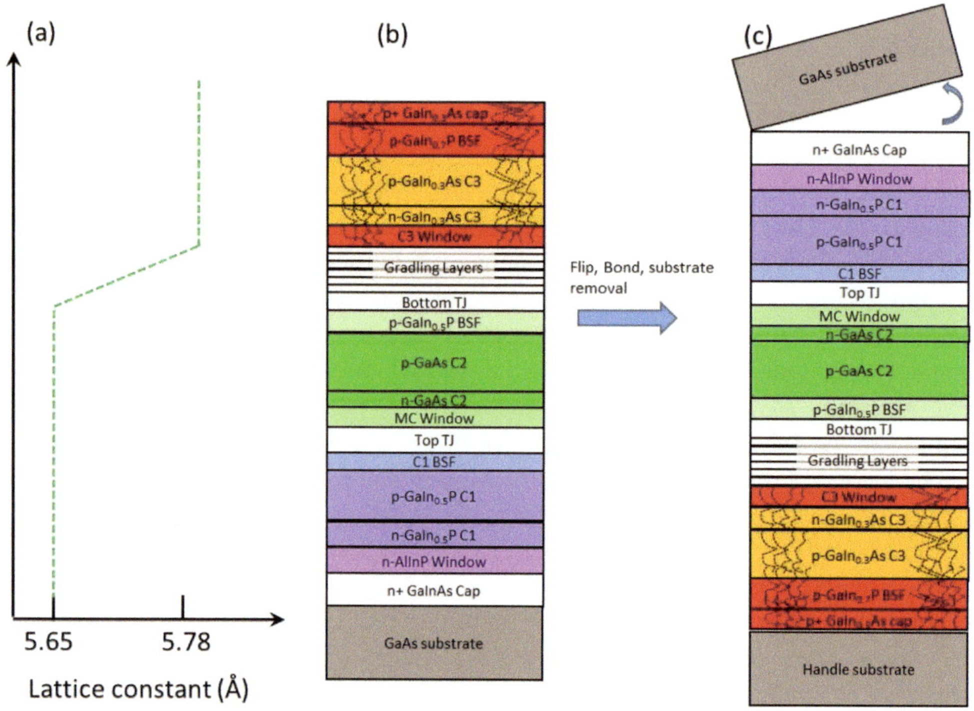

Figure 4.25 (A) Lattice constant change as a function of thickness, (B) as grown layer schematic, and (C) layer schematic after processing for an IMM 3J cell. *Source: author.*

first be indirectly bonded to a handle substrate. Previous indirect bonding methods have included the use of metal eutectic bonds [31] or adhesive bonds [32]. The handle substrate can be chosen to decrease the weight or increase the flexibility of the fabricated cell, providing a potential design advantage for the IMM cell. Second, the original epitaxial substrate must be either chemically removed or lifted off. The advantage of the liftoff technique is that it allows the potential reuse of the substrate for cost mitigation [33,34]. Chemical etching is a relatively simpler and less time consuming process, but it entirely wastes the original growth substrate. Following the additional processing steps, the IMM 3J can be processed similar to that of the LM 3J cell.

After accounting for the bonding and substrate removal steps, the IMM 3J has exhibited excellent BOL efficiency results. Even with the use of CGB layers, the literature reports TDD values between 10^6 and 5×10^6 cm^{-2} for an $In_{0.3}GaAs$ subcell 3 [35,36]. The resulting W_{oc} values average around 0.44 V, with the exact W_{oc} value depending on CGB quality. Even with a W_{oc} of 0.44, the absolute value of a 0.97-eV C3 is still 0.53 V, resulting in a subcell 3 voltage gain of about 300 mV over that of a Ge cell in the LM 3J case. Moreover, due to the inverted design of the IMM 3J growth, the high

TDD only affects C3 and leaves the top junctions of the IMM 3J cell lattice matched and of quality mirroring the LM 3J. Due to the high quality of subcells 1 and 2, and the voltage gain of C3 due to the higher C3 bandgap, the IMM 3J cell has achieved BOL efficiency near 33%, which is appreciably higher than even the best LM 3J cells. In 2010, Spectrolab reported a 32.8% IMM 3J cell [37]. In 2009, Solearo reported a high-quality IMM 3J at 32%. With improved current matching between the C1 and C2, the projection for that cell would be similar to Spectrolab's 32.9% [36]. The increase in efficiency is attributed to the total cell voltage of the IMM 3J, which varies between 2.98 and 3.05 V and is significantly higher than that of contemporary LM 3J cells around 2.7 V.

While the cell performance of the IMM 3J does provide a gain relative to the LM 3J cell at BOL, these gains are completely lost at EOL. The most important feature of the IMM 3J is that it replaces the Ge cell with a 0.97 eV $In_{0.3}GaAs$ cell. While this serves to improve BOL efficiency through improved current matching and higher cell voltage, this replacement also exchanges a radiation-hard Ge material with a radiation-soft $In_{0.3}GaAs$ material. In Table 4.1, it is apparent that $In_{0.3}GaAs$ has a K_L value that is at least 10^5 times lower than that of Ge. The EOL LIV modeling shows the effect of the change in radiation damage coefficient on the retention of the bottom subcell. Comparing the subcell retentions in Figs. 4.20 and 4.26, we see that since the materials used for subcell 1 and 2 are unchanged, unsurprisingly the radiation retention for these subcells is also unchanged. However, the retention for subcell 3 drops dramatically from 0.94 to 0.57. This decrease in subcell 3 retention reduces the overall 3J retention from 0.85 to 0.79. If we multiply the retention by the BOL efficiency, we see that the EOL efficiency for the IMM 3J cell of 26.2% is equal to the earlier generation LM 3J cell and is 3% less than the latest LM 3J cell.

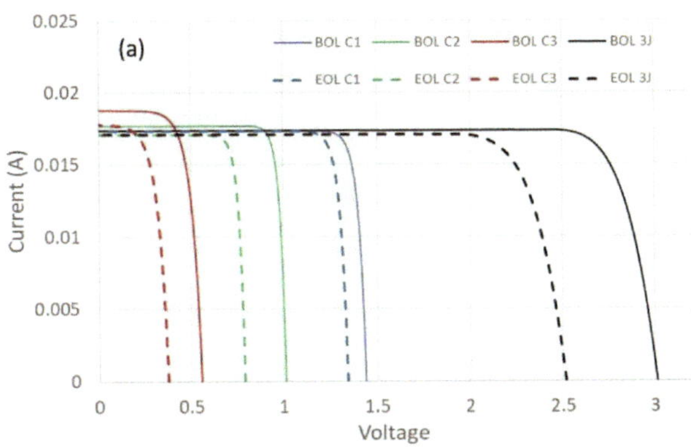

(b)

Subcell	Material	NP_{mp}
1	InGaP	0.93
2	$In_{0.01}GaAs$	0.73
3	$In_{0.3}GaAs$	0.57
3J	-	0.790

Figure 4.26 (A) Modeled MJ LIV curve for IMM 3J cell shown in Fig. 4.25. (B) Summary of power retention factors for subcells and MJ cell. *Source: author.*

Similar to the UMM 3J, the IMM 3J cell does not provide any advantage in EOL performance or EOL $/W advantage over the LM 3J cell. The lack of EOL efficiency gain for the IMM 3J is attributed to a significantly lower radiation retention factor from a weak $In_{0.3}GaAs$ third junction. Moreover, the IMM 3J technology is far more expensive to produce than both the UMM 3J and LM 3J cells. Growth of the IMM 3J cell requires growth of CGB layers and the $In_{0.3}GaAs$ junction itself. Recall that in the case of the LM 3J cell, the third junction is realized with little additional epitaxial growth. Conversely, the epitaxial growth thickness for an IMM 3J is more than double that of a typical LM 3J cell. More significantly, the IMM cell requires two additional process steps including wafer bonding and substrate removal. There is potential to recover some of this additional cost through substrate reuse, but the economics of the substrate reuse process are not clear at this time.

4.8 Lattice-mismatched quadruple (four)-junction solar cells

Quadruple or four-junction (4J) cells offer an alternative method of increasing cell efficiency above that of the LM 3J cell. Conceptually, a 4J cell can be achieved simply starting with an LM 3J cell and adding an ~1 eV junction between cell 2 and cell 3 of the LM 3J stack. This results in a 4J cell with a 1.9/1.4/1.0/0.7 eV bandgap combination. The addition of the 1 eV junction converts the nearly two times excess current of the Ge cell into additional voltage. While the 1.9/1.4/1.0/0.7 bandgap combination is not exactly equal to the optimum 4J bandgaps described in Fig. 4.12A, it is only shifted by about 0.1 eV from the optimum for all four junctions. Therefore, current matching between all four junctions is still well preserved. A high-quality 4J cell with this bandgap combination is modeled with ~35%–37% efficiency at BOL. The UMM 4J and IMM 4J technologies are two contrasting methods to implement the 1.9/1.4/1.0/0.7 eV combination. This section of the chapter elaborates on these two methods by which the 4J is implemented and the resulting performance at BOL and EOL.

The cell structure in Fig. 4.27 shows how the UMM 4J achieves the target 4J bandgap combination. The structure begins with a standard Ge diffused junction. Next a CGB layer takes the lattice constant from 5.65 to 5.75 Å, with a 1.7% lattice mismatch for the final CGB layer relative to the Ge substrate. Next, the three upper junctions are grown with a 1.7% lattice mismatch, all with relatively high density of threading dislocations resulting from that lattice mismatch. The first of these upper junctions to be grown is C3: an $In_{0.23}GaAs$ alloy with a 1.1-eV bandgap. Next is C2, an $AlIn_{0.23}GaAs$ alloy with a 1.4-eV bandgap. 23% indium is required in the quaternary alloy to maintain lattice matching to the top layer of the CGB and C3. However, the indium in the alloy reduces bandgap, requiring the addition of an equal measure of aluminum in the alloy to achieve a 1.4-eV bandgap. Finally, C1 is an AlInGaP alloy with 73% In and 23% Al in the group III lattice.

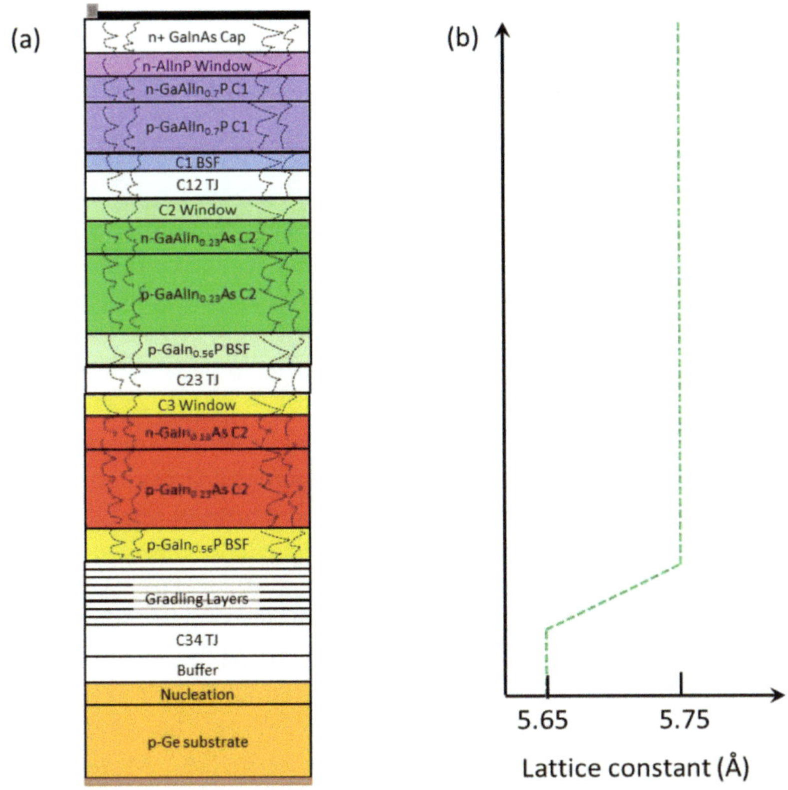

Figure 4.27 (A) Layer schematic and (B) lattice constant change as function of thickness for a UMM 4J cell. *Source: author.*

The aluminum in this phosphide quaternary is again needed to compensate for the 23% indium needed for lattice matching to the grade.

The quality of the top three subcells in the UMM 4J structure is challenged simultaneously by two factors. All three high bandgap junctions are grown with 1.7% lattice mismatch. At 1.7% lattice mismatch, the TDD increases to $\sim 4.4 \times 10^6$ [38,39]. Since threading dislocations are SRH defects, there is an impact to the W_{oc} for all three top junctions. The W_{oc} of C3, which is only affected by TDD, varies from 0.42 to 0.44 V. Subcells 1 and 2 are not only affected by TDD, but also the inclusion of 23% aluminum in their respective quaternary alloys. As discussed in Section 4.5, Al-containing materials have a tendency to incorporate C and O contamination from the source and reactor, leading to a further increase in SRH defects. The evidence of this combined material degradation is again found in Fig. 4.24, where W_{oc} values around 0.48 V were measured for the top two junctions. The use of an AlInGaP cell for C1 is particularly challenging due to its implications on cell sheet resistance and blue response. The emitter of C1 absorbs the majority of photons in the blue portion of the spectrum from 350 to 500 nm. The C1 emitter is also responsible for moving carriers laterally from any point on the cell toward the gridlines. Maintaining sufficient material quality to meet both requirements is difficult for AlInGaP emitters because both the majority carrier and minority carrier mobility decrease with Al incorporation. The loss of minority carrier mobility reduces blue response that is clearly evident in published QE spectra for the UMM 4J [29,40]. The reduction in majority carrier mobility reduces sheet resistance that affects total cell series resistance and FF. The loss of overall cell quality due to the high TDD in C1−C3 and Al based defects in C1−C2 results in a relatively poor UMM 4J BOL efficiency. Two of the latest reported BOL efficiency values for the UMM 4J cell are shown in Table 4.3. Spectrolab has reported a 30.5% cell, with a clear path to 31% with improved current balancing [41]. Azur has introduced a UMM 4J product with a stated 31.8% BOL efficiency [42]. Both results are substantially less than the 35%−37% that is possible for high-quality 4J cells with bandgaps achieved by the UMM 4J cell.

In sharp contrast to the relatively poor BOL performance of the UMM 4J relative to the predicted optimum, the UMM 4J cell has excellent radiation resistance. EOL modeling of the UMM 4J structure is shown in Fig. 4.28. The retention factors for subcells 1, 3, and 4 are similar to that observed for subcells 1−3 of LM 3J cells. The main difference between the retention factor of UMM 4J and LM 3J is due to AlInGaAs cell 2, which has a subcell retention factor of 0.85. The retention factor of cell 2 is improved relative to GaAs or $In_{0.23}GaAs$ for two reasons. It has a slightly better damage coefficient that $In_{0.23}GaAs$ or GaAs due to the incorporation of Al [43]. More importantly, for current matching purposes, cell 2 in the UMM 4J cell (1 μm) is much thinner than cell 2 in a 3J (~3 μm). The reduced thickness plays a key role in maintaining a low W_p/L_e ratio at EOL. The higher subcell retention factor for cell 2 reduces the effect of the relatively poor subcell retention and raises the power-weighted average for the overall 4J NP_{mp} to a modeled value of 0.86. This is close to that reported by Azur for their UMM 4J product

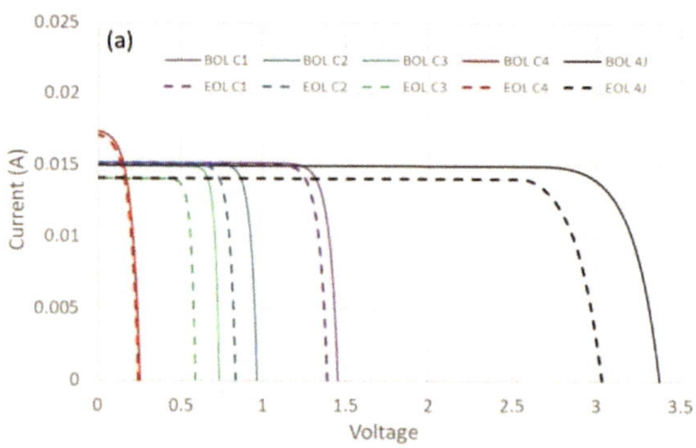

(b)

Subcell	Material	NP_{mp}
1	$AlIn_{0.7}GaP$	0.95
2	$AlIn_{0.23}GaAs$	0.85
3	$In_{0.23}GaAs$	0.73
4	Ge	0.93
4J	-	0.86

Figure 4.28 (A) Modeled MJ LIV curve for UMM 4J cell shown in Fig. 4.27. (B) Summary of power retention factors for subcells and MJ cell. *Source: author.*

where the NP_{mp} is reported at ~0.87 (calculated without European Cooperation for Space Standards annealing) [42]. The corresponding EOL efficiency for the UMM 4J cell reaches 27.7% at 1×10^{15}, 1 MeV e^-/cm^2. This is approximately 1.7% better than the EOL efficiency of the latest LM 3J cell.

The IMM 4J cell structure provides a second method of obtaining the target 1.9/1.4/1.0/0.7 eV bandgap combination. The layer schematic for the IMM 4J cell structure shown in Fig. 4.29B is similar to the IMM 3J cell structure described in the previous section. The major difference is that the IMM 4J structure has an additional set of CGB layers to increase the lattice constant from 5.78 to 5.89 Å. An $In_{0.5}GaAs$ ternary (C4) is grown lattice matched to the final CGB layer. The total lattice mismatch between the C4 and the substrate is 4.2%. The large lattice mismatch results in a high TDD in C4. Through careful engineering of CGB layers between C3 and C4, the TDD values for C4 are only slightly higher or similar to that observed for C3. TDD values between 3×10^6 and 5×10^6 cm^{-2} have been previously reported [44]. As such, W_{oc} values for C4 are also

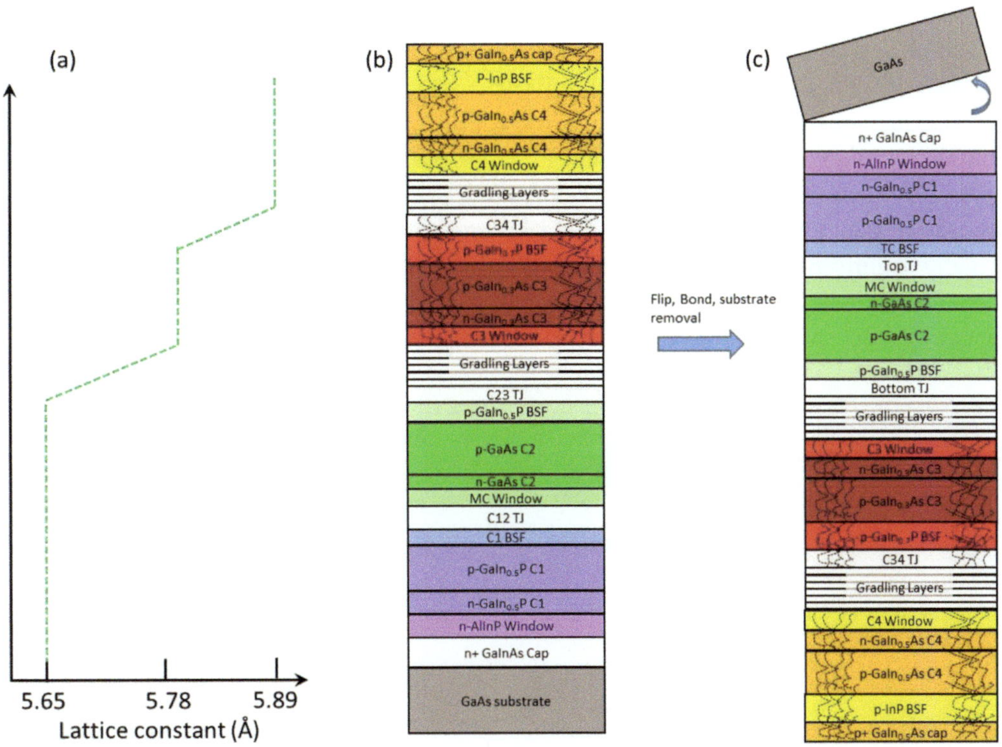

Figure 4.29 (A) Lattice constant change as a function of thickness, (B) as grown layer schematic, and (C) layer schematic after processing for an IMM 4J cell. *Source: author.*

comparable to that achieved for C3 in an IMM 3J. From Fig. 4.24, we see that W_{oc} values for C4 are near 0.45 V, resulting in a C4 voltage of 0.25 V. It should be noted that this subcell 4 voltage is nearly identical to that reported for Ge subcells in LM 3J and UMM structures.

The addition of C4 in an IMM 4J structure increases the cell voltage by 0.25 V without effect on the performance of subcells 1−3, resulting in excellent BOL cell efficiency. Similar to the IMM 3J cell, the IMM 4J takes advantage of the inverted growth structure to confine high TDD within C3 and C4 metamorphic junctions. The top two junctions remain lattice matched and of high quality. As a result, the IMM 4J cell has achieved some impressive BOL efficiency values. In particular Solaero and Spectrolab have both reported BOL efficiencies of 33.6% [45] and 33.3% [46], respectively; this is still short of the ideal target efficiency for a 4J cell between 35% and 37%. We attribute the discrepancy to the high TDD so lower quality of the bottom two junctions. A W_{oc} value of 0.45 V for the bottom two junctions falls short of the value ideal by 0.1 and 0.15 V for C3 and C4, respectively. Nonetheless, the IMM 4J still exhibits a marked improvement in BOL efficiency over the UMM 4J cell, mainly due to the high-quality LM top junctions. Moreover, the IMM 4J also has a higher BOL efficiency relative to the LM 3J cell, largely resulting from the reduction of current mismatch losses. In fact, there is a 4%−5% relative increase in BOL efficiency over the latest LM 3J cell.

Although the IMM 4J exhibits a marked improvement in BOL efficiency over the LM 3J cell, the radiation retention and EOL efficiency remain points of weakness. The EOL LIV modeling for the IMM 4J cell is shown in Fig. 4.30. The retention factors for subcells 1, 2, and 3 are 0.94, 0.80, and 0.75, respectively. All subcell retention factors are actually slightly higher than that of C1−C3 in IMM 3J, largely due to a reduction in subcell thickness to improve current matching to subcell 4. However, IMM 4J again replaces a radiation-hard Ge cell with a radiation-weak $In_{0.5}GaAs$ cell. Recall that the Ge

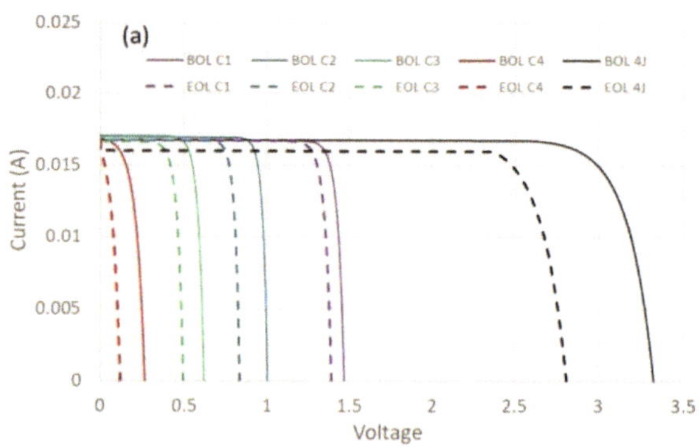

(b)

Subcell	Material	NP_{mp}
1	InGaP	0.94
2	$In_{0.01}GaAs$	0.80
3	$In_{0.3}GaAs$	0.75
4	$In_{0.5}GaAs$	0.31
4J	-	0.82

Figure 4.30 (A) Modeled MJ LIV curve for IMM 4J cell shown in Fig. 4.29. (B) Summary of power retention factors for subcells and MJ cell. *Source: author.*

C4 in UM M4J structure has a subcell retention factor of 0.93 compared to 0.31 for the $In_{0.5}GaAs$ C4 in the IMM 4J structure. The extremely poor subcell retention factor for C4 is sufficient to overcome any subtle gains in subcells 1−3, resulting in an NP_{mp} of the overall IMM 4J cell of 0.82. This is lower than that of a typical LM 3J cell by 3%. Even with the high BOL efficiency of the IMM 4J cell is 4%−5% higher than the LM 3J cell, the 3% relative loss in retention factor largely negates the BOL gain. The EOL efficiency at standard GEO fluence is 27.6% and about 1.5% higher than that of the latest LM 3J cell.

In contrast to their MM 3J counterparts, both the UMM 4J and IMM 4J cells have achieved an increased EOL efficiency relative to the LM 3J cell. However, the improvement in both cases is only 1.5% and comes with significant tradeoffs. The UMM 4J and IMM 4J cells are more complex and expensive than the MM 3J counterparts, and even more so relative to the LM 3J cell. In the case of the UMM 4J cell, the total thickness of the cell is about two times that of an LM 3J. Subcell 3 and CGB layers account for nearly 6 μm in total thickness that is not present in LM 3J cells. Substrate costs are also higher due to the need for greater thickness to counteract the bow from the 1.7% lattice mismatch. The IMM4 J is about four times thicker than a LM3J cell. The IMM 4J requires the growth of two sets of CGB layers, and the growth of subcells and 3 and 4. This accounts for nearly 12 μm in total thickness that is not present in LM 3J cells. In addition to the higher growth costs, the inverted layer structure requires additional fabrication processes including wafer bonding and substrate removal. For the MM 4J cells, it is at best unclear that a 1.5% increase in EOL efficiency outweighs the inevitable increase in costs associated with growth, substrates, and fabrication. Without an EOL $/W reduction, adopters of the MM 4J technologies will need to prioritize other metrics that could be favorable to MM 4J cells. These include a modest increase in EOL efficiency and increased cell voltage/lower cell current to reduce electrostatic discharge risk. In the case of the IMM 4J technology, the potential to bond to a lightweight or flexible substrate could have applications in higher W/kg flexible blanket arrays.

4.9 Lattice mismatched quintuple (five)-junction solar cells

The final cell structure to be discussed in this chapter is a LM quintuple or five-junction (5J) cell with a 2.2/1.7/1.4/1.05/0.73 bandgap combination. Although the bandgaps of all subcells are higher than those listed in Fig. 4.12A, the shift is approximately 0.1 eV higher for all subcells, resulting in excellent current balance in all five junctions. The LM 5J cell is achieved by performing two epitaxial growths on two substrates. More specifically, the top three junctions are grown inverted on GaAs substrates. As shown in Fig. 4.31B, the top three junctions are composed of InAlGaP, AlGaAs, and GaAs alloys respectively lattice matched to the GaAs lattice constant. The bottom junctions with bandgaps are grown upright on InP substrates. The importance of

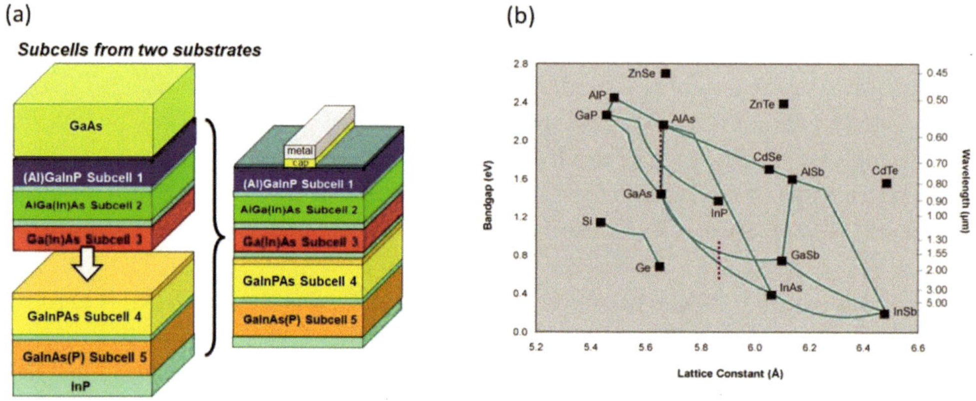

Figure 4.31 (A) Layer structure and process for a single-bonded (SBT) 5J cell. (B) Bandgap lattice constant diagram for the SBT 5J cell. *Black line* highlights the top three junctions' growth at the GaAs lattice constant. *Purple line* highlights the bottom two junctions grown at the InP lattice constant. *Source: author.*

the growth on InP is that high-quality InGaAsP and $In_{0.5}GaAs$ alloys with 1.05 and 0.73 eV bandgaps, respectively, are lattice matched to InP. As depicted in Fig. 4.31A, the inverted top three junctions are then direct bonded to the bottom two junctions. Direct semiconductor bonding technology (SBT) is an elegant bonding approach that maintains both high conductivity and low optical transmission loss [47]. The GaAs substrate on the top three junctions can then be removed by either direct chemical etching or epitaxial liftoff. The remaining cell fabrication steps are conserved with respect to LM 3J cells.

The main advantage of the SBT approach is that all five junctions can be grown lattice matched to their respective substrates. The benefit of the LM SBT approach is most apparent in the performance of the bottom two junctions. LIV measurements of the bottom two junctions alone indicate a V_{oc} of 1.08 V [48]. This equates to an average W_{oc} of 0.335V, which is close to the ideal values shown in Fig. 4.6. By comparison, the average W_{oc} for the bottom junctions of an IMM 4J is more than 100 mV higher. The strong contrast in W_{oc} values for LM and mismatched materials demonstrates the importance of low TDD values on cell voltage and performance. LIV measurements of the top three junctions indicate a V_{oc} of 3.68 V.

Due to primarily to the high material quality of InP-based bottom junctions, SBT cells have achieved the highest reported BOL space efficiencies to date. As shown in Fig. 4.32, a 35.8% efficiency 5J SBT cell has been reported and independently confirmed by the National Renewable Energy Lab (NREL) [49]. The V_{oc} of the 5J cell of 4.76 V is equal to the sum of the top three junction components grown on GaAs and bottom two junction components on InP, as would be expected for a direct bond without parasitic junction formation. The SBT 5J BOL performance represents a significant improvement

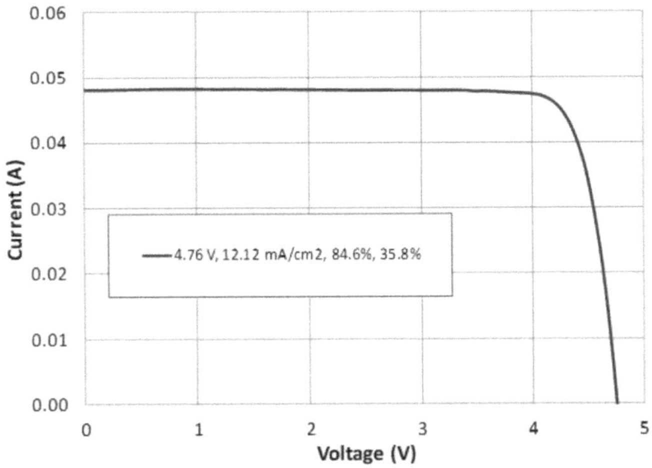

Figure 4.32 (A) Measured LIV curve to the best SBT 5J cell fabricated to date. *Source: author.*

over previously discussed MJ cell technologies. It is 11.5% and 6.5% higher than the latest respective LM 3J cell and best IMM 4J reported to date.

While the 5J SBT cell does achieve an impressive BOL efficiency, it is short of the ideal target for the given bandgap combination by approximately 3%. Although the top three junctions are lattice matched to GaAs, the average W_{oc} values for the top three junctions (0.53 V) are still relatively high compared with the bottom two junctions. The higher W_{oc} value is primarily attributed to the inclusion of aluminum in the top two junctions. More specifically, increasing aluminum content in the InAlGaP C1 leads to a steadily increasing W_{oc} value, with small gains in V_{oc} relative to aluminum-free InGaP [50]. We recall from previous sections that aluminum-containing materials are susceptible to oxygen- and silicon-related deep-level traps that increase the W_{oc}. In fact, the relatively high W_{oc} that varies between 0.5 and 0.6 V value [51,52] for high-bandgap materials with $E_g > 2$ eV is a challenge limiting any multijunction cell design where the number of junctions is more than five. Improvements in the W_{oc} of the C1 InAlGaP toward its ideal value would in and of itself increase the 5J cell efficiency by 6%.

In the case of the SBT 5J cell, we do not offer any predictions of NP_{mp} or EOL efficiency due to the lack of experimental data on the radiation damage coefficients for the 1.05 eV InGaAsP cell. However, even if we bound the NP_{mp} at a level similar to that of the LM 3J, the SBT 5J cell technology still faces similar challenges in terms in cell cost and EOL $/W as the IMM-based cell structures. The total cell thickness of the SBT cell is approximately two times that of a LM 3J cell. The technology is also based on the concept of two separate epitaxial growths, which require the overhead of two separate heat-ups and cool-downs in the growth tool. The two separate growths also require an extra InP substrate, which is typically 10 times more expensive than either GaAs or Ge substrates. In terms of fabrication, the SBT cell also requires additional process steps similar to the IMM cell including cell bonding and substrate removal. The multitude of cell cost increases associated with the SBT 5J cell concept are likely to outweigh the gains in EOL efficiency, resulting in an increase in EOL $/W. Similar to the IMM cell, adoption of the SBT 5J would require missions that emphasize cell efficiency over the EOL $/W metric.

4.10 Conclusions

Over the past 30 years, the BOL cell efficiency of III-V-based solar cells has increased steadily from 19% for the first SJ GaAs cell to 32% for the latest state-of-the-art upright 3J solar cell (InGaP/GaAs/Ge). EOL efficiency has also increased at the same pace due to a slight increase in radiation retention (NP_{mp}) of these technologies over time. Consequentially, these increases in EOL efficiency have come with either relatively small or no increase in cell cost, resulting in significant reduction in EOL $/W. The concurrent increase in EOL efficiency and decrease in EOL $/W have led to an uninterrupted and rapid adoption of all technologies between SJ GaAs to the latest 3J solar cell.

To increase BOL efficiency above that achieved by the upright 3J cell, multiple alternative technologies have been proposed and implemented. The main efficiency limitation of the upright 3J cell is poor current matching between the top two junctions and the bottom Ge junction. Two alternative 3J structures that improve the current matching include the UMM 3J and IMM 3J by decreasing the bandgap of the top junctions and increasing the bandgap of the bottom junction, respectively. However, the BOL efficiency of the UMM 3J structure is encumbered by higher defect density from the use of CGB layers. The EOL efficiency of the IMM 3J structure is hindered by poor radiation retention from the bottom InGaAs junction. In both UMM3 and IMM 3J structures, the EOL efficiency does not even exceed that of the upright 3J structure.

An additional junction(s) can increase BOL and EOL in the case of the UMM 4J and IMM 4J structure. In fact, the UMM 4J structure has achieved slightly higher (+1.7% relative) EOL efficiency relative to the upright 3J structure. The IMM 4J and SBT 5J technologies have achieved impressive BOL efficiencies of near 34% and 36%, respectively. However, all four- and five-junction cell technologies evaluated in this chapter are accompanied with a significant increase in cell cost that greatly exceeds any increases in EOL cell efficiency. This increase in cost is due to higher substrate costs in the case of the SBT 5J cell technology, thicker structures leading to higher epitaxial growth costs for all 4–5J technologies, and higher processing costs (substrate removal and bonding) for the IMM 4J and SBT 5J cases. Adoption of these advanced technologies in future missions will likely need to prioritize EOL cell efficiency and other advantageous metrics including specific power and stowage volume over the EOL $/W.

References

[1] R.L. Easton, M.J. Votaw, Vanguard I IGY satellite (1958 beta), Rev. Sci. Instrum. 30 (2004) 70, https://doi.org/10.1063/1.1716492.

[2] 5 Minutes for 5 Questions with Boeing Satellite Systems International, Viasat.Com. (n.d.). https://www.viasat.com/about/newsroom/blog/5-minutes-for-5-questions-with-boeing-satellite-systems/ (accessed March 28, 2021).

[3] L.M. Hague, K.J. Metcalf, G.M. Shannon, R.C. Hill, C.-Y. Lu, Performance of International Space Station electric power system during station assembly, in: IECEC 96 Proc. 31st Intersoc. Energy Convers. Eng. Conf, vol 1, 1996, pp. 154–159, https://doi.org/10.1109/IECEC.1996.552863.

[4] S. Dawson, P. Stella, W. McAlpine, B. Smith, JUNO photovoltaic power at Jupiter, in: 10th Int. Energy Convers. Eng. Conf, American Institute of Aeronautics and Astronautics, 2012, https://doi.org/10.2514/6.2012-3833.

[5] S.R. Messenger, G.P. Summers, E.A. Burke, R.J. Walters, M.A. Xapsos, Modeling solar cell degradation in space: A comparison of the NRL displacement damage dose and the JPL equivalent fluence approaches, Prog. Photovoltaics Res. Appl. 9 (2001) 103–121, https://doi.org/10.1002/pip.357.

[6] A. Luque, S. Hegedus (Eds.), Handbook of Photovoltaic Science and Engineering, second ed., Wiley, Chichester, West Sussex, U.K, 2011.

[7] S.M. Sze, Physics of Semiconductor Devices, second ed., Wiley-Interscience, New York, 1981.

[8] 2000 ASTM Standard Extraterrestrial Spectrum Reference E-490-00 | Grid Modernization | NREL, (n.d.). https://www.nrel.gov/grid/solar-resource/spectra-astm-e490.html (accessed January 13, 2021).

[9] M.A. Green, Solar Cells: Operating Principles, Technology and System Applications, University of New South Wales, Kensington, NSW, 1986.

[10] B.E. Anspaugh, GaAs Solar Cell Radiation Handbook, CreateSpace Independent Publishing Platform, 2014.

[11] G.B. Stringfellow, Organometallic Vapor-phase Epitaxy: Theory and Practice: Stringfellow, Gerald B.: 9780126738421: Amazon.Com: Books, second ed., Academic Press, 1998. https://www.amazon.com/Organometallic-Vapor-Phase-Epitaxy-Theory-Practice/dp/0126738424. (Accessed 31 December 2020).

[12] P. Chiu, 32% Efficient 3J Space Solar Cells for Targeted Space Environments, 2018.

[13] J.I. Pankove, Optical Processes in Semiconductors, 2nd Revised ed. edition, Dover Publications, Mineola NY, 2010.

[14] M. Taguchi, A. Yano, S. Tohoda, K. Matsuyama, Y. Nakamura, T. Nishiwaki, K. Fujita, E. Maruyama, 24.7% record efficiency HIT solar cell on thin silicon wafer, IEEE J. Photovoltaics 4 (2014) 96–99, https://doi.org/10.1109/JPHOTOV.2013.2282737.

[15] K. Yoshikawa, H. Kawasaki, W. Yoshida, T. Irie, K. Konishi, K. Nakano, T. Uto, D. Adachi, M. Kanematsu, H. Uzu, K. Yamamoto, Silicon heterojunction solar cell with interdigitated back contacts for a photoconversion efficiency over 26%, Nat. Energy 2 (2017) 1–8, https://doi.org/10.1038/nenergy.2017.32.

[16] M. Yamaguchi, K. Lee, K. Araki, N. Kojima, Y. Okuno, M. Imaizumi, Analysis for radiation degradation of advanced Si space solar cells, in: 2019 IEEE 46th Photovolt. Spec. Conf, PVSC, 2019, pp. 2377–2380, https://doi.org/10.1109/PVSC40753.2019.8981219.

[17] S. Subramanian, S. Anand, B.M. Arora, Y.C. Lu, E. Bauser, Charge state of the DX center in aluminum gallium arsenide from photo-Hall measurements, Phys. Rev. B 48 (1993) 8757–8760, https://doi.org/10.1103/PhysRevB.48.8757.

[18] M.J. Ludowise, R.A. LaRue, P.G. Borden, P.E. Gregory, W.T. Dietze, High-efficiency organometallic vapor phase epitaxy AlGaAs/GaAs monolithic cascade solar cell using metal interconnects, Appl. Phys. Lett. 41 (1982) 550–552, https://doi.org/10.1063/1.93592.

[19] B.-C. Chung, G.F. Virshup, S. Hikido, N.R. Kaminar, 27.6% efficiency (1 sun, air mass 1.5) monolithic Al 0.37 Ga 0.63 As/GaAs two-junction cascade solar cell with prismatic cover glass, Appl. Phys. Lett. 55 (1989) 1741–1743, https://doi.org/10.1063/1.102204.
[20] R.B. Capaz, B. Koiller, Partial-ordering effects in InGaP, Phys. Rev. B 47 (1993) 4044–4047, https://doi.org/10.1103/PhysRevB.47.4044.
[21] J.K. Shurtleff, R.T. Lee, C.M. Fetzer, G.B. Stringfellow, Band-gap control of GaInP using Sb as a surfactant, Appl. Phys. Lett. 75 (1999) 1914–1916, https://doi.org/10.1063/1.124869.
[22] J.M. Olson, W.E. McMahon, S. Kurtz, Effect of Sb on the properties of GaInP top cells, in: 2006 IEEE 4th World Conf. Photovolt. Energy Conf., IEEE, Waikoloa, HI, 2006, pp. 787–790, https://doi.org/10.1109/WCPEC.2006.279574.
[23] J.M. Olson, S.R. Kurtz, A.E. Kibbler, P. Faine, A 27.3% efficient Ga0.5In0.5P/GaAs tandem solar cell, Appl. Phys. Lett. 56 (1990) 623–625, https://doi.org/10.1063/1.102717.
[24] K.A. Bertness, S.R. Kurtz, D.J. Friedman, A.E. Kibbler, C. Kramer, J.M. Olson, 29.5%-efficient GaInP/GaAs tandem solar cells, Appl. Phys. Lett. 65 (1994) 989–991, https://doi.org/10.1063/1.112171.
[25] P.T. Chiu, D.C. Law, C.M. Fetzer, M. Haddad, S. Mesropian, R. Cravens, P.H. Hebert, J.H. Ermer, J.P. Krogen, Qualification of 32% BOL and 28% EOL efficient XTE solar cells, in: 2019 IEEE 46th Photovolt. Spec. Conf, PVSC, 2019, pp. 1506–1509, https://doi.org/10.1109/PVSC40753.2019.8981370.
[26] P. Chiu, Qualification and Production of 32%-Efficient XTJ Targeted Environment (XTE) 3J Space Solar Cells, 2021.
[27] C.M. Fetzer, R.R. King, P.C. Colter, K.M. Edmondson, D.C. Law, A.P. Stavrides, H. Yoon, J.H. Ermer, M.J. Romero, N.H. Karam, High-efficiency metamorphic GaInP/GaInAs/Ge solar cells grown by MOVPE, J. Cryst. Growth 261 (2004) 341–348, https://doi.org/10.1016/j.jcrysgro.2003.11.026.
[28] R.R. King, D. Bhusari, A. Boca, D. Larrabee, X.-Q. Liu, W. Hong, C.M. Fetzer, D.C. Law, N.H. Karam, Band gap-voltage offset and energy production in next-generation multijunction solar cells, Prog. Photovoltaics Res. Appl. 19 (2011) 797–812, https://doi.org/10.1002/pip.1044.
[29] X.Q. Liu, C. Fetzer, P. Chiu, M. Haddad, X. Zhang, R. Cravens, D. Law, J. Ermer, J. Krogen, S. Sharma, J. Hanley, Large area multijunction III-V space solar cells over 31% efficiency, in: 2017 IEEE 44th Photovolt. Spec. Conf, PVSC, 2017, pp. 2094–2098, https://doi.org/10.1109/PVSC.2017.8366024.
[30] W. Guter, F. Dunzer, l. Ebel, K. Hillerich, Space Solar Cells – 3G30 and Next Generation Radiation Hard Products, ResearchGate, 2017, p. 03005, https://doi.org/10.1051/e3sconf/20171603005.
[31] C.C. Lee, C.Y. Wang, G. Matijasevic, Au-In bonding below the eutectic temperature, IEEE Trans. Compon. Hybrids Manuf. Technol. 16 (1993) 311–316, https://doi.org/10.1109/33.232058.
[32] F. Niklaus, G. Stemme, J.-Q. Lu, R.J. Gutmann, Adhesive wafer bonding, J. Appl. Phys. 99 (2006) 031101, https://doi.org/10.1063/1.2168512.
[33] X.Y. Lee, M. Goertemiller, M. Boroditsky, R. Ragan, E. Yablonovitch, Thin film GaAs solar cells on glass substrates by epitaxial liftoff, AIP Conf. Proc. 394 (1997) 719–727, https://doi.org/10.1063/1.52872.
[34] R. Tatavarti, G. Hillier, G. Martin, A. Wibowo, R. Navaratnarajah, F. Tuminello, D. Hertkorn, M. Disabb, C. Youtsey, D. McCallum, N. Pan, Lightweight, low cost InGaP/GaAs dual-junction solar cells on 100 Mm Epitaxial Liftoff (ELO) Wafers, in: 2009 34th IEEE Photovolt. Spec. Conf, PVSC, 2009, pp. 002065–002067, https://doi.org/10.1109/PVSC.2009.5411455.
[35] R.M. France, J.F. Geisz, M.A. Steiner, B. To, M.J. Romero, W.J. Olavarria, R.R. King, Reduction of crosshatch roughness and threading dislocation density in metamorphic GaInP buffers and GaInAs solar cells, J. Appl. Phys. 111 (2012) 103528, https://doi.org/10.1063/1.4721367.
[36] M. Stan, D. Aiken, B. Cho, A. Cornfeld, J. Diaz, V. Ley, A. Korostyshevsky, P. Patel, P. Sharps, T. Varghese, Very high efficiency triple junction solar cells grown by MOVPE, J. Cryst. Growth 310 (2008) 5204–5208, https://doi.org/10.1016/j.jcrysgro.2008.07.024.
[37] J. Boisvert, D. Law, R. King, D. Bhusari, X. Liu, A. Zakaria, W. Hong, S. Mesropian, D. Larrabee, R. Woo, A. Boca, K. Edmondson, D. Krut, D. Peterson, K. Rouhani, B. Benedikt, N.H. Karam,

Development of advanced space solar cells at spectrolab, in: 2010 35th IEEE Photovolt. Spec. Conf, 2010, pp. 000123–000127, https://doi.org/10.1109/PVSC.2010.5614522.

[38] F. Dimroth, U. Schubert, A.W. Bett, 25.5% efficient Ga/sub 0.35/In/sub 0.65/P/Ga/sub 0.83/In/sub 0.17/As tandem solar cells grown on GaAs substrates, IEEE Electron. Device Lett. 21 (2000) 209–211, https://doi.org/10.1109/55.841298.

[39] D.C. Law, C.M. Fetzer, R.R. King, P.C. Colter, H. Yoon, T.D. Isshiki, K.M. Edmondson, M. Haddad, N.H. Karam, Multijunction solar cells with subcell materials highly lattice-mismatched to germanium, in: Conf. Rec. Thirty-First IEEE Photovolt. Spec. Conf. 2005, 2005, pp. 575–578, https://doi.org/10.1109/PVSC.2005.1488196.

[40] W. Guter, F. Dunzer, L. Ebel, K. Hillerich, W. Köstler, T. Kubera, M. Meusel, B. Postels, C. Wächter, Space solar cells – 3G30 and next generation radiation hard products, E3S Web Conf. 16 (2017) 03005, https://doi.org/10.1051/e3sconf/20171603005.

[41] D. Law, X. Liu, P. Chiu, Ultra High Efficiency Multijunction Space Solar Cell Structures Approaching 37% AM0 Efficiency, Spectrolab, 2019.

[42] 32% Quadruple Junction GaAs Solar Cell, Azur Space, 2019.

[43] M. Yamaguchi, Radiation resistance of compound semiconductor solar cells, J. Appl. Phys. 78 (1995) 1476–1480, https://doi.org/10.1063/1.360236.

[44] R.M. France, I. García, W.E. McMahon, A.G. Norman, J. Simon, J.F. Geisz, D.J. Friedman, M.J. Romero, Lattice-mismatched 0.7-eV GaInAs solar cells grown on GaAs using GaInP compositionally graded buffers, IEEE J. Photovoltaics 4 (2014) 190–195, https://doi.org/10.1109/JPHOTOV.2013.2281724.

[45] M. Stan, D. Aiken, B. Cho, A. Cornfeld, V. Ley, P. Patel, P. Sharps, T. Varghese, High-efficiency quadruple junction solar cells using OMVPE with inverted metamorphic device structures, J. Cryst. Growth 312 (2010) 1370–1374, https://doi.org/10.1016/j.jcrysgro.2009.10.059.

[46] D. Law, Recent Progress in MJ Solar Cell Development and Production, 2015.

[47] D. Bhusari, D. Law, R. Woo, J. Boisvert, S. Mesropian, D. Larrabee, W. Hong, N. Karam, Direct Semiconductor Bonding Technology (SBT) for high efficiency III-V multi-junction solar cells, in: 2011 37th IEEE Photovolt. Spec. Conf, 2011, pp. 001937–001940, https://doi.org/10.1109/PVSC.2011.6186332.

[48] P.T. Chiu, D.C. Law, R.L. Woo, S.B. Singer, D. Bhusari, W.D. Hong, A. Zakaria, J. Boisvert, S. Mesropian, R.R. King, N.H. Karam, Direct semiconductor bonded 5J cell for space and terrestrial applications, IEEE J. Photovoltaics 4 (2014) 493–497, https://doi.org/10.1109/JPHOTOV.2013.2279336.

[49] P.T. Chiu, D.C. Law, R.L. Woo, S.B. Singer, D. Bhusari, W.D. Hong, A. Zakaria, J. Boisvert, S. Mesropian, R.R. King, N.H. Karam, 35.8% space and 38.8% terrestrial 5J direct bonded cells, in: 2014 IEEE 40th Photovolt. Spec. Conf, PVSC, 2014, pp. 0011–0013, https://doi.org/10.1109/PVSC.2014.6924957.

[50] S. Tomasulo, K. Yaung, M. Lee, Metamorphic GaAsP and InGaP solar cells on GaAs, photovolt, IEEE J. Of. 2 (2012) 56–61, https://doi.org/10.1109/JPHOTOV.2011.2177640.

[51] L. Hongbo, X. li, Z. Wei, Z. Dayong, S. Mengqi, S. Lijie, C. Kaijian, A 2.05 eV AlGaInP sub-cell used in next generation solar cells, J. Semiconduct. 35 (2014) 094010, https://doi.org/10.1088/1674-4926/35/9/094010.

[52] A.B. Cornfeld, P. Patel, J. Spann, D. Aiken, J. McCarty, Evolution of a 2.05 eV AlGaInP top sub-cell for 5 and 6J-IMM applications, in: 2012 38th IEEE Photovolt. Spec. Conf, 2012, pp. 002788–002791, https://doi.org/10.1109/PVSC.2012.6318171.

[53] H. Yoon, K. Edmondson, G. Kinsey, R. King, P. Hebert, B.T. Cavicchi, N. Karam, Minority carrier lifetime and radiation damage coefficients of germanium, in: IEEE Conference Publication, IEEE, 2005. https://ieeexplore.ieee.org/abstract/document/1488264. (Accessed 6 December 2020).

[54] F. Zhuge, Z. Zheng, P. Luo, L. Liang, Y. Huang, H. Li, T. Zhai, Nanostructured Materials and Architectures for Advanced Infrared Photodetection, Adv. Mater. Technol. 2 (2017) 1700005, https://doi.org/10.1002/admt.201700005.

CHAPTER FIVE

Perovskite solar cells: Background and prospects for space power applications

Lyndsey McMillon-Brown and Timothy J. Peshek
NASA Glenn Research Center, Photovoltaic & Electrochemical Systems Branch, Cleveland, OH United States

5.1 Introduction

The next generation of space exploration will require substantial power. Scientific exploration, Earth observations, telecommunications, electric propulsion systems, and human life support systems are driving the increasing power requirements [1]. The aerospace industry will rely on solar panels to meet this growing energy demand. There is great interest in operating high-voltage systems (300—600 V), but we currently lack the capabilities required for long-duration high-voltage power supply systems. As high-voltage systems are implemented, it is ideal to eliminate conversion units and several kilometers of associated wiring that increase the resulting systems volume and weight. Ideally, the resulting power system would operate at a lower current, reducing the power system volume and weight and thus reducing the system cost. There is great interest in high-efficiency perovskite thin-film solar cells for implementation in space [2]. Perovskite-structured solar cells are promising candidates for aerospace due to their exceptional optoelectronic properties. Here we review the recent progress of perovskite devices for aerospace applications and highlight the challenges and opportunities that this technology must overcome before being broadly implemented as a high-voltage energy solution.

5.1.1 Perovskite materials

The name perovskite(s) applies to the crystal structure; the ABX_3 stoichiometry is shared with the perovskite mineral calcium titanium oxide ($CaTiO_3$). Perovskite-structured materials can display insulating, semiconducting, or superconducting behavior. Semiconductor materials with perovskite structure are commonly a hybrid organic-inorganic lead or tin halide—based material. The structure is nearly cubic, following the general stoichiometry ABX_3 where the A-site cation is in the center of the lattice, the B-site cation is on the corners of the lattice, and the X is a halide (Fig. 5.1A).

Perovskites can be processed from solution at room temperature, and they exhibit exceptional optoelectronic properties with performance levels similar to polycrystalline

Photovoltaics for Space
ISBN 978-0-12-823300-9, https://doi.org/10.1016/B978-0-12-823300-9.00018-2

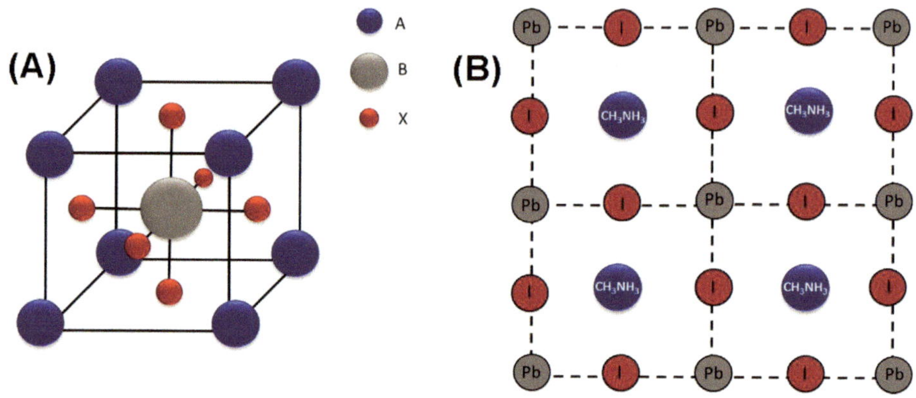

Figure 5.1 (A) General schematic of the perovskite crystal structure. (B) Schematic perovskite crystal structure of $MAPbI_3$. *Credit: First author's original work, as a US Government work product; this is in the public domain.*

Si. These characteristics have led researchers to consider perovskites for many optoelectronic applications. Initially implemented as a light sensitizer in mesoporous dye cells, perovskites also function as an absorber and transport layer in solid-state dye cell architectures. Recently, perovskites have been employed as the bulk material in a standard planar thin-film solar cell [3]. Perovskite materials have also been employed to produce light-emitting electrochemical cells [4] and diodes [5,6]. These aforementioned achievements have been enabled by rapid materials development, through constant improvement of deposition techniques informed by a deeper understanding of the perovskite crystallization process. Crystallization mechanisms and film morphology are impacted by deposition technique (solution vs. vapor) as well as the order that device layers are deposited, and the chemical composition of those layers [7]. Crystallization mechanisms are also impacted by the selection of solvents, their concentrations, and any additives [8,9].

5.1.2 Recent advances in terrestrial perovskite photovoltaics

Recently, solar cells based on hybrid perovskites have become increasingly attractive for low-cost photovoltaic applications since the demonstration of viable devices ($\sim$10% efficiency in 2012) [10,11]. Perovskite solar cells have now reached 24% single-junction efficiency [12]. Perovskites are promising candidates for photovoltaic applications due to their favorable optoelectronic properties [13–15], potential for high power conversion efficiency [16], and low-temperature solution or vapor-based fabrication methods [17–19]. Perovskites also exhibit a superior light absorption coefficient ($\sim 10^5\ cm^{-1}$), which creates a high density of photoexcited charges and a smaller absorption length that requires only a submicrometer thickness of perovskite for sufficient light harvesting

[15,20–22]. Finally, perovskites have long electron and hole diffusion lengths in thin-film (>1 μm) and single-crystal (>175 μm) specimens, suppressing the recombination of photoexcited charges [13,14,20,23]. Despite rapid success and growth in performance, poor stability has slowed widespread commercial integration of perovskites [24].

Thin films enable use of lightweight devices that can be processed via solution or vapor phase deposition methods. When these thin-film devices are deposited onto flexible substrates, they can result in a flexible solar module with low mass and low storage volume. Perovskites also offer opportunities for multijunction devices. Increased efficiencies can be achieved by exploiting perovskite-perovskite tandem structures [25]. Also, the tunable bandgaps from 1.2 up to 2.3 eV of perovskites make them attractive options for top cells, especially when combined with crystalline silicon or copper indium gallium selenide bottom cells [26,27].

5.2 Characteristics of perovskite solar cells

The most popular hybrid perovskite active layer material for photovoltaics is methylammonium lead iodide ($CH_3NH_3PbI_3$ or $MAPbI_3$ where $MA = CH_3NH_3^+$). The MA is a positively charged organic cation at the center of a lead iodide cage structure (Fig. 5.1B). The particular process that any solar cell uses to separate an electron-hole pair divides existing solar cell technologies into two distinct classes: conventional cells, like silicon and III-V devices; and excitonic solar cells, which tend to be organic based solar cells made from polymers or dye-sensitized cells.

In conventional solar cells, light absorption directly and immediately results in a free electron-hole pair whose collection is guided by the energetics of the cell, leading electrons to the cathode and holes to the anode by making other routes energetically unfavorable, or impossible. In the case of excitonic solar cells, the absorption of light results in a bound electron-hole pair that is not yet free. It must diffuse to an interface where a localized electric field will assist the exciton separation; only then can an electron travel to the cathode, facilitating current. These excitonic solar cells have an added challenge because the excitons have short diffusion lengths—in some cases just 10 nm before they recombine—so many clever architectures (like mixing donor and acceptor materials to shorten the distance to an interface) have been implemented to introduce many interfaces to facilitate this exciton separation. Perovskite solar cells fall in the class of conventional solar cells. This is unique because, although an inorganic-organic hybrid that can be solution processed—as well as thin and flexible like the excitonic solar cells—they do not fall prey to challenges of charge extraction endemic to excitonic cells.

5.2.1 Defect tolerance

Perovskite crystals exhibit a remarkable tolerance for defects. Defect properties are critical to the performance of an optoelectronic material, especially those of intrinsic point defects, often determining their performance. Above a certain density, defects deteriorate performance and limit the ability to control material properties [28]. Intrinsic defects determine the doping limit, and carrier mobility, carrier lifetime, and recombination rate are all greatly affected by defects [29]. Defect tolerance indicates either a low density of defects in the material or a relative insensitivity of functional properties to defects. Defect tolerance does not equate to defect immunity, however. It is necessary to have an accurate understanding of the defect formation process to control material stability [28]. The generation of defects within the conduction and valence bands is not without consequence. The impact of defects is mitigated in $MAPbI_3$, preserving electrical properties, yielding robust performance. This is because the defects primarily form within the valence band or conduction band, so they do not lead to nonradiative recombination since states in the gap are avoided.

5.2.2 Radiation hardness

Radiation in space is characterized by high intensity (fluence ~ 1.5 millisievert/day) and has deleterious effects on materials; it consists of both ionizing and nonionizing radiation [29]. Solar flares release a large amount of X-rays, gamma rays, and streams of electrons and photons with kinetic energy ranging from keV to GeV [30]. The Van Allen radiation belt is a zone of charged particles, mostly originating from the solar wind, that are held around a planet by that planet's magnetic field. In the Earth's case, when the inner Van Allen belt expands, it includes the orbits of the International Space Station (ISS) and many other satellites, providing further impetus for space photovoltaics having radiation-resistant characteristics. AIAA S-111 standards require solar cells to withstand 1×10^{16} electrons per square cm (e^-/cm^2) fluence at 1 MeV and 1×10^{13} protons per square cm (p^+/cm^2) fluence at 3 MeV [31]. To date, experiments have been conducted to determine the tolerance of perovskite solar cells under electron [32–34], proton [32–40], neutron [41], and γ–ray irradiations [42–45].

When materials experience high-energy electron and proton irradiations, atomic displacements generate lattice defects in semiconductors (Fig. 5.2). These lattice defects manifest as vacancies, interstitials, and complex defects that all act as recombination centers or majority- and minority-trapping centers. The recombination centers result in a decrease in the output power of solar cells [46]. To date, many experiments have been carried out to investigate perovskite solar cell tolerance to high-energy radiation [32,34–38,41,47].

Initial investigations of perovskite radiation hardness to electron radiation yield favorable results. Huang et al. exposed a formamidinium (FA)-based perovskite solar cell

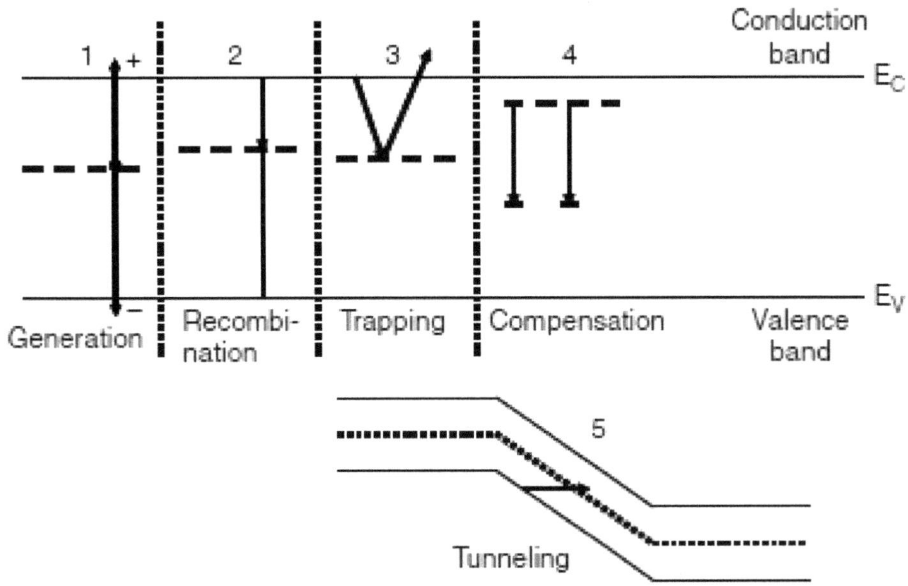

Figure 5.2 Potential effects of radiation-induced defect levels on charge transport in a solar cell. *Reproduced with permission from A. Reinders, P. Verlinden, W. van Sark, A. Freundlich,* Photovoltaic Solar Energy: From Fundamentals to Applications, *copyright (2017) John Wiley & Sons.*

to 1 MeV electrons at fluences between 10^{12} to 10^{16} cm^{-2} at room temperature under a vacuum of 10^{-5} torr. The solar cell architecture was ITO/TiO_2/$FAPbI_3$/Spiro-MeOTAD/Ag on a quartz substrate. They irradiated the cells through the Ag contact to ensure that the quartz did not serve as a shield to the perovskite active layer. There was no detectable degradation up to 10^{15} cm^{-2} fluence, and just 10% reduction in photovoltaic performance (FF and PCE) and spectral response at 10^{16} cm^{-2} fluence. No significant changes in the morphology and crystal phase of the perovskites were observed [34]. Miyazawa et al. similarly reported no detectable degradation of perovskite solar cell architecture ITO/mp-TiO_2/$MAPbI_{3\text{-}x}Cl_x$/P3HT/Au performance at 10^{16} cm^{-2} fluence at 1 MeV energies [33]. In a separate study, Miyazawa et al. exposed encapsulated mixed cation/halide perovskites [ITO/mp-TiO_2/$Cs_xFA_{0.85}MA_{0.15}Pb(I_{0.85}Br_{0.15})_3$/P3HT/Au ($x < 0.05$)] to a 1-MeV electron beam for an accumulated fluence of 10^{16} cm^{-2}. After irradiation, short circuit current (J_{sc}), open circuit voltage (V_{oc}), and power conversion efficiency (PCE) for the cells maintained $99 \pm 4\%$, $97 \pm 3\%$, and $93 \pm 13\%$, respectively, of the initial values. The fill factor (FF) in the devices did drop but the external quantum efficiency (EQE) did not show any degradation [32]. All aforementioned perovskite devices demonstrate high durability against high-fluence electron irradiation, even when compared to Si and III-V multijunction solar cells. It is critical to highlight that the Huang [34] and Miyazawa [32,33] investigations used quartz substrates instead of soda-lime glass, because soda-lime becomes darkened by radiation. This substrate degradation may otherwise affect the evaluation of radiation tolerance of the devices.

Perovskite devices under proton radiation have retained over 90% of their initial performance after exposure to high proton fluences of 10^{12} p^+/cm^2 [38] and 10^{14} p^+/cm^2 [32] with proton energies ranging from 0.05 to 68 MeV [39]. In perovskites, electronic scattering of the incident proton ionizes the target material and can lead to the disruption of C—H and N—H bonds in the organic and hybrid perovskite layers in the device stacks [39,48,49]. Nuclear scattering of the incident proton causes the target nuclei to recoil and be displaced, which instigates a cascade of damage events that generate defects in the material [39]. Lang et al. assessed the radiation hardness of perovskite tandem devices in operando, and their perovskite/CIGS tandems retain over 85% of their initial performance after 68 MeV proton irradiation and a dose of $\Phi = 2 \times 10^{12}\, p^+/\text{cm}^2$. Further investigation revealed that the open circuit voltage of the perovskite top cell was unaffected after high-dose proton irradiation [39]. This performance is comparable to conventional GaInP/GaAs/Ge absorbers that retain approximately 82% of their initial performance at identical displacement damage [50]. These irradiation conditions correspond to more than 50 years in space at the ISS orbit [39]. Brus et al. demonstrated that MA perovskite cells are durable under proton fluence up to 10^{13} cm^{-2} with high energy 68 MeV protons [36]. However, it is possible that such high-energy protons penetrate the perovskite active layer without causing a collision event, which makes it difficult to correctly assess proton tolerance [32]. The effect of proton irradiation on perovskite solar cells varies by the stopping position of the protons within the device [33].

The ISS receives a fluence of approximately 2.8×10^{11} neutrons cm^{-2} each year, varying in energies from 10^{-1} to 10^{11} eV. Fast neutrons—neutrons with energies >10 MeV—are secondary particles generated upon the interaction of primary cosmic rays (~90% protons) with the atmosphere and spacecraft components resulting in damage to electronic devices. Paternò et al. conducted the first in operando study of illuminated and neutron-irradiated devices. Encapsulated devices [ITO/PEDOT:PSS/$CH_3NH_3PbI_{3-x}Cl_x$/PCBM/Al] were irradiated for up to 1.5×10^9 e^- $cm^{-2}s^{-1}$ simulating approximately 80 years of fast neutron exposure on the ISS. Fast neutron irradiation led to the formation of permanent defects that likely originate from atomic displacements within the active layer. The investigators suspect that the atomic displacement can promote the formation of shallow-traps (Frenkel defects) that act as dopants and contribute to a decreased leakage current. Ultimately, perovskites demonstrate that they can be resilient to neutron radiation [41].

Gamma rays have the highest penetrating ability, and they cannot be stopped by regular encapsulation materials [45]. After 20 years of operation in space, a typical solar cell is expected to accumulate approximately 1000 krad of gamma radiation [30]. Despite the significance of gamma radiation in space [51], the stability of perovskite devices under gamma radiation remains largely unexplored. K. Yang et al. investigated the stability of a hysteresis-free high-efficiency encapsulated perovskite solar cell [ITO/PTAA/ $Cs_{0.5}FA_{0.81}MA_{0.14}PbI_{2.55}Br_{0.45}/C_{60}$/BCP/Cu] under gamma rays and visible light simultaneously. Their experiments indicate that hybrid perovskites are stable after exposure to gamma-ray irradiation for 1535 h with an accumulated radiation dose of 2.3 Mrad, under continuous illumination. The transmittance of the soda-lime glass substrates used in this experiment reduced from 90% over the entire visible range to about 50%–75% after gamma radiation. After correcting for the substrate darkening, Yang et al. resolved that the perovskite retained 96.8% of its original PCE [42]. S. Yang et al. irradiated [ITO/SnO_2/$FA_{0.945}MA_{0.025}Cs_{0.03}Pb(I_{0.975}Br_{0.025})_3$/Spiro-MeOTAD/Ag] with gamma rays spanning 100 krad, 200 krad, and 500 krad. All cells retained their V_{oc} and FF, but the J_{sc} did decrease gradually with the increased radiation dose, indicating a decrease of photocurrent induced by irradiation [39]. The perovskite films did not exhibit a color change after irradiation. Existing X-ray diffraction peaks of the film remain unchanged, but a new peak emerges and is attributed to the δ-phase $FAPbI_3$, which is likely caused by phase transition of α-phase $FAPbI_3$. This phase transition results in decreased absorbance of the perovskite film, resulting in reduced J_{sc} as irradiation dose increases.

K. Yang et al. also observed that with increased gamma-ray irradiation dose [42], the glass substrate deepens in color, as shown in Fig. 5.3A. A gamma-irradiated structure thus produces structural defects and destroys the stable potential field in the cell interior, which causes changes in the electronic energy state. Boldryreva et al. systematically investigated the radiation stability of different components of a triple-cation perovskite device, as shown in Fig. 5.3B. This permitted the specific investigation of the impact of gamma-ray irradiation on the hole-collecting bottom electrode, triple-cation perovskite absorber material, and electron transport layer, respectively. The radiation dosage was 10 krad, 30 krad, and 50 krad, which are dosages likely to be experienced by satellites in low and middle Earth orbit. They found that doses up to 50 krad do not damage the hole-collecting bottom electrode [glass/ITO/PEDOT:PSS], although the radiation did significantly impact the photovoltaic performance of the triple-cation perovskite absorber and the electron transport layers. These losses were due to significant photocurrent loss (25%–30%) that were caused by halide phase segregation [44]. Boldryeva et al. later conducted a comparative study of gamma-ray stability of various perovskite absorber layers including $MAPbI_3$, $Cs_{0.15}FA_{0.85}PbI_3$, $CS_{0.1}MA_{0.15}FA_{0.75}PbI_3$, $CsPbI_3$, and $CsPbBr_3$ with doses ranging from 10 to 500 krad. They observed that the composition of the perovskite strongly impacts the gamma radiation stability of the material, and they concluded that $MAPbI_3$-based cells were the most resistant to gamma radiation, as perovskite underwent rapid self-healing. $MAPbI_3$ solar cells were shown to resist a 1000-krad gamma radiation dose.

5.2.3 UV sensitivity

Exposure to ultraviolet (UV) rays can damage the molecular makeup of certain materials and cause molecular chains to break down, causing both physical and chemical changes.

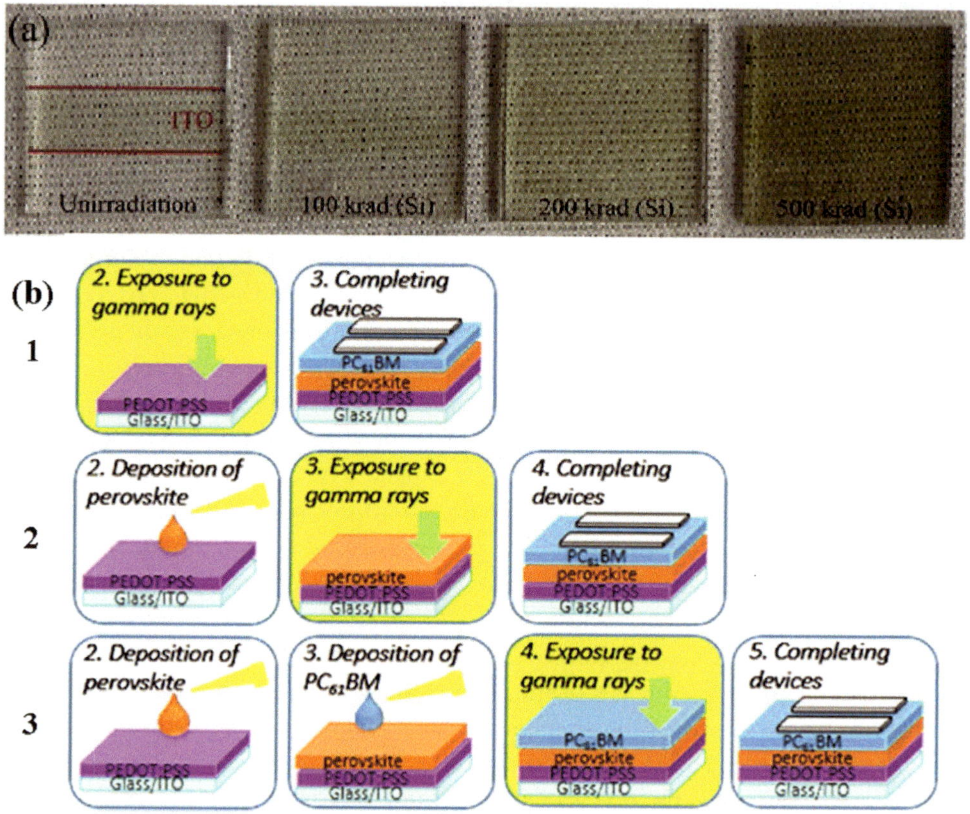

Figure 5.3 (A) Photographs of glass/ITO substrates before and after exposure to gamma-ray irradiation dose at 100, 200, and 500 krad. (B) Architecture of investigated samples and general methodology of the experiment on assessment of the radiation stability of different components of the device: (1) glass/ITO/PEDOT:PSS, (2) glass/ITO/PEDOT:PSS/perovskite, and (3) glass/ITO/PEDOT:PSS/perovskite/PC61BM. *Reproduced with permission from: (A) K. Yang, K. Huang, X. Li, S. Zheng, P. Hou, J. Wang, H. Guo et al., Radiation tolerance of perovskite solar cells under gamma ray, Org. Electron. 71 (2019) 79–84, copyright (2019) Elsevier; (B) A.G. Boldyreva, A.F. Akbulatov, S.A. Tsarev, S.Yu. Luchkin, I.S. Zhidkov, E.Z. Kurmaev, K.J. Stevenson, V.G. Petrov, P.A. Troshin, γ-Ray-Induced Degradation in the Triple-Cation Perovskite Solar Cells, J. Phys. Chem. Lett. 10 (2019) 813–818, copyright (2019) American Chemical Society.*

The solar spectrum contains UV (10%), and most of the UV radiation from the Sun is absorbed in Earth's atmosphere. Approximately 33% of all UV radiation penetrates the atmosphere and reaches Earth's surface. Thus, UV-induced degradation is a greater concern for photovoltaic devices operating in space. The underlying degradation process for perovskite photovoltaics exposed to UV radiation can occur in the perovskite absorber and originate from interfaces and other layers within the device, resulting in a cumulative phenomenon [52]. For example, the perovskite absorber can experience metal infiltration from the top contact [53], chemical bonds between the hole transport layer and the electrode can degrade [54], and the perovskite absorber can exhibit photoinduced performance degradation [55].

Kim et al. investigated the specific impact of UV light on perovskites by illuminating $MAPbI_3$ devices with 365-nm UV light in an Ar glovebox (<0.5 ppm average humidity) at 25°C for 1000 h at open circuit voltage [56]. The device structure was [FTO/TiO_2/mp-TiO_2/$MAPbI_3$/Spiro-MeOTAD/Au]. X-ray diffraction (XRD) verified that $MAPbI_3$ decomposed to PbI_2 continuously with increasing UV exposure. Device absorption also decreased, and transmission increased, as UV exposure exceeded 200 h. This absorption change was attributed to the degraded ability of perovskites to absorb light. A postdegradation light soak at one sun, however, yielded recovery of FF and PCE, which went from 12.6% initially, 1.36% after UV exposure, and recovered to 10.4% after the light soak, as shown in Fig. 5.4A.

Results of a postdegradation light soak vary, suggesting that longer UV exposure can change the UV degradation recovery and mechanism. Domanski et al. investigate the effect of UV light on unencapsulated devices in inert conditions compared to devices illuminated by a UV-free light-emitting diode lamp [57]. Both sets of devices display a two-regime degradation behavior, initially exhibiting exponential loss followed by linear loss. The devices aged under UV light suffer accelerated degradation, but the observed

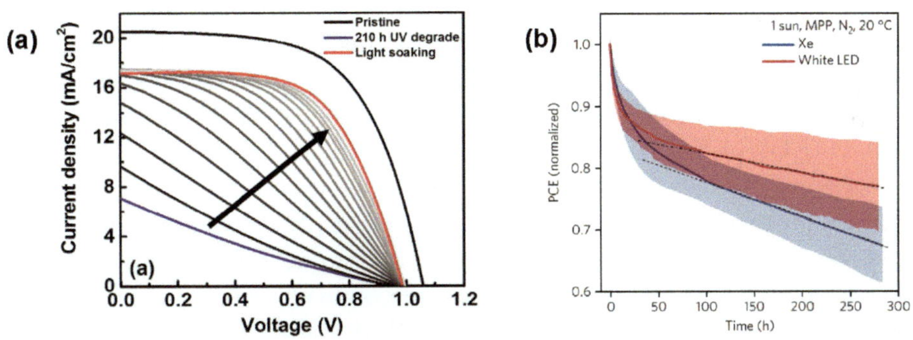

Figure 5.4 (A) Light I-V curve of pristine device, 210 h UV exposed, and during 15 min light soaking. (B) The effect of ultraviolet light on the stability of perovskite solar cells. Comparison of aging under an unfiltered Xe and an UV-free white LED lamp. Values are normalized to the initial values. Solid lines represent averaged behavior of the devices, while the shaded area represents standard deviation. Dashed lines serve to guide the eye. *(A) Reproduced from: S.-W. Lee, S. Kim, S. Bae, K. Cho, T. Chung, L.E. Mundt, S. Lee et al., UV Degradation and Recovery of Perovskite Solar Cells, Sci. Rep. 6 (2016) 38150, this work is licensed under a Creative Commons 4.0 International License, see: http://creativecommons.org/licenses/by/4.0/ (B) reproduced with permission from: K. Domanski, E.A. Alharbi, A. Hagfeldt, M. Grätzel, W. Tress, Systematic investigation of the impact of operation conditions on the degradation behavior of perovskite solar cells, Nat. Energy. 3 (2018) 61–67, copyright (2018) Springer Nature.*

differences were not statistically significant, as shown in Fig. 5.4B. From this investigation they conclude that UV degradation is not catastrophic for cells under inert conditions, but several reports exist of severe UV degradation in the presence of O_2 [58,59]. This result makes UV light an extrinsic degradation factor that can be viewed analogously to O_2 and H_2O and can similarly be prevented via encapsulation. Further, perovskite device operation in space will be void of O_2 and H_2O, so the UV degradation will be further minimized and can likely be addressed through materials selection and device encapsulation.

It has been shown that the use of UV light–sensitive TiO_2 as an electron transport material results in poor operational stability of organic metal lead halide perovskite devices as a result of the degradation of the perovskite absorber layer at the perovskite/TiO_2 contact interface [60–63]. Snaith et al. found that TiO_2-based perovskite solar cells, encapsulated in nitrogen atmosphere, displayed rapid decay in PCE [64]. Their unencapsulated devices exhibited better stability. The presence of oxygen was deemed necessary for these devices, because it often removed the surface states of TiO_2. UV light is absorbed close to the interface of perovskite/TiO_2, resulting in meta-stable trap states and charge accumulation close to the mp-TiO_2/perovskite interface. The presence of oxygen decreases the number of empty deep trap sites that form by adsorbing onto TiO_2 and pacifying these states [63].

The UV irradiation–induced deterioration of perovskite solar cells can be mitigated by modifying or replacing TiO_2 with more UV-resilient materials. Snaith et al. demonstrated enhanced UV stability with a UV filter or upon substitution of mesoporous TiO_2 with Al_2O_3-based meso-superstructured perovskite solar cells [65]. Guo et al. introduced a silicon oxide (SiO_x) interfacial modifier for the electron transport layer in TiO_2-based planar heterojunction perovskite devices [FTO/TiO_2/SiO_x/$MAPbI_3$/Spiro-MeOTAD/Au] [66]. The SiO_x modifier was placed between the perovskite and planar TiO_2 and was shown to passivate the trap states of the perovskite absorber, significantly suppressing the charge recombination of the perovskite solar cells. The devices with a SiO_x modifier also demonstrated improved UV stability, as they maintained over 74% of their original efficiencies after 65 mW/cm^2 UV light soaking.

Regarding approaches designed to mitigate UV degradation by preventing the TiO_2 from absorbing the incident UV light by adding a UV filter, Komarala et al. implemented a thin-film down-shifting nano-phosphor layer (doped yttrium vanadate YVO_4) that absorbs incident UV light and emits red visible light, which can be absorbed by the perovskite active layer, as shown in Fig. 5.5 [67]. These modified deices exhibited an improvement in device stability, retaining more than 50% of their efficiency after prolonged illumination, control devices degraded to 35% of their initial efficiency. The down-converted (UV to red) photons also resulted in an 8.5% photocurrent enhancement.

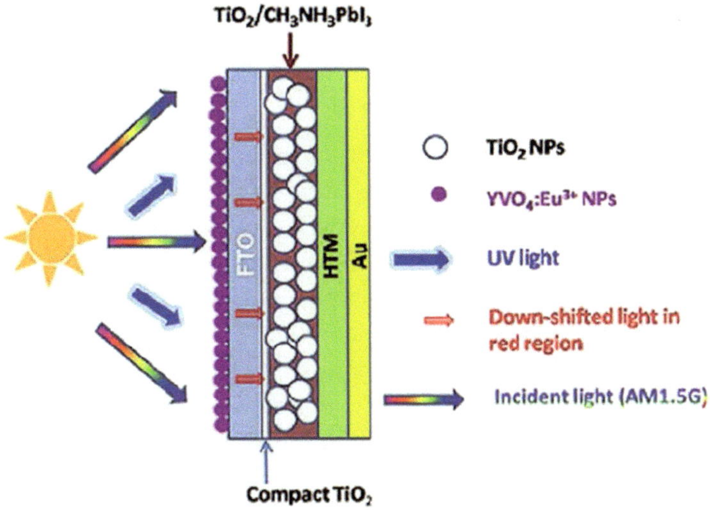

Figure 5.5 Perovskite solar cell device structure with down-shifting nano-phosphor layer on FTO glass superstrate. *Reproduced with permission from N. Chander, A.F. Khan, P.S. Chandrasekhar, E. Thouti, S.K. Swami, V. Dutta, V.K. Komarala, Reduced ultraviolet light induced degradation and enhanced light harvesting using $YVO_4:Eu^{3+}$ down-shifting nano-phosphor layer in organometal halide perovskite solar cells, Appl. Phys. Lett. 105 (2014) 033904, copyright (2014) American Institute of Physics.*

Farooq et al. studied the impact of electron transport layer by developing $MAPbI_3$ devices with SnO_2 electron transport layer (ETL) and various synthesized TiO_2 layers, including compact, nanoparticle, e-beam treated and mesoporous varieties [52]. The control devices included some wrapped in Al foil to experience the same heat but no illumination, and others under a long-pass UV filter with 400-nm cutoff but otherwise illuminated. The experimental devices were exposed to UV-B (310–317 nm) and UV-A (360–380 nm) filtered light. All devices exposed to UV-B radiation had significantly deteriorated J_{sc} values, regardless of ETL. The devices based on TiO_2 nanoparticles exposed to the UV-A lost up to 64% of their J_{sc} over 1750 h of testing. This degradation was driven by the decomposition of the perovskite active layer. All remaining devices exposed to UV-A were largely unchanged. Farooq et al. recommend filtration of the photons with UV-B wavelengths to increase the stability of perovskite devices, sacrificing 0.53 mA/cm^2 photocurrent as a tradeoff for extended device operation.

Ultimately, the literature investigations performed thus far have varied the UV spectra and illumination levels. Some investigations have applied light stress at one sun, whereas others only apply UV radiation. It is not yet established which UV wavelengths and intensities impact device stability. It is accepted that typically the presence of O_2 and H_2O intensify UV-induced degradation, so there is strong potential that this challenge can be overcome for in space operation. Regardless, materials selection is critical to thwarting these degradation pathways.

5.2.4 Oxygen and moisture sensitivity

Perovskite devices are easily decomposed in the presence of humidity due to the hygroscopic amine salts. Upon exposure to elevated temperature and humidity, the moisture-induced decomposition of the perovskite crystal is accelerated, resulting in significant device degradation [68,69]. In addition to humidity, improved stability against oxygen exposure is critical to terrestrial implementation [69]. Two components in ambient air affect the chemical stability of perovskite solar cells, resulting in the degradation of perovskite with the film color changing from dark brown-black to yellow.

This challenge is being addressed by employing moisture-resistant layers (e.g., carbon nanotubes) and other encapsulation techniques [70–72]. In early studies, the degradation of perovskite solar cells was attributed to hydrolysis reactions wherein water molecules reacted with organic species in the photoactive layer. The degradation of $MAPbI_3$ (where MA = $CH_3NH_3^+$) was further studied via XRD by Niu et al.; the degradation process can be explained as follows (Eqs. 5.1–5.4), from Ref. [59]:

$$CH_3NH_3PbI_3(s) \overset{H_2O}{\leftrightarrow} CH_3NH_3I\ (aq) + PbI_2(s) \tag{5.1}$$

$$CH_3NH_3I(aq) \leftrightarrow CH_3NH_2(aq) + HI(aq) \tag{5.2}$$

$$4HI(aq) + O_2 \leftrightarrow 2I_2(s) + 2H_2O(l) \tag{5.3}$$

$$2HI(aq) \overset{h\nu}{\leftrightarrow} H_2 \uparrow\ + I_2(s) \tag{5.4}$$

In the first reaction, in the presence of H_2O, $CH_3NH_3PbI_3$ will decompose into CH_3NH_3I solution and PbI_2. In the second reaction, CH_3NH_3I solution will decompose into CH_3NH_2 solution and HI solution. In the third reaction, there are two pathways for HI to react. Reaction 5.3 represents the redox reaction, which has a negative standard Gibbs free energy indicating that this reaction can move forward easily. Reaction 5.4 is a photochemical reaction where, under light conditions, HI can easily decompose into H_2 and I_2. The consumption of HI drives the decomposition process and causes the decrease in perovskite solar cell efficiency [24]. Frost et al. propose two reversible reaction mechanisms for decomposition with a combination of H_2O and $MAPbI_3$.

Aluminum oxide (Al_2O_3) has been employed as a postmodification material in $CH_3NH_3PbI_3$ ($MAPbI_3$) sensitized solar cells. Al_2O_3 can serve two roles in these solar cells. One role is to serve as an insulator between TiO_2 and the HTM (spiro-MeOTAD) to impede electron recombination. The other role of Al_2O_3 is to protect the $MAPbI_3$ and the sensitized films from moisture and sunlight-induced degradation. Niu et al. exposed sensitized $MAPbI_3$ films to air with 60% humidity at 35°C for 18 h (Fig. 5.6); devices prepared with an Al_2O_3 modification layer exhibited more stable photo-conversion efficiency under moisture [59]. The untreated $MAPbI_3$ film absorption between 520 and 800 nm sharply decreases. The absorption at $\lambda < 520$ nm is attributed to PbI_2 verifying that $MAPbI_3$ is quite sensitive to moisture and corrodes to PbI_2. The film postmodified with Al_2O_3 maintains the main absorption of $MAPbI_3$. The Al_2O_3 effectively shields the $MAPbI_3$ layer from air, preventing oxygen- and moisture-induced degradation. The device prepared with an Al_2O_3 modification layer maintained 48% of its efficiency, whereas the control device maintained only 20% of its original efficiency [59].

Space environments, however, are devoid of oxygen and water, making perovskites an attractive alternative to current state-of-the-art space photovoltaics such as III-V and Si solar cells. The lack of problematic degradation mechanisms, coupled with perovskites' radiation hardness, further enhance their feasibility for space applications [35]. The final four subsections will address further considerations that are critical concerns for operation in the space environment.

5.2.5 Thermal cycling

$MAPbI_3$ has an orthorhombic-to-tetragonal transition occurring at 160 K with a stable cubic phase that exists up to approximately 330 K [2]. Miyazawa et al. prepared thermally stable mixed cation/halide perovskites $FAMAPb(IBr)_3$ and $MAPbI_3$ to temperatures of 100 and −80°C, which correspond to the temperatures of satellite orbit with and without solar illumination, respectively. Some cells used P3HT as an HTL and others had spiro-MeOTAD. All P3HT-based encapsulated cells mostly maintained performance without losing J_{sc} and V_{oc} at the full thermal test rate for up to 16.7 h. Unencapsulated cells lost >50% of their J_{sc} and had yellow deposits due to PbI_2. These initial results

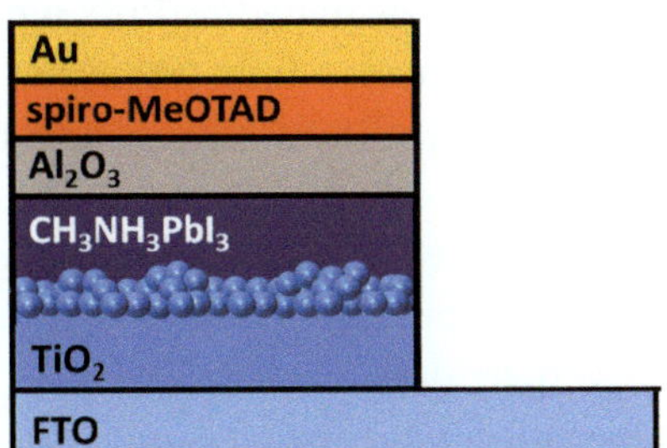

Figure 5.6 Schematic of solar cell implemented for degradation studies, exposed for 18 h to air at 60% humidity at 35°C. *First author's original work, as a US Government work product; this is in the public domain.*

suggest that encapsulation will be necessary to prevent the evaporation of MA and FA at high temperatures under vacuum. Miyazawa et al. also tested encapsulated spiro-MeOTAD-based $MAPbI_3$ devices under the same conditions. Upon heating to 100°C the devices showed an 80% reduction in PCE within 2 h; all spiro-MeOTAD cells had drastically degraded or became inoperable after 16.7 h. This rapid degradation can be contributed to spiro-MeOTAD thermal degradation, making spiro-MeOTAD unlikely to be a suitable material for incorporation in perovskite solar cells for space.

5.2.6 Vacuum stability

Perovskite devices exhibit unique challenges when exposed to vacuum. Perovskite films undergo significantly more degradation under ultrahigh vacuum ($\sim 10^{-9}$ torr) than at normal pressure [73,74]. Vacuum exposure accelerates the degradation of perovskite absorber layers that are composed of volatile materials. The perovskite absorber degradation begins at grain boundaries and is accompanied by defect formation and outgassing [75]. The defects result in accelerated ion migration across the device. Outgassing is the liberation of gas. In the case of perovskites, liberated organic gases are likely to deposit on the surface of a space vehicle causing contamination, thereby impacting spacecraft performance and altering the properties of sensitive optics. ASTM-E1559 standards have been established as a means to qualify materials for operation in space by quantifying the total loss of mass and volatile condensable materials to predict the outgassing characteristics of those materials [76].

Device encapsulation can reduce outgassing [32], but encapsulation does not fully eliminate it [77]. Ultimately, the vacuum stability of a perovskite device depends on the composition of the absorber, the absorber processing method, and the device encapsulation. Triple-cation perovskite degrades less than $MAPbI_3$ when exposed to the same vacuum and illumination conditions [74,78]. The triple-cation stability superiority is attributed to its higher thermal stability. The method of film preparation is significant to the vacuum stability of the resulting film. For example, a $MAPbI_3$ film prepared with dimethyl formamide (DMF) precursor degraded to PbI_2 when it was stored in vacuum without illumination for 90 days, whereas film fabricated with DMF and dimethyl sulfoxide (DMSO) did not exhibit significant degradation under the same conditions [79]. Outgassing rates are accelerated when perovskite devices are illuminated in vacuum, as they would be in normal operation. Gunasekaran et al. conclude that illumination is not necessary for vacuum exposure to result in structural degradation and phase segregation in perovskites. Even in dark vacuum, metallic traps and mobile I_2^- ions decompose from PbI_2 and are critical to the self-degradation of perovskite films. The I_2^- ions that diffuse to the surface of the perovskites react with the metal electrode to form AuI_2^- ions. The resulting surface traps at the interface degrade the perovskite stability

[79]. It is crucial that future perovskite devices designed for operation in space have minimal ion migration and outgassing issues. Perovskites for space must also incorporate encapsulation techniques that can substantially thwart vacuum-induced degradation.

5.2.7 Photocurrent density voltage hysteresis

The photocurrent density voltage (*J-V*) responses of perovskite solar cells often exhibit a hysteresis rarely observed in other photovoltaic technologies [60]. Remarkably, perovskite devices often exhibit slow relaxation to steady state following an extrinsic excitation, which most notably manifests as *J-V* hysteresis where the forward-scanned *J-V* curve is different from the reverse scanned *J-V* curve. Typically, the reverse scan from V_{oc} to J_{sc} results in a higher measured power conversion efficiency than a forward scan from J_{sc} to V_{oc} condition. Thus, the reverse scan overestimates the PCE and the forward scan underestimates PCE [60,80–86]. Hysteretic *J-V* behavior demonstrates an anomalous dependence on the solar cell voltage scanning rate, direction, voltage range, precondition, and device architecture [20,87,88]. Mobile ionic charge is a fundamental contributor to hysteretic behavior. In $MAPbI_3$ perovskites, for example, the I^- anions and MA^+ cations are likely mobile and various other photo-induced ions are suspected within the device [88,89]. It is certain that the migration of ions from one side to another results in changes to charge distribution, capacitance, and internal electric field within the device, causing changes in *J-V* curves [88]. It is highly probable that the influence of ions on the conduction of holes and electrons in the perovskite absorber is a source of hysteresis [90]. This as-yet unexplained phenomenon translates to the inability for researchers to compare measurements from one laboratory to another, which ultimately impacts the perceived reproducibility of perovskite cell characteristics.

In Fig. 5.7A we see the efficiency under reverse scan increases with increased scan rate; that trend reverses, however, with an increased scan rate (and efficiency under forward scan decreases) [91]. Hysteretic behavior becomes more pronounced when scan rate is increased. The hysteresis may be eliminated if a sufficiently slow scan rate is implemented. The impact of scan range on *J-V* hysteresis is apparent in Fig. 5.7B and C. A negative scan with a more positive initial bias can increase the device performance (Fig. 5.7B) [92]. Alternatively, a forward scan with a more negative initial bias (from −0.2 to −1 V) deteriorates the measured efficiency, and the efficiency discrepancy increases from approximately 2.5%–7%, respectively (Fig. 5.7C) [93].

Preconditioning a device with light soaking at varying biases can have a significant impact on measured *J-V* performance [87]. After storage in the dark and prior to light soaking, Unger et al. observed that *J-V* curves for planar perovskite devices exhibit a slight S-shape and low performance. A light soak under far-forward bias conditions ($J > 0$) results in a significantly improved PCE via dramatically increased fill factor and photocurrent. This resulted in a PCE increase from 2% to 4%. Further device

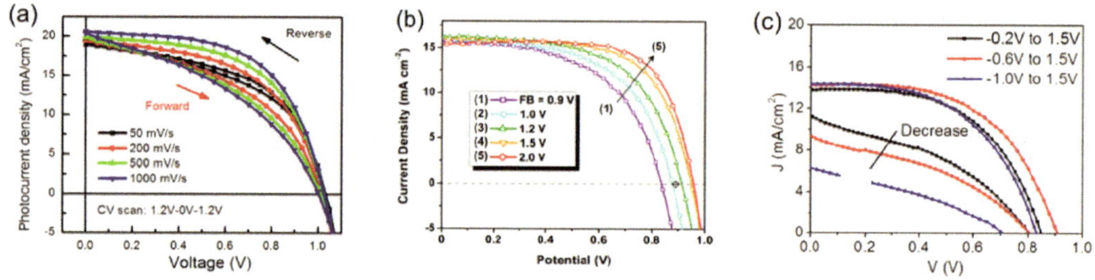

Figure 5.7 Dependence of the *J-V* hysteresis behavior on (A) scan rate, (B) positive starting bias under reverse scan, and (C) negative starting bias under forward scan. *Reproduced with permission from: (A) B. Chen, M. Yang, X. Zheng, C. Wu, W. Li, Y. Yan, J. Bisquert, G. Garcia-Belmonte, K. Zhu, S. Priya, Impact of capacitive effect and ion migration on the hysteretic behavior of perovskite solar cells, J. Phys. Chem. Lett. 6 (2015) 4693–4700, copyright (2015) American Chemical Society; (B) J.A. Christians, J.S. Manser, P.V. Kamat, Best practices in perovskite solar cell efficiency measurements. Avoiding the error of making bad cells look good, J. Phys. Chem. Lett. 6 (2015) 852–857, copyright (2015) American Chemical Society; (C) J. Wei, Y. Zhao, H. Li, G. Li, J. Pan, D. Xu, Q. Zhao, D. Yu, Hysteresis analysis based on the ferroelectric effect in hybrid perovskite solar cells, J. Phys. Chem. Lett. 5 (2014) 3937–3945, copyright (2014) American Chemical Society.*

improvement (PCE up to 7% for the device shown in Fig. 5.7D) was realized by cycling the device between forward bias and 0 V under illumination multiple times before performing the *J-V* measurement. Light soaking under a large reverse bias, flowing the current in the reverse direction, has proven to suppress photovoltaic PCE. Unger et al. also discovered that they could repeatedly improve or diminish device efficiency by cycling the voltage bias conditions during light soaking, indicating these processes are reversible. However, the results did not translate to mesoporous titania-based devices that performed independently of illumination and bias conditions prior to *J-V* measurements.

One meaningful alternative measurement technique is to compare the performance of perovskite absorber solar cells by measuring the power generated by the device over time [87]. In this method the efficiency can be derived by measuring the current over time at a constant potential close to the maximum power point determined from *I-V* measurements at slow scan rates. Unger et al. conducted this experiment on both mesoporous titania (mp-TiO_2) $MAPbI_3$ devices and planar thin-film $MAPbI_3$ devices on compact TiO_2 architectures. During cyclic illumination the mp-TiO_2 devices had a stable efficiency of 12%, which is comparable to the device performance data obtained from *I-V* measurements at long delay times of 5 s after each voltage step. A delay time of 5 s after each voltage step is equivalent to a scan rate of 10 mV/s, and the positive and negative device scans converge at an efficiency of 8.5% ± 0.1%, demonstrating suppression of hysteretic behavior. The cyclic illumination of thin-film devices gave a fairly consistent power conversion efficiency of close to 8%.

Electrical circuits exhibit transient responses when a sudden change in input alters the steady-state equilibrium of the circuit. The transient time is the response period in which the circuit settles from nonsteady state to a steady-state condition. Transient responses typically persist on a time scale on the order of 1 ms. In the specific case of a solar cell, the transient disturbance is caused by a change from illuminated to nonilluminated after a short light pulse. Once the light is switched off, the exponential decay of charge carriers occurs over time. The appropriate scan rate for *J*-*V* characterization of a solar cell depends on the time constant of transient effects in perovskite solar cells. Transient effects can lead to an overestimation of the device performance, resulting in a lower apparent photocurrent for the positive scan direction and a higher apparent photocurrent for the negative scan direction, if delay times are too short. The 10 mV/s deemed suitable for the aforementioned experimental devices [87] may need to be adjusted for application in other perovskite solar cells. For example, there is evidence suggesting that the hysteretic effects observed for thin-film planar perovskite devices hysteretic effects are caused by different mechanisms than the mp-TiO_2-based devices. Planar perovskite devices on compact TiO_2 often exhibit more severe hysteresis than perovskite solar cells on a mp-TiO_2 scaffold [20,87,93]. Mesoporous devices exhibit faster transients, likely because the titania scaffolding limits the growth and size of perovskite crystals and provides efficient pathways for electron extractions. The transient processes linked to hysteretic effects likely originate from processes in the bulk perovskite absorber; contacts, however, can influence the size and direction of the transients and hysteresis [87].

Recently, researchers suggested the presence of electronic defects also contribute to the changes in *J*-*V* response. Five energy states have been experimentally detected in CdTe cells at the *n*-CdTe/metal interface resulting in a ladder of defect levels [94]; this same phenomenon is expected in perovskites, but more defects at higher concentrations are expected due to low-temperature processing of perovskite materials. These defects can be in material layers and at interfaces. Differences in crystal lattice constants and electronic properties result in a higher concentration of defects at metal/semiconductor interfaces (i.e., contacts) than at semiconductor/semiconductor interfaces (Fig. 5.8) [88]. The location of these defect levels has not yet been identified for perovskites but the presence of slowly responding defects causes variability in *J*-*V* curves during measurements. More research needs to be conducted to fully understand the locations of electronic defects, transient processes, and their resulting hysteretic effects of perovskite solar cells toward the goal of accurate measurement and the long-term reliability of perovskite solar devices.

5.2.8 Arcing and charging

Up to this point in the chapter, we have primarily discussed the impacts of space on the cell-level design considerations; however, to scale up to arrays of cells, a primary

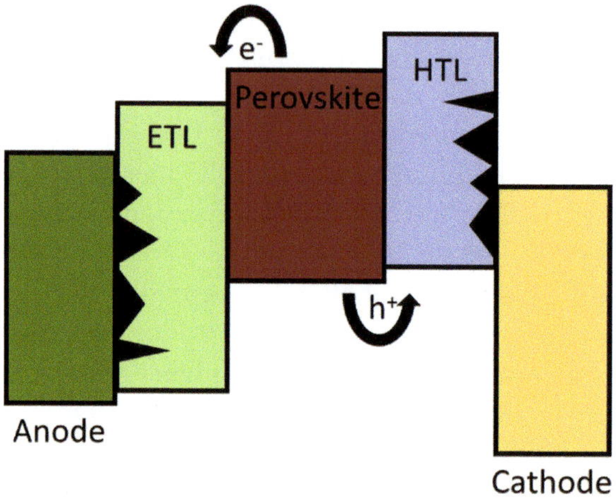

Figure 5.8 Energy diagram for a general perovskite solar cell showing possible ladders of defect levels, not yet experimentally observed, at the contact interfaces in the device. *Adapted from I.M. Dharmadasa, Y. Rahaq, A.E. Alam, Perovskite solar cells: Short lifetime and hysteresis behavior of current–voltage characteristics, J. Mater. Sci. Mater. Electron. 30 (2019) 12851–12859; this article is distributed under the terms of the Creative Commons Attribution 4.0 International License (http://creativecommons.org/licenses/by/4.0/).*

consideration in solar array design is the local charging environment to which it will be exposed. Most notably, the solar array voltage bus is determined by the short-circuit current and the charging environment of the intended application. The most significant and most thoroughly studied charging environments are low-Earth orbit (LEO) and geostationary (or geosynchronous) Earth orbit (GEO) [95]. In LEO the charged particle environment consists of a moderate density of particles ($\sim 10^6/\text{cm}^3$) in a quasineutral plasma that is thermalized with the environment and a mean energy of approximately 100 meV [96]. The plasma consists mostly of electrons and oxygen ions. The solar array is illuminated and creates a positive voltage, V_{BUS}, with respect to the chassis, and electrons are attracted to the array, while ions are swept up by the spacecraft chassis, generating a current to the plasma. Due to the much higher mobility of electrons compared to ions, the local floating potential of the solar array will become strongly negative (approximately 90% V_{BUS}), yet the insulating cover glass just 10 mils removed from that surface will be held at the plasma potential because of the density of the plasma [97]. The plasma potential is typically only about -1 V with respect to the plasma neutral voltage. Therefore, a differential voltage across the insulating gap of the solar cell exists, and it has been observed that potential differences of approximately 75 V lead to arc inception [98].

In GEO, the situation differs significantly; the plasma environment is nonneutral and contains very few ions [96]. The electrons are far less dense than in GEO, but significantly more energetic. There is a significant population density of electrons in the region of energies that are above the unity yield of secondary electron emission, thereby causing electrons to be emitted from the surface of the cover glasses, and leaving a more positive surface than the underlying solar array. The steady-state floating potential is determined where $I_e + I_i + I_{see}$ is equal to zero. In this *inverted gradient* condition, the differential voltage between the cover glass and the array may be hundreds of volts [99].

The arcs that are initiated by the high differential voltages on solar arrays create a very high-density plasma for a short period of time, locally. This local plasma will expand and drain static charge from a potentially wide region, the equivalent of a few microfarads of capacitance, and it generally lasts for a few tens of microseconds [100]. If this plasma region includes solar cells that are at a voltage difference to each other so the local electric field between them is coupled to this plasma, then a secondary arc can be formed between the two cells, and that arc is a DC arc that is fed by the photogenerated current of the solar cell. This powerful arc may have a current sufficient to cause pyrolysis of any insulator separating the two cells and will therefore not cease until the cells have been destroyed. This phenomena is known as a "sustained arc," and it is the dominant concern for LEO and GEO solar arrays in terms of designing the V_{BUS}, cell spacings, and configuration of the cell interconnects [101].

Much of the design constraints for solar arrays is dependent upon the rigid wafer-based geometry of the solar cells that are sourced to construct the arrays. W_e contrast this situation with the potential of perovskite thin-film solar arrays. Roll-to-roll fabrication of arrays has been demonstrated where the cells are deposited globally onto the substrate material and then cells and cell-level interconnects are formed by laser scribing processes [102]. In this scenario there are multiple benefits of scribed thin-film solar arrays to charging concerns: 1) the overall architecture of the cell is extremely thin and uniformly smooth, minimizing changes in the local electric field that leads to interconnect related sustained arcing; 2) the only geometric constraint on the system is the requirement that cells in the string source the same current; and 3) the shape of the cells can vary to accommodate any structural elements and provide designers freedom to avoid neighboring cells that exist at high potential to each other.

The thin-film solar array paradigm may also have disadvantages with respect to charging. Although it seems that sustained arcing can be avoided by judicious cell scribing, it is unclear if primary arcing can be avoided, particularly in LEO. Generally, primary arcs are considered nondamaging for existing solar arrays, but the reason for this consideration is unclear. It is likely that the energy involved in the discharge does not locally heat a wafer-based technology high enough to damage the *p-n* junction. A thin-film device on a light polymeric substrate, however, will have a very different thermal capacity and response to the discharge energy. Therefore, it is imperative that charge control and dissipation be included in any thin-film design. At NASA Glenn Research Center, we engage in charging and electrostatic discharge tests of space solar arrays in accordance with ISO 11221:2011 [74], and we can perform these tests on thin-film arrays.

5.3 Use of perovskites for space power: Issues and opportunities for improvement

Reproducibility, although imperative for widespread implementation of a novel energy conversion technology, is not yet resolved for perovskites. High reproducibility would be considered a narrow scatter of power conversion efficiency and solar cell parameters of perovskite solar cells that are processed under the same conditions [7]. Variation in film quality and device performance is often observed depending on the synthesis method and processing temperature [103,104]. Statistics are necessary to determine the reproducibility and device performance. Histograms of a large number of devices (30+) with device-to-device and batch-to-batch reproducibility statistics should be required to prove the desired device reproducibility [7].

As noted previously, perovskite materials are known to decompose via outgassing when exposed to vacuum. $MAPbI_3$ undergoes reversible and irreversible photodecomposition reactions. Organic gas components like $CH_3NH_2 + HI$ (reversible), $CH_3I + NH_3$ (irreversible), and $Pb + I_2$ (reversible photodecomposition) are liberated from $CH_3NH_3PbI_3$ ($MAPbI_3$) powder [105]. $MAPbI_3$ was found to liberate I_2 independent of light conditions. The effect of vacuum on the operational stability of perovskite solar cells remains largely uninvestigated. Selective contacts should be chemically inert toward the volatile released products, and a greater understanding of the reversible degradation routes, which can be leveraged as self-healing, is needed to prolong perovskite solar cell device lifetimes.

Sellers et al. conducted the first energy-independent investigation of perovskite devices irradiated with proton energies ranging from highly nuclear to highly ionizing. They illustrate the inherent resilience of perovskite devices compared to traditional photovoltaic technologies as degradation was not observed until a fluence of 1×10^{12} p/cm^2, mostly as result of decreased V_{oc} [106]. Kirmani et al. report that low-energy protons (0.05–0.15 MeV) are an optimum radiation source for creating atomic vacancies and are thus best suited for probing radiation effects in perovskites [107].

5.4 Conclusion

In this chapter, we have reviewed the challenges and opportunities for perovskite solar cell implementation in space. The primary challenges are its instabilities in vacuum, ultraviolet illumination, and thermal cycling. UV illumination instability can be addressed via implementation of UV filters or down conversion layers. Down conversion materials convert incident high-energy UV photons to visible light, which improves device power conversion efficiency and reduces resulting UV-induced damage [108,109]. Materials selection for perovskite devices will be critical to managing both thermal and vacuum stability. Device materials that are "resistant" to ion migration and phase separation will be crucial to achieving long-term operation of perovskites in space.

Presently, cesium-based perovskite absorbers are favored because they possess the best thermal stability and acceptable phase stability at low and high temperature extremes. The absorber selection will have to be resolved with considerations for radiation tolerance, and other device performance considerations relevant to long duration operation in space. There is room for development in perovskite device architectures for long duration operation in space.

Perovskites are also uniquely positioned as strong candidates for implementation in tandem devices. The combination of excellent photovoltaic properties, tunable bandgap in a wide range, and low-temperature processability make perovskite materials attractive for multijunction tandem cells. Thirty percent power conversion efficiency is anticipated for tandem devices with a radiative efficiency limit of around 42% determined for perovskite tandems under AM1.5 illumination. Wide-bandgap perovskite top cells and low-bandgap bottom cells including another perovskite (typically mixed tin-lead perovskite) or a crystalline silicon or copper indium gallium selenide bottom cell are all suitable materials for realizing the goal of low-cost tandems suitable for space. There are some technical challenges presented by the current technology that must be overcome to ensure their integration into such devices. These include understanding and mitigating the tendency for the metal contact to dope the perovskite or undergo redox chemistry. The interfacial layers need to ideally be defect free, compatible with the process methods of the rest of the device, and have the appropriate band alignment. Similarly, recombination layers need to have low optical and electronic losses and also exhibit solvent compatibility to processing techniques and be stable. The absorber layers sometimes fall prey to ion migration, phase segregation, or environmental-induced decomposition. It is attainable for the perovskite and space photovoltaic community to overcome these challenges to realize high-efficiency, stable perovskite tandem cells.

Demonstration of perovskite solar cells in space is ongoing. In 2021, McMillon-Brown and Peshek et al. launched perovskite solar cells for a 6-month flight in LEO via the 15th Materials International Space Station Experiment (MISSE-15). This is the first long-duration flight of a perovskite solar cell in LEO. McMillon-Brown and Peshek et al. have previously flown a $MAPbI_3$ thin-film encapsulated in DC 93−500 aboard MISSE-13. After 10 months of exposure to LEO, the sample exhibited no significant changes from exposure to the extreme environment of LEO [110]. Tu et al. flew an encapsulated perovskite solar cell with a UV filter for a 2-h high-altitude flight; the device retained 95% of its power conversion efficiency [111]. Cardinaletti et al. launched perovskite solar cells to 32 km altitude, and their performance was measured in situ. The cells did exhibit severe performance loss attributed to damaged encapsulation layers [112].

Perovskite devices demonstrate immense promise for large-area, high-voltage arrays and outer planetary SmallSat or CubeSat missions under low-intensity, low-temperature conditions. Preliminary investigations indicate that carefully selected and treated perovskite materials are capable of surviving in the space environment; in fact, it may be feasible to manufacture perovskite solar cells in space [113]. Therefore, perovskite materials are a promising technology solution to address the growing demand for low-cost, high-voltage, capable solar arrays for long-duration operation in space.

References

[1] A. Bermudez-Garcia, P. Voarino, Olivier Raccurt, Environments, needs and opportunities for future space photovoltaic power generation: A review, Appl. Energy 290 (2021) 116757.

[2] L. McMillon-Brown, K.M. Crowley, K.T. VanSant, T.J. Peshek, Prospects for perovskites in space, in: 2021 IEEE 48th Photovolt. Spec. Conf. PVSC, 2021, pp. 0222−0225, https://doi.org/10.1109/PVSC43889.2021.9518700.

[3] J.M. Frost, K.T. Butler, F. Brivio, C.H. Hendon, M. van Schilfgaarde, A. Walsh, Atomistic origins of high-performance in hybrid halide perovskite solar cells, Nano Lett. 14 (2014) 2584−2590, https://doi.org/10.1021/nl500390f.

[4] M.F. Ayguler, M.D. Weber, B.M.D. Puscher, D.D. Medina, P. Docampo, R.D. Costa, Light-emitting electrochemical cells based on hybrid lead halide perovskite nanoparticles, J. Phys. Chem. C 119 (2015) 12047−12054.

[5] J. Wang, N. Wang, Y. Jin, J. Si, Z.-K. Tan, H. Du, L. Cheng, et al., Interfacial control toward efficient and low-voltage perovskite light-emitting diodes, Adv. Mater. 27 (2015) 2311−2316, https://doi.org/10.1002/adma.201405217.

[6] A. Sadhanala, S. Ahmad, B. Zhao, N. Giesbrecht, P.M. Pearce, F. Deschler, R.L.Z. Hoye, et al., Blue-green color tunable solution processable organolead chloride−bromide mixed halide perovskites for optoelectronic applications, Nano Lett. 15 (9) (2015) 6095−6101. https://pubs.acs.org/doi/abs/10.1021/acs.nanolett.5b02369.

[7] M.L. Petrus, J. Schlipf, C. Li, T.P. Gujar, N. Giesbrecht, P. Müller-Buschbaum, M. Thelakkat, T. Bein, S. Hüttner, P. Docampo, Capturing the sun: A review of the challenges and perspectives of perovskite solar cells, Adv. Energy Mater. 7 (2017) 1700264, https://doi.org/10.1002/aenm.201700264.

[8] P. Docampo, F.C. Hanusch, S.D. Stranks, M. Döblinger, J.M. Feckl, M. Ehrensperger, N.K. Minar, M.B. Johnston, H.J. Snaith, T. Bein, Solution deposition-conversion for planar heterojunction mixed halide perovskite solar cells, Adv. Energy Mater. 4 (2014) 1400355, https://doi.org/10.1002/aenm.201400355.

[9] N.J. Jeon, J.H. Noh, Y.C. Kim, W.S. Yang, S. Ryu, S.I. Seok, Solvent engineering for high-performance inorganic−organic hybrid perovskite solar cells, Nat. Mater. 13 (2014) 897−903, https://doi.org/10.1038/nmat4014.

[10] M.M. Lee, J. Teuscher, T. Miyasaka, T.N. Murakami, H.J. Snaith, Efficient hybrid solar cells based on meso-superstructured organometal halide perovskites, Science 338 (2012) 643−647.

[11] H.-S. Kim, C.-R. Lee, J.-H. Im, K.-B. Lee, T. Moehl, A. Marchioro, S.-J. Moon, R. Humphry-Baker, J.-H. Yum, J.E. Moser, Lead iodide perovskite sensitized all-solid-state submicron thin film mesoscopic solar cell with efficiency exceeding 9%, Sci. Rep. 2 (2012) 591.

[12] M. Grätzel, The light and shade of perovskite solar cells, Nat. Mater. 13 (2014) 838.

[13] G. Xing, N. Mathews, S. Sun, S.S. Lim, Y.M. Lam, M. Grätzel, S. Mhaisalkar, T.C. Sum, Long-range balanced electron- and hole-transport lengths in organic-inorganic $CH_3NH_3PbI_3$, Science 342 (6156) (October 2013) 344−347. https://www.science.org/doi/abs/10.1126/science.1243167.

[14] S.D. Stranks, G.E. Eperon, G. Grancini, C. Menelaou, M.J.P. Alcocer, T. Leijtens, L.M. Herz, A. Petrozza, H.J. Snaith, Electron-hole diffusion lengths exceeding 1 micrometer in an organometal trihalide perovskite absorber, Science 342 (2013) 341−344, https://doi.org/10.1126/science.1243982.

[15] W.-J. Yin, T. Shi, Y. Yan, Unique properties of halide perovskites as possible origins of the superior solar cell performance, Adv. Mater. 26 (2014) 4653−4658, https://doi.org/10.1002/adma.201306281.

[16] K. Bush, K. Frohna, R. Prasanna, R. Beal, T. Leijtens, S. Swifter, M. McGehee, Compositional engineering for efficient wide band gap perovskites with improved stability to photoinduced phase segregation, ACS Energy Lett. 3 (2018) 428−435.

[17] H. Gaonkar, J. Zhu, R. Kottokkaran, B. Bhageri, M. Noack, V. Dalal, Thermally stable, efficient, vapor deposited inorganic perovskite solar cells, ACS Appl. Energy Mater. 3 (2020) 3497−3503, https://doi.org/10.1021/acsaem.0c00010.

[18] R. Kottokkaran, H.A. Gaonkar, H.A. Abbas, M. Noack, V. Dalal, Performance and stability of co-evaporated vapor deposited perovskite solar cells, J. Mater. Sci. Mater. Electron. 30 (2019) 5487−5494, https://doi.org/10.1007/s10854-019-00842-y.

[19] L.E. Polander, P. Pahner, M. Schwarze, M. Saalfrank, C. Koerner, K. Leo, Hole-transport material variation in fully vacuum deposited perovskite solar cells, Apl. Mater. 2 (2014) 081503, https://doi.org/10.1063/1.4889843.

[20] B. Chen, M. Yang, S. Priya, K. Zhu, Origin of J−V hysteresis in perovskite solar cells, J. Phys. Chem. Lett. 7 (2016) 905−917, https://doi.org/10.1021/acs.jpclett.6b00215.

[21] T. Chien Sum, N. Mathews, Advancements in perovskite solar cells: Photophysics behind the photovoltaics, Energy Environ. Sci. 7 (2014) 2518−2534, https://doi.org/10.1039/C4EE00673A.

[22] S. Sun, T. Salim, N. Mathews, M. Duchamp, C. Boothroyd, G. Xing, T. Chien Sum, Y. Ming Lam, The origin of high efficiency in low-temperature solution-processable bilayer organometal halide hybrid solar cells, Energy Environ. Sci. 7 (2014) 399−407, https://doi.org/10.1039/C3EE43161D.

[23] Q. Dong, Y. Fang, Y. Shao, P. Mulligan, J. Qiu, L. Cao, J. Huang, Electron-hole diffusion lengths > 175 μm in solution-grown $CH_3NH_3PbI_3$ single crystals, Science 347 (2015) 967−970, https://doi.org/10.1126/science.aaa5760.

[24] X. Zhao, N.-G. Park, Stability issues on perovskite solar cells, Photonics 2 (2015) 1139−1151, https://doi.org/10.3390/photonics2041139.

[25] M.T. Hörantner, T. Leijtens, M.E. Ziffer, G.E. Eperon, M.G. Christoforo, M.D. McGehee, H.J. Snaith, The potential of multijunction perovskite solar cells, ACS Energy Lett. 2 (2017) 2506−2513, https://doi.org/10.1021/acsenergylett.7b00647.

[26] T.C.-J. Yang, P. Fiala, Q. Jeangros, C. Ballif, High-bandgap perovskite materials for multijunction solar cells, Joule 2 (2018) 1421−1436, https://doi.org/10.1016/j.joule.2018.05.008.

[27] M. Filipič, P. Löper, B. Niesen, S. De Wolf, J. Krč, C. Ballif, M. Topič, CH_3NH_3PbI_3 perovskite/silicon tandem solar cells: Characterization based optical simulations, Opt Express 23 (2015) A263, https://doi.org/10.1364/OE.23.00A263.

[28] J. Berry, T. Buonassisi, D.A. Egger, G. Hodes, L. Kronik, Y.-L. Loo, I. Lubomirsky, S.R. Marder, Y. Mastai, J.S. Miller, D.B. Mitzi, Y. Paz, A.M. Rappe, I. Riess, B. Rybtchinski, O. Stafsudd, V. Stevanovic, M.F. Toney, D. Zitoun, A. Kahn, D. Ginley, D. Cahen, Hybrid organic−inorganic perovskites (HOIPs): Opportunities and challenges, Adv. Mater. 27 (2015) 5102−5112, https://doi.org/10.1002/adma.201502294.

[29] J. Kang, L.-W. Wang, High defect tolerance in lead halide perovskite CsPbBr3, J. Phys. Chem. Lett. 8 (2017) 489−493, https://doi.org/10.1021/acs.jpclett.6b02800.

[30] E.J. Daly, G. Drolshagen, A. Hilgers, H.D.R. Evans, Space Environment Analysis: Experience and Trends, European Space Agency, (Special Publication) ESA SP Issue 392, 1996, pp. 15−22.

[31] Standard: Qualification and Quality Requirements for Space Solar Cells (AIAA S-111A-2014), American Institute of Aeronautics and Astronautics, Inc., Washington, DC, 2014, https://doi.org/10.2514/4.102806.

[32] Y. Miyazawa, M. Ikegami, H.-W. Chen, T. Ohshima, M. Imaizumi, K. Hirose, T. Miyasaka, Tolerance of perovskite solar cell to high-energy particle irradiations in space environment, iScience 2 (2018) 148−155, https://doi.org/10.1016/j.isci.2018.03.020.

[33] Y. Miyazawa, M. Ikegami, T. Miyasaka, T. Ohshima, M. Imaizumi, K. Hirose, Evaluation of radiation tolerance of perovskite solar cell for use in space, in: Proc. IEEE 42nd Photovolt. Spec. Conf, 2015, pp. 1−4, https://doi.org/10.1109/PVSC.2015.7355859.

[34] J. Huang, M.D. Kelzenberg, P. Espinet-González, C. Mann, D. Walker, A. Naqavi, N. Vaidya, E. Warmann, H.A. Atwater, Effects of electron and proton radiation on perovskite solar cells for space solar power application, in: Proc. IEEE 44th Photovolt. Spec. Conf. PVSC, 2017, pp. 1248−1252, https://doi.org/10.1109/PVSC.2017.8366410.

[35] F. Lang, N.H. Nickel, J. Bundesmann, S. Seidel, A. Denker, S. Albrecht, V.V. Brus, J. Rappich, B. Rech, G. Landi, H.C. Neitzert, Radiation hardness and self-healing of perovskite solar cells, Adv. Mater. 28 (2016) 8726−8731, https://doi.org/10.1002/adma.201603326.

[36] V.V. Brus, F. Lang, J. Bundesmann, S. Seidel, A. Denker, B. Rech, G. Landi, H.C. Neitzert, J. Rappich, N.H. Nickel, Defect dynamics in proton irradiated $CH_3NH_3PbI_3$ perovskite solar cells, Adv. Electron. Mater. 3 (2017) 1600438, https://doi.org/10.1002/aelm.201600438.

[37] J. Barbé, D. Hughes, Z. Wei, A. Pockett, H.K.H. Lee, K.C. Heasman, M.J. Carnie, T.M. Watson, W.C. Tsoi, Radiation hardness of perovskite solar cells based on aluminum-doped zinc oxide electrode under proton irradiation, Sol. RRL. 3 (2019) 1900219, https://doi.org/10.1002/solr.201900219.

[38] F. Lang, M. Jošt, J. Bundesmann, A. Denker, S. Albrecht, G. Landi, H.-C. Neitzert, J. Rappich, N.H. Nickel, Efficient minority carrier detrapping mediating the radiation hardness of triple-cation perovskite solar cells under proton irradiation, Energy Environ. Sci. 12 (2019) 1634−1647, https://doi.org/10.1039/C9EE00077A.

[39] F. Lang, M. Jošt, K. Frohna, E. Köhnen, A. Al-Ashouri, A.R. Bowman, T. Bertram, et al., Proton radiation hardness of perovskite tandem photovoltaics, Joule 4 (2020) 1054−1069, https://doi.org/10.1016/j.joule.2020.03.006.

[40] O. Malinkiewicz, A. Yella, Y.H. Lee, G.M. Espallargas, M. Graetzel, M.K. Nazeeruddin, H.J. Bolink, Perovskite solar cells employing organic charge-transport layers, Nat. Photonics 8 (2014) 128−132, https://doi.org/10.1038/nphoton.2013.341.

[41] G.M. Paternò, V. Robbiano, L. Santarelli, A. Zampetti, C. Cazzaniga, V.G. Sakai, F. Cacialli, Perovskite solar cell resilience to fast neutrons, Sustain. Energy Fuels 3 (2019) 2561−2566, https://doi.org/10.1039/C9SE00102F.

[42] S. Yang, Z. Xu, S. Xue, P. Kandlakunta, L. Cao, J. Huang, Organohalide lead perovskites: more stable than glass under gamma-ray radiation, Adv. Mater. 31 (2019) 1805547, https://doi.org/10.1002/adma.201805547.

[43] K. Yang, K. Huang, X. Li, S. Zheng, P. Hou, J. Wang, H. Guo, et al., Radiation tolerance of perovskite solar cells under gamma ray, Org. Electron. 71 (2019) 79−84, https://doi.org/10.1016/j.orgel.2019.05.008.

[44] A.G. Boldyreva, A.F. Akbulatov, S.A. Tsarev, S.Y. Luchkin, I.S. Zhidkov, E.Z. Kurmaev, K.J. Stevenson, V.G. Petrov, P.A. Troshin, γ-Ray-Induced degradation in the triple-cation perovskite solar cells, J. Phys. Chem. Lett. 10 (2019) 813−818, https://doi.org/10.1021/acs.jpclett.8b03222.

[45] A.G. Boldyreva, L.A. Frolova, I.S. Zhidkov, L.G. Gutsev, E.Z. Kurmaev, B.R. Ramachandran, V.G. Petrov, K.J. Stevenson, S.M. Aldoshin, P.A. Troshin, Unravelling the material composition effects on the gamma ray stability of lead halide perovskite solar cells: $MAPbI_3$ breaks the records, J. Phys. Chem. Lett. 11 (2020) 2630−2636, https://doi.org/10.1021/acs.jpclett.0c00581.

[46] M. Yamaguchi, Radiation-resistant solar cells for space use, Sol. Energy Mater. Sol. Cells 68 (2001) 31−53, https://doi.org/10.1016/S0927-0248(00)00344-5.

[47] O. Malinkiewicz, M. Imaizumi, S.B. Sapkota, T. Ohshima, S. Öz, Radiation effects on the performance of flexible perovskite solar cells for space applications, Emergent Mater 3 (2020) 9−14, https://doi.org/10.1007/s42247-020-00071-8.

[48] R.A. Street, J.E. Northrup, B.S. Krusor, Radiation induced recombination centers in organic solar cells, Phys. Rev. B 85 (2012) 205211, https://doi.org/10.1103/PhysRevB.85.205211.

[49] F. Lang, O. Shargaieva, V.V. Brus, H.C. Neitzert, J. Rappich, N.H. Nickel, Influence of radiation on the properties and the stability of hybrid perovskites, Adv. Mater. 30 (2018) 1702905, https://doi.org/10.1002/adma.201702905.

[50] S.R. Messenger, E.A. Burke, R.J. Walters, J.H. Warner, G.P. Summers, Using SRIM to calculate the relative damage coefficients for solar cells, Prog. Photovoltaics Res. Appl. 13 (2005) 115–123, https://doi.org/10.1002/pip.608.

[51] E.R. Benton, E.V. Benton, Space radiation dosimetry in low-Earth orbit and beyond, Nucl. Instrum. Methods Phys. Res. Sect. B Beam Interact. Mater. Atoms 184 (2001) 255–294, https://doi.org/10.1016/S0168-583X(01)00748-0.

[52] A. Farooq, I.M. Hossain, S. Moghadamzadeh, J.A. Schwenzer, T. Abzieher, B.S. Richards, E. Klampaftis, U.W. Paetzold, Spectral dependence of degradation under ultraviolet light in perovskite solar cells, ACS Appl. Mater. Interfaces 10 (2018) 21985–21990, https://doi.org/10.1021/acsami.8b03024.

[53] S. Guarnera, A. Abate, W. Zhang, J.M. Foster, G. Richardson, A. Petrozza, H.J. Snaith, Improving the long-term stability of perovskite solar cells with a porous Al2O3 buffer layer, J. Phys. Chem. Lett. 6 (3) (February 2015) 432–437. https://pubs.acs.org/doi/abs/10.1021/jz502703p.

[54] D. Wei, T. Wang, J. Ji, M. Li, P. Cui, Y. Li, G. Li, J.M. Mbengue, D. Song, Photo-induced degradation of lead halide perovskite solar cells caused by the hole transport layer/metal electrode interface, J. Mater. Chem. 4 (2016) 1991–1998, https://doi.org/10.1039/C5TA08622A.

[55] A.K. Jena, A. Kulkarni, M. Ikegami, T. Miyasaka, Steady state performance, photo-induced performance degradation and their relation to transient hysteresis in perovskite solar cells, J. Power Sources 309 (2016) 1–10, https://doi.org/10.1016/j.jpowsour.2016.01.094.

[56] S.-W. Lee, S. Kim, S. Bae, K. Cho, T. Chung, L.E. Mundt, S. Lee, et al., UV degradation and recovery of perovskite solar cells, Sci. Rep. 6 (2016) 38150, https://doi.org/10.1038/srep38150.

[57] K. Domanski, E.A. Alharbi, A. Hagfeldt, M. Grätzel, W. Tress, Systematic investigation of the impact of operation conditions on the degradation behaviour of perovskite solar cells, Nat. Energy 3 (2018) 61–67, https://doi.org/10.1038/s41560-017-0060-5.

[58] W. Li, W. Zhang, S.V. Reenen, R.J. Sutton, J. Fan, A.A. Haghighirad, M.B. Johnston, L. Wang, H.J. Snaith, Enhanced UV-light stability of planar heterojunction perovskite solar cells with caesium bromide interface modification, Energy Environ. Sci. 9 (2016) 490–498, https://doi.org/10.1039/C5EE03522H.

[59] G. Niu, W. Li, F. Meng, L. Wang, H. Dong, Y. Qiu, Study on the stability of $CH_3NH_3PbI_3$ films and the effect of post-modification by aluminum oxide in all-solid-state hybrid solar cells, J. Mater. Chem. A. 2 (2013) 705–710, https://doi.org/10.1039/C3TA13606J.

[60] H.J. Snaith, A. Abate, J.M. Ball, G.E. Eperon, T. Leijtens, N.K. Noel, S.D. Stranks, J.T.-W. Wang, K. Wojciechowski, W. Zhang, Anomalous hysteresis in perovskite solar cells, J. Phys. Chem. Lett. 5 (2014) 1511–1515, https://doi.org/10.1021/jz500113x.

[61] Y. Shao, Z. Xiao, C. Bi, Y. Yuan, J. Huang, Origin and elimination of photocurrent hysteresis by fullerene passivation in CH 3 NH 3 PbI 3 planar heterojunction solar cells, Nat, Commun. Now. 5 (2014) 5784, https://doi.org/10.1038/ncomms6784.

[62] C. Qin, T. Matsushima, T. Fujihara, W.J. Potscavage, C. Adachi, Degradation mechanisms of solution-processed planar perovskite solar cells: Thermally stimulated current measurement for analysis of carrier traps, Adv. Mater. 28 (2016) 466–471, https://doi.org/10.1002/adma.201502610.

[63] G. Niu, X. Guo, L. Wang, Review of recent progress in chemical stability of perovskite solar cells, J. Mater. Chem. A. 3 (2015) 8970–8980, https://doi.org/10.1039/C4TA04994B.

[64] S.K. Pathak, A. Abate, P. Ruckdeschel, B. Roose, K.C. Gödel, Y. Vaynzof, A. Santhala, S.-I. Watanabe, D.J. Hollman, N. Noel, A. Sepe, U. Wiesner, R. Friend, H.J. Snaith, U. Steiner, Performance and stability enhancement of dye-sensitized and perovskite solar cells by Al doping of TiO_2, Adv. Funct. Mater. 24 (2014) 6046–6055, https://doi.org/10.1002/adfm.201401658.

[65] T. Leijtens, G.E. Eperon, S. Pathak, A. Abate, M.M. Lee, H.J. Snaith, Overcoming ultraviolet light instability of sensitized TiO_2 with meso-superstructured organometal tri-halide perovskite solar cells, Nat. Commun. 4 (2013) 2885, https://doi.org/10.1038/ncomms3885.

[66] J. Ren, Q. Luo, Q. Hou, H. Chen, T. Liu, H. He, J. Wang, Q. Shao, M. Dong, S. Wu, N. Wang, J. Lin, Z. Guo, Suppressing charge recombination and ultraviolet light degradation of perovskite solar cells using silicon oxide passivation, Chemelectrochem 6 (2019) 3167–3174, https://doi.org/10.1002/celc.201900688.

[67] N. Chander, A.F. Khan, P.S. Chandrasekhar, E. Thouti, S.K. Swami, V. Dutta, V.K. Komarala, Reduced ultraviolet light induced degradation and enhanced light harvesting using YVO_4:Eu^{3+} down-shifting nano-phosphor layer in organometal halide perovskite solar cells, Appl. Phys. Lett. 105 (2014) 033904, https://doi.org/10.1063/1.4891181.

[68] Z. Song, S.C. Watthage, A.B. Phillips, M.J. Heben, Pathways toward high-performance perovskite solar cells: review of recent advances in organo-metal halide perovskites for photovoltaic applications, J. Photon. Energy 6 (2) (2016) 022001, https://doi.org/10.1117/1.JPE.6.022001.

[69] J.H. Noh, S.H. Im, J.H. Heo, T.N. Mandal, S.I. Seok, Chemical management for colorful, efficient, and stable inorganic–organic hybrid nanostructured solar cells, Nano Lett. 13 (2013) 1764–1769, https://doi.org/10.1021/nl400349b.

[70] S.N. Habisreutinger, T. Leijtens, G.E. Eperon, S.D. Stranks, R.J. Nicholas, H.J. Snaith, Carbon nanotube/polymer composites as a highly stable hole collection layer in perovskite solar cells, Nano Lett. 14 (2014) 5561–5568, https://doi.org/10.1021/nl501982b.

[71] X. Li, M. Tschumi, H. Han, S.S. Babkair, R.A. Alzubaydi, A.A. Ansari, S.S. Habib, M.K. Nazeeruddin, S.M. Zakeeruddin, M. Grätzel, Outdoor performance and stability under elevated temperatures and long-term light soaking of triple-layer mesoporous perovskite photovoltaics, Energy Technol. 3 (2015) 551–555, https://doi.org/10.1002/ente.201500045.

[72] T. Leijtens, G.E. Eperon, N.K. Noel, S.N. Habisreutinger, A. Petrozza, H.J. Snaith, Stability of metal halide perovskite solar cells, Adv. Energy Mater. 5 (2015) 1500963, https://doi.org/10.1002/aenm.201500963.

[73] J. Yang, Q. Bao, L. Shen, L. Ding, Potential applications for perovskite solar cells in space, Nano Energy 76 (2020) 105019, https://doi.org/10.1016/j.nanoen.2020.105019.

[74] J. Yang, Q. Hong, Z. Yuan, R. Xu, X. Guo, S. Xiong, X. Liu, et al., Unraveling photostability of mixed cation perovskite films in extreme environment, Adv. Opt. Mater. 6 (2018) 1800262, https://doi.org/10.1002/adom.201800262.

[75] Y. Jiang, S.-C. Yang, Q. Jeangros, S. Pisoni, T. Moser, S. Buecheler, A.N. Tiwari, F. Fu, Mitigation of vacuum and illumination-induced degradation in perovskite solar cells by structure engineering, Joule 4 (2020) 1087–1103, https://doi.org/10.1016/j.joule.2020.03.017.

[76] B.D. Green, Satellite Contamination and Materials Outgassing Knowledgebase: An Interactive Database Reference, National Aeronautics and Space Administration, Marshall Space Flight Center, March 2001. Contract NAS8-98215, NASA/CR-2001-210909.

[77] M.M. Finckenor, K. de Groh, A researcher's guide to space environmental effects, Natl. Aeronaut. Space Adm. Int. Space Stn. Res. Guide Ser. NP-2015–03-015-ISC (2017) 15.

[78] E. Smecca, Y. Numata, I. Deretzis, G. Pellegrino, S. Boninelli, T. Miyasaka, A.L. Magna, A. Alberti, Stability of solution-processed $MAPbI_3$ and $FAPbI_3$ layers, Phys. Chem. Chem. Phys. 18 (2016) 13413–13422, https://doi.org/10.1039/C6CP00721J.

[79] R.K. Gunasekaran, D. Chinnadurai, A.R. Selvaraj, R. Rajendiran, K. Senthil, K. Prabakar, Revealing the self-degradation mechanisms in methylammonium lead iodide perovskites in dark and vacuum, ChemPhysChem 19 (2018) 1507–1513, https://doi.org/10.1002/cphc.201800002.

[80] S. Meloni, T. Möhl, W. Tress, M. Franckevičius, M. Saliba, Y.H. Lee, P. Gao, Ionic polarization-induced current–voltage hysteresis in CH3NH3PbX3 perovskite solar cells, Nature Commun 7 (2016) 10334. https://www.nature.com/articles/ncomms10334.

[81] C. Xia, W.-D. Song, C.-Z. Zhang, S.-Y. Yuan, W.-X. Hu, P. Qin, R.-P. Wang, L.-L. Zhao, X.-F. Wang, M. He, S.-T. Li, Alleviating hysteresis and improving device stability of perovskite solar cells via alternate voltage sweeps, Chin. Phys. B 26 (2017) 018401, https://doi.org/10.1088/1674-1056/26/1/018401.

[82] K. Ankireddy, A.H. Ghahremani, B. Martin, G. Gupta, T. Druffel, Rapid thermal annealing of $CH_3NH_3PbI_3$ perovskite thin films by intense pulsed light with aid of diiodomethane additive, J. Mater. Chem. A. 6 (2018) 9378–9383, https://doi.org/10.1039/C8TA01237G.

[83] M.F. Aygüler, A.G. Hufnagel, P. Rieder, M. Wussler, W. Jägermann, T. Bein, V. Dyakonov, M.L. Petrus, A. Baumann, P. Docampo, Influence of fermi level alignment with tin oxide on the hysteresis of perovskite solar cells, ACS Appl. Mater. Interfaces 10 (14) (2018) 11414–11419. https://pubs.acs.org/doi/abs/10.1021/acsami.8b00990.

[84] H.-S. Kim, N.-G. Park, Parameters affecting I–V hysteresis of $CH_3NH_3PbI_3$ perovskite solar cells: Effects of perovskite crystal size and mesoporous TiO_2 layer, J. Phys. Chem. Lett. 5 (17) (2014). https://pubs.acs.org/doi/abs/10.1021/jz501392m.

[85] S.-H. Turren-Cruz, M. Saliba, M.T. Mayer, H. Juárez-Santiesteban, X. Mathew, L. Nienhaus, W. Tress, et al., Enhanced charge carrier mobility and lifetime suppress hysteresis and improve efficiency in planar perovskite solar cells, Energy Environ. Sci. 11 (2018) 78–86, https://doi.org/10.1039/C7EE02901B.

[86] W. Tress, J.P. Correa Baena, M. Saliba, A. Abate, M. Grätzel, Inverted current–voltage hysteresis in mixed perovskite solar cells: Polarization, energy barriers, and defect recombination, Adv. Energy Mater. 6 (19) (2016) 1600396. https://onlinelibrary.wiley.com/doi/abs/10.1002/aenm.201600396.

[87] E.L. Unger, E.T. Hoke, C.D. Bailie, W.H. Nguyen, A.R. Bowring, T. Heumüller, M.G. Christoforo, M.D. McGehee, Hysteresis and transient behavior in current–voltage measurements of hybrid-perovskite absorber solar cells, Energy Environ. Sci. 7 (2014) 3690–3698, https://doi.org/10.1039/C4EE02465F.

[88] I.M. Dharmadasa, Y. Rahaq, A.E. Alam, Perovskite solar cells: Short lifetime and hysteresis behaviour of current–voltage characteristics, J. Mater. Sci. Mater. Electron. 30 (2019) 12851–12859, https://doi.org/10.1007/s10854-019-01759-2.

[89] C. Earnes, J.M. Frost, P.R.F. Barnes, B.C. O'Regan, A. Walsh, M.S. Islam, Ionic transport in hybrid lead iodide perovskite solar cells, Nat. Commun. 6 (June 2015) 7497, https://doi.org/10.1038/ncomms8497.

[90] D. Walter, A. Fell, Y. Wu, T. Duong, C. Barugkin, N. Wu, T. White, K. Weber, Transient photovoltage in perovskite solar cells: Interaction of trap-mediated recombination and migration of multiple ionic species, J. Phys. Chem. C 122 (2018) 11270–11281, https://doi.org/10.1021/acs.jpcc.8b02529.

[91] B. Chen, M. Yang, X. Zheng, C. Wu, W. Li, Y. Yan, J. Bisquert, G. Garcia-Belmonte, K. Zhu, S. Priya, Impact of capacitive effect and ion migration on the hysteretic behavior of perovskite solar cells, J. Phys. Chem. Lett. 6 (2015) 4693–4700, https://doi.org/10.1021/acs.jpclett.5b02229. Available from:.

[92] J.A. Christians, J.S. Manser, P.V. Kamat, Best practices in perovskite solar cell efficiency measurements. Avoiding the error of making bad cells look good, J. Phys. Chem. Lett. 6 (2015) 852–857, https://doi.org/10.1021/acs.jpclett.5b00289.

[93] J. Wei, Y. Zhao, H. Li, G. Li, J. Pan, D. Xu, Q. Zhao, D. Yu, Hysteresis analysis based on the ferroelectric effect in hybrid perovskite solar cells, J. Phys. Chem. Lett. 5 (2014) 3937–3945, https://doi.org/10.1021/jz502111u.

[94] I.M. Dharmadasa, Recent developments and progress on electrical contacts to CdTe, CdS and ZnSe with special reference to barrier contacts to CdTe, Prog. Cryst. Growth Char. Mater. 36 (1998) 249–290, https://doi.org/10.1016/S0960-8974(98)00010-2.

[95] D.E. Hastings, M. Cho, H. Kuninaka, Arcing rates for high voltage solar arrays - theory, experiment, and predictions, J. Spacecraft Rockets 29 (1992) 538–554, https://doi.org/10.2514/3.25497. Avaiable from:.

[96] D.C. Ferguson, Solar array arcing in plasmas, in: Proc. Third Annual Workshop on Space Operations Automation and Robotics, NASA Johnson Space Center, Houston, TX, USA, CP-N90-25558, 1990, pp. 509–513, available at: https://ntrs.nasa.gov/citations/19900016242.

[97] D.C. Ferguson, G.B. Hillard, New NASA SEE LEO Spacecraft Charging Design Guidelines: How to Survive in LEO rather than GEO, NASA/TM—2003-212737, December 2003 available at: https://ntrs.nasa.gov/citations/20040026971.

[98] D.E. Hastings, G. Weyl, D. Kaufman, Threshold voltage for arcing on negatively biased solar arrays, J. Spacecraft Rockets 27 (1990) 539–544, https://doi.org/10.2514/3.26177.

[99] D. Ferguson, D. Snyder, B. Vayner, J. Galofaro, Array arcing in orbit - from LEO to GEO, in: 37th Aerosp. Sci. Meet. Exhib, American Institute of Aeronautics and Astronautics, Reno, NV, USA, 1999, https://doi.org/10.2514/6.1999-218.

[100] B. Vayner, Physical Principles of Flashover Current Pulse Formation, American Institute of Aeronautics and Astronautics, 2011, https://doi.org/10.2514/6.2011-3976.

[101] B. Vayner, J.T. Galofaro, Inception and prevention of sustained discharges on solar arrays, IEEE Trans. Plasma Sci. 40 (2012) 388–393, https://doi.org/10.1109/TPS.2011.2177480.

[102] B. Dou, J.B. Whitaker, K. Bruening, D.T. Moore, L.M. Wheeler, J. Ryter, N.J. Breslin, et al., Roll-to-roll printing of perovskite solar cells, ACS Energy Lett. 3 (10) (2018) 2558–2565, https://doi.org/10.1021/acsenergylett.8b01556.

[103] F. Brivio, A.B. Walker, A. Walsh, Structural and electronic properties of hybrid perovskites for high-efficiency thin-film photovoltaics from first-principles, Apl. Mater. 1 (2013) 042111, https://doi.org/10.1063/1.4824147.

[104] Z. Cheng, J. Lin, Layered organic–inorganic hybrid perovskites: Structure, optical properties, film preparation, patterning and templating engineering, CrystEngComm 12 (2010) 2646–2662, https://doi.org/10.1039/C001929A.

[105] E.J. Juarez-Perez, L.K. Ono, M. Maeda, Y. Jiang, Z. Hawash, Y. Qi, Photodecomposition and thermal decomposition in methylammonium halide lead perovskites and inferred design principles to increase photovoltaic device stability, J. Mater. Chem. 6 (2018) 9604–9612, https://doi.org/10.1039/C8TA03501F.

[106] B. Durant, H. Afshari, S. Singh, B. Rout, G. Eperon, I. Sellers, Tolerance of perovskite solar cells to targeted proton irradiation and electronic ionization induced healing, ACS Energy Lett. 6 (2021) 2362–2368, https://doi.org/10.1021/acsenergylett.1c00756.

[107] A. Kirmani, B. Durant, J. Grandidier, N. Haegel, M. Kelzenberg, Y. Lao, M. McGehee, L. McMillon-Brown, D. Ostrowski, T. Peshek, B. Rout, I. Sellers, M. Steger, D. Walker, D. Wilt, K. VanSant, J. Luther, Countdown to perovskite space launch: Guidelines to performing relevant radiation-hardness experiments, Joule 6 (2022) 1–17, https://doi.org/10.1016/j.joule.2022.03.004.

[108] S.R. Pathipati, M.N. Shah, X. Pan, Interfacial engineering and down-conversion of ultraviolet light for efficient perovskite solar cells, Sol. Energy 197 (2020) 363–370, https://doi.org/10.1016/j.solener.2020.01.001.

[109] J. Jia, J. Dong, J. Lin, Z. Lan, L. Fan, J. Wu, Improved photovoltaic performance of perovskite solar cells by utilizing down-conversion $NaYF_4:Eu^{3+}$ nanophosphors, J. Mater. Chem. C 7 (2019) 937–942, https://doi.org/10.1039/C8TC05864D.

[110] T.J. Peshek, K.M. Crowley, W. Delmas, S. Ghosh, J.W. Williams, L. McMillon-Brown, On the performance of $MAPbI_3$ in the space environment, in: 2021 IEEE 48th Photovolt. Spec. Conf. PVSC, 2021, pp. 2611–2613, https://doi.org/10.1109/PVSC43889.2021.9518552.

[111] Y. Tu, G. Xu, X. Yang, Y. Zhang, Z. Li, R. Su, D. Luo, W. Yang, Y. Miao, R. Cai, Mixed-cation perovskite solar cells in space, Sci. China Phys. Mech. Astron. 62 (2019) 1–4.

[112] I. Cardinaletti, T. Vangerven, S. Nagels, R. Cornelissen, D. Schreurs, J. Hruby, J. Vodnik, et al., Organic and perovskite solar cells for space applications, Sol. Energy Mater. Sol. Cells 182 (2018) 121–127, https://doi.org/10.1016/j.solmat.2018.03.024.

[113] L. McMillon-Brown, J. Luther, T. Peshek, What would it take to manufacture perovskite solar cells in space? ACS Energy Lett. 7 (2022) 1040–1042, https://doi.org/10.1021/acsenergylett.2c00276.

CHAPTER SIX

Photovoltaics and nuclear energy conversion for space power: Background and issues

Mark Antonio Prelas[1], Tariq Rizvi Alam[2] and Modeste Tchakoua Tchouaso[3]

[1]Electrical Engineering, University of Missouri, Columbia, MO, United States
[2]Nuclear Environments and System Assessments, Applied Research Associates, Santa Barbara, CA, United States
[3]Physics, North Carolina A&T State University, Greensboro, NC, United States

6.1 Introduction

There are several areas of concern in the design of nuclear energy conversion systems that use *p-n* junctions as transducers [1,2]. One of the most important issues is radiation damage; semiconductors are highly susceptible to ionizing radiation [3,4]. A second area of concern is the availability of radioisotopes for nuclear energy conversion systems [5]. This chapter discusses radiation damage and radioisotope supplies. A most promising approach for providing energy for space exploration is nuclear reactors [6]. It is feasible to integrate photovoltaics with operating nuclear reactors using photon intermediate direct energy conversion (PIDEC) since this technology can provide shielding for the semiconductor [7]. Concepts for nuclear reactor energy conversion are discussed in another chapter in this book [8].

6.2 Radiation damage

Radiation damage is the most significant concern in the design of nuclear energy conversion systems using photovoltaics. The issues stem from the interaction of ionizing radiation with the photovoltaic cell, which causes damage to its structure. Photovoltaic cells are typically made from a solid. Solids are composed of tightly bound atoms that on a microscopic level are either a crystalline solid (with a regular geometrical pattern in the lattice, which includes metals, semiconductors, ice, etc.) or an amorphous solid (non-crystalline solid, which includes glass, plastic, gel, amorphous forms of silicon (Si), amorphous forms of carbon, etc.).

Neutrons, ions (e.g., fission fragments, alpha particles (He^{2+} nuclei), protons (H^+), etc.), beta particles (electrons (β^-), and positrons (β^+)) and gamma rays (γ) are forms of ionizing radiation that are created in nuclear reactions. There are four basic effects caused by ionizing radiation that can produce damage in a solid: 1) transmutation of nuclei into

Photovoltaics for Space
ISBN 978-0-12-823300-9, https://doi.org/10.1016/B978-0-12-823300-9.00016-9

other nuclei through neutron capture; 2) atom displacement that damages the structure of the material; 3) ionization; 4) localized heating that can change material properties. The most significant of these effects on photovoltaics is atom displacement. Displacements create vacancies in the material that are specifically harmful to semiconductors. Displacements are a result of nonionizing energy loss (NIEL).

In the operation of betavoltaics and alphavoltaics, the desired effect of the ionizing radiation is to create electron-hole pairs [9]. Electron-hole pair production in a *p-n* junction-based transducer is responsible for the current flow in the direct energy conversion device. However, NIEL, which can generate recoil atoms and form vacancies, is a competing effect. Vacancies create microscopic damage to the atomic structure of the semiconductor.

Neutrons interact with matter through elastic and inelastic collisions with the nuclei of atoms in the structure. The two primary reactions that create damage to semiconductors are 1) neutron capture leading to transmutation of the atom and 2) a neutron colliding with a nucleus to create a recoil ion, known as a primary knock-on atom (PKA), leading to the formation of a Frankel pair. A Frankel pair is defined as a vacancy in the structure of a solid and the free atom from that vacancy residing in an interstitial position. Both transmutation and a PKA damage the structure of the solid. However, the most important damage mechanism is a PKA that is a NIEL; NIEL models for radiation damage in a solid have been experimentally verified [3].

The minimum energy for neutrons or protons, which have a mass of 1 atomic mass unit (amu), to create a Frankel pair in crystalline Si is 110 eV. The threshold energy for an electron to create a Frankel pair in Si is 260 keV. The difference in the two cases is that the electron is much lighter than either a neutron or proton (0.0005486 vs. 1 amu). Thus, it is much more probable in an elastic collision for a heavy particle to generate a PKA. Fig. 6.1 shows the calculation of NIEL for different ionizing radiations (electrons, muons, pions, neutrons, protons, deuterons, and helium ions) where a limited set of radiation damage data is available. These calculations do an excellent job in predicting the radiation damage for the various types of ionizing radiation in semiconductor devices.

Another model that is useful for predicting the effect of ion energy and mass on creating PKAs is SRIM/TRIM [10,11]. As an example, in Fig. 6.2, the displacements (25 displacements per ion) created in a Si crystal by 1 MeV protons are shown. As the proton energy increases to 10 MeV the displacements per ion increases to 91 and the range increases, as shown in Fig. 6.3. A higher mass ion of the same energy will create more displacements (991 per ion) over a much shorter range, as shown in Fig. 6.4 for a 1 MeV carbon ion interacting with a Si substrate. In Fig. 6.5, as the energy of the carbon ion increases to 10 MeV, the range increases as well as the displacements per ion (1316).

Neutrons create ions by a recoil reaction. The recoil energy of a PKA can be extremely high depending on the energy of the incident neutron. The recoil ion can

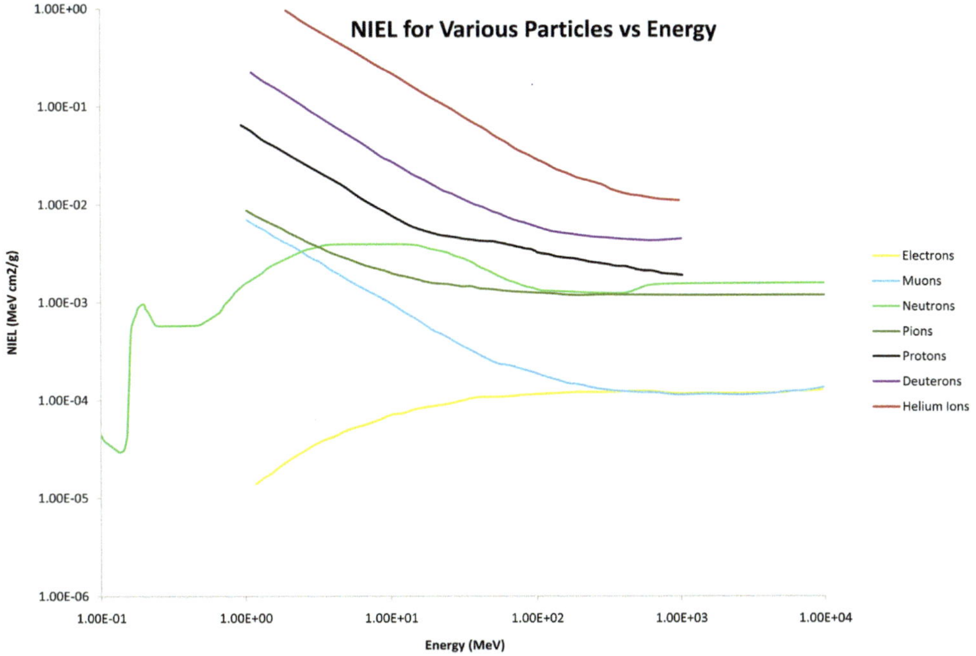

Figure 6.1 Theoretical values of NIEL in silicon for various ionizing radiation particles as a function of particle energy. Experimental data show a relationship between NIEL and displacement damage. *Adapted from C. Claeys, E. Simoen, Radiation Effects in Advanced Semiconductor Materials and Devices, Springer-Verlag, Berlin, Germany (2002) 426 pp, https://doi.org/10.1007/978-3-662-04974-7.*

knock out other atoms from the lattice giving rise to a PKA cascade. About half of the damage occurs when the recoil ion loses its final 10 keV of energy. In a PKA cascade, it is possible to form clusters of Frankel pairs. A high percentage of Frankel pairs recombine (about 90%) with interstitials, resulting in no net damage to the crystal structure. Some of the clusters of vacancies can form multivacancy defects, and the interstitial atoms diffuse through the crystal and can interact with other defects or impurity atoms to form stable complexes.

Electrons have a much different and not well understood mechanism for creating Frankel pairs because of their small mass compared to the mass of the nucleus of the atoms with which they interact. A direct knockout collision is not very probable even though NIEL occurs. However, an alternate view will be discussed that may help explain the long-term slow radiation damage rate in *p-n* junction-based transducers exposed to low-energy beta particles. Electrons are efficient in transporting energy to the bound electrons in atoms that make up the target to create ions. Thus, the electrons that form covalent bonds in a solid can be kicked out. The probability that a covalent bond is broken on a specific atom at a given instant in time by a beta is P (which is the power

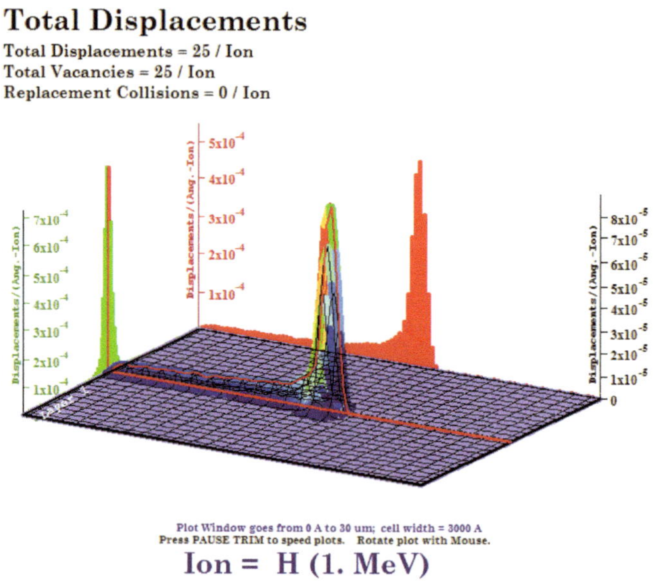

Figure 6.2 Total displacements in a silicon lattice created by 1 MeV protons. *Adapted from J.F. Ziegler, M.D. Ziegler, J.P. Biersack, SRIM—The stopping and range of ions in matter, Nuclear Instruments and Methods in Physics Research Section B: Beam Interactions with Materials and Atoms 268 (11) (2010) 1818–23 and plotted using software on website: www.SRIM.org, accessed October 2, 2021.*

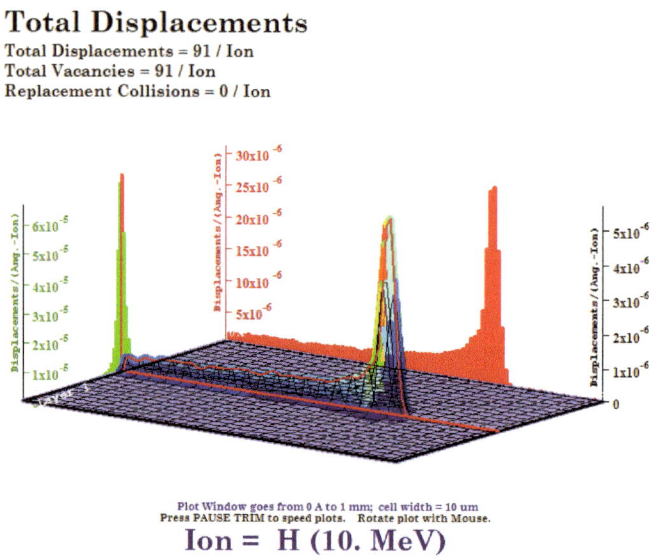

Figure 6.3 Displacements in a silicon lattice created by 10 MeV protons. *Adapted from J.F. Ziegler, M.D. Ziegler, J.P. Biersack, SRIM—The stopping and range of ions in matter, Nuclear Instruments and Methods in Physics Research Section B: Beam Interactions with Materials and Atoms 268 (11) (2010) 1818–23 and plotted using software on website: www.SRIM.org, accessed October 2, 2021.*

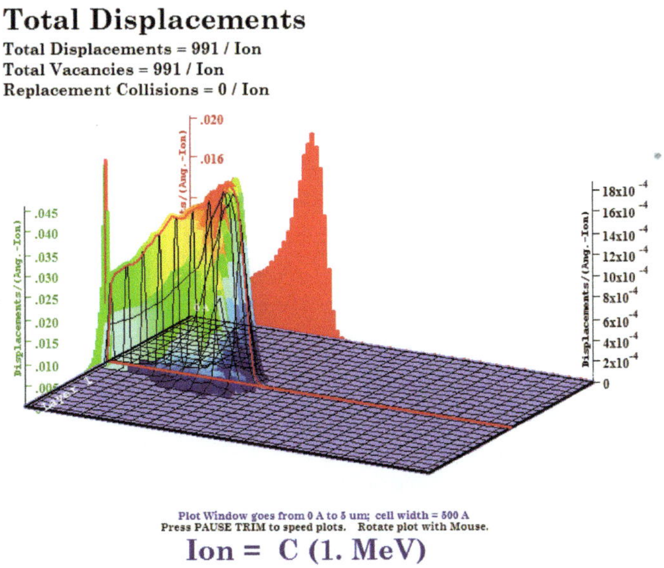

Figure 6.4 Shown are displacements in a silicon lattice created by 1 MeV carbon ions. *Adapted from J.F. Ziegler, M.D. Ziegler, J.P. Biersack, SRIM—The stopping and range of ions in matter, Nuclear Instruments and Methods in Physics Research Section B: Beam Interactions with Materials and Atoms 268 (11) (2010) 1818–23 and plotted using software on website: www.SRIM.org, accessed October 2, 2021.*

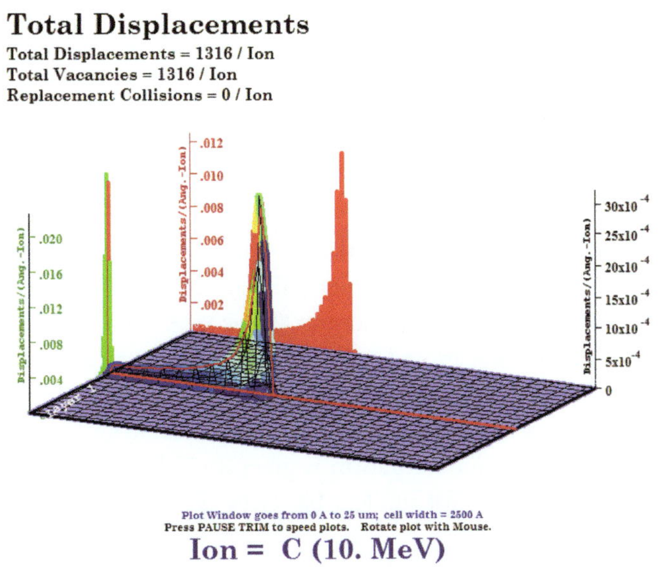

Figure 6.5 Shown are displacements in a silicon lattice created by 10 MeV carbon ions. *Adapted from J.F. Ziegler, M.D. Ziegler, J.P. Biersack, SRIM—The stopping and range of ions in matter, Nuclear Instruments and Methods in Physics Research Section B: Beam Interactions with Materials and Atoms 268 (11) (2010) 1818–23 and plotted using software on website: www.SRIM.org, accessed October 2, 2021.*

density that the betas deposit in the material and the atomic density of the atoms in the crystal lattice). If the atom has "n" covalent bonds, the probability that all the covalent bonds are broken simultaneously is P^n. Once the covalent bonds that hold an atom in place are broken, the atom can drift away and create a Frankel pair. The probability function does have a dependence on the maximum beta energy (because it directly factors into the power density) and is important to the concept of the damage energy threshold (E_{dth}).

The radiation damage threshold energy (E_{dth}) for the various semiconductors used in betavoltaic energy conversion has been reported for Si (about 200 keV), GaAs (about 225 keV), and Ge (about 350 keV) [12]. Many beta sources have maximum energies that exceed E_{dth}. Thus, nuclear battery studies have focused on radioisotopes with lower beta energies such as tritium (maximum energy of 18 keV), Ni-63 (maximum energy of 67 keV), and Pm-147 (maximum energy of 230 keV). Even though low beta energies minimize radiation damage, these particles still have enough energy to break bounds. Betavoltaic cells using low-energy beta particles still degrade over time because ionizing radiation above 1 keV has enough energy to break covalent bonds in a solid [9]. The probability function (P) predicts that there is a probability that all the bonds binding an atom to a lattice can be broken simultaneously (P^n), which can create a Frankel pair. As power density increases, P^n increases. As P^n increases, damage in the crystal lattice increases. Power density can be increase by using radioisotopes that emit higher energy beta particles and/or by using radioisotopes with shorter half-lives.

The production of Frankel pairs is harmful for p-n junctions. Vacancies and multi-vacancy defects can lead to traps that limit the lifetime of carriers. In binary semiconductors, vacancies could be filled with the wrong atom thus causing a structural defect in the crystal. Another consequence is that the p and n type impurities in the depletion region can be reordered in a way that damages the junction.

6.3 Radioisotopes

The nuclear power source most associated with space exploration is radioisotopes. There are three types of radioisotopes: natural, cosmogenic, and manmade. Natural radioisotopes come from the decay of uranium (U)-238, U-235, and thorium (Th)-232 (Table 6.1) or from the earth's crust and seas (Table 6.2). Cosmogenic radioisotopes come from the interactions of cosmic rays with the elements on earth (Table 6.3). Manmade radioisotopes are produced by nuclear reactors or accelerators; a partial listing of radioisotopes of interest for space exploration is given in Table 6.4. Finally, a summary of other potentially useful isotopes for nuclear batteries available from spent nuclear fuel is shown in Table 6.5.

Tables 6.1–6.5 demonstrate that there is a limited supply of suitable isotopes. The most promising source of isotopes is manmade. Isotope costs have been discussed

Table 6.1 Estimated world supply of natural decay chains' radioisotopes.

Series	Isotope	Half-life (yr)	World supply (kg)	Total activity (Ci)	Type(s) of radiation (see Section 6.2)
Uranium	U-238	4.47E+09	7.843E+09	2.636E+06	α, γ
	Th-234	6.60E-02	1.139E-01	2.636E+06	β, γ
	Pa-234m	2.20E-06	3.796E-06	2.631E+06	β, γ
	U-234	2.46E+05	4.237E+05	2.636E+06	α, γ
	Th-230	7.54E+04	1.279E+05	2.636E+06	α, γ
	Ra-226	1.60E+03	2.667E+03	2.636E+06	α, γ
	Rn-222	1.05E-02	1.714E-02	2.636E+06	α, γ
	Po-218	5.89E-06	9.469E-06	2.636E+06	α
	Pb-214	5.10E-05	8.039E-05	2.636E+06	β, γ
	Bi-214	3.78E-05	5.971E-05	2.636E+06	β, α
	Po-214	5.21E-12	8.214E-12	2.636E+06	α, γ
	Pb-210	2.22E+01	3.437E+01	2.636E+06	β, γ
	Bi-210	1.37E-02	2.124E-02	2.636E+06	β, α
	Po-210	3.79E-01	5.865E-01	2.636E+06	α, γ
	At-218	4.75E-08	1.528E-11	5.270E+02	α
	Pa-234	7.64E-04	4.349E-06	8.692E+03	β
Actinium	U-235	7.04E+08	5.691E+07	1.214E+05	α, γ
	Th-231	2.91E-03	2.284E-04	1.214E+05	β, γ
	Pa-231	3.28E+04	2.570E+03	1.214E+05	α, γ
	Ac-227	2.18E+01	1.679E+00	1.214E+05	β, α, γ
	Th-227	5.11E-02	3.889E-03	1.197E+05	α, γ
	Ra-223	3.13E-02	2.370E-03	1.214E+05	α, γ
	Rn-219	1.25E-07	9.333E-09	1.214E+05	α, γ
	Po-215	5.64E-11	4.121E-12	1.214E+05	α, β
	Pb-211	6.86E-05	4.918E-06	1.214E+05	β, γ
	Bi-211	4.07E-06	2.915E-07	1.214E+05	α, β, γ
	Tl-207	9.07E-06	6.356E-07	1.211E+05	β, γ
	Po-211	1.64E-08	3.278E-12	3.397E+02	α, γ
	Fr-223	4.18E-05	4.372E-08	1.675E+03	β, α, γ
Thorium	Th-232	1.400E+10	5.385E+09	5.927E+05	α, γ
	Ra-228	5.750E+00	2.174E+00	5.927E+05	β
	Ac-228	7.016E-04	2.652E-04	5.927E+05	β, γ
	Th-228	1.912E+00	7.227E-01	5.927E+05	α, γ
	Ra-224	9.944E-03	3.693E-03	5.927E+05	α, γ
	Rn-220	1.762E-06	6.427E-07	5.927E+05	α, γ
	Po-216	4.595E-09	1.646E-09	5.927E+05	α
	Pb-212	1.214E-03	4.266E-04	5.927E+05	β, γ
	Bi-212	1.151E-04	4.046E-05	5.927E+05	α, β, γ
	Po-212	9.475E-15	2.134E-15	3.797E+05	α
	Tl-208	5.805E-06	7.191E-07	2.129E+05	β, γ

Sources: adapted from the literature and https://www.nndc.bnl.gov/.

Table 6.2 Estimated world supply of nonseries-primordial radioisotopes.

Radioisotope	Relative abundance	Half-life (years)	Supply Earth's crust (kg)	Supply sea (kg)	Total activity EC (Ci)	Total activity sea (Ci)	Type(s) of radiation[a]
K-40	0.0117	1.248E+09	5.771E+16	7.327E+13	4.228E+14	5.368E+11	β, ε
V-50	0.25	2.10E+17	7.080E+15	1.009E+10	2.366E+05	3.371E-01	ε, β
Rb-87	27.85	4.81E+10	5.915E+17	5.394E+13	5.144E+13	4.691E+09	β
In-115	96.67	4.41E+14	5.704E+15	3.121E+13	4.027E+07	2.203E+05	β
Te-123	0.87	9.20E+16	2.053E+11	n/a	6.253E+00	n/a	ε
La-138	0.089	1.02E+11	8.192E+14	4.884E+06	2.067E+10	1.232E+02	ε, β
Ce-142	11.7	5.00E+16	1.836E+17	2.266E+08	9.370E+06	1.156E-02	β
Nd-144	23.8	2.29E+15	2.331E+17	1.076E+09	2.523E+08	1.164E+00	α
Sm-147	15.1	1.06E+11	2.512E+16	1.097E+08	5.635E+11	2.460E+03	α
Sm-148	11.35	7.00E+15	1.888E+16	8.244E+07	6.414E+06	2.800E-02	α
Gd-152	0.205	1.08E+14	3.000E+14	2.316E+06	6.314E+06	4.876E-02	α
Hf-174	0.163	2.00E+15	1.154E+14	1.842E+07	1.156E+05	1.844E-02	α
Lu-176	2.588	3.76E+10	4.886E+14	6.266E+06	2.655E+10	3.405E+02	β
Re-187	62.93	4.33E+10	1.040E+13	4.063E+09	4.609E+08	1.801E+05	β
Pt-190	0.0127	6.50E+11	1.499E+10	n/a	4.225E+04	n/a	α
Pb-204	1.4	1.40E+17	4.626E+15	6.779E+08	5.701E+04	8.355E-03	α

[a]- See 6.2; ε = electron capture.
Source: https://www.nndc.bnl.gov/.

Table 6.3 Cosmogenic isotope inventory.

Radioisotope	Half-life (years)	Inventory (MCi)		Type of radiation[c]
		O'Brien[a]	Masarik[b]	
Be-10	1.39E+06	3.5E+00	2.5E+00	β^-
Al-26	7.17E+05	5.6E-03	1.6E-03	β^+, ε
Cl-36	3.01E+05	1.2E-01	2.7E-01	β^-, ε
Kr-81	229,000	-	4.9E-05	ε
C-14	5700	2.4E+02	3.0E+02	β^-
H-3	12.32	3.5E+01	3.5E+01	β^-
Na-22	2.603	5.0E-03	—	ε
S-35	87.37 days	8.0E-02	—	β^-
Be-7	53.24 days	7.7E+00	4.9E+00	ε
P-33	25.35 days	4.5E-02	—	β^-
P-32	14.26 days	4.7E-02	—	β^-
Mg-28	20.92 h	2.3E-03	—	β^-
Na-24	15 h	9.7E-03	—	β^-
S-38	170.3 min	2.0E-03	—	β^-
Si-31	157.3 min	2.2E-02	—	ε
F-18	109.77 min	4.2E-03	—	ε
Cl-39	56.2 min	1.3E-01	—	β^-
Cl-38	37.24 min	8.0E-02	—	β^-
Cl-34m	32.00 min	5.4E-03	—	ε, IT
Al-29	6.56 min	1.4E-03	—	β^-
S-37	5.05 min	5.5E-03	—	β
Ne-24	3.38 min	7.2E-04	—	β
P-30	2.45 min	5.8E-03	—	ε
Al-28	2.24 min	2.0E-02	—	β

[a]K. O'Brien, Secular variations in the production of cosmogenic isotopes in Earth's atmosphere, Journal of Geophysical Research: Space Physics 84 (A2) (February 1979) 423—431.
[b]J. Masarik, M. Frank, J.M. Schäfer, R. Wieler, Correction of in situ cosmogenic nuclide production rates for geomagnetic field intensity variations during the past 800,000 years, Geochimica et Cosmochimica Acta 65 (17) (September 2001) 2995—3003.
[c]β^-, electron; β^+, positron; ε, electron capture; IT, isomeric transition; see Section 6.2 for more details.
Sources: public domain and literature cited in table.

Table 6.4 Estimated world supplies of Kr-85, Pu-238, Pu-241, and Am-241: a partial listing of radioisotopes of interest for space exploration.

Radioisotope	Half-life (yr)	Year of estimation	Total activity (Ci)	World supply (kg)
Kr-85[a]	10.752	2009	1.46E+08	3.73E+02
Pu-238	87.7	End of 2014	6.11E+07	3.57E+03
Pu-241	14.325	End of 2014	1.34E+09	1.29E+04
Am-241	432.6	End of 2004	2.33E+08	6.80E+04

[a]Global atmospheric content.

Table 6.5 Estimation of radioisotope supply from spent fuel in the United States including potential useful isotopes for nuclear batteries.

Radioisotope	Half-life (yr)	Total activity (Ci)	Supply (g)
Cs-137	30.08	4.62E+09	5.31E+07
Sr-90	28.9	2.98E+09	2.11E+07
Cm-244	18.1	3.03E+08	3.74E+06
Kr-85	10.752	1.69E+08	4.29E+05
Am-241	432.6	1.58E+08	4.61E+07
Eu-154	8.6	1.01E+08	3.73E+05
Pm-147	2.6234	4.05E+07	4.37E+04
Sm-151	90	1.88E+07	7.16E+05
Cs-134	2.065	1.53E+07	1.18E+04
Eu-155	4.753	1.10E+07	2.23E+04
Sb-125	2.76	3.23E+06	3.08E+03
Cm-243	29.1	1.32E+06	2.56E+04
Sn-121m	43.9	7.31E+05	1.36E+04
Eu-152	13.528	1.34E+05	7.59E+02
Ru-106	1.02	3.92E+04	1.18E+01
Cd-113m	14.1	1.15E+04	5.14E+01
Th-228	1.912	1.53E+03	1.86E+00
U-232	68.9	1.51E+03	6.82E+01
Pm-146	5.53	1.33E+03	2.99E+00
Pu-236	2.858	2.84E+02	5.43E-01
Cf-250	13.08	2.39E+01	2.19E-01
Tm-171	1.92	4.80E+00	4.40E-03
Cf-252	2.645	1.12E+00	2.08E-03
Ac-227	21.772	9.10E-02	1.26E-03
Po-210	0.379	8.71E-04	1.94E-07
Pb-210	22.2	8.71E-04	1.14E-05
Ra-228	5.75	1.13E-05	4.14E-08

Source: https://www.nndc.bnl.gov/.

elsewhere; they are expensive ($K/g) [13]. Producing large quantities of isotopes using neutron capture with existing high-power reactors or accelerators is costly. The most cost-effective approach appears to be processing of spent nuclear fuel. But this would require a worldwide commitment to the recycling of spent nuclear fuel, which is not likely to happen given concerns about nuclear nonproliferation.

6.4 Energy conversion technologies

For energy conversion applications, there are other issues to consider. In Tables 6.1–6.5, the estimated activity in Ci (curies) is shown for each isotope. This number is important because the isotope activity is proportional to the available power output, as shown in Table 6.6 [13]. In the eighth column (Power W/gm), the total power produced by a gram of the isotope is shown. In the 10th column (Power W/Ci), the total power produced by a Ci of the isotope is shown.

6.4.1 Nuclear batteries

The utility of the data in Table 6.6 can be demonstrated by the following example:

Example 1. If a Radioisotope Thermoelectric Generator (RTG) uses 1 kg of Pu-238, using the figure of 0.556 W/gm from column 8 of Table 6.6 and multiplying by 1000 results in a figure of 556 W of thermal power. If the RTG produces electricity with 7% efficiency, then the electrical power produced is 38.9 W.

In the example, the Pu-238 produces heat and the heat is converted to electricity using a thermoelectric converter. The power density of the device has minimal dilution of the isotope due to the compact structure of a thermoelectric converter. Betavoltaics and alphavoltaics are far more complicated in that there is a significant dilution of the isotope's power density due to the transducer diffuse structure (Fig. 6.6) [9,13].

A nuclear battery is composed of layers of materials. The two most important layers are the source and the transducer. The efficiency of a nuclear battery (η_{NB}) is defined as the transducer efficiency ($\eta_{transducer}$) times the fraction of power deposited in the transducer or the power deposition efficiency $\eta_{pd} = P_4 \big/ (P_1 + P_2 + P_3 + P_4)$ [9,13]. Example two calculates the thermal power of 1 kilogram of tritium in the form of T_2O:

Example 2. To find the thermal power of 1 kg of tritium, using the figure of 0.360 W/gm from column 8 of Table 6.6 and multiplying by 1000 produces a result of 360 W of thermal power.

Tritium is typically stored as T_2O, one of its most dense chemical forms. Using T_2O as a guide, it is possible to simply illustrate one of the shortcomings of betavoltaics. The density of T_2O is 0.331 gm/cm^3 so the volume of 1 kilogram of T_2O is 3021 cm^3. In constructing a device, the average density of tritium (for example as part of a polymer coating) in a betavoltaic cell structure is typically reduced by a factor of 1000 compared to T_2O [13]. So, a large array of betavoltaic cells containing 1 kilogram of tritium will have a volume of approximately 3,021,000 cm^3. Assuming a theoretical maximum efficiency for a betavoltaic of ~3% [14], the power density of the betavoltaic array will be approximately 3.6 μW/cm^3. This power density is too low to compete with an RTG. Low power density is not only a problem for a tritium source, it is an overall weakness for betavoltaics and alphavoltaics regardless of the isotope used [13]. However, there are low-power applications, such as nanosats, where betavoltaics are appropriate [15]; other applications or missions are discussed in a chapter in the final section of the book [8].

Table 6.6 Properties of some potential isotopes for nuclear batteries: power is calculated for beta and alpha particles only.

Nuclide	Decay energy (MeV)	Half-life (yr)	Decay	β_{max} (MeV)	Other emission (units in MeV)	Activity/gm	Mass specific power (W/gm)	Energy stored (J/gm)	Power W/Ci	Energy stored (J/Ci)
H-3	0.019	12.33	Beta	0.019		3.559 E+14	0.361	1.97×10^8	3.75 E−05	2.10×10^4
Kr-85	0.67	10.755	Beta	0.67 MeV 99.6%, 0.15 MeV 0.4%	0.4% yield of 0.514 γ	1.449 E+13	0.518	2.45×10^8	0.00132	6.27×10^5
Pu-238	5.593	87.74	Alpha (100%), fis (1.85E-7%)		α mainly 5.456 (28.98%) and 5.499 (70.91%)	6.336 E+11	0.556	2.19×10^9	0.0331	1.28×10^8
Pu-241	0.021	14.35	Beta (99.998%), alpha (0.00245)	0.0208	α mainly 4.853 (12.2%), 4.896 (83.2%)	3.826 E+12	2.856	8.13×10^6	0.000109	7.86×10^4
Am-241	5.638	432.2	Alpha (100%), fission (4.3E-10%)		α mainly 5.442 (13%) and 5.485 (84.5%) with 0.05954 (35.9%) gammas	1.270 E+11	0.109	2.18×10^9	0.0316	6.36×10^8

Adapted from literature and https://www.nndc.bnl.gov/.

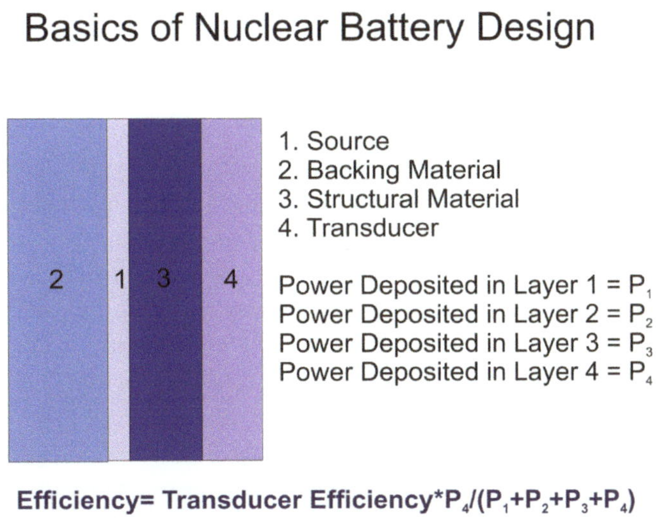

Figure 6.6 A basic nuclear battery design. *Source: the authors.*

6.4.2 Photon intermediate direct energy conversion

It is possible to overcome the radiation damage of photovoltaic cells by implementing an indirect energy conversion method known as PIDEC [16–22]. With PIDEC, ionizing radiation interacts with a fluorescer that is not prone to radiation damage, and photons are then transported to a shielded photovoltaic cell using a waveguide or protective barrier. As described in the references, PIDEC can be implemented with nuclear fission, fusion, and radioisotopes. There are potential isotopes that can benefit from PIDEC. One of the most favorable is Kr-85 (Mark A. Prelas and Tchouaso, 2018) [19].

Due to isotope costs and shortages, NASA has initiated the Kilopower Reactor Using Stirling Technology (KRUSTY) program [23,24]. Nuclear reactors may enable future human colonies on the moon and Mars as well as to provide a power source for deep space missions. PIDEC could play a role in future KRUSTY designs since the photovoltaic convertor can act as a high-temperature topping power cycle and still produce heat for a Stirling engine [19].

6.5 Conclusions

Major drawbacks of using radioisotopes for space power generation are cost and availability. The most attractive radioisotopes are those that are produced as a byproduct of nuclear fission. The cost involved in reprocessing spent nuclear fuel is prohibitive and only makes sense if it is part of a larger infrastructure designed for nuclear fuel recycling. Currently the only source of separated isotopes come from the inventory of reprocessed nuclear fuel from the nuclear weapons programs. These inventories are nearly depleted and have required a substantial investment to produce additional supplies of Pu-238 from a small inventory of Np-237 [25].

If the problem with supplies of beta-emitting isotopes can be solved, there are additional considerations. Radiation damage to nuclear batteries using *p-n* junction transducers is mainly focused on pure beta emitters that have low energy, below the radiation damage threshold of the semiconductors used. Although these types of batteries have inherently low power densities, recent advancements in nanotechnology are promising for deployment of betavoltaics for space applications [26]. A brief description of nuclear batteries, different types of nuclear battery technologies, detailed end-to-end design principles of betavoltaics, and direct energy conversion systems for nuclear reactors are discussed in other chapters in this book [8,26].

References

[1] F. Rahmani, H. Khosravini, Optimization of Silicon parameters as a betavoltaic battery: Comparison of Si *p-n* and Ni/Si Schottky barrier, Radiat. Phys. Chem. 125 (August 2016) 205–212.

[2] R. Zheng, J. Lu, Y. Liu, X. Li, X. Xu, R. He, Z. Tao, Y. Gao, Comparative study of GaN betavoltaic battery based on *p-n* junction and Schottky barrier diode, Radiat. Phys. Chem. 168 (March 2020) 108595.

[3] C. Claeys, E. Simoen, Radiation Effects in Advanced Semiconductor Materials and Devices, Springer-Verlag, Berlin, Germany, 2002, p. 426, https://doi.org/10.1007/978-3-662-04974-7.

[4] S.J. Pearton, A. Aitkaliyeva, M. Xian, F. Ren, et al., Review-radiation damage in wide and ultra-wide bandgap semiconductors, ECS J. Solid State Sci. Technol. 10 (5) (2021) 055008.

[5] J.F. Zakrajsek, J. Hamley, D. Cairns-Gallimore, T.J. Sutliff, T. Bishop, C.E. Sandifer, P. McCallum, M. McCune, Radioisotope Power Systems Program Status and Expectations, AIAA 2017-4609, 15th International Energy Conversion Engineering Conference, 10–12 July 2017, Atlanta, GA USA, https://doi.org/10.2514/6.2017-4609.

[6] G.L. Bennett, Radioisotope Power: Historical Review, in: Encyclopedia of Nuclear Energy, E. Greenspan (Ed.), Elsevier Inc., 2021, pp. 174–190, https://doi.org/10.1016/B978-0-12-819725-7.00104-5.

[7] M.A. Prelas, F.P. Boody, E.J. Charlson, G.H. Miley, A two-step photon-intermediate technique for the production of electricity, chemicals or lasers in nuclear energy conversion, Prog. Nucl. Energy 23 (3) (1990) 223–240.

[8] M.T. Tchouaso, T.R. Alam, M.A. Prelas, Space nuclear power: Radioisotopes, technologies, and the future, in: S.G. Bailey, A.F. Hepp, D.C. Ferguson, R.P. Raffaelle, S.M. Durbin (Eds.), Space PV, Elsevier, Cambridge, MA USA, 2022, pp. 443–488.

[9] M.A. Prelas, C.L. Weaver, M.L. Watermann, E.D. Lukosi, R.J. Schott, D.A. Wisniewski, A review of nuclear batteries, Prog. Nucl. Energy 75 (August 2014) 117–148.

[10] J.F. Ziegler, M.D. Ziegler, J.P. Biersack, SRIM—The stopping and range of ions in matter, Nucl. Instrum. Methods Phys. Res. Sect. B Beam Interact. Mater. Atoms 268 (11) (2010) 1818–1823.

[11] Software and references in: www.SRIM.org, accessed October 2, 2021.

[12] S.T. Revankar, T.E. Adams, Advances in betavoltaic power sources, J. Energy Power Sources 1 (6) (2014) 321–329.

[13] M. Prelas, M. Boraas, F. De La Torre Aguilar, J.D. Seelig, M.T. Tchouaso, D. Wisniewski, Nuclear batteries and radioisotopes (Lecture Notes in Energy - Volume 56), Springer International Publishing, Switzerland, 2016, 350pp.

[14] K. Oh, M.A. Prelas, J.B. Rothenberger, E.D. Lukosi, J. Jeong, D.E. Montenegro, D.A. Wisniewski, Theoretical maximum efficiencies of optimized slab and spherical betavoltaic systems utilizing sulfur-35, strontium-90, and yttrium-90, Nucl. Technol. 179 (2) (August 2012) 234–242, https://doi.org/10.13182/NT12-A14095.

[15] J. Gonzalo, D. Domínguez, D. López, On the challenge of a century lifespan satellite, Prog. Aero. Sci. 70 (October 2014) 28–41.
[16] M.A. Prelas, A potential UV fusion light bulb for energy conversion, Bull. Am. Phys. Soc. 26 (1) (1981) 1045.
[17] M.A. Prelas, F.P. Boody, G.H. Miley, J.F. Kunze, Nuclear driven flashlamps, Laser Part. Beams 6 (February 1988) 25–62, https://doi.org/10.1017/S0263034600003803.
[18] M.A. Prelas, E.J. Charlson, E.M. Charlson, J. Meese, G. Popovici, T. Stacy, Diamond photovoltaic energy conversion, in: M. Yoshikawa, M. Murakawa, Y. Tzeng, W.A. Yarbrough (Eds.), Proceedings of Second International Conference on the Application of Diamond Films and Related Materials, MYU Tokyo, Japan, 1993, pp. 329–335.
[19] M.A. Prelas, M.T. Tchouaso, High efficiency dual-cycle conversion system using Kr-85, Appl. Radiat. Isot. 139 (2018) 70–80.
[20] R.J. Schott, Photon Intermediate Direct Energy Conversion Using a Sr-90 Beta Source, Ph.D. Thesis, University of Missouri, 2012, http://hdl.handle.net/10355/33103.
[21] C.L. Weaver III, PIDECα: Photon Intermediate Direct Energy Conversion Using the Alpha Emitter Polonium-210, Ph.D. thesis, University of Missouri, 2012, http://hdl.handle.net/10355/15908.
[22] C.L. Weaver, R.J. Schott, M.A. Prelas, D.A. Wisniewski, J.B. Rothenberger, E.D. Lukosi, K. Oh, Radiation resistant PIDECα cell using photon intermediate direct energy conversion and a 210Po source, Appl. Radiat. Isot. 132 (2018) 110–115.
[23] R. Sanchez, D. Hayes, Kilowatt reactor using Stirling technology (Krusty) experiment update, Trans. Am. Nucl. Soc. 117 (November 2017) 825–828.
[24] R. Sanchez, T. Grove, D. Hayes, et al., Kilowatt Reactor Using Stirling TechnologY (KRUSTY) component-critical experiments, Nuclear Technology, in: Nuclear Technology, 206, June 2020, pp. S56–S67, https://doi.org/10.1080/00295450.2020.1722553.
[25] National Isotope Development Center, https://isotopes.gov, accessed October 3, 2021.
[26] T.R. Alam, M.T. Tchouaso, M.A. Prelas, Summary of the design principles of betavoltaics and space applications, in: S.G. Bailey, A.F. Hepp, D.C. Ferguson, R.P. Raffaelle, S.M. Durbin (Eds.), Photovoltaics for Space, Elsevier, Cambridge, MA USA, 2022, pp. 293–345.

PART TWO

New materials technologies and advanced processing

CHAPTER SEVEN

Perovskite solar cells on the horizon for space power systems

Brandon K. Durant[1], Ian R. Sellers[1] and Bibhudutta Rout[2]
[1]Homer L. Dodge Department of Physics and Astronomy, University of Oklahoma, Norman, OK, United States
[2]Department of Physics, University of North Texas, Denton, TX, United States

7.1 Background of perovskite solar cells

Perovskite solar cells (PSCs) have garnered an unprecedented level of research activity since a breakthrough efficiency of >10% by Lee et al. in 2012 [1]. PSCs have now surpassed other polycrystalline thin-film technologies in record-certified cell efficiencies of 25.5% [2], eclipsing more mature thin-film technologies like copper indium gallium selenide (CIGS) and cadmium telluride (CdTe) within the last decade. Issues with scaling up to large areas still need to be realized for commercialization as well as proven long-term stability with regard to moisture, thermal cycling, and illumination. However, the solution-processed methods of record cells offer hope for rapid deployment, low processing costs, and flexible architectures with a high specific power.

The prototypical crystal structure for perovskite (named after the mineral perovskite, $CaTiO_3$) is composed of an ABX_3 composition with corner-sharing lead or tin halide octahedra with an organic cation within the cavities. Stability with the methylammonium cations originally used for early PSCs has led to the replacement with larger organic cations formamidinium as well as inorganic cations of cesium. The valence and conduction band extrema are determined by the overlap of the Pb^{2+} or Sn^{2+} and halide antibonding orbitals, and therefore the bonding angle and bond length determine the bandgap of these materials. Mixed tin—lead iodide compounds offer the narrowest bandgaps showing a bowing parameter [2—4], while the pure lead halide compounds can be tuned from 1.6 to 2.3 eV increasing from I^- to Br^- [5] as well as further increase with Cl^- incorporation [6], and mixed-halide compositions in between. Although the orbitals of the A site cation do not directly participate in the conduction or valence bands, the size does affect the sterics and spin-orbit coupling of the lead halide network so have consequences on the bandgap as well, offering further tunability [7,8].

Photovoltaics for Space
ISBN 978-0-12-823300-9, https://doi.org/10.1016/B978-0-12-823300-9.00011-X

7.2 Defect tolerance and ion mobility

One of the impressive qualities of PSCs is the long carrier lifetimes of photogenerated carriers. These lifetimes coupled with carrier mobilities result in long diffusion lengths of carriers, an order of magnitude longer than the absorber thickness for high-efficiency devices. The origin of this is the defect tolerance that appears to be more than other polycrystalline thin-film materials such as CdTe and CIGS, significantly contributing to the rapid enhancement of conversion efficiencies over a relatively short period of time. The resulting transition energy levels due to various defects that have been calculated for $MAPbI_3$ cubic crystal structure show that many of the low-formation defects do not reside within the bandgap [9]. Calculations have gone further to explain why halide-deficient films exhibited the best performing devices by inhibiting interstitial iodide [10]. Although the low annealing temperatures offer to lower costs and increase throughput for processing of PSCs, the volatility of organic and halide constituents makes the synthesized absorber layers very sensitive to processing conditions and annealing times [11] (Fig. 7.1). Additionally, this sensitivity limits the options and deposition techniques for remaining layers to complete device stacks.

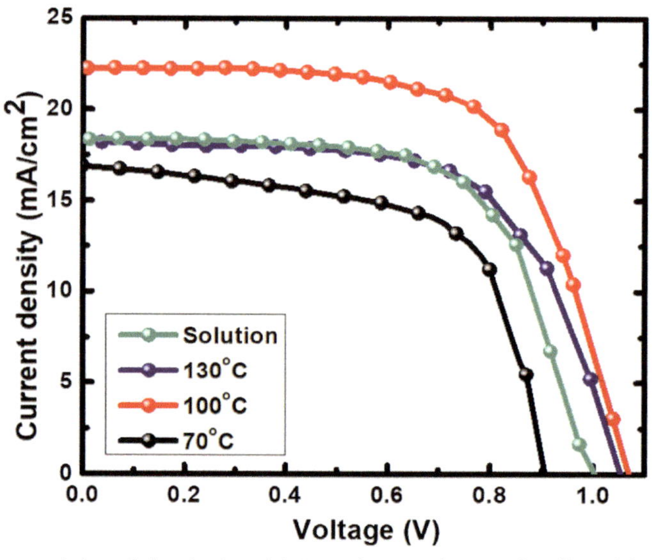

Figure 7.1 I-V characteristics of the devices fabricated using the $MAPbI_3$ films fabricated at different temperatures using a quasi-vapor deposition (QVC) or solution processing method, for comparison with (power conversion efficiency) 70°C (10.2), 100°C (15.6), 130°C (12.3), and solution (12.0). *Reproduced with permission from R.P. Vijendar, S. Maniyarasu, D.K. Reshma, R.K. Battula, P.U. Bhaskar, E. Ramasamy, G. Veerappan, Temperature dependence of $MAPbI_3$ films by quasi-vapor deposition technique and impact on photovoltaic performance and stability of perovskite solar cells, Journal of Alloys and Compounds 888 (2021) 161448, copyright (2021) Elsevier.*

A frequently observed phenomenon of PSC is the hysteresis observed in the current density-voltage sweeps, characterized by a nonsuperposition of the forward and reverse scans [12–16]. Frequent speculation for the origin of this property has been extrinsic ion migration [13], mobility of ions within the lattice [12], and/or charging and discharging of interfaces. These two possibilities should also not be considered mutually exclusive considering the ionicity and charged species that can promote redox reactions [17] or trap state formation/filling when accumulated at an interface. For instance, it has been shown the vacancies at interfaces provide recombination centers [10]. The depletion of A site cations and volatility of organic halide compounds [18] leads to PbI_2 formation [19], and reduction of Pb^{2+} to metallic Pb has been shown to form at interfaces [20]. Additionally, problems exist with I^- expulsion [21] and I_2 catalyzed degradation [22]. Tin-based perovskites suffer from redox reactions at interfaces due to the facile oxidation of Sn^{2+} to Sn^{4+} and reduction of Pb^{2+} to metallic Pb [23–25].

Efforts have been successful in reducing or practically eliminating the hysteresis by optimizing the composition and stoichiometry of the perovskite layer [26]. Additionally, mixed-halide compositions exhibit segregation of lower bandgap I-rich regions under illumination producing a type I band alignment with the neighboring mixed or I-poor regions, characterized by enhanced luminescence from the I-rich regions and detrimental photovoltaic (PV) performance. The force or forces causing this segregation are not well understood yet, but the entropic forces are sufficient to provide "remixing" of the halides in the dark at room temperature and recovery of performance [27–29]. The issue of halide segregation, and the prevention, has been a key research endeavor within the community [30,31] and has so far led to the new state-of-the-art mixed-cation, mixed-halide composition that minimizes the strain within the lattice [32–36] and shows superior longevity to illumination even at high intensities compared with earlier compositions. The mobility of the halide ions is the key aspect of the segregation phenomena and the inherent tolerance of PSCs to particle irradiation, as will be discussed later.

7.3 Particle radiation tolerance

Inherent properties of PSCs that enable the remarkable tolerance to high-energy particle radiation are their high absorption and therefore relatively thin absorber depth, defect tolerance, remarkably long carrier lifetimes and diffusion lengths compared to absorber depth, and ion (specifically halide) mobility. For missions into space, devices will encounter hostile conditions with respect to temperature fluctuations, high-energy particles of protons and electrons trapped in the magnetosphere, as well as protons, neutrons, electrons, X-rays, and gamma rays ejected from solar events. The fluence and spectra of energetic protons and electrons can be predicted based on models adapted from data collected from instrumentation coupled with spacecraft trajectories and orbits.

Protons with their much larger mass compared to electrons and therefore greater damage a crystalline lattice are a concern for most PV technologies used to power spacecraft due to the inherent need to be exposed for power generation.

Particles interact with matter, and energy is transferred via two forms: ionizing and nonionizing events. The nonionizing events have shown to be the dominant mechanism for degradation in incumbent PV technologies. Energetic protons traversing a lattice can collide with nuclei causing displacement of atoms that produce defect centers that typically result in parasitic trap states and recombination centers, reducing the photovoltaic power output. A simplistic result of an atomic displacement would be the formation of a vacancy and interstitial defect pair, or a Frenkel defect. However, multiple proton—nuclei interactions, and nuclei—nuclei collisions can also result in defect clusters. The degree of interaction depends on the energy of the incident protons and the effective cross-section of the nuclei. Defects produced within the active regions in PV devices are the most sensitive to reduced performance. The interaction, and therefore the number of atomic displacements, is maximum for particle energies sufficient to traverse encapsulation and conductive layers, slowing down but with enough remaining energy compared to the displacement energy threshold of host ions within the lattice to produce displacements. This eventually leads to the protons "stopping" within the active regions. Higher energies have a lower degree of interaction with the absorber regions as the effective cross-section of the nuclei is smaller, and most protons traverse the lattice without as much interaction and will stop deeper in the device, substrate, or supporting apparatus, or will simply pass through if the amount of material is thin enough. The so-called nonionizing energy loss (NIEL) value of materials is a calculation that represents the rate of energy transfer from the particles to the host lattice nuclei.

A benefit that PSCs have compared to traditional technologies is they utilize a thin active region, typically on the order of hundreds of nanometers due to their high absorption coefficient. With a shortened interaction length with incident particles, although displacement rates may be somewhat similar, this inherently lowers the number of displacements. Additionally, the amount of energy transferred throughout the absorber layer is reduced, thereby lowering the lattice interactions as protons traverse.

To reconcile the use of monoenergetic particles irradiated normal to a solar cell surface using laboratory accelerators for the effects of the poly energetic omnidirectional particles encountered on space missions, the Naval Research Laboratory introduced a displacement damage dose analysis [37] to shorten testing methodology compared to the JPL method of equivalent fluences approach. The latter involves rigorous testing of four electron and eight proton energies to determine damage coefficients for a particular solar cell. The former offers the use of NIEL values of a material as a function of energy that can be theoretically calculated and integrating the NIEL with the differential spectra of protons and electrons accounting for shielding to produce a single dose (energy per mass) value. This method condenses the testing considerably to two electron and one proton energies. The two electron energies are performed to calculate an exponential term that calculates an "effective 1 MeV electron displacement damage dose" that collapses the electron energy data into a single curve.

The displacement damage dose analysis has proven appropriate for traditional silicon and III-V-based technologies through validation by years of empirical evidence. This large amount of data so far does not exist for the nascent perovskite technologies that do not yet have a standard structure. Fig. 7.2 is the calculated NIEL values as a function of proton energy irradiated normal to the material surface [38]. What is clear is that the values of low displacement energy for hydrogen contribute to low energy peaks for protons and electrons extending to nearly 10 eV and 10 keV, respectively. Shielding however can lower the fluence of these energies significantly, and the penetration depth is on the order of several nanometers. Compared to Si the NIEL values are significantly less at all energies and lower for high energies compared with III-V InGaP, which is considered a more radiation-tolerant PV material, contributing at least theoretically to a greater tolerance for all perovskite compositions. Additionally, these calculations if used for the displacement damage dose method are assumed to be irreversible, which considering the mobility of halides and A site cations would not necessarily be accurate at least at room temperature. Thus, there is strong evidence that at room temperature, defects formed in perovskites self-heal [27,39,40].

To further exacerbate the differences, there is no foreseeable solution to produce a purely unencapsulated PSC device, simply by thinning of encapsulation layers. Finally,

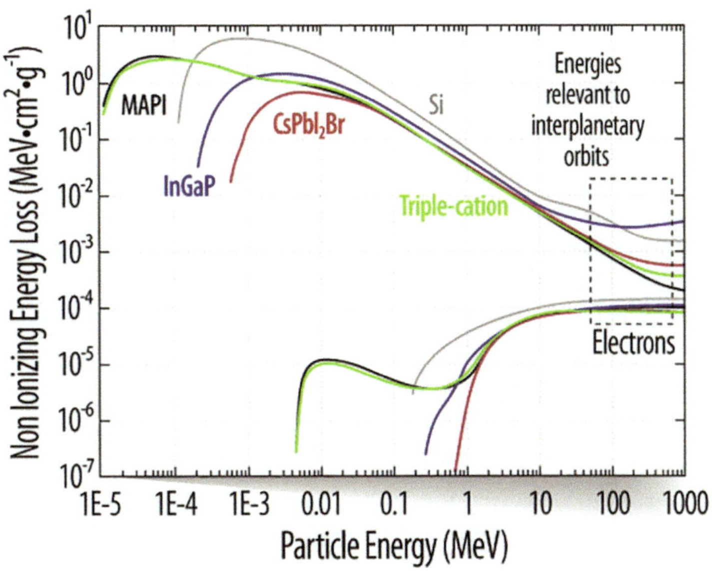

Figure 7.2 Nonionizing energy loss (NIEL) calculations of various photovoltaic absorber materials. *Reproduced with permission from A.R. Kirmani, B.K. Durant, J. Grandidier, N.M. Haegel, M.D. Kelzenberg, Y.M. Lao, M.D. McGehee, et al., Countdown to perovskite space launch: Guidelines to performing relevant radiation-hardness experiments, Joule 6 (5) (2022) 1015-1031, copyright (2022) Elsevier.*

the use of organic cations and transport layers in the highest efficiency PSCs, although showing a more robust performance for terrestrial applications, has not been thoroughly investigated for their susceptibility and degradation mechanisms from high-energy particle irradiation. The understanding of inorganic materials that comprise the traditional technologies would not be expected to translate fully to the inherent covalent characteristics of the organic components. A better theoretical understanding needs to be realized to provide insights into the effects of radiation on not only PSCs but also organic solar cells for the accurate assessment of space mission environments on these systems. Interpretation from the existing inorganic technologies should be treated with caution.

The first reports of irradiation tolerance by Miyawaza et al. [41] and following reports by Lang et al. [3,42] and Huang et al. [42] provided the early investigations of PSCs stability to high-energy protons and electrons. In these reports, the consistent observation was that the PSCs were more tolerant to both types of radiation when compared to traditional Si and III-V technologies. Although the proton energies of 68 MeV were used, which did not provide a high degree of interaction with the thin absorber layer, the in situ characterization with the comparison of a reference Si cell demonstrated a remarkable resilience. Additionally, these reports illustrated the ability of the cells to "self-heal" when annealed [41] or left in the dark for several days [40].

A study reported by Lang et al. assesses the radiation hardness of the widely used triple cation—based perovskite absorber material ($Cs_{0.05}MA_{0.17}FA_{0.83}Pb(I_{0.83}Br_{0.17})_3$) employing 20 and 68 MeV proton irradiation [43]. In situ measurements of the degradation of the proton-induced current as well as the PV performance during proton irradiation are used as two independent metrics. The optimized triple cation—based perovskite-based space solar cells have measured AM0 efficiencies of 18.8% and maintain 95% of their initial efficiency even after irradiation with protons at an energy of 68 MeV and a total dose of 10^{12} p/cm^2; performance degradation under 20-MeV proton irradiation is even less. Despite the negligible impact on solar cell device performance, this study identifies that proton irradiation is changing the recombination kinetics under low excitation densities profoundly. Dark capacitance-voltage and current-voltage characteristics, photoluminescence (PL), and open-circuit voltage (V_{oc}) decays are analyzed in depth.

Also apparent is the difference in energy loss mechanisms describing the inelastic energy transfer to electrons (ionizing energy loss) and elastic energy transfer to nuclei (nonionizing energy loss). These figures of merit are calculated using stopping and range of ions in matter (SRIM) methods [44], which account for incident particle and energy, material stoichiometry, density, and displacement energy of host atoms. Additionally, these calculations can be converted into energy loss per fluence and vacancy generation rate of constituent atoms. Apparent in these calculations is that the ionization energy losses are many orders of magnitude greater than the recoil energy losses, but typically they are not considered for the majority of the degradation in incumbent inorganic technologies. However, the organic cations within most high-efficiency PSCs will be susceptible to covalent bond breaking via ionization, and the detailed consequences of this are not yet realized. The displacement energy of each atom for these calculations assumes a purely inorganic type crystal structure, making the organic cation contributions somewhat convoluted. Additionally, the ionized electron—phonon interactions within the perovskite material can lead to localized heating via thermalization, a point that will be discussed later.

Remarkably with in situ monitoring of the proton-induced external quantum efficiency, measuring the ion beam—induced current, the perovskite device showed less apparent degradation than a silicon carbide (SiC) reference device; SiC is well known to have impressive radiation hardness. The prolonged PL and V_{oc} decay lifetimes, coupled with the decrease in PL intensity were attributed to the production of trap states for minority carriers (calculated as the electrons) likely due to iodide interstitials and associated vacancies. These trap states, calculated to be 310 meV below the conduction band, underwent relatively fast release of carriers (detrapping), which did not reduce performance at full sunlight intensities. Similar results of V_{oc} decay lifetimes were reported by Barbé where the main source of device degradation appeared to be related to the organic Spiro-OMeTAD hole transport layer and its interface with the perovskite absorber [45].

Although the mixed-cation, mixed-halide, lead-based perovskites have a wider bandgap than optimal for single-junction PV technologies (~1.4 eV) [46], the stability of these compositions are so far superior. As such, possessing a bandgap at 1.7 eV makes them ideal for tandem double-junction PVs, increasing the theoretical efficiency from 32% for a 1.1-eV c-Si bandgap to 46% [47]. Practical record efficiencies of monolithic perovskite/Si cells have now reached 29.5% [2,48], surpassing the 26.7% record for silicon single-junction devices. Other narrow bandgap materials such as tin-based perovskites [49—56] and CIGS [57—59] have also been investigated as rear junction material. Incorporation onto highly optimized and rapidly manufactured silicon-based panels by adding a cheap high-throughput perovskite top junction would increase efficiencies while not greatly increasing weight or manufacturing costs. Recently, attention has been paid to evaluating the proton irradiation tolerance of perovskite/Si and perovskite/CIGS tandem devices [60], which boasted specific powers of 0.42 and 7.4 W/g, respectively, with the latter utilizing a flexible form factor as well (Fig. 7.3).

While the perovskite/CIGS device retained 85% of power conversion performance after 2×10^{12} p/cm^2 (68 MeV) fluence, the perovskite/Si silicon heterojunction (SHJ) device retained only 1% after 1×10^{11} p/cm^2, showing superior tolerance of the thin-film device to proton irradiation, mainly due to loss of FF and V_{oc}. While the perovskite/Si SHJ device showed an almost complete loss in performance, quasi-Fermi-level separation measurements (QFLS) alluded to the loss as dominated by the proton-induced defects in the relatively thick (hundreds of micrometers) Si SHJ bottom cell rather than the perovskite top cell, although a modest decrease in luminescence intensity was

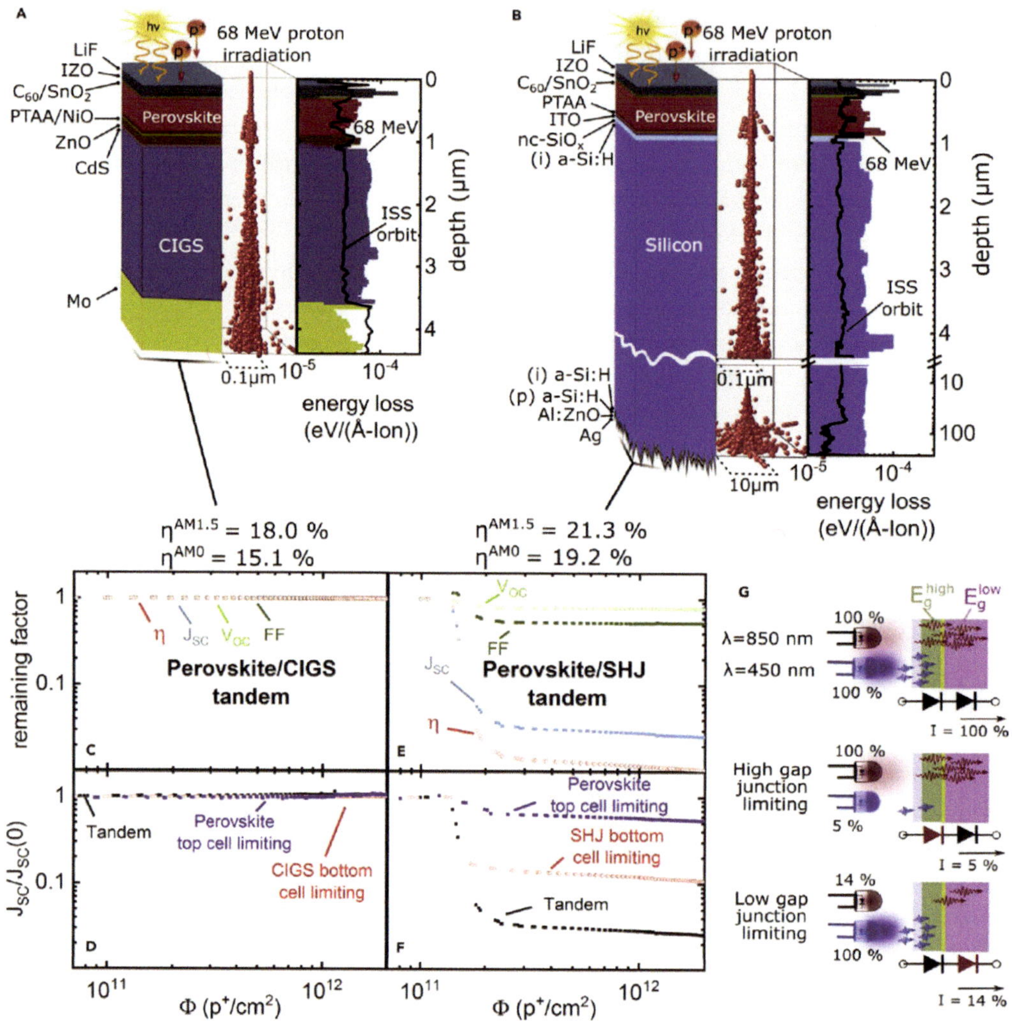

Figure 7.3 (A and B) 3D scatter plots of the straggling of 68 MeV protons within the perovskite/CIGS (A) and perovskite/SHJ (B) tandem solar cells. The corresponding energy loss of the incident 68 MeV protons to recoils is plotted as a function of depth based on SRIM simulations with a total of 5×10^7 protons. The damage of a real space environment at the orbit of the International Space Station is shown as black line considering polyenergetic and omnidirectional proton irradiation (see reference for further details). (C and E) Operando measurements of V_{oc}, J_{sc}, FF, and η of the investigated perovskite/CIGS (C) and perovskite/SHJ (E) tandem solar cell as a function of the accumulated proton dose (Φ). All values are normalized to their initial value. The proton energy amounted to 68 MeV. (D–G) Normalized short-circuit current of perovskite/CIGS (D) and perovskite/SHJ (F) tandem solar cell under illumination with near-infrared ($\lambda = 850$ nm) and blue LEDs ($\lambda = 450$ nm) that were alternatingly set to either 100% or 5/14% (see reference for further details) to mimic current matching under AM0 or forcing one subcell into limitation as illustrated in (G). *Reproduced from F. Lang, M. Jošt, K. Frohna, E. Köhnen, A. Al-Ashouri, A.R. Bowman, T. Bertram et al., Proton radiation hardness of perovskite tandem photovoltaics, Joule 4 (5) (2020) 1054–1069, published by Elsevier under terms of a Creative Commons (4.0 BY-CC) license.*

observed for the perovskite layer. The perovskite/CIGS losses were attributed to the reduced shunt resistance from the CIGS bottom cell and contact layers, as well as modest decrease in luminescence from the CIGS. Additionally, it appeared to have negligible effects on the perovskite bulk, with an increase in photoluminescence due to deterioration of the charge extraction layers and their interfaces with the perovskite.

Although considered less impactful on device performance for incumbent technologies, the electronic ionization energy losses have shown to play a remarkable role with respect to the effects of proton irradiation. Using a constant fluence of 1×10^{12} p/cm^2, the dependence of device performance on energy was investigated (Fig. 7.4) [39]. SRIM calculations showed that the low-energy (50 keV) protons stopped and caused a significant amount of atomic displacements (recoil energy loss) within the back transparent contact, electron selective transport layers, as well as the active perovskite due to the very thin encapsulation. This, again, showed remarkable resilience to irradiation compared to traditional technologies with the degradation coming from a decrease in V_{oc} and FF rather than J_{sc}, indicating that any decrease in carrier diffusion lengths was not dictating the reduced performance output. As energy increased, although a significant rate of vacancy generation was still produced within the active region, no degradation was observed. As the energy was increased further and protons stopped past the active region, V_{oc} and power conversion efficiency (PCE) increased relative to preirradiated parameters of the same devices. When integrating the electronic ionization due to inelastic energy transfer mechanisms within the lattice of the back contact and absorber layers, it is apparent that at 300-keV energies, the local heating due to ionizing energy loss is maximized, correlating well with the increase in performance.

As electrons are excited by energy transfer above the conduction band edge from the incident protons, they will thermalize by phonon-mediated processes. The same is true for the holes generated below the valence band edge. This behavior results in a heating of the lattice, which would provide the energy required for ions to migrate into energetically favorable and electrically benign positions in the lattice, at grain boundaries, and interfaces. Here it is shown not only to negate displacements and vacancies generated by the proton's elastic nuclear collisions, but defects present prior to irradiation, resulting in improved V_{oc} and power output. Additionally, perovskites have shown to have thermal conductivities several orders of magnitude lower than III-V materials [61], and as such the heat dissipation is significantly lower, and would presumably amplify this effect.

While both the works described previously with respect to perovskite tandems suggest the superior performance of the perovskite subcell (with respect to their partner in these tandems, silicon and CIGS, respectively), care must be taken in this hypothesis since the termination of the 68 MeV protons—and the more prohibitive and destructive nuclear interactions—occur solely in the nonperovskite layers by virtue of geometry in these reports. Indeed, CIGS, in particular, has been observed previously to display considerable radiation hardness at higher electron and proton fluences compared to more

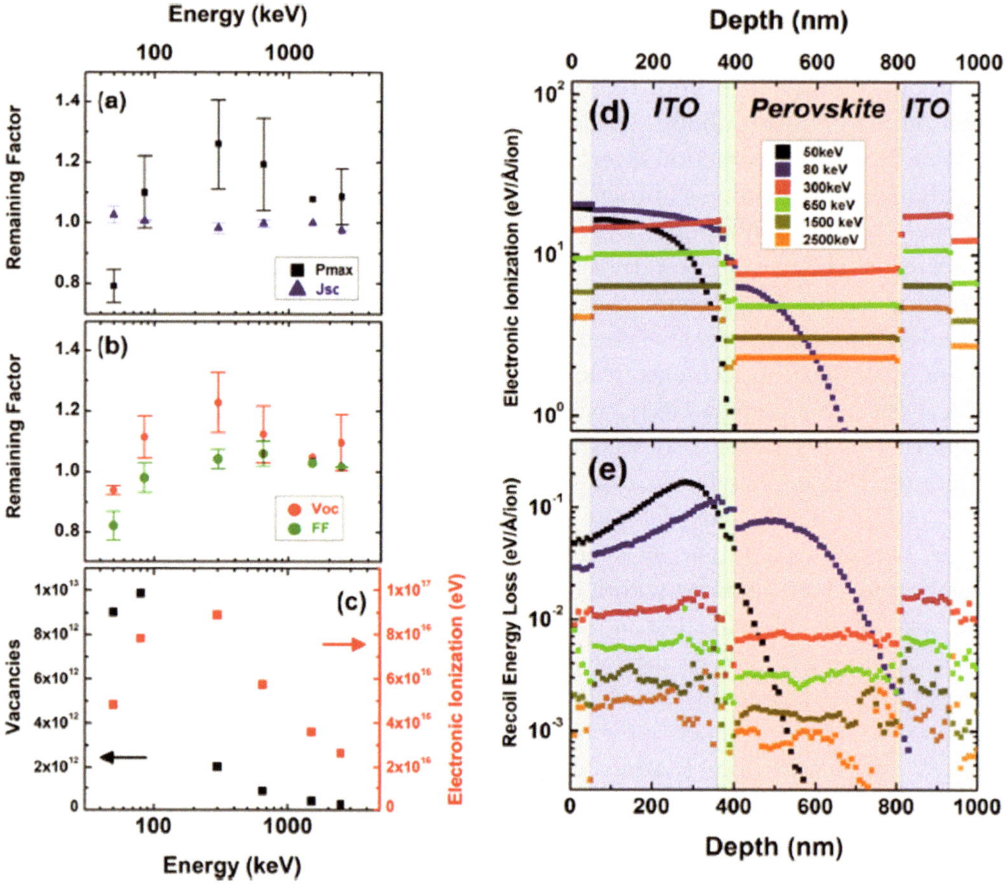

Figure 7.4 Energy-dependent remaining factor (final value/initial value) of JV AM0 reverse parameters (A) P_{max} and J_{sc} and (B) V_{oc} and FF with respect to prior to proton irradiation, and (C) the cumulative vacancies (black, left) and electronic ionization (red, right) generated within the indium tin oxide (ITO) back contact and perovskite layers. The highest vacancy generation with 50 and 80 keV protons results in the greatest decrease in performance, while a relative improvement in V_{oc} corresponds to the higher electronic ionization (heating) at higher energies. Devices were irradiated with a proton fluence of 1×10^{12} p/cm^2 for all energies. Stopping and range of ions in matter (SRIM) calculated electronic energy loss (electronic ionization) (D) and nuclear (recoil) energy loss (E) for 50 keV (black), 80 keV (blue), 300 keV (red), 650 keV (green), 1500 keV (yellow), and 2500 keV (orange) protons. *Reproduced with permission from B.K. Durant, H. Afshari, S. Singh, B. Rout, G.E. Eperon, I.R. Sellers, Tolerance of perovskite solar cells to targeted proton irradiation and electronic ionization induced healing, ACS Energy Letters 5 (7) (2021) 2362–2368, copyright (2021) American Chemical Society .*

traditional III-V solar cells [62]. Recently, a comparison of the relative radiation tolerance of a mixed Sn–Pb-based perovskite solar cell was compared to that of commercially available flexible CIGS [63,64]. While a direct comparison requires unencapsulated devices or devices with the same cover glass, and the same structure and absorber layer thickness, here care was taken to choose proton energies and fluences that generated similar defect densities in each solar cell absorber.

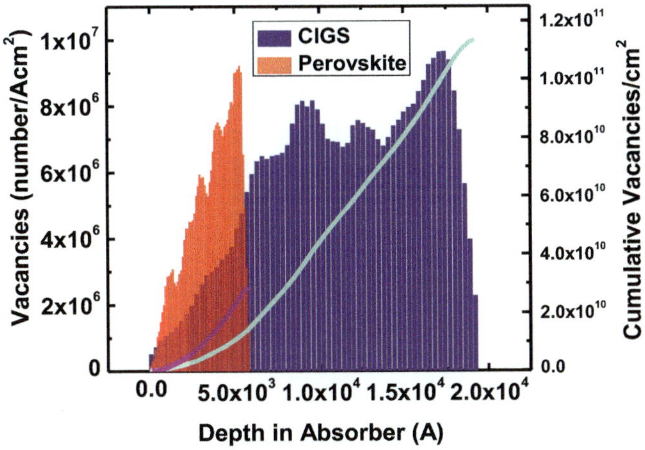

Figure 7.5 Stopping and range of ions in matter (SRIM) simulated damages of $(FASn)_{0.6}(MAPb)_{0.4}I_3$ perovskite layer (red) and CIGS (blue) irradiated with 3.7 MeV protons through the back contact at a fluence of 1×10^{11} ions/cm^2, and 1.5 MeV protons through the front at a fluence of 1×10^{11} ions/cm^2, respectively. Cumulative vacancies by integrating vacancy generation in represented perovskite (magenta) and CIGS (cyan) is plotted against right *y*-axis. *Reproduced with permission from B.K. Durant, H. Afshari, S. Sourabh, V. Yeddu, M.T. Bamidele, S. Singh, B. Rout, G.E. Eperon, D.Y. Kim, I.R. Sellers, Radiation stability of mixed tin–lead halide perovskites: Implications for space applications, Sol. Energy Mater. Sol. Cells 230 (2021) 111232, copyright (2021) Elsevier.*

Fig. 7.5 shows the SRIM simulation for the two devices investigated in Durant et al., illustrating the similar levels of proton irradiation *and* vacancy creation experienced by the two cells at the energies and fluence levels selected to consider the difference in the structure and encapsulation. The CIGS was irradiated through the thin front contact as opposed to the back for the perovskite device, while the absorber thickness of the CIGS was also considerably thicker, consequently increasing the interaction length for proton damage and (inherently) the number of defects produced in comparison to the perovskite cell studied. Despite this, the authors suggest the irradiation tolerance for perovskites still appears qualitatively superior to that of the CIGS since the vacancy generation per depth is similar for both materials while the device performance is *quite different*. As such, the authors suggest that the $(FASn)_{0.6}(MAPb)_{0.4}I_3$ perovskite assessed [63], in general, outperforms the commercial CIGS under high levels (fluence) of proton irradiation.

This is illustrated in Fig. 7.6, which compares the current density-voltage dependence of both the (a) CIGS and (b) $(FASn)_{0.6}(MAPb)_{0.4}I_3$ solar cells at AM0. While perovskites display relatively stableperformance under proton irradiation of 1×10^{14} protons/cm^2, levels in excess of 1×10^{13} protons/cm^2 result in a performance reduction for thin-film CIGS. Specifically, the CIGS shown in Fig. 7.6A lose 14.3% of V_{oc}, 22.6% J_{sc}, and 16.4% FF, resulting in an overall relative loss in PCE of 44.6%. This is compared to 4.0% *gain* in V_{oc}, 7.7% loss in J_{sc}, and 0.6% loss in FF, totaling a mere 4.6% relative loss in PCE for the $(FASn)_{0.6}(MAPb)_{0.4}I_3$ perovskite shown in Fig. 7.6B.

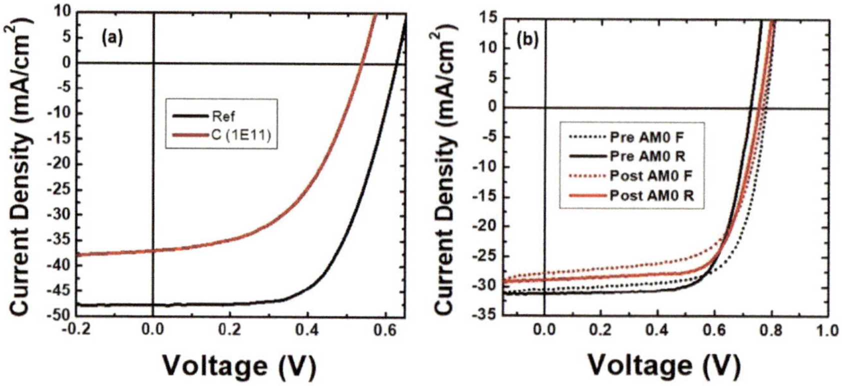

Figure 7.6 Current density-voltage (JV) characteristic curves of under AM0 intensity at room temperature of (A) CIGS and (B) $(FASn)_{0.6}(MAPb)_{0.4}I_3$ perovskite devices prior to proton irradiation (black), and irradiated with 1×10^{11} ions/cm^2 proton fluence (red). F and R refer to forward (negative voltage to positive voltage) and reverse (positive voltage to negative voltage), respectively. *Reproduced with permission from B.K. Durant, H. Afshari, S. Sourabh, V. Yeddu, M.T. Bamidele, S. Singh, B. Rout, G.E. Eperon, D.Y. Kim, I.R. Sellers, Radiation stability of mixed tin–lead halide perovskites: Implications for space applications, Sol. Energy Mater. Sol. Cells 230 (2021) 111232, copyright (2021) Elsevier.*

The authors attribute the relative stability of the $(FASn)_{0.6}(MAPb)_{0.4}I_3$ perovskite solar cell and the more deleterious behavior of the CIGS performance under irradiation to the nature of the defects formed in the perovskite relative to the CIGS that are attributed to halide segregation and/or Sn^{4+}-based defects in the $(FASn)_{0.6}(MAPb)_{0.4}I_3$, which are relatively shallow or reside at grain boundaries in the perovskites that have a less parasitic effect than deeper Cu-interstitials that are suggested to significantly reduce the voltage in CIGS upon irradiation [65]. Several recent works have also demonstrated that the dynamic ionic motion present in perovskite or "softness" also serves to enable rapid self-healing, further limiting any losses associated with the effects of halide defects, which dominate in these systems as shown below.

Elemental analysis of defects induced by proton irradiation in the $(FASn)_{0.6}(MAPb)_{0.4}I_3$ solar cell are presented in Fig. 7.7; similar analysis was also performed by Durant et al. in triple-halide perovskites [39]. Fig. 7.7A shows the SRIM simulation results for the distribution of the total vacancy created due to the collision events between incident protons (energy of 3.7 MeV and fluence of 10^{11} H^+/cm^2) and host atoms in the various layers that constitute the $(FASn)_{0.6}(MAPb)_{0.4}I_3$ solar cell. The defect density gradually increases as protons penetrate and slow down deeper into the perovskite absorber layer. Initially, the higher energy protons that pass through the upper layers of the solar cell create predominately electron ionization events and heat, while deeper in the devices as the protons slow down and cause greater nuclear damage.

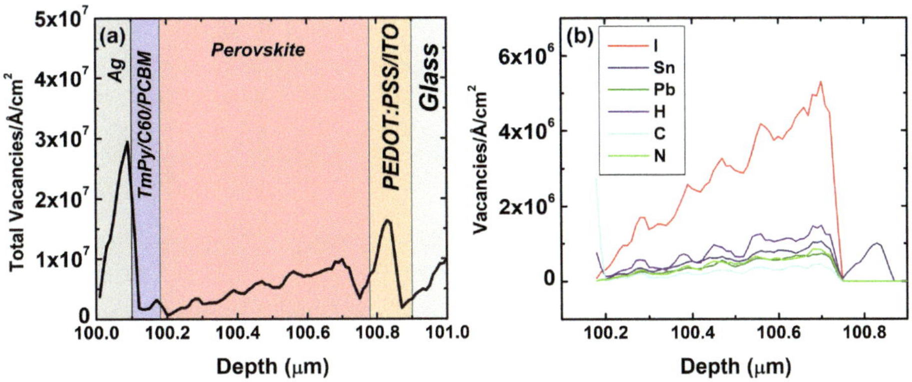

Figure 7.7 (A) SRIM simulated damages in the target layer due to the irradiation of 3.7 MeV protons with a fluence of 1×10^{11} ions/cm^2. The target was encapsulated with 100-μm cover glass. (B) Individual contributions of vacancies associated with the individual elements I (black), H (green), Sn (blue), C (green), N (purple), and Pb (red) in the perovskite layer are also shown. *Reproduced with permission from B.K. Durant, H. Afshari, S. Sourabh, V. Yeddu, M.T. Bamidele, S. Singh, B. Rout, G.E. Eperon, D.Y. Kim, I.R. Sellers, Radiation stability of mixed tin–lead halide perovskites: Implications for space applications, Sol. Energy Mater. Sol. Cells 230 (2021) 111232, copyright (2021) Elsevier.*

Fig. 7.7B shows the individual contributions of the vacancies associated with the constituent elements in the perovskite absorber, which are dominated by halide vacancies with iodine contributing ~50% of the defects in the $(FASn)_{0.6}(MAPb)_{0.4}I_3$, with the metal cations of Sn and Pb contributing <10% of the total vacancies. The H atoms present in the layer due to the organic cations contribute slightly above 10% of the total vacancies in the perovskite. These results illustrate that significant defects are generated upon high energy irradiation, which might be expected from a soft material like perovskites, yet these materials show remarkable tolerance to these defects in their PV parameters. This unique ability to withstand defect generation is attributed to a combination of the dynamic nature of the perovskites, the tendency for ionic motion, and therefore the subsequent self-healing evident in the materials.

One of the consistent results from proton irradiation studies has been an unexpected decrease in dark current. With regards to mature technologies, the creation of defects within the lattice increases leakage pathways for carriers, thereby increasing the dark current, increasing nonradiative recombination, and therefore decreasing the open-circuit voltage. Brus et al. provided detailed device characterization of 68 MeV 1×10^{13} cm^{-2} irradiated devices (reproduced in Fig. 7.8A) [3] and concluded that the proton-induced formation of Frenkel defects compensated the photo degradation of $CH_3NH_3^+$ molecules and reduced nonradiative recombination losses. This effect was also observed by Durant et al. (Fig. 7.8B) and associated with high electronic energy loss, potentially annealing interfaces, where the V_{oc} also increased [39]. Interestingly, observations of the dark current of devices without perovskite active layers also decreased after

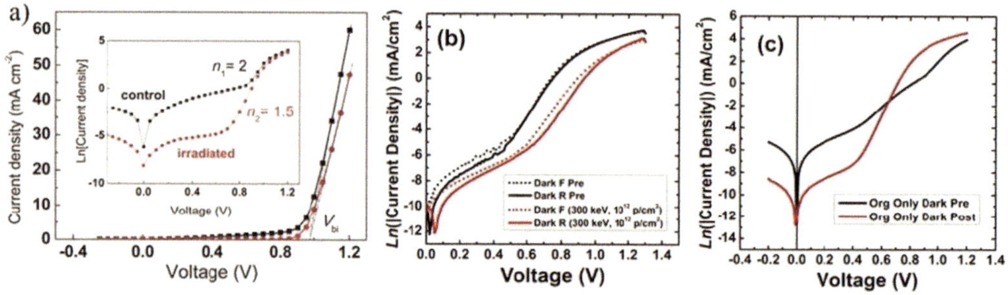

Figure 7.8 (A) J-V curves of the control (black rectangles) and irradiated (red circles) solar cells. *Vbi* denotes the build-in voltage. The inset shows a semilogarithmic plot of the data. Dark J-V characteristic curves (B) of the device before (pre) and after 1×10^{12} p/cm^2, 300 keV proton irradiation of forward (F) and reverse (R) scans, and (C) dark J-V curves of organic-only device (without perovskite active layer) before (black) and after (red) 1×10^{12} p/cm^2, 3.7 MeV proton irradiation. *Reproduced with permission from: (a) V.V. Brus, F. Lang, J. Bundesmann, S. Seidel, A. Denker, B. Rech, G. Landi, H.C. Neitzert, J. Rappich, N.H. Nickel, Defect dynamics in proton irradiated $CH_3NH_3PbI_3$ perovskite solar cells, Advanced Electronic Materials 3 (2) (2017) 1600438, copyright (2017) Wiley-VCH; (b) B.K. Durant, H. Afshari, S. Singh, B. Rout, G.E. Eperon, I.R. Sellers, Tolerance of perovskite solar cells to targeted proton irradiation and electronic ionization induced healing, ACS Energy Letters 5 (7) (2021) 2362–2368, copyright (2021) American Chemical Society; (c) B.K. Durant, A. Hadi, S. Sourabh, V. Yeddu, M.T. Bamidele, S. Singh, B. Rout, G.E. Eperon, D.Y. Kim, I.R. Sellers, Proton Radiation Tolerance of Wide and Narrow Band Gap Perovskite Solar Cells, in: 2021 IEEE 48th Photovoltaic Specialists Conference (PVSC), 2021, pp. 1111–1114, https://doi.org/10.1109/PVSC43889.2021.9518848, copyright (2021) IEEE.*

proton irradiation (Fig. 7.8C) [64], leading to speculation that the effects at least partially also arise from a contribution of the increased shunt resistance and improvement in diode quality of the organic transport layers and interfaces.

The results of proton irradiation studies on PSC consistently indicate that the degradation in performance is not related to the bulk material but rather the interfaces or simply instability of the devices evident in reference samples. These findings are consistent with carrier lifetime measurements of high-performance devices and isolated perovskite films [66], where the latter shows long lifetimes and high QFLS that are reduced substantially by interfacial recombination in the former. It can therefore be conjectured that further improvements in device performance with interface engineering would lead to greater performance and irradiation tolerance. Defect migration to the interfaces, however, could be potentially masking the effects in the bulk material, but passivated interfaces should help alleviate this accumulation to an extent.

7.4 Thermal stability

A unique property of perovskites is the increase in bandgap with increasing temperature, which is opposite of the vast majority of semiconducting materials [61,67].

This property has been shown to be beneficial at higher temperatures with a very small negative temperature coefficient [68]. However, devices on space missions will undergo rapid thermal changes and Earth orbits usually encompass periods without illumination. Only a few reports so far describe the effects of these large (colder) temperature regimes [63,69,70]. What is evident is the phase of the tin-based perovskite is stable down to 100 K, while the mixed-cation, mixed-halide, lead-based perovskite retains phase stability at all low temperatures, unlike the pure $MAPbI_3$ [71] and $FAPbI_3$ [72] or $CsPbI_3$ [73] compositions. However, the structures used for all of these devices do not perform well at temperatures below −10 [69], −40 [70], and −173°C [63] mainly due to a large decrease in fill factor. This has been attributed to a barrier to photogenerated carrier collection only circumvented by thermionic emission at higher temperatures or transport within the carrier selective contacts. This detrimental performance, however, was shown to be recoverable at low intensities of outer planetary missions where these temperatures would be most relevant during operation [69].

Other issues do exist for temperatures exceeding 90°C for long durations, which is of concern for terrestrial applications as well [74–76]. Namely, it is the volatility of organic halide salts leading to the decomposition of the absorber material, which is exacerbated in a vacuum. This can be alleviated by effective encapsulation or by using an all-inorganic perovskites, which have not yet reached the performance of mixed organic–inorganic systems. Recent data have also suggested unrecoverable V_{oc} loss with higher temperatures even with state-of-the-art devices incorporating effective encapsulation for terrestrial applications. Furthermore, the American Institute of Aeronautics and Astronautics (AIAA) specifications for space PVs [77] also require stable operation in high humidity conditions (95% humidity at 42°C for 60 days) as well as high temperatures (100°C) in vacuum that to date are particularly challenging to current perovskite solar technologies. Humidity exposure is known to result in rapid irreversible loss of performance and segregation of both the perovskite and organic transporting layers, suggesting that encapsulation and perhaps alternative transporting layers will be critical in the practical realization of PSCs for space power applications with respect to thermal considerations, in general.

7.5 Conclusion and outlook

Metal halide perovskites have demonstrated tremendous stability to the high-energy particle irradiation that solar cells might experience in space environments. Indeed, while the protocols for terrestrial space testing require further consideration—and considerably more empirical data from space missions—so laboratories around the world are assessing these systems in a manner that truly and consistently establishes their radiation tolerance, recent trends suggest that the unique interaction of high-energy particles with these systems is beneficial for long-term performance. In particular, the softness and dynamic nature of these materials make them remarkably radiation tolerant. Numerous reports now indicate that PSCs, while experiencing defect generation at levels constant with other technologies under irradiation, suffer much less degradation in their performance upon this exposure due to a combination of the shallow nature of the predominantly halide-based defects created, the tendency for such defects to propagate to grain boundaries, and the mobile nature of the ions in these systems that facilitate rapid self-healing at ambient temperatures [39,40,42,63]. How such processes might affect the PSCs in deep space or in cooler temperature regimes is still under discussion with only limited work under (for example) low-intensity low-temperature (LILT) conditions available in the literature [69].

In this recent work from Brown et al., evidence for inhibited performance at $T < 200$ K was presented and attributed to parasitic barriers to carrier transport at low temperatures and AM0 solar irradiance [69]. However, these authors also noted that at lower levels of solar irradiance such as those at Saturn or Jupiter that the carrier extraction was recovered due to the lower generation rate and therefore effective thermionic extraction under these conditions. Similar results were found for $(FASn)_{0.6}(MAPb)_{0.4}I_3$-based solar cells with a loss of fill factor at low temperatures and AM0 intensities [63]. While these data suggest potential applications for perovskite PVs in LILT missions, they also suggest the solar cell architectures used for terrestrial applications require further development specifically for space applications, along with careful consideration of the encapsulates, transport layers, and substrate and cover glass for optimum performance and stability in space [63].

These final points also speak to the current or major bottleneck to the practical implementation of perovskite PVs in space, their thermal stability. To date, the state-of-the-art perovskites being developed by the terrestrial PV community cannot withstand the AIAA high-temperature thermal cycling and humidity protocols [77] without significant and irreversible degradation. Although the lack of moisture and oxygen in space environments has been suggested as favorable for the stability of perovskites, effective encapsulation and material stability at low and high temperatures in humid environments must still be assessed for terrestrial conditions prior to deployment as well as prevention of the outgassing of organic constituents. While this is not a priori a fundamental limitation to the practical implementation of PSCs for space, it requires significant investment in research and development targeting space-specific solar cell architectures and encapsulations in these systems and the development of more stable and unique transporting layer for these applications. This might include for example NiO_X or other inorganic materials that are currently under investigation by the community and are significantly more stable than the current organic-based transporting layers used in high-efficiency perovskites currently. Additionally, efforts to develop high-efficiency perovskite compositions with lower organic content may provide a route to greater thermal and vacuum stability in these systems.

References

[1] M.M. Lee, J. Teuscher, T. Miyasaka, T.N. Murakami, H.J. Snaith, Efficient hybrid solar cells based on meso-superstructured organometal halide perovskites, Science 338 (6107) (2012) 643–647.

[2] U.S. DOE National Renewable Energy Laboratory. Research Cell Record Efficiency Chart, Best Research-Cell Efficiencies. https://www.nrel.gov/pv/cell-efficiency.html. accessed June 4, 2021.

[3] V.V. Brus, F. Lang, J. Bundesmann, S. Seidel, A. Denker, B. Rech, G. Landi, H.C. Neitzert, J. Rappich, N.H. Nickel, Defect dynamics in proton irradiated $CH_3NH_3PbI_3$ perovskite solar cells, Adv. Electron. Mater. 3 (2) (2017) 1600438.

[4] E.L. Unger, L. Kegelmann, K. Suchan, D. Sörell, L. Korte, S. Albrecht, Roadmap and roadblocks for the band gap tunability of metal halide perovskites, J. Mater. Chem. 5 (23) (2017) 11401–11409.

[5] G.E. Eperon, S.D. Stranks, C. Menelaou, M.B. Johnston, L.M. Herz, H.J. Snaith, Formamidinium lead trihalide: A broadly tunable perovskite for efficient planar heterojunction solar cells, Energy Environ. Sci. 7 (3) (2014) 982–988.

[6] L. Wang, K. Wang, G. Xiao, Q. Zeng, B. Zou, Pressure-induced structural evolution and band gap shifts of organometal halide perovskite-based methylammonium lead chloride, J. Phys. Chem. Lett. 7 (24) (2016) 5273–5279.

[7] A. Amat, E. Mosconi, E. Ronca, C. Quarti, P. Umari, M.K. Nazeeruddin, M. Grätzel, F. De Angelis, Cation-induced band-gap tuning in organohalide perovskites: Interplay of spin–orbit coupling and octahedra tilting, Nano Lett. 14 (6) (2014) 3608–3616.

[8] R. Prasanna, A. Gold-Parker, T. Leijtens, B. Conings, A. Babayigit, H.G. Boyen, M.F. Toney, M.D. McGehee, Band gap tuning via lattice contraction and octahedral tilting in perovskite materials for photovoltaics, J. Am. Chem. Soc. 139 (32) (2017) 11117–11124.

[9] W.J. Yin, T. Shi, Y. Yan, Unusual defect physics in $CH_3NH_3PbI_3$ perovskite solar cell absorber, Appl. Phys. Lett. 104 (6) (2014), 063903.

[10] D. Meggiolaro, S.G. Motti, E. Mosconi, A.J. Barker, J. Ball, C.A.R. Perini, F. Deschler, A. Petrozza, F. De Angelis, Iodine chemistry determines the defect tolerance of lead-halide perovskites, Energy Environ. Sci. 11 (3) (2018) 702–713.

[11] R.P. Vijendar, S. Maniyarasu, D.K. Reshma, R.K. Battula, P.U. Bhaskar, E. Ramasamy, G. Veerappan, Temperature dependence of $MAPbI_3$ films by quasi-vapor deposition technique and impact on photovoltaic performance and stability of perovskite solar cells, J. Alloys Compd. 888 (2021) 161448.

[12] J. Liu, M. Hu, Z. Dai, W. Que, N.P. Padture, Y. Zhou, Correlations between electrochemical ion migration and anomalous device behaviors in perovskite solar cells, ACS Energy Lett. 6 (3) (2021) 1003–1014.

[13] Z. Li, C. Xiao, Y. Yang, S.P. Harvey, D.H. Kim, J.A. Christians, M. Yang, P. Schulz, S.U. Nanayakkara, C.S. Jiang, J.M. Luther, J.J. Berry, M.C. Beard, M.M. Al-Jassim, K. Zhu, Extrinsic ion migration in perovskite solar cells, Energy Environ. Sci. 10 (5) (2017) 1234–1242.

[14] Y. Yuan, J. Huang, Ion migration in organometal trihalide perovskite and its impact on photovoltaic efficiency and stability, Acct. Chem. Res. 49 (2) (2016) 286–293.

[15] C. Eames, J.M. Frost, P.R.F. Barnes, B.C. O'Regan, A. Walsh, M.S. Islam, Ionic transport in hybrid lead iodide perovskite solar cells, Nat. Commun. 6 (1) (2015) 7497.

[16] L. Krückemeier, B. Krogmeier, Z. Liu, U. Rau, T. Kirchartz, Understanding transient photoluminescence in halide perovskite layer stacks and solar cells, Adv. Energy Mater. 11 (19) (2021) 2003489.

[17] C.C. Boyd, R.C. Shallcross, T. Moot, R. Kerner, L. Bertoluzzi, A. Onno, S. Kavadiya, C. Chosy, E.J. Wolf, J. Werner, J.A. Raiford, C. de Paula, A.F. Palmstrom, Z.J. Yu, J.J. Berry, S.F. Bent, Z.C. Holman, J.M. Luther, E.L. Ratcliff, N.R. Armstrong, M.D. McGehee, Overcoming redox reactions at perovskite-nickel oxide interfaces to boost voltages in perovskite solar cells, Joule 4 (8) (2020) 1759–1775.

[18] L. Shi, M.P. Bucknall, T.L. Young, M. Zhang, L. Hu, J. Bing, D.S. Lee, J. Kim, T. Wu, N. Takamure, D.R. McKenzie, S. Huang, M.A. Green, A.W.Y. Ho-Baillie, Gas chromatography–mass spectrometry analyses of encapsulated stable perovskite solar cells, Science 368 (6497) (2020) aba2412.

[19] A. Dualeh, N. Tétreault, T. Moehl, P. Gao, M.K. Nazeeruddin, M. Grätzel, Effect of annealing temperature on film morphology of organic—inorganic hybrid pervoskite solid-state solar cells, Adv. Funct. Mater. 24 (21) (2014) 3250—3258.
[20] W. Zhang, S. Pathak, N. Sakai, T. Stergiopoulos, P.K. Nayak, N.K. Noel, A.A. Haghighirad, V.M. Burlakov, D.W. deQuilettes, A. Sadhanala, W. Li, L. Wang, D.S. Ginger, R.H. Friend, H.J. Snaith, Enhanced optoelectronic quality of perovskite thin films with hypophosphorous acid for planar heterojunction solar cells, Nat. Commun. 6 (1) (2015) 10030.
[21] P.S. Mathew, G.F. Samu, C. Janáky, P.V. Kamat, Iodine (I) expulsion at photoirradiated mixed halide perovskite interface. Should I stay or should I go? ACS Energy Lett. 5 (6) (2020) 1872—1880.
[22] S. Wang, Y. Jiang, E.J. Juarez-Perez, L.K. Ono, Y. Qi, Accelerated degradation of methylammonium lead iodide perovskites induced by exposure to iodine vapour, Nat. Energy 2 (1) (2016) 16195.
[23] L.E. Mundt, J. Tong, A.F. Palmstrom, S.P. Dunfield, K. Zhu, J.J. Berry, L.T. Schelhas, E.L. Ratcliff, Surface-activated corrosion in tin-lead halide perovskite solar cells, ACS Energy Lett. 5 (11) (2020) 3344—3351.
[24] L. Lanzetta, T. Webb, N. Zibouche, X. Liang, D. Ding, G. Min, R.J.E. Westbrook, B. Gaggio, T.J. Macdonald, M.S. Islam, S.A. Haque, Degradation mechanism of hybrid tin-based perovskite solar cells and the critical role of tin (IV) iodide, Nat. Commun. 12 (1) (2021) 2853.
[25] R. Prasanna, T. Leijtens, S.P. Dunfield, J.A. Raiford, E.J. Wolf, S.A. Swifter, J. Werner, G.E. Eperon, C. dePaula, A.F. Palmstrom, C.C. Boyd, M.F.A.M. van Hest, S.F. Bent, G. Teeter, J.J. Berry, M.D. McGehee, Design of low bandgap tin—lead halide perovskite solar cells to achieve thermal, atmospheric and operational stability, Nat. Energy 4 (11) (2019) 939—947.
[26] J. Xu, C.C. Boyd, Z.J. Yu, A.F. Palmstrom, D.J. Witter, B.W. Larson, R.M. France, J. Werner, S.P. Harvey, E.J. Wolf, W. Weigand, S. Manzoor, M.F.A.M. van Hest, J.J. Berry, J.M. Luther, Z.C. Holman, M.D. McGehee, Triple-halide wide—band gap perovskites with suppressed phase segregation for efficient tandems, Science 367 (6482) (2020) 1097—1104.
[27] T. Elmelund, B. Seger, M. Kuno, P.V. Kamat, How interplay between photo and thermal activation dictates halide ion segregation in mixed halide perovskites, ACS Energy Lett. 5 (1) (2020) 56—63.
[28] T. Elmelund, R.A. Scheidt, B. Seger, P.V. Kamat, Bidirectional halide ion exchange in paired lead halide perovskite films with thermal activation, ACS Energy Lett. 4 (8) (2019) 1961—1969.
[29] W. Peng, C. Aranda, O.M. Bakr, G. Garcia-Belmonte, J. Bisquert, A. Guerrero, Quantification of ionic diffusion in lead halide perovskite single crystals, ACS Energy Lett. 3 (7) (2018) 1477—1481.
[30] A.J. Knight, L.M. Herz, Preventing phase segregation in mixed-halide perovskites: A perspective, Energy Environ. Sci. 13 (7) (2020) 2024—2046.
[31] W. Mao, C.R. Hall, S. Bernardi, Y.B. Cheng, A. Widmer-Cooper, T.A. Smith, U. Bach, Light-induced reversal of ion segregation in mixed-halide perovskites, Nat. Mater. 20 (1) (2021) 55—61.
[32] A.J. Knight, J. Borchert, R.D.J. Oliver, J.B. Patel, P.G. Radaelli, H.J. Snaith, M.B. Johnston, L.M. Herz, Halide segregation in mixed-halide perovskites: Influence of A-site cations, ACS Energy Lett. 5 (3) (2021) 799—808.
[33] Y. Zhao, P. Miao, J. Elia, H. Hu, X. Wang, T. Heumueller, Y. Hou, G.J. Matt, A. Osvet, Y.T. Chen, M. Tarrago, D. de Ligny, T. Przybilla, P. Denninger, J. Will, J. Zhang, X. Tang, N. Li, C. He, A. Pan, A.J. Meixner, E. Spiecker, D. Zhang, C.J. Brabec, Strain-activated light-induced halide segregation in mixed-halide perovskite solids, Nat. Commun. 11 (1) (2020) 6328.
[34] Y. An, J. Hidalgo, C.A.R. Perini, A.F. Castro-Méndez, J.N. Vagott, K. Bairley, S. Wang, X. Li, J.P. Correa-Baena, Structural stability of formamidinium- and cesium-based halide perovskites, ACS Energy Lett. 5 (6) (2021) 1942—1969.
[35] Y. Liu, Y. Zhang, X. Zhu, Z. Yang, W. Ke, J. Feng, X. Ren, K. Zhao, M. Liu, M.G. Kanatzidis, S. Liu, Inch-sized high-quality perovskite single crystals by suppressing phase segregation for light-powered integrated circuits, Sci. Adv. 7 (7) (2021) abc8844. https://www.science.org/doi/10.1126/sciadv.abc8844.
[36] J. Cho, P.V. Kamat, How chloride suppresses photoinduced phase segregation in mixed halide perovskites, Chem. Mater. 32 (14) (2020) 6206—6212.

[37] S.R. Messenger, G.P. Summers, E.A. Burke, R.J. Walters, M.A. Xapsos, Modeling solar cell degradation in space: A comparison of the NRL displacement damage dose and the JPL equivalent fluence approaches, Prog. Photovoltaics Res. Appl. 9 (2) (2001) 103–121.

[38] A.R. Kirmani, B.K. Durant, N. Haegel, M.D. Kelzenberg, Y.M. Lao, M.D. McGehee, L. McMillon-Brown, D.P. Ostrowski, T.J. Peshek, B. Rout, I.R. Sellers, M. Steger, D. Walker, D.M. Wilt, K.T. Vansant, J.M. Luther, Countdown to Perovskite Space Launch: Guidelines to Performing Relevant Radiation-Hardness Experiments, Joule, 6 (5) (2022) 1015–1031. In this issue, https://www.sciencedirect.com/science/article/pii/S2542435122001313.

[39] B.K. Durant, H. Afshari, S. Singh, B. Rout, G.E. Eperon, I.R. Sellers, Tolerance of perovskite solar cells to targeted proton irradiation and electronic ionization induced healing, ACS Energy Lett. 5 (7) (2021) 2362–2368.

[40] F. Lang, N.H. Nickel, J. Bundesmann, S. Seidel, A. Denker, S. Albrecht, V.V. Brus, J. Rappich, B. Rech, G. Landi, H.C. Neitzert, Radiation hardness and self-healing of perovskite solar cells, Adv. Mater. 28 (39) (2016) 8726–8731.

[41] Y. Miyazawa, M. Ikegami, T. Miyasaka, T. Ohshima, M. Imaizumi, K. Hirose, In Evaluation of radiation tolerance of perovskite solar cell for use in space, in: 2015 IEEE 42nd Photovoltaic Specialist Conference (PVSC), 2015, pp. 1–4.

[42] J. Huang, M.D. Kelzenberg, P. Espinet-González, C. Mann, D. Walker, A. Naqavi, N. Vaidya, E. Warmann, H.A. Atwater, In effects of electron and proton radiation on perovskite solar cells for space solar power application, in: 2017 IEEE 44th Photovoltaic Specialist Conference (PVSC), 2017, pp. 1248–1252.

[43] F. Lang, M. Jošt, J. Bundesmann, A. Denker, S. Albrecht, G. Landi, H.C. Neitzert, J. Rappich, N.H. Nickel, Efficient minority carrier detrapping mediating the radiation hardness of triple-cation perovskite solar cells under proton irradiation, Energy Environ. Sci. 12 (5) (2019) 1634–1647.

[44] J.F. Ziegler, M.D. Ziegler, J.P. Biersack, SRIM — the stopping and range of ions in matter, Nucl. Instrum. Methods Phys. Res. Sect. B Beam Interact. Mater. Atoms 268 (11) (2010) 1818–1823.

[45] J. Barbé, D. Hughes, Z. Wei, A. Pockett, H.K.H. Lee, K.C. Heasman, M.J. Carnie, T.M. Watson, W.C. Tsoi, Radiation hardness of perovskite solar cells based on aluminum-doped zinc oxide electrode under proton irradiation, Solar RRL 3 (12) (2019) 1900219.

[46] W. Shockley, H.J. Queisser, Detailed balance limit of efficiency of p-n junction solar cells, J. Appl. Phys. 32 (3) (1961) 510–519.

[47] T. Leijtens, K.A. Bush, R. Prasanna, M.D. McGehee, Opportunities and challenges for tandem solar cells using metal halide perovskite semiconductors, Nat. Energy 3 (10) (2018) 828–838.

[48] Oxford PV Hits New World Record for Solar Cell 21 December, 2020. https://www.oxfordpv.com/news/oxford-pv-hits-new-world-record-solar-cell. (Accessed 8 October 2021).

[49] G.E. Eperon, T. Leijtens, K.A. Bush, R. Prasanna, T. Green, J.T.W. Wang, D.P. McMeekin, G. Volonakis, R.L. Milot, R. May, A. Palmstrom, D.J. Slotcavage, R.A. Belisle, J.B. Patel, E.S. Parrott, R.J. Sutton, W. Ma, F. Moghadam, B. Conings, A. Babayigit, H.G. Boyen, S. Bent, F. Giustino, L.M. Herz, M.B. Johnston, M.D. McGehee, H.J. Snaith, Perovskite-perovskite tandem photovoltaics with optimized band gaps, Science 354 (6314) (2016) 861–865.

[50] D. Zhao, C. Chen, C. Wang, M.M. Junda, Z. Song, C.R. Grice, Y. Yu, C. Li, B. Subedi, N.J. Podraza, X. Zhao, G. Fang, R.G. Xiong, K. Zhu, Y. Yan, Efficient two-terminal all-perovskite tandem solar cells enabled by high-quality low-bandgap absorber layers, Nat. Energy 3 (12) (2018) 1093–1100.

[51] D. Zhao, C. Wang, Z. Song, Y. Yu, C. Chen, X. Zhao, K. Zhu, Y. Yan, Four-Terminal all-perovskite tandem solar cells achieving power conversion efficiencies exceeding 23%, ACS Energy Lett. 3 (2) (2018) 305–306.

[52] R. Lin, K. Xiao, Z. Qin, Q. Han, C. Zhang, M. Wei, M.I. Saidaminov, Y. Gao, J. Xu, M. Xiao, A. Li, J. Zhu, E.H. Sargent, H. Tan, Monolithic all-perovskite tandem solar cells with 24.8% efficiency exploiting comproportionation to suppress Sn(II) oxidation in precursor ink, Nat. Energy 4 (10) (2019) 864–873.

[53] A.F. Palmstrom, G.E. Eperon, T. Leijtens, R. Prasanna, S.N. Habisreutinger, W. Nemeth, E.A. Gaulding, S.P. Dunfield, M. Reese, S. Nanayakkara, T. Moot, J. Werner, J. Liu, B. To, S.T. Christensen, M.D. McGehee, M.F.A.M. van Hest, J.M. Luther, J.J. Berry, D.T. Moore, Enabling flexible all-perovskite tandem solar cells, Joule 3 (9) (2019) 2193–2204.

[54] J. Tong, Z. Song, D.H. Kim, X. Chen, C. Chen, A.F. Palmstrom, P.F. Ndione, M.O. Reese, S.P. Dunfield, O.G. Reid, J. Liu, F. Zhang, S.P. Harvey, Z. Li, S.T. Christensen, G. Teeter, D. Zhao, M.M. Al-Jassim, M.F.A.M. van Hest, M.C. Beard, S.E. Shaheen, J.J. Berry, Y. Yan, K. Zhu, Carrier lifetimes of > 1 μs in Sn-Pb perovskites enable efficient all-perovskite tandem solar cells, Science 364 (6439) (2019) 475–479.

[55] B. Abdollahi Nejand, I.M. Hossain, M. Jakoby, S. Moghadamzadeh, T. Abzieher, S. Gharibzadeh, J.A. Schwenzer, P. Nazari, F. Schackmar, D. Hauschild, L. Weinhardt, U. Lemmer, B.S. Richards, I.A. Howard, U.W. Paetzold, Vacuum-assisted growth of low-bandgap thin films (FA0.8MA0.2Sn0.5Pb0.5I3) for all-perovskite tandem solar cells, Adv. Energy Mater. 10 (5) (2020) 1902583.

[56] A.R. Bowman, F. Lang, Y.H. Chiang, A. Jiménez-Solano, K. Frohna, G.E. Eperon, E. Ruggeri, M. Abdi-Jalebi, M. Anaya, B.V. Lotsch, S.D. Stranks, Relaxed current matching requirements in highly luminescent perovskite tandem solar cells and their fundamental efficiency limits, ACS Energy Lett. 6 (2) (2021) 612–620.

[57] Q. Han, Y.T. Hsieh, L. Meng, J.L. Wu, P. Sun, E.P. Yao, S.Y. Chang, S.H. Bae, T. Kato, V. Bermudez, Y. Yang, High-performance perovskite/Cu(In,Ga)Se_2 monolithic tandem solar cells, Science 361 (6405) (2018) 904–908.

[58] A. Al-Ashouri, A. Magomedov, M. Roß, M. Jošt, M. Talaikis, G. Chistiakova, T. Bertram, et al., Conformal monolayer contacts with lossless interfaces for perovskite single junction and monolithic tandem solar cells, Energy Environ. Sci. 12 (11) (2019) 3356–3369.

[59] M. Langenhorst, B. Sautter, R. Schmager, J. Lehr, E. Ahlswede, M. Powalla, U. Lemmer, B.S. Richards, U.W. Paetzold, Energy yield of all thin-film perovskite/CIGS tandem solar modules, Prog. Photovolt. 27 (4) (2019) 290–298.

[60] F. Lang, M. Jošt, K. Frohna, E. Köhnen, A. Al-Ashouri, A.R. Bowman, T. Bertram, et al., Proton radiation hardness of perovskite tandem photovoltaics, Joule 4 (5) (2020) 1054–1069, https://doi.org/10.1016/j.joule.2020.03.006. Access at.

[61] L.D. Whalley, J.M. Skelton, J.M. Frost, A. Walsh, Phonon anharmonicity, lifetimes, and thermal transport in $CH_3NH_3PbI_3$ from many-body perturbation theory, Phys. Rev. B 94 (22) (2016) 220301.

[62] A. Boden, D. Braunig, J. Klaer, F.H. Karg, B. Hosselbarth, G.L. Roche, Proton-irradiation of Cu(In,Ga)Se_2 and $CuInS_2$ thin-film solar cells, in: Conference Record of the Twenty-Eighth IEEE Photovoltaic Specialists Conference - 2000 (Cat. No.00CH37036), 2000, pp. 1038–1041.

[63] B.K. Durant, H. Afshari, S. Sourabh, V. Yeddu, M.T. Bamidele, S. Singh, B. Rout, G.E. Eperon, D.Y. Kim, I.R. Sellers, Radiation stability of mixed tin–lead halide perovskites: Implications for space applications, Sol. Energy Mater. Sol. Cells 230 (2021) 111232.

[64] B.K. Durant, A. Hadi, S. Sourabh, V. Yeddu, M.T. Bamidele, S. Singh, B. Rout, G.E. Eperon, D.Y. Kim, I.R. Sellers, Proton radiation tolerance of wide and narrow band gap perovskite solar cells, in: 2021 IEEE 48th Photovoltaic Specialists Conference, PVSC), 2021, pp. 1111–1114, https://doi.org/10.1109/PVSC43889.2021.9518848.

[65] H. Afshari, B.K. Durant, C.R. Brown, K. Hossain, D. Poplavskyy, B. Rout, I.R. Sellers, The role of metastability and concentration on the performance of CIGS solar cells under low-intensity-low-temperature conditions, Sol. Energy Mater. Sol. Cells 212 (2020) 110571.

[66] M. Stolterfoht, C.M. Wolff, J.A. Márquez, S. Zhang, C.J. Hages, D. Rothhardt, S. Albrecht, P.L. Burn, P. Meredith, T. Unold, D. Neher, Visualization and suppression of interfacial recombination for high-efficiency large-area pin perovskite solar cells, Nat. Energy 3 (10) (2018) 847–854.

[67] A. Francisco-López, B. Charles, O.J. Weber, M.I. Alonso, M. Garriga, M. Campoy-Quiles, M.T. Weller, A.R. Goñi, Equal footing of thermal expansion and electron–phonon interaction in the temperature dependence of lead halide perovskite band gaps, J. Phys. Chem. Lett. 10 (11) (2019) 2971–2977.

[68] T. Moot, J.B. Patel, G. McAndrews, E.J. Wolf, D. Morales, I.E. Gould, B.A. Rosales, C.C. Boyd, L.M. Wheeler, P.A. Parilla, S.W. Johnston, L.T. Schelhas, M.D. McGehee, J.M. Luther, Temperature coefficients of perovskite photovoltaics for energy yield calculations, ACS Energy Lett. 6 (5) (2021) 2038–2047.

[69] C.R. Brown, G.E. Eperon, V.R. Whiteside, I.R. Sellers, Potential of high-stability perovskite Solar Cells for low-intensity–low-temperature (LILT) outer planetary space missions, ACS Appl. Energy Mater. 2 (1) (2019) 814–821.

[70] J. Barbé, A. Pockett, V. Stoichkov, D. Hughes, H.K.H. Lee, M. Carnie, T. Watson, W.C. Tsoi, In situ investigation of perovskite solar cells' efficiency and stability in a mimic stratospheric environment for high-altitude pseudo-satellites, J. Mater. Chem. C 8 (5) (2020) 1715–1721.

[71] R.L. Milot, G.E. Eperon, H.J. Snaith, M.B. Johnston, L.M. Herz, Temperature-dependent charge-carrier dynamics in $CH_3NH_3PbI_3$ perovskite thin films, Adv. Funct. Mater. 25 (39) (2015) 6218–6227.

[72] A.D. Wright, C. Verdi, R.L. Milot, G.E. Eperon, M.A. Pérez-Osorio, H.J. Snaith, F. Giustino, M.B. Johnston, L.M. Herz, Electron–phonon coupling in hybrid lead halide perovskites, Nat. Commun. 7 (1) (2016) 11755.

[73] Z. Li, M. Yang, J.S. Park, S.H. Wei, J.J. Berry, K. Zhu, Stabilizing perovskite structures by tuning tolerance factor: Formation of formamidinium and cesium lead iodide solid-state alloys, Chem. Mater. 28 (1) (2016) 284–292.

[74] G. Divitini, S. Cacovich, F. Matteocci, L. Cinà, A. Di Carlo, C. Ducati, In situ observation of heat-induced degradation of perovskite solar cells, Nat. Energy 1 (2) (2016) 15012.

[75] N.K. Kim, Y.H. Min, S. Noh, E. Cho, G. Jeong, M. Joo, S.W. Ahn, Y. Kang, H.S. Lee, D. Kim, Investigation of thermally induced degradation in $CH_3NH_3PbI_3$ perovskite solar cells using in-situ synchrotron radiation analysis, Sci. Reports 7 (1) (2017) 4645.

[76] A.F. Akbulatov, V.M. Martynenko, L.A. Frolova, N.N. Dremova, I. Zhidkov, S.A. Tsarev, S.Y. Luchkin, E.Z. Kurmaev, S.M. Aldoshin, K.J. Stevenson, P.A. Troshin, Intrinsic thermal decomposition pathways of lead halide perovskites $APbX_3$, Sol. Energy Mater. Sol. Cells 213 (2020) 110559.

[77] AIAA Solar Cells and Solar Panels Committee on Standards, Qualification and Quality Requirements for Space Solar Cells, AIAA S-111A-2014 (Revision of AIAA S-111-2005), American Institute of Aeronautics and Astronautics, Reston, VA USA, June 2014, pp. 25.

CHAPTER EIGHT

Thermophotovoltaic energy conversion in space

Alejandro Datas[1] and Donald L. Chubb[2]
[1]Instituto de Energía Solar – Universidad Politécnica de Madrid, Madrid, Spain
[2]Retired from NASA Glenn Research Center, Cleveland, OH, United States

8.1 Introduction

Thermophotovoltaics (TPV) is the direct conversion of radiant heat into electricity through the photovoltaic (PV) effect [1–4]. Photons radiated by a high temperature body (or emitter) are directed toward a closely spaced infrared-sensitive PV cell (or TPV cell) that subsequently produces electricity (Fig. 8.1). The proximity between the photon emitter and the TPV cell enables tuning the radiative flux between them and minimizes the net transfer of out-band radiation, i.e., photons with energies lower than that of the TPV cell bandgap, which do not produce electron-hole pairs and contribute to increased heat losses (Fig. 8.2). Using strategies of this kind, current state-of-the-art TPV cells have reached conversion efficiencies over 30% at emitter temperatures of

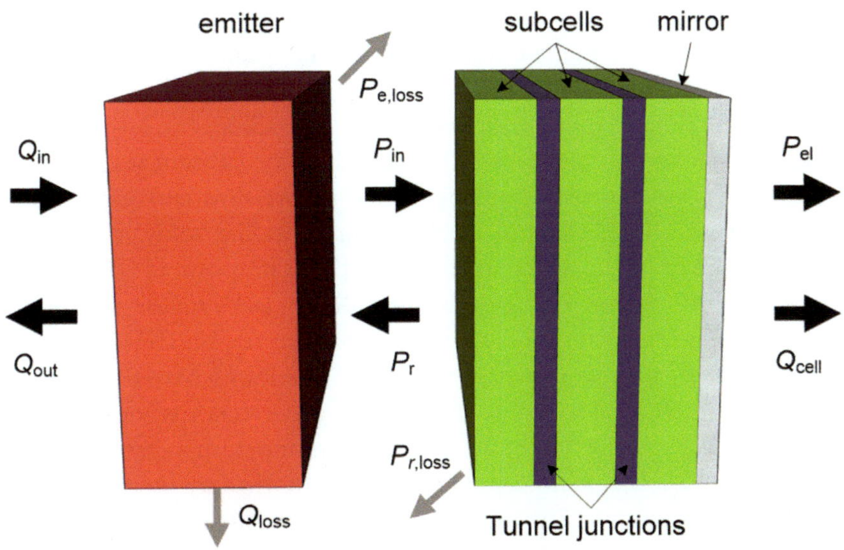

Figure 8.1 Thermophotovoltaic energy conversion. *Adapted from A. Datas, Optimum semiconductor bandgaps in single junction and multijunction thermophotovoltaic converters, Sol. Energy Mater. Sol. Cells 134 (Mar. 2015) 275–290, https://doi.org/10.1016/j.solmat.2014.11.049.*

Photovoltaics for Space
ISBN 978-0-12-823300-9, https://doi.org/10.1016/B978-0-12-823300-9.00005-4

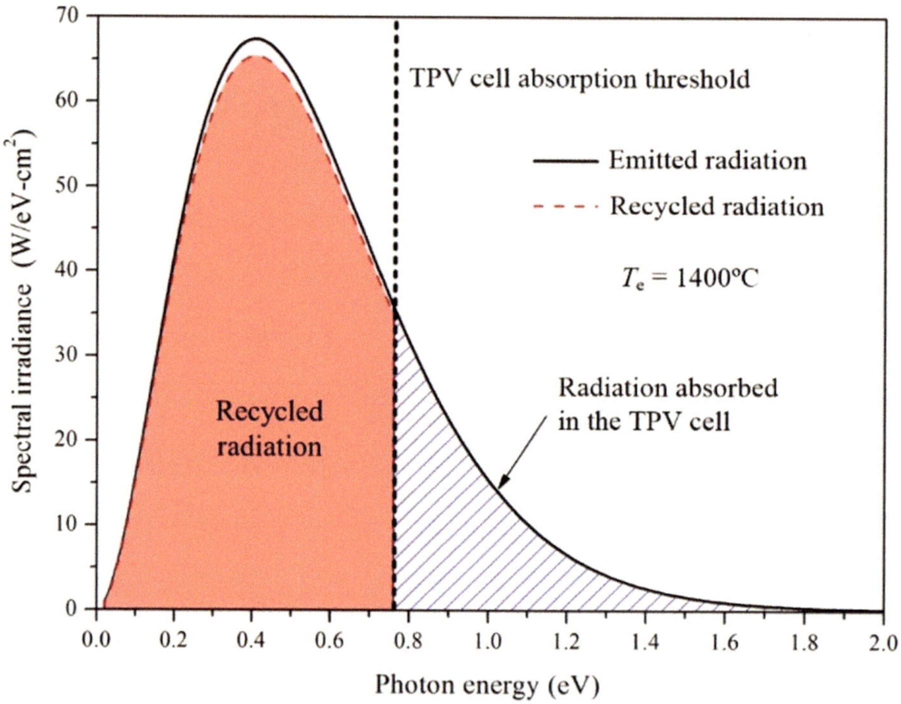

Figure 8.2 Spectral irradiances in a thermophotovoltaic converter. *Adapted from A. Datas, Optimum semiconductor bandgaps in single junction and multijunction thermophotovoltaic converters, Sol. Energy Mater. Sol. Cells 134 (Mar. 2015) 275–290, https://doi.org/10.1016/j.solmat.2014.11.049.*

~1200°C [5,6]. This is the highest conversion efficiency among all kinds of solid-state thermal-to-electric energy converters, including thermoelectrics [7] and thermionics [8], whose record efficiencies are in the range of 10%–15%.

The high conversion efficiency of TPV added to the advantages of all solid-state converters (lack of moving parts, independence of pressure and gravitational force, direct production of DC electricity) results in an appealing solution for thermal-to-electric energy conversion in space applications [9]. The possible sources of thermal energy in space include nuclear (radioisotope decay and nuclear fission) and solar thermal, but only nuclear systems have flown in space so far. Nuclear power generators are indicated for deep-space or planetary settlement missions where the solar resource is too weak or intermittent, and subsequently solar PV generators are not suitable. TPV has been primarily considered for the replacement of thermoelectric generators (TEGs) in small power (~100 W class) radioisotope power systems that are used in deep space and planetary settlement missions. The low rejection temperature of TPV, which is limited by the maximum cell temperature, implies a large radiator area that makes its use difficult in higher power applications. The high efficiency of TPV enables higher specific power

(W/kg) than TEG, but the main concern is the unproven reliability, especially regarding the possible contamination of the TPV optical elements by condensation of materials that readily evaporate at operational temperatures.

In this chapter, we review the state of the art and historical development of TPV technology in the frame of space power research programs. Section 8.2 reviews the state of the art of thermal-to-electric energy conversion in space applications. Section 8.3 describes the state of the art of TPV technology. Finally, Section 8.4 describes specific TPV developments targeting space power applications.

8.2 Thermal-to-electric energy conversion in space

Thermal-to-electric energy conversion has been widely developed for space power applications. Both solar and nuclear thermal inputs have been considered, but only nuclear-based systems have been flown in space so far [11]. Nuclear-powered generators are well suited for missions where the solar resource is too weak or intermittent, such as deep-space or planetary settlements, and where day/night cycling or settling of dust is problematic for solar PV systems. Two main kinds of nuclear power generators have been flown in space: radioisotope power systems (RPSs) and nuclear fission power systems (FPSs), targeting small ($\sim$100 W_{el}-class) and high (>1 kW_{el}-class) power outputs, respectively. In what follows we briefly describe the main developments related to each kind of system. A number of reviews can be found elsewhere that focus on thermal-to-electric energy conversion in space applications [9,11] or on the application of particular technologies including thermoelectrics [12], thermionics [13], dynamic engines [14,15], or TPV [9]. Fig. 8.3 illustrates the main features (efficiency, hot-side temperature, specific power, and nominal output power) for each of the technologies that are described subsequently.

Fission power systems were flown in space in the 1960–80s mostly by the former Soviet Union, which launched 35 missions of this kind, most of them (32) for radar ocean reconnaissance applications [12]. These reactors (BUK and ROMASHKA) were primarily using 1.5%–3% efficient SiGe TEGs to produce low output power (<3 kW_{el}) and were designed with lifetimes less than $\sim$6 months. The United States flew only one nuclear reactor in space (SNAP-10A, launched in 1965) and used $\sim$1.5% efficient SiGe TEGs to produce over 500 W_{el}. Two additional nuclear reactors were flown in space by the former Soviet Union based on thermionic energy converters (TOPAZ and ENISEY) [16,17]. These systems operated at extreme reactor core temperatures ($\sim$1800 K), lasted up to 1 year in orbit, and produced over 5 kW_{el} [18]. A nuclear in-core thermionic generator was also built by the United States at General Atomics (TRIGA) between 1962 and 1973. This system operated at core temperatures as high as 1900 K for more than 12,500 h, but it was never flown in space. Other early FPS research programs at NASA (SNAP-1, SNAP-2, SNAP-8, and SNAP-50) focused

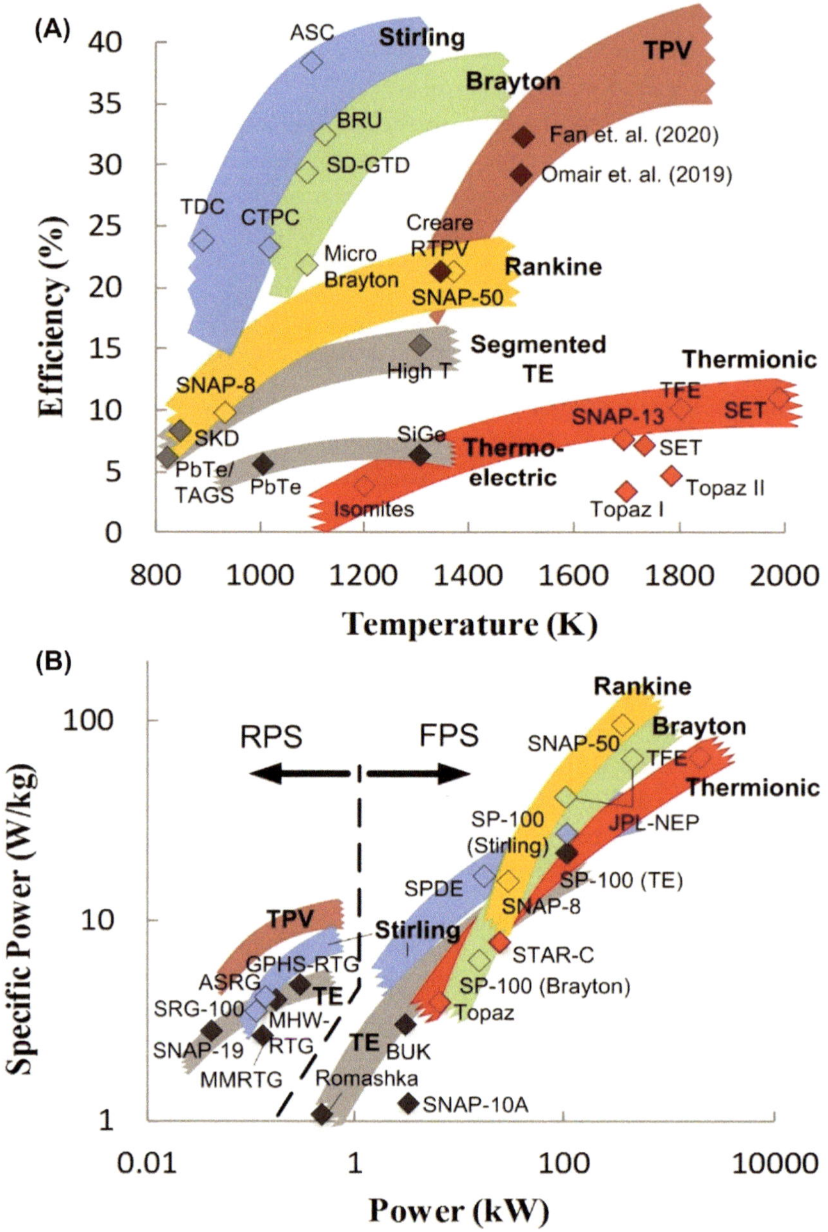

Figure 8.3 Efficiency and specific power of thermal-to-electric energy conversion devices as a function of (A) heat source temperature and (B) output nominal power. *Adapted and updated from A. Datas, A. Martí, Thermophotovoltaic energy in space applications: Review and future potential, Sol. Energy Mater. Sol. Cells 161 (Mar. 2017) 285–296, https://doi.org/10.1016/j.solmat.2016.12.007.*

on liquid metal—cooled Rankine engines based on mercury [19,20] or potassium [21] working fluids, and they were designed to produce a higher electric output power in the range of 4—350 kWel [11].

Since the last FPS-based missions were launched by the Soviet Union in the late 1980s, most of the research on FPS has focused on the development of intermediate (kW-class) and high (~100 kW-class) power generators for future science and human exploration missions. For instance, the US program SP-100 (1983—93) aimed at developing ~100 kW_{el} nuclear reactors using different energy conversion technologies. Among them, the Thermionic Fuel Element verification project investigated the lifetime issues of thermionic generators for very high power (>100 kWe) in-core nuclear reactors, validating the potential of this technology to meet mission life ranging from 3 to 7 years. High temperature (~1077°C) SiGe TEG were also considered, along with Stirling generators with power outputs in the range of 12—25 kW_{el} [11]. More recently, the Fission Surface Power program (starting in 2008) has developed ~12 kW_{el} Brayton [22] and Stirling [23] engines for nuclear fission power generation in Moon and Mars settlement missions. Stirling generators have recently become the baseline technology for power conversion in a new 1-kWe class of nuclear reactors that aim to address the gap above current RPS and below large-scale FPS [24—26]. A system of this kind, called Kilopower Reactor Using Stirling Technology (KRUSTY), was successfully tested by NASA in 2018 [27].

Radioisotope power systems have been continuously flown in space since 1961, mostly by the United States [12,28]. All of these RPS implementations use TEG for thermal-to-electric energy conversion and Pu-238 as a fuel. Early designs of these radioisotope thermoelectric generators (RTGs) used 5%—6% efficient PbTe TEGs to produce less than 70 W_{el}. In 1976, PbTe was replaced by SiGe to build the multihundred-watt RTG (MHW-RTG) units that enable a higher heat source temperature (900°C), conversion efficiency (6.6%), and output power (160 W_{el}). Since 1989, the predominant technology used general purpose heat source (GPHS) modules containing SiGe TEGs. Each RTG contains 18 GPHS modules each having a specific power of 5.1 W_{el}/kg and a nominal efficiency of 6.6% to generate a total of ~290 W_{el}. More recently, the GPHS-RTG has been replaced by the so-called multimission RTG (MMRTG), designed to operate both in vacuum and oxidizing atmospheres. It is composed of 8 GPHS modules that operate at lower temperatures (535°C) and utilizes PbTe/TAGS (tellurides of antimony, germanium, and silver) TEG (efficiency of 6.3%) to generate 125 W_{el}. This system shows a specific power of 2.8 W/kg and has been recently used in Mars missions (Curiosity and Mars 2020). Next-generation developments on RTGs at NASA include the replacement of the current PbTe/TAGS devices by more efficient skutterudite (~8%) or segmented (~15%) thermoelectric couples [29] to reach specific powers in excess of 3.6 and 8 W/kg, respectively.

Dynamic engine cycles, such as Stirling, Brayton, or Rankine [30], are also being considered for replacement of thermoelectrics in future RPS. Remarkably, the US Stirling Radioisotope Generator (SRG) (SRG110 program, 1997–2006) produced more than 20 Stirling engines with a nominal output power of ~60 W_{el}, demonstrating more than 26,000 h of operation. More recent designs have demonstrated 38% conversion efficiency for a hot-end (cold-end) temperature of 1123 K (362 K) that enable an overall radioisotope power generator specific power of ~4.1 W/kg [31]. Future advanced Stirling radioisotope generators (ASRGs) target a specific power over 8 W_{el}/kg and an output power of ~160 W_{el} [32]. Micro-machined Brayton engines have been also developed for RPS applications with expected efficiencies of ~21% at a turbine inlet temperature of 1055 K without noticeable degradation [30].

Early US developments (SNAP-3C1, 3D1, and SNAP-13) also included small thermionic converters for RPS applications [13]. The systems included both vacuum (SNAP-3C1, 3D1) and ignited-mode cesium-filled (SNAP-13) devices, and they demonstrated up to 6.6% conversion efficiency at hot-end temperatures of 1680 K. Also, milliwatt-class vacuum thermionic converters—so-called isomites—were developed in the early 1970s to produce ~1 W with an overall efficiency of ~4% at an emitter temperature of 1200 K.

Besides nuclear-powered systems, solar thermal power generators have been also investigated for space applications. Perhaps the earliest developments started within the US Solar Energy Technology (SET) program (1961 to early 1970s) and focused on thermionic energy converters. More than 135 devices were tested at 1700–2000 K emitter temperatures for more than 68,000 h, demonstrating efficiencies between 7% and 11% and power densities between 17 and 25 W/cm^2. A few decades later, the solar dynamic ground test demonstration (SD-GTD) project (1994–98) developed a 2 kW_{el} solar-powered prototype that used Brayton generators [33]. This prototype used mini-Brayton rotating unit engines (0.5–2 kW_{el}) [15] to successfully demonstrate 29% heat-to-electricity conversion efficiency (17% solar-to-electric) and over 800 h of operation (372 cycles). The system was also designed to incorporate thermal energy storage in LiF–CaF_2 phase change materials (PCMs) [34]. Despite the fact that complete solar thermal power generators have not yet been flown in space, experiments on the use of molten lithium fluoride PCMs (melting point of ~1050°C) have been already tested in space during the Space Shuttle Thermal Energy Storage flight experiments in 1996 [35].

In summary, only solid-state (static) thermal-to-electric energy converters (TEG and thermionic) have been flown in space so far, and only TEGs are still in use today. Primarily used in RPS, the main advantages of TEGs are their robustness and reliability, having demonstrated lifetime periods over 40 years (e.g., Voyager 1 and 2). Alternative technologies being considered for future space missions are thermionics, TPV, and dynamic generators (Stirling, Brayton, and Rankine). Thermionic generators are especially appealing for FPS due to their ability to operate inside the nuclear core. This eliminates the necessity of a heat transfer fluid, thereby reducing the overall weight of the converter [16]. In addition, the approach has a high rejection temperature (several hundred kelvin) that enables the use of a small, low-weight heat rejection radiator. Dynamic systems are a highly efficient alternative to TEG that could enable higher specific powers and a significant reduction of the fuel consumption. The latter is especially relevant for RPS applications that use the extremely scarce Pu-238 fuel. Due to the higher conversion efficiency and high rejection temperatures, dynamic power generators also enable larger output power generation ($>1\ kW_{el}$). However, dynamic systems face several barriers for their adoption, including a lack of lifetime qualification and their not being as robust against externally applied dynamic loads compared to RTGs [36]. TPV is the most efficient solid-state (static) converter, and the only one that has not yet been flown in space. TPV could theoretically enable high reliability, high specific power, and lower fuel consumption, but the lack of lifetime qualification is presently a barrier for its adoption.

8.3 Overview of thermophotovoltaic energy conversion

In TPV converters, thermally radiated photons are converted into electricity using infrared-sensitive PV cells, and so there have been strong links to the development of low-bandgap semiconductors such as gallium antimonide (GaSb, E_g ~0.73 eV) and indium gallium arsenide (InGaAs, E_g ~0.6–0.74 eV). TPV research dates to the early 1960s, mostly in the frame of military research programs to develop silent portable power generators [37]. At that time, silicon and germanium were the main semiconductors available, and it was not until the mid-1990s that high-quality and low-bandgap III-V semiconductors enabled the fabrication of highly efficient infrared PV devices. Originally designed as bottom subcells for multijunction tandem solar cells, InGaAs and GaSb have traditionally been the most widely used contemporary TPV materials. Developments in the field eventually resulted in the first demonstration of over 20% TPV cell conversion efficiency in early 2000s [38]. A key factor enabling this result was the inclusion of a back-surface reflector (BSR) at the rear side of a 0.6-eV InGaAs monolithic interconnected module (MIM) [39,40]. The MIM was comprised of 30 small monolithically series-interconnected 0.6-eV InGaAs PV cells within a large area ($2 \times 2\ cm^2$) device: thus, enabling a high output voltage (~12 V) and low output current (~0.4 A). The BSR reflected the low energy photons that do not contribute to photogeneration in the cell back to the heat source. Therefore, these photons are not considered to be lost since they contribute to reduce heat losses (Fig. 8.2). Very recently (2019–20), the combination of very thin film TPV cell structures with improved BSR have led to record TPV cell conversion efficiencies of 29% [6] and 32% [5]. Making thin-film TPV cells reduces free carrier absorption of the subbandgap photons, enabling subbandgap broadband reflectivities of ~95% [6]. Improved BSR designs include a dielectric spacer (or just air) between the cell and the reflector that enable reaching broadband reflectivities in the range of 96%–99% [5,41].

Thermophotovoltaic cell conversion efficiencies are typically defined as the ratio between generated electric power (P_{el}) and the heat absorbed by the cell ($Q_{abs} = P_{in} - P_r$). However, this definition does not consider other possible paths for the heat to flow from the emitter to the environment that exist in real TPV systems (Fig. 8.1). These losses may include nonidealities in the thermal insulation of the emitter (Q_{loss}), back-end heat losses in the emitter (Q_{out}), or front-end view factor optical losses ($P_{e,loss}$ and $P_{r,loss}$). Many different intermediate efficiency definitions exist in the TPV literature to account for these losses [1]. Some examples are the thermal efficiency (i.e., total thermal power into the emitter divided by the total input power to the system) or the spectral cavity efficiency (i.e., radiation absorbed in the TPV cell within the useful spectral range for PV generation divided by the total thermal power into the emitter). For the sake of simplicity, in what follows, we refer only to the TPV cavity and TPV system efficiencies, defined in this chapter as $\eta_{cavity} = P_{el}/(Q_{in} - Q_{out})$ and $\eta_{system} = P_{el}/Q_{in}$, respectively [3].

Most types of TPV systems use a heat source that is exogenous to the system, e.g., sunlight or hydrocarbon fuels, so they require a channel for the heat to flow into the system. This channel represents a path for a large amount of heat losses (Q_{out}) that deteriorate the system efficiency ($\eta_{system} < \eta_{cavity}$). Higher system efficiencies can be potentially achieved if the heat source is integrated within the system and it is very well insulated from the environment, i.e., $Q_{out} = 0$ and $\eta_{system} = \eta_{cavity}$. Radioisotope TPV (RTPV) generators (described in the next section) are a type of such a closed-TPV system that have demonstrated system conversion efficiencies of 17%–20% [42], which is only slightly lower than that of the individual TPV cells used in the system (~24% [38]). However, open-TPV systems, such as solar- or combustion-powered systems, have yet not reached high system conversion efficiencies, currently limited to typically less than 10% [4,43] due to the large heat leakages linked to the external heating process (absorber reemission in solar TPV or rejection of hot exhaust gases in combustion-TPV).

A typical strategy to reduce the heat rejection losses (Q_{out}) of open-TPV systems is to increase the fraction of power that is radiated toward the TPV cell with respect to the one that is emitted back to the heat source. For instance, in solar-powered systems this can be done by using spectrally selective absorbers that minimize back emission losses [44] or by increasing the emitter/cell area, thereby increasing the amount of radiation that is incident on the cells [45]. Alternatives to large emitter/cell areas include light-pipes [46], near-field TPV arrangements [47–50], hybrid devices like thermionic-photovoltaic [51,52] or thermoelectric-thermophotovoltaics [53], and thermophotonic devices [54], all of which are intended to increase the power density (W/cm^2) of the TPV generator. To a lesser extent, the use of multijunction TPV cells [10,55] also enables an increase in the output power density and reduces the sensitivity to spectral control inefficiencies. These high power density alternatives do not only address the challenge of rejection heat losses in open-TPV systems, but they also enable a higher specific power (W/kg), which is particularly relevant for space power applications.

Spectrally selective emitters, which are designed to preferably emit in-band thermal radiation and suppress out-band (subbandgap) emission, can reduce thermalization losses and improve the spectral efficiency of individual TPV cells [4,56,57]. Moreover, by restricting the emission to a narrower range of wavelengths, the view factor optical cavity losses ($P_{e,loss}$ and $P_{r,loss}$ in Fig. 8.2) can also be significantly reduced without affecting the in-band radiative power impinging the TPV cells. Different kinds of selective emitters have been developed for TPV applications, including intrinsically selective materials (e.g., rare earth materials and transition metal-doped oxides) and structurally tunable materials (e.g., photonic crystals and metamaterials) [57]. Some of these approaches have been already tested in system demonstrations [56], and others have demonstrated high temperature stability operating at temperatures near or beyond 1000°C for over 100 h [4,56]. Individual spectral efficiencies of selective emitters (defined as the ratio of in-band radiative energy flux to the total radiative energy flux when paired with a black body [4]) range from less than 30% in the case of structured materials to nearly 50% for transition metal oxide emitters, compared to individual spectral efficiencies of over 60% for TPV cells having front side filters and a BSR [4]. Selective emitters and filters can be paired with TPV cells that incorporate a BSR to reach spectral efficiencies of over 80% [58].

8.4 Thermophotovoltaic systems for space applications

Research on TPV energy conversion for space applications began in the 1980s at the NASA Lewis Research Center, later renamed the NASA Glenn Research Center. A demonstration by the Gillette Corporation of a combustion-driven TPV energy converter created an interest in developing TPV for space power systems. That prototype converter employed a Coleman lantern as the thermal source, a neodymium oxide (Nd_2O_3) photon emitter, and a silicon (Si) PV cell. The initial TPV research program at NASA investigated the potential of rare earth aluminum garnets ($R_3Al_5O_{12}$), where R represents a rare earth. Thin (0.1−1 mm) layers of erbium, holmium, and thulium aluminum garnets were studied. The photon emitter samples were backed with a very small emittance platinum foil to block the radiation originating from the source that heats the sample. Theoretical spectral emittances were calculated, and experimental emittances were measured for these materials. Also in the 1980s the development of InGaAs PV cells began at NASA, including the development of MIM devices.

Eventually, research on TPV for space applications focused on the development of RPS. Radioisotope TPV (RTPV) was proposed in mid-1990s [59] as an alternative to

RTG and SRG with the potential to achieve specific powers beyond 15 W/kg [42,60–62]. This is three to five times greater than the SRG110 (3.6 W/kg), MMRTG (2.8 W/kg), and GPHS-RTG (5.1 W/kg) systems described before. Experimental work to develop RTPV generators started in 2003, when NASA issued contracts to EDTEK and Creare to develop an RTPV system using an electrically heated thermal source to emulate a 250-Wth GPHS unit [60] (Fig. 8.4). The TPV converter used 0.6-eV InGaAs MIMs [39,40] fabricated by Emcore and interference/plasma filters made by Rugate. A 10 × 10 cm TPV array containing 16 MIMs, each of them comprised of 25 series-

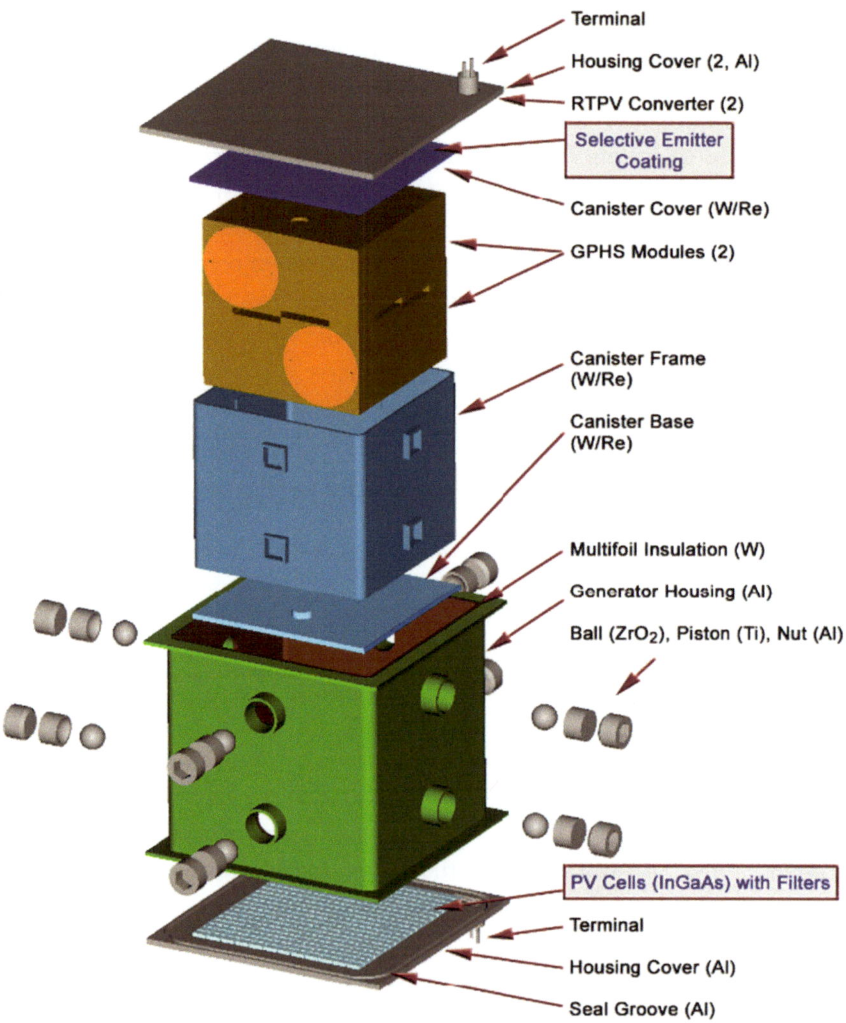

Figure 8.4 RTPV system concept developed by Creare. *Reproduced with permission from D. Wolford, D.L. Chubb, Theoretical performance of a radioisotope thermophotovoltaic (RTPV) power system, Presented at the 7th International Energy Conversion Engineering Conference, Denver, Colorado, Aug. 2009, https://arc.aiaa.org/doi/10.2514/6.2009-4655.*

interconnected InGaAs subcells, was fabricated to provide ~50 W_{el} at a heat source temperature of 1077°C, demonstrating conversion efficiencies of ~20% [42]. Although specific powers of up to 17 W/kg were initially predicted for this system [60], more recent studies have yielded more conservative estimates, typically in the range of 6–8 W_{el}/kg [62,63]. A low rejection temperature (<80°C) is presently the main bottleneck of RTPV to increase the specific power. Smaller radiators would enable lower weight at the expense of higher cell temperature and lower output power and conversion efficiency [60]. For this reason, state-of-the-art 0.6-eV InGaAs MIM devices were tested at 90°C during more than 4000 h, demonstrating no relevant degradation. The TPV cell degradation under neutron radiation from radioisotope fuel was also investigated by the Naval Research Laboratory, ultimately predicting a 20% degradation by the end of a 14-year mission [60], compared to 14% for the MMRTG system.

One important conclusion resulting from this seminal RTPV research was that a significant amount of radiation "leaks" out in the gap between the photon emitter and the PV array, even though this gap is comparatively small (2 mm) [1]. To reduce such leakage, thin reflecting radiation shields could be used to block the radiation loss [62]. Also, a significant conclusion from modeling work at NASA was the importance of reflecting back to the emitter most of the photons that cannot be converted to electricity; to obtain high system efficiency (>20%), the reflectance for the photons that cannot be converted must be greater than 95%.

More recently, the micro-isotope power system program produced the first mW-scale RTPV system [64,65], where additional degradation tests were performed on 0.6-eV InGaAs MIMs using both Pu-238 and Cm-244 radioisotope sources. The MIMs were exposed to Cm-244 over 120 days, which is equivalent to more than 100 years of operation under Pu-238. The MIMs showed less than 1% of degradation per year, which includes the temperature decay of the heat source and cell degradation due to neutron damage. Also, the system was tested using an enriched Pu-238 heat source during more than 500 days, corroborating less than 1% degradation per year due to neutron damage. Higher power versions (up to ~3 W_{el}) of hermetically sealed TPV systems have been also developed by General Atomics to demonstrate a TPV autonomous power system conversion efficiency of ~8%. The lower efficiency with respect to earlier NASA developments is attributed to the scaling issues of TPV for such small output powers.

Recent works have also reported several RTPV system designs that incorporate improved spectral control strategies [66], such as two-dimensional micropatterned photonic crystal (PhC) emitters [67–70]. Experimentally validated models [67] predict that an efficiency of 7.8% would be attainable by RTPV that use 0.55-eV InGaAsSb TPV cells to produce an output power of 4.7 W_{el}. This result agrees with the one reported by General Atomics for a similar power output. The same authors later predicted that RTPV efficiencies of 8.6% and specific powers of ~10 W/kg could be attainable by pairing such InGaAsSb TPV cells to 2D-PhCs [68]. According to that study, more efficient InGaAs MIMs could achieve ~18% efficiency and specific powers in the range of 13.6–21 W/kg, which is similar to the initial predictions made for Creare's RTPV system [60].

Solar TPV (STPV) [44,45,71] has been also proposed for space applications, the main advantage being the possibility of integrating extremely energetically dense thermal energy storage [72–75]. Thermal energy storage is particularly advantageous for solar thermal propulsion (STP) systems, which are being investigated to replace conventional chemical and electrical rockets [76]. In these systems, thermal energy storage could enable STP during eclipses. Experimental research has been already conducted to evaluate the feasibility of using silicon and boron PCMs for latent heat energy storage to augment STP in microsatellites operating in low-Earth orbits [77–81]. Silicon and boron have remarkably high latent heats of 1800 J/g and 4650 J/g, respectively, and high melting temperatures (1410°C and 2540°C, respectively). Due to these extreme temperatures, TPV has been considered a better choice than TEG for power generation. STPV may also offer some advantages for missions in high radiation environments [60], such as in mid-Earth orbits coinciding with the Van Allen radiation belts, where PV systems degrade significantly in short periods of time. TPV cells are shielded by the thermal emitter, which can be made resistant to radiation and thus withstand such extreme environmental conditions. Finally, STPV could be also used in missions very near to the Sun, where the solar flux is too intense for direct incidence onto solar cells [82,83]. For instance, the NASA Parker Solar Probe mission [84], launched in 2018, incorporates a 2.7-m-diameter, 17-cm-thick carbon–carbon composite shield that will reach temperatures of nearly 1400°C at the closest distances to the Sun. TPV was considered during the initial design phase of this mission for converting the radiant heat emitted by the thermal shield into electricity [60].

Early experimental research on STPV was carried out by McDonnell Douglas in the mid-1990s [85] and by EDTEK in the early 2000s [86] for both space and terrestrial applications, demonstrating high emitter temperatures (up to 1360°C) and power densities (~0.9 W/cm^2) under real outdoor solar irradiance conditions. However, no experimental system efficiencies were reported until the early 2010s, when solar-to-electricity STPV conversion efficiencies of ~1% [87] were measured. This result was progressively improved during the 2010s [88–90] to finally reach solar-to-electricity conversion efficiencies in the range of ~7%–8% [91–93].

8.5 Conclusions

Thermophotovoltaic devices have been developed in the frame of space power research programs since the 1980s. With record cell efficiencies already over 30%, TPV is the most efficient solid-state thermal-to-electric energy converter technology developed so far, and the only one that has not yet been flown in space. TPV has been investigated primarily as an alternative to TEGs in RPSs. Radioisotope TPV could be used in deep space and planetary settlement missions, where solar irradiance is too weak, precluding the use of other solar PV technologies. The higher efficiency could enable a reduction of over 60% in Pu-238 fuel consumption and a threefold specific power enhancement with respect to current systems using TEGs. The main drawbacks are unproven reliability and the requirement of low cell temperatures, which preclude application in high-power systems. Preliminary reliability tests have concluded that TPV cells can withstand relatively high temperatures (~90°C) for more than 4000 h without relevant degradation and have an acceptable degradation under neutron radiation (20% degradation by the end of a 14-year mission).

References

[1] D.L. Chubb, Fundamentals of Thermophotovoltaic Energy Conversion, Elsevier, 2007, p. 515.
[2] T. Bauer, Thermophotovoltaics: Basic Principles and Critical Aspects of System Design, Springer, 2011, 222 pp.
[3] A. Datas, R. Vaillon, Chapter 11—thermophotovoltaic energy conversion, in: A. Datas (Ed.), Ultra-High Temperature Thermal Energy Storage, Transfer and Conversion, Woodhead Publishing, 2021, pp. 285—308.
[4] T. Burger, C. Sempere, B. Roy-Layinde, A. Lenert, Present Efficiencies and Future Opportunities in Thermophotovoltaics, Joule, Jul. 2020, https://doi.org/10.1016/j.joule.2020.06.021.
[5] D. Fan, T. Burger, S. McSherry, B. Lee, A. Lenert, S.R. Forrest, Near-perfect photon utilization in an air-bridge thermophotovoltaic cell, Nature (Sep. 2020) 1—5, https://doi.org/10.1038/s41586-020-2717-7.
[6] Z. Omair, G. Scranton, L.M. Pazos-Outón, T.P. Xiao, et al., Ultraefficient thermophotovoltaic power conversion by band-edge spectral filtering, Proc. Natl. Acad. Sci. 116 (31) (Jul. 2019) 15356—15361, https://doi.org/10.1073/pnas.1903001116.
[7] M. Zebarjadi, K. Esfarjani, M.S. Dresselhaus, Z.F. Ren, G. Chen, Perspectives on thermoelectrics: from fundamentals to device applications, Energy Environ. Sci. 5 (1) (2012) 5147—5162, https://doi.org/10.1039/C1EE02497C.
[8] K.A.A. Khalid, T.J. Leong, K. Mohamed, Review on thermionic energy converters, IEEE Trans. Electron. Dev. 63 (6) (Jun. 2016) 2231—2241, https://doi.org/10.1109/TED.2016.2556751.
[9] A. Datas, A. Martí, Thermophotovoltaic energy in space applications: review and future potential, Sol. Energy Mater. Sol. Cells. 161 (Mar. 2017) 285—296, https://doi.org/10.1016/j.solmat.2016.12.007.
[10] A. Datas, Optimum semiconductor bandgaps in single junction and multijunction thermophotovoltaic converters, Sol. Energy Mater. Sol. Cells. 134 (Mar. 2015) 275—290, https://doi.org/10.1016/j.solmat.2014.11.049.
[11] A. Mazzetti, M. Gianotti Pret, G. Pinarello, L. Celotti, M. Piskacev, A. Cowley, Heat to electricity conversion systems for moon exploration scenarios: a review of space and ground technologies, Acta Astronaut. 156 (Mar. 2019) 162—186, https://doi.org/10.1016/j.actaastro.2018.09.025.
[12] D.M. Rowe, Thermoelectrics Handbook: Macro to Nano, CRC Press, New York, 2005.
[13] G. Miskolczy, D.P. Lieb, Radioisotope thermionic converters for space applications, in: Energy Conversion Engineering Conference, 1990. IECEC-90. Proceedings of the 25th Intersociety, vol. 1, Aug. 1990, pp. 222—226, https://doi.org/10.1109/IECEC.1990.716874.
[14] L.S. Mason, J.G. Schreiber, A Historical Review of Brayton and Stirling Power Conversion Technologies for Space Applications, Nov. 2007. Accessed: Feb. 11, 2016. [Online]. Available: http://ntrs.nasa.gov/search.jsp?R=20070038168.

[15] L.S. Mason, A Summary of Closed Brayton Cycle Development Activities at NASA, Apr. 29, 2009. Accessed: Feb. 12, 2016. [Online]. Available: http://ntrs.nasa.gov/search.jsp?R=20120016632.
[16] Thermionics Quo Vadis?: An Assessment of the DTRA's Advanced Thermionics Research and Development Program, National Academies Press, Washington, DC, 2001.
[17] G.N. Hatsopoulos, E.P. Gyftopoulos, Thermionic Energy Conversion, 2 vols, The MIT Press, Cambridge, Massachusetts, 1979.
[18] Y. Deng, et al., Performance evaluation and efficiency enhancement of a space thermionic fuel element for thermal energy conversion and utilization, Appl. Therm. Eng. 173 (Jun. 2020) 115237, https://doi.org/10.1016/j.applthermaleng.2020.115237.
[19] R.T. Lahey, V. Dhir, Research in Support of the Use of Rankine Cycle Energy Conversion Systems for Space Power and Propulsion, Jul. 2004 [Online]. Available: http://ntrs.nasa.gov/search.jsp?R=20040086725. (Accessed 31 May 2016).
[20] R.L. Cataldo, G.L. Bennett, U.S. Space Radioisotope Power Systems and Applications: Past, Present and Future, 2011. Accessed: Jun. 02, 2016. [Online]. Available: http://ntrs.nasa.gov/search.jsp?R=20120000731.
[21] G.L. Yoder, Technology Development Program for an Advanced Potassium Rankine Power Conversion System Compatible with Several Space Reactor Designs, ORNL/TM-2004/214, 2005. Accessed: Jun. 10, 2016. [Online]. Available: http://web.ornl.gov/~webworks/cppr/y2001/rpt/121399.pdf.
[22] R.L. Fuller, Closed Brayton Cycle Power Conversion Unit for Fission Surface Power Phase I Final Report, Jun. 2010. Accessed: Feb. 12, 2016. [Online]. Available: http://ntrs.nasa.gov/search.jsp?R=20100026654.
[23] M. Briggs, M. Gibson, S. Geng, J. Pearson, T. Godfroy, Development Status of the Fission Power System Technology Demonstration Unit, Jul. 2012, https://doi.org/10.2514/6.2012-3711.
[24] L.S. Mason, M.A. Gibson, D. Poston, Kilowatt-Class Fission Power Systems for Science and Human Precursor Missions, Sep. 2013. Accessed: Apr. 22, 2016. [Online]. Available: http://ntrs.nasa.gov/search.jsp?R=20140010823.
[25] M.A. Gibson, L.S. Mason, C.L. Bowman, D.I. Poston, P.R. McClure, J. Creasy, C. Robinson, Development of NASA's Small Fission Power System for Science and Human Exploration, Jun. 2015. Accessed: Apr. 22, 2016. [Online]. Available: http://ntrs.nasa.gov/search.jsp?R=20150011642.
[26] S. Wilson, Overview of stirling technology research at NASA Glenn Research Center, in: 13th International Energy Conversion Engineering Conference, American Institute of Aeronautics and Astronautics, 2015. Available at: https://ntrs.nasa.gov/citations/20150021865.
[27] A. Barco, R.M. Ambrosi, H.R. Williams, K. Stephenson, Radioisotope power systems in space missions: overview of the safety aspects and recommendations for the European safety case, J. Space Saf. Eng. 7 (2) (Jun. 2020) 137–149, https://doi.org/10.1016/j.jsse.2020.03.001.
[28] R.G. Lange, W.P. Carroll, Review of recent advances of radioisotope power systems, Energy Convers. Manag. 49 (3) (Mar. 2008) 393–401, https://doi.org/10.1016/j.enconman.2007.10.028.
[29] NASA, NASA Technology Roadmaps. TA 3: Space Power and Energy Storage, 2015. Accessed: Jun. 10, 2016. [Online]. Available: https://www.nasa.gov/sites/default/files/atoms/files/2015_nasa_technology_roadmaps_ta_3_space_power_energy_storage_final.pdf.
[30] D. Anderson, W. Wong, K. Tuttle, An overview and status of NASA's radioisotope power conversion technology NRA, in: 3rd International Energy Conversion Engineering Conference, American Institute of Aeronautics and Astronautics, Available at: https://doi.org/10.2514/6.2005-5713.
[31] NASA, Advanced Stirling Radioisotope Generator (ASRG), NF-2013-07-568-HQ. https://solarsystem.nasa.gov/rps/docs/APP%20ASRG%20Fact%20Sheet%20v3%209-3-13.pdf.
[32] S.M. Oriti, Advanced stirling radioisotope generator engineering unit 2 (ASRG EU2) final assembly, in: Presented at the Nuclear and Emerging Technologies for Space 2015, Albuquerque, NM, United States, Feb. 2015. Accessed: Feb. 18, 2016. [Online]. Available: http://ntrs.nasa.gov/search.jsp?R=20150004104.
[33] D. Alexander, 2 kWe Solar Dynamic Ground Test Demonstration Project, 1997. Accessed: Dec. 25, 2020. [Online]. Available: https://ntrs.nasa.gov/citations/19970012688.

[34] C.A. Hall, E.K. Glakpe, J.N. Cannon, T.W. Kerslake, Modeling cyclic phase change and energy storage in solar heat receivers, in: Presented at the Thermodynamics, 23-25 Jun. 1997, United States, Jun. 1997. Accessed: Feb. 11, 2016. [Online]. Available: http://ntrs.nasa.gov/search.jsp?R=19970025060.

[35] C. Tolbert, Jr F. Robinson, Experimental results from the thermal energy storage-2 (TES-2) flight experiment, in: 36th AIAA Aerospace Sciences Meeting and Exhibit, American Institute of Aeronautics and Astronautics.

[36] E.S. Brummel, P. Ostdiek, D. Woerner, K. Hibbard, et al., Identifying and Mitigating Barriers to the Adoption of Dynamic Radioisotope Power Systems for Space Flight, in: 2019 IEEE Aerospace Conference, Mar. 2019, pp. 1−16, https://doi.org/10.1109/AERO.2019.8742171.

[37] R.E. Nelson, A brief history of thermophotovoltaic development, Semicond. Sci. Technol. 18 (2003) S141−S143.

[38] B. Wernsman, R.R. Siergiej, S.D. Link, R.G. Mahorter, et al., Greater than 20% radiant heat conversion efficiency of a thermophotovoltaic radiator/module system using reflective spectral control, IEEE Trans. Electron. Dev. 51 (3) (Mar. 2004) 512−515, https://doi.org/10.1109/TED.2003.823247.

[39] A. Datas, P.G. Linares, Monolithic interconnected modules (MIM) for high irradiance photovoltaic energy conversion: a comprehensive review, Renew. Sustain. Energy Rev. 73 (Suppl. C) (Jun. 2017) 477−495, https://doi.org/10.1016/j.rser.2017.01.071.

[40] D. Wilt, R. Wehrer, M. Palmisiano, M. Wanlass, C. Murray, Monolithic interconnected modules (MIMs) for thermophotovoltaic energy conversion, Semicond. Sci. Technol. 18 (5) (2003) p. PII S0268-1242(03)60134-3.

[41] T. Burger, D. Fan, K. Lee, S.R. Forrest, A. Lenert, Thin-film architectures with high spectral selectivity for thermophotovoltaic cells, ACS Photonics 5 (7) (Jul. 2018) 2748−2754, https://doi.org/10.1021/acsphotonics.8b00508.

[42] C.J. Crowley, N.A. Elkouh, S. Murray, D.L. Chubb, Thermophotovoltaic converter performance for radioisotope power systems, in: AIP Conference Proceedings, vol. 746, Feb. 2005, pp. 601−614, https://doi.org/10.1063/1.1867178.

[43] C. Ferrari, F. Melino, M. Pinelli, P.R. Spina, M. Venturini, Overview and status of thermophotovoltaic systems, Energy Proc. 45 (2014) 160−169, https://doi.org/10.1016/j.egypro.2014.01.018.

[44] Z. Zhou, E. Sakr, Y. Sun, P. Bermel, Solar thermophotovoltaics: reshaping the solar spectrum, Nanophotonics 0 (0) (Jan. 2016), https://doi.org/10.1515/nanoph-2016-0011.

[45] A. Datas, C. Algora, Global optimization of solar thermophotovoltaic systems, Prog. Photovoltaics Res. Appl. (2012), https://doi.org/10.1002/pip.2202 vol. Published Online.

[46] D.L. Chubb, Light Pipe Thermophotovoltaics (LTPV), vol. 890, Feb. 2007, pp. 297−316, https://doi.org/10.1063/1.2711748.

[47] A. Fiorino, L. Zhu, D. Thompson, R. Mittapally, P. Reddy, E. Meyhofer, Nanogap near-field thermophotovoltaics, Nat. Nanotechnol. (Jun. 2018) 1, https://doi.org/10.1038/s41565-018-0172-5.

[48] T. Inoue, T. Koyama, D.D. Kang, T. Asano, S. Noda, S. Noda, One-chip integrated near-field thermophotovoltaic devices using intermediate transparent substrates, in: Conference on Lasers and Electro-Optics (2019), Paper SM1J.7, May 2019, p. SM1J.7, https://doi.org/10.1364/CLEO_SI.2019.SM1J.7.

[49] C. Lucchesi, D. Cakiroglu, J.-P. Perez, T. Taliercio, E. Tournié, P.-O. Chapuis, R. Vaillon, Harnessing near-field thermal photons with efficient photovoltaic conversion, arXiv:1912.09394 [physics] (Dec. 2019). May 23, 2020. [Online]. Available: http://arxiv.org/abs/1912.09394.

[50] G.R. Bhatt, B. Zhao, S. Roberts, I. Datta, et al., Integrated near-field thermo-photovoltaics for heat recycling, Nat. Commun. 11 (1) (May 2020), https://doi.org/10.1038/s41467-020-16197-6. Art. no. 1.

[51] A. Datas, Hybrid thermionic-photovoltaic converter, Appl. Phys. Lett. 108 (14) (Apr. 2016) 143503, https://doi.org/10.1063/1.4945712.

[52] A. Datas, R. Vaillon, Thermionic-enhanced near-field thermophotovoltaics, Nano Energy 61 (Jul. 2019) 10−17, https://doi.org/10.1016/j.nanoen.2019.04.039.

[53] D.L. Chubb, B.S. Good, A combined thermophotovoltaic-thermoelectric energy converter, Sol. Energy 159 (Jan. 2018) 760–767, https://doi.org/10.1016/j.solener.2017.11.030.

[54] N.-P. Harder, M.A. Green, Thermophotonics, Semicond. Sci. Technol. 18 (5) (2003) S270, https://doi.org/10.1088/0268-1242/18/5/319.

[55] R.R. Siergiej, S. Sinharoy, T. Valko, R.J. Wehrer, et al., InGaAsP/InGaAs tandem TPV device, in: AIP Conference Proceedings, vol. 738, Nov. 2004, pp. 480–488, https://doi.org/10.1063/1.1841927.

[56] R. Sakakibara, V. Stelmakh, W.R. Chan, M. Ghebrebrhan, et al., Practical emitters for thermophotovoltaics: a review, JPE 9 (3) (Feb. 2019) 032713, https://doi.org/10.1117/1.JPE.9.032713.

[57] N.A. Pfiester, T.E. Vandervelde, Selective emitters for thermophotovoltaic applications, Phys. Status Solidi A 214 (1) (2017) 1600410, https://doi.org/10.1002/pssa.201600410.

[58] P.M. Fourspring, D.M. DePoy, J.F Beausang, E.J. Gratrix, et al., Thermophotovoltaic spectral control, in: AIP Conference Proceedings, vol. 738, Nov. 2004, pp. 171–179, https://doi.org/10.1063/1.1841892.

[59] A. Schock, C.T. Or, V. Kumar, Modified design of radioisotope thermophotovoltaic generator to mitigate adverse effect of measured cell voltage, in: Energy Conversion Engineering Conference, 1996. IECEC 96., Proceedings of the 31st Intersociety, vol. 2, Aug. 1996, pp. 979–986, https://doi.org/10.1109/IECEC.1996.553832, vol. 2.

[60] D. Wilt, D. Chubb, D. Wolford, P. Magari, C. Crowley, Thermophotovoltaics for space power applications, in: AIP Conference Proceedings, vol. 890, Feb. 2007, pp. 335–345, https://doi.org/10.1063/1.2711751.

[61] C.S. Murray, C.J. Crowley, S. Murray, N.A. Elkouh, R.W. Hill, D.E. Chubb, Thermophotovoltaic converter design for radioisotope power systems, in: AIP Conference Proceedings, vol. 738, Dec. 2004, pp. 123–132, https://doi.org/10.1063/1.1841887.

[62] D. Wolford, D.L. Chubb, Theoretical performance of a radioisotope thermophotovoltaic (RTPV) power system, in: Presented at the 7th International Energy Conversion Engineering Conference, Denver, Colorado, Aug. 2009. Accessed: Apr. 19, 2016. [Online]. Available: https://arc.aiaa.org/doi/10.2514/6.2009-4655.

[63] L.S. Mason, Realistic specific power expectations for advanced radioisotope power systems, J. Propul. Power 23 (5) (2007) 1075–1079, https://doi.org/10.2514/1.26444.

[64] J. Strauch, A. Klein, P. Charles, C. Murray, M. Du, Radioisotope fueled thermophotovoltaic power systems for space applications, in: Proceedings of Nuclear & Emerging Technologies for Space, NETS), 2015, 5145 pp. Accessed: May 10, 2016. [Online]. Available: http://anstd.ans.org/wp-content/uploads/2015/07/Proceedings-of-NETS-2015.pdf.

[65] J.E. Strauch, A. Klein, P. Charles, C. Murray, M. Du, General Atomics radioisotope fueled thermophotovoltaic power systems for space applications, in: 13th International Energy Conversion Engineering Conference, 0 vols, American Institute of Aeronautics and Astronautics, 2015.

[66] H. Wang, X. Tang, Y. Liu, Z. Xu, Z. Yuan, K. Liu, Z. Zhang, T. Jiang, Thermal emission-enhanced and optically modulated radioisotope thermophotovoltaic generators, Energy Technol. 8 (3) (2020) 1901170, https://doi.org/10.1002/ente.201901170.

[67] X. Wang, W. Chan, V. Stelmakh, I. Celanovic, P. Fisher, Toward high performance radioisotope thermophotovoltaic systems using spectral control, Nucl. Instrum. Methods Phys. Res. Sect. A Accel. Spectrom. Detect. Assoc. Equip. 838 (Dec. 2016) 28–32, https://doi.org/10.1016/j.nima.2016.09.028.

[68] X. Wang, R. Liang, P. Fisher, W. Chan, J. Xu, Radioisotope thermophotovoltaic generator design methods and performance estimates for space missions, J. Propul. Power 36 (4) (2020) 593–603, https://doi.org/10.2514/1.B37623.

[69] J. Lee, S. Cheon, S. Hong, Y. Nam, A radioisotope thermophotovoltaic converter with nanophotonic emitters and filters, Int. J. Heat Mass Tran. 108 (May 2017) 1115–1125, https://doi.org/10.1016/j.ijheatmasstransfer.2016.12.049.

[70] S.J. Cheon, S.G. Hong, J.H. Lee, Y.S. Nam, Design and performance analysis of a 500-W heat source for radioisotope thermophotovoltaic converters, Int. J. Energy Res. 42 (2) (2018) 817–829, https://doi.org/10.1002/er.3889.

[71] N.-P. Harder, P. Würfel, Theoretical limits of thermophotovoltaic solar energy conversion, Semicond. Sci. Technol. 18 (2003) S151–S157.
[72] D.L. Chubb, B.S. Good, R.A. Lowe, Solar Thermophotovoltaic (STPV) System with Thermal Energy Storage, 1995, pp. 181–198.
[73] A. Datas, D.L. Chubb, A. Veeraragavan, Steady state analysis of a storage integrated solar thermophotovoltaic (SISTPV) system, Sol. Energy 96 (2013) 33–45, https://doi.org/10.1016/j.solener.2013.07.002.
[74] A. Veeraragavan, L. Montgomery, A. Datas, Night time performance of a storage integrated solar thermophotovoltaic (SISTPV) system, Sol. Energy 108 (0) (Oct. 2014) 377–389, https://doi.org/10.1016/j.solener.2014.07.005.
[75] A. Datas, A. Ramos, A. Martí, C. del Cañizo, A. Luque, Ultra high temperature latent heat energy storage and thermophotovoltaic energy conversion, Energy 107 (Jul. 2016) 542–549, https://doi.org/10.1016/j.energy.2016.04.048.
[76] M.R. Gilpin, Chapter 12—ultra-high temperature space power applications, in: A. Datas (Ed.), Ultra-High Temperature Thermal Energy Storage, Transfer and Conversion, Woodhead Publishing, 2021, pp. 309–330.
[77] M.R. Gilpin, D.B. Scharfe, M.P. Young, A.P. Pancotti, Molten boron phase-change thermal energy storage to augment solar thermal propulsion systems, in: Presented at the 47th AIAA Joint Propulsion Conference, San Diego, CA (US), 2011. Available at: https://doi.org/10.2514/6.2011-5986.
[78] M.R. Gilpin, D.B. Scharfe, M.P. Young, Phase-change thermal energy storage and conversion: development and analysis for solar thermal propulsion, in: Presented at the 48th AIAA/ASME/SAE/ASEE Joint Propulsion Conference & Exhibit, Atlanta, GE (US), 2012.
[79] M.R. Reid, D.B. Scharfe, R.N. Webb, Computational evaluation of a latent heat energy storage system, Sol. Energy 95 (Sep. 2013) 99–105, https://doi.org/10.1016/j.solener.2013.06.010.
[80] M.R. Gilpin, High Temperature Latent Heat Thermal Energy Storage to Augment Solar Thermal Propulsion for Microsatellites, University of Southern California, 2015.
[81] T.R. Amundson, K.M. Webb, R.N. Webb, Numerical power output predictions for low-bandgap thermophotovoltaic cells coupled with a latent-heat energy storage system, J. Energy Storage 6 (May 2016) 204–212, https://doi.org/10.1016/j.est.2016.04.008.
[82] E. Marsch, R. Marsden, R. Harrison, R. Wimmer-Schweingruber, B. Fleck, Solar Orbiter—mission profile, main goals and present status, Adv. Space Res. 36 (8) (2005) 1360–1366, https://doi.org/10.1016/j.asr.2004.11.012.
[83] J. van Casteren, M. Novara, BepiColombo mission, Memor. Soc. Astronom. Ital. 82 (2011) 394.
[84] Solar Probe Plus: Report of the Science and Technology Definition Team, NASA/TM—2008-214161, 2008. Accessed: Apr. 22, 2016. [Online]. Available: http://solarprobe.jhuapl.edu/mission/docs/SolarProbe_STDT2008.pdf.
[85] K.W. Stone, E.F. Leingang, S.M. Kusek, On-sun Test Results of McDonnell Douglas' Prototype Solar Thermophotovoltaic Power System, 1994, pp. 2010–2013.
[86] E. Horne, Hybrid Thermophotovoltaic Power Systems, EDTEK, Inc., 2002.
[87] A. Datas, C. Algora, Development and experimental evaluation of a complete solar thermophotovoltaic system, Prog. Photovoltaics Res. Appl. (2012), https://doi.org/10.1002/pip.2201 vol. Published Online.
[88] A. Lenert, D.M. Bierman, Y. Nam, W.R. Chan, I. Celanović, M. Soljačić, E.N. Wang, A nanophotonic solar thermophotovoltaic device, Nat. Nanotechnol. 9 (2) (Feb. 2014) 126–130.
[89] C. Ungaro, S.K. Gray, M.C. Gupta, Solar thermophotovoltaic system using nanostructures, Opt. Express 23 (19) (Sep. 2015) A1149, https://doi.org/10.1364/OE.23.0A1149.
[90] A. Kohiyama, M. Shimizu, H. Yugami, Unidirectional radiative heat transfer with a spectrally selective planar absorber/emitter for high-efficiency solar thermophotovoltaic systems, APEX 9 (11) (Oct. 2016) 112302, https://doi.org/10.7567/APEX.9.112302.
[91] D.M. Bierman, A. Lenert, W.R. Chan, B. Bhatia, I. Celanović, M. Soljačić, E.N. Wang, Enhanced photovoltaic energy conversion using thermally based spectral shaping, Nat. Energy 1 (6) (May 2016) 16068, https://doi.org/10.1038/nenergy.2016.68.

[92] R. Bhatt, I. Kravchenko, M. Gupta, High-efficiency solar thermophotovoltaic system using a nanostructure-based selective emitter, Sol. Energy 197 (Feb. 2020) 538–545, https://doi.org/10.1016/j.solener.2020.01.029.

[93] A. Kohiyama, M. Shimizu, K. Konno, T. Furuhashi, H. Yugami, Effective photon recycling in solar thermophotovoltaics using a confined cuboid emitter, Opt. Express 28 (26) (Dec. 2020) 38567–38578, https://doi.org/10.1364/OE.412764.

CHAPTER NINE

Thin-film materials for space power applications

Ina T. Martin[1], Kyle Crowley[2] and Aloysius F. Hepp[3]
[1]Department of Materials Science & Engineering, Case Western Reserve University, Cleveland, OH, United States
[2]NASA Glenn Research Center, Cleveland, OH, United States
[3]Nanotech Innovations LLC, Oberlin, OH, United States

9.1 Introduction

Photovoltaics are an ideal method for extraterrestrial power production as the panels are self-contained, do not require replenishing with external fuel, and do not have emissions. This introductory section details differences between classic crystalline and thin-film solar cells (TFSCs), summarizes the types of materials considered for TFSCs, and discusses the difference between space (air mass zero, AM0) and terrestrial (AM1.0 or AM1.5) solar radiation. TFSCs are differentiated from classic crystalline technologies by their thinner absorber layers, which range from hundreds of nanometers to tens of microns across numerous semiconductor materials. Numerous thin-film PV technologies have been proposed for space. Absorber materials encompass organic and inorganic materials, but most work has been focused on inorganic thin films. The efficiency of the solar cell is sensitive to the power and energy of the radiative light source; thus, different light sources are used for standard measurements of cells used in terrestrial versus extraterrestrial applications.

9.1.1 Technology background

Thin-film materials discussed in this chapter range in structure from amorphous to polycrystalline and require a solid support on which to initiate growth and impart mechanical stability [1,2]. TFSCs can have significantly reduced mass compared to classic crystalline Si and III-V semiconductors, resulting in the advantageous potential for high mass specific power (power per unit mass) and stowability (power per unit volume). Thin films with low temperature processing will also allow for deposition on lightweight, flexible substrates, further reducing the module mass, and increasing the potential applications to include flexible blankets and solar sails, for example [3,4]. The quality and performance of a thin-film absorber depends on the nature of the material (i.e., band structure, bandgap, absorption coefficient, crystallinity, defects). These qualities, in turn, are sensitive to deposition parameters, layer thickness, postdeposition treatments, and the layers on which they are grown.

Photovoltaics for Space
ISBN 978-0-12-823300-9, https://doi.org/10.1016/B978-0-12-823300-9.00015-7

Photovoltaic devices encounter unique stressors in the space environment that vary with location [5]. Thermal cycling in space has greater extremes than those encountered by terrestrial installations, making delamination a particular concern. A range of electromagnetic and ionizing radiation is present that can result in displacement and ionization damage. The Van Allen belts and other solar wind regions are mostly composed of protons and electrons. High-energy protons, alpha particles, and heavier particles make up the galactic cosmic rays; solar flare activity results in eruptions of electromagnetic radiation and particle (electrons, protons, heavier ions) acceleration. Solar panels will be exposed to different types of radiation depending on location. Generally, device performance will degrade as a function of the differential flux spectrum and the total ionizing dose [3–6]. Both thermal and radiation stressors can result in interdiffusion of elements through the devices, resulting in changes to materials and device performance. Lastly, atomic oxygen is an abundant species in low-Earth orbit (LEO); its reactivity with carbon makes carbon-based polymers susceptible to oxidation and thinning and silicon-based polymers susceptible to oxidation and contamination from the resulting deposits [7].

A recent comprehensive review provides a valuable analysis that can help spacecraft designers during the development of photovoltaic assemblies for space applications [5]. A critical outcome includes findings that point toward optimal appropriate materials, device technology, and key environmental challenges (Table 9.1) to overcome for future mission concepts and needs. When focusing on TFSCs, in particular, consideration of application-specific external stressors must be considered in the design and encapsulation of TFSCs [3–7].

9.1.2 Air mass standards for terrestrial and space photovoltaics

The solar irradiation that powers solar cells in nonterrestrial environments encompasses a broad parameter space across missions: Earth orbiting, near-Earth, near-Sun, and outer planetary missions will all have different exposures to sunlight, which are coupled with varying power needs based on the particular mission requirements and lengths. Terrestrial solar cells are qualified using air mass (AM) 1.0 or 1.5, whereas space efficiencies are qualified using AM0, as shown in Fig. 9.1. The air mass provides a quantitative accounting for the absorption of sunlight by Earth's atmosphere, and the number refers to the path length the light takes through the atmosphere relative to the shortest possible path length (when the Sun is directly overhead). For example, AM1.0 is the solar constant at the surface of the earth at noon on a completely clear day (1.000 kW/m^2), while AM1.5 corresponds to light passing through 1.5 atm thickness, representative of the annual average of mid-latitude locations, such as those in the continental United States, and scaled to a specific integrated power density [8]. As most solar cells do not operate under ideal conditions, AM1.5 has been the standardized value used by the solar industry since the 1970s.

Table 9.1 Summary of the impact of the space environment on solar cells.

Environment factors	Effects
Solar irradiance	Power conversion dependence
Temperature	Efficiency degradation
	Degradation mechanisms (carrier freeze-out and thermal barriers to conduction)
	Thermo-elastic stress cycles (e.g., cracks in solder joints of the interconnects)
	Electric resistances
Vacuum	Contamination (degassing)
	Pressure differentials (decompression)
Plasmas	Surface charging, electrostatic discharge, and dielectric breakdown
	Enhanced sputtering and reattraction of contamination
	Increased leakage current
Energetic particle radiation	Total ionizing dose effects (electronic degradation)
	Displacement damage
	Single event effects (upset, latch-up, burnout)
	Degradation in optical properties (e.g., cover glass, optics, etc.)
Electrically neutral particles	Mechanical effects (aerodynamic drag, physical sputtering)
	Chemical effects (ATOX, spacecraft flow)
Ultraviolet and X-ray radiation	Degradation of thermoelectric properties
	Degradation of optical properties (e.g., cover glass, optics, etc.)
	Structural damage(s)
Micrometeoroids and debris	Damage to cell active area and interconnects
	Damage of optical systems caused by hypervelocity impacts (cover glass, lenses, mirrors)
	Increased cell shunt resistance

Reproduced with permission from Ref. [5], copyright (2021) Elsevier.

Space efficiencies are calculated using AM0; the lack of atmospheric absorption results in a higher power (1.367 kW/m^2), with significantly higher irradiance in the near UV and short wavelength regions that are typically filtered out in terrestrial atmospheric conditions. As solar cell efficiency is a ratio of power out to power in, devices commonly have a lower efficiency at AM0 than under terrestrial conditions, due to poorer performance at lower wavelengths [3,4]. Correction factors can be used to predict AM0 efficiency values from AM1.5 efficiency measurements; as these depend on the spectral response of the cell, they vary for different technologies. For materials with 1−1.5 eV bandgaps such as c-Si and chalcogenides, a 15%−20% efficiency reduction is reported. Due to its higher bandgap, amorphous Si:H AM1.5 efficiency values are multiplied by a factor of 1.25 to estimate the AM0 values [9].

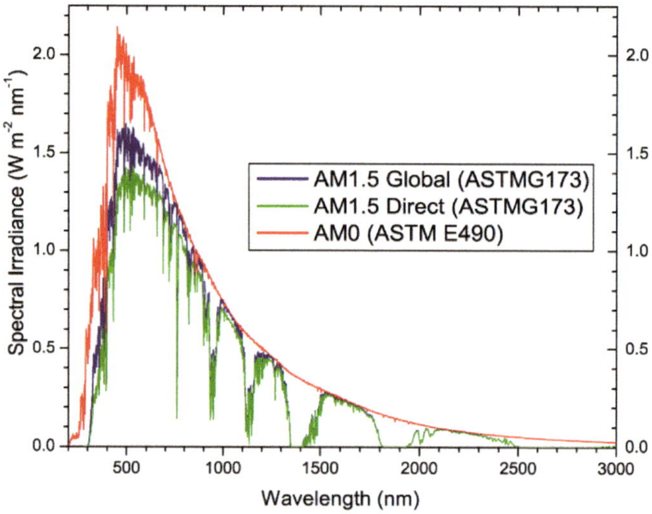

Figure 9.1 Standard solar spectra for space and terrestrial use. *Reproduced with permission from https://www.pveducation.org/pvcdrom/appendices/standard-solar-spectra.*

9.1.3 Overview of materials technologies

Fig. 9.2 illustrates the number of published items per year that contain references to solar cells used for space applications. The search was refined to include citations that included specific solar cell absorber materials, or the terms thin film or multijunction. Publications on thin films for space started in the 1990s, mostly on CdTe and amorphous Si:H. Publications for most materials have increased over time, with a surge in 2021. Perovskites for space first appeared in 2015 and currently comprise a similar number of publications to multijunctions.

The inorganic absorbers comprise binary, ternary, and quaternary semiconductors, including amorphous Si (a-Si:H), CdTe, CdS, quantum dot solar cells, $CuIn_{1-x}Ga_xSe_2$ (CIGS) and its variations, e.g., Cu_2ZnSnS_4 (CZTS), and most recently, perovskite solar cells (PSCs) [3,4,10–13]. The most studied perovskite absorbers are organic/inorganic hybrid materials, which are ionic semiconductors in contrast to the covalently bonded materials above. Their rapid progress since the first published devices [14] and their potential for space applications are covered in several chapters of this book [15,16]. Considerations for each material technology include cost, how deposition on space-appropriate substrates affects the device performance, and the robustness of performance under extraterrestrial stressors. Device cost is materials dependent, both in terms of general production and whether small batches need to be customized for extraterrestrial applications.

This chapter primarily reviews three inorganic TFSC materials and devices that have been theorized, tested, and launched into space, and how they each perform under the

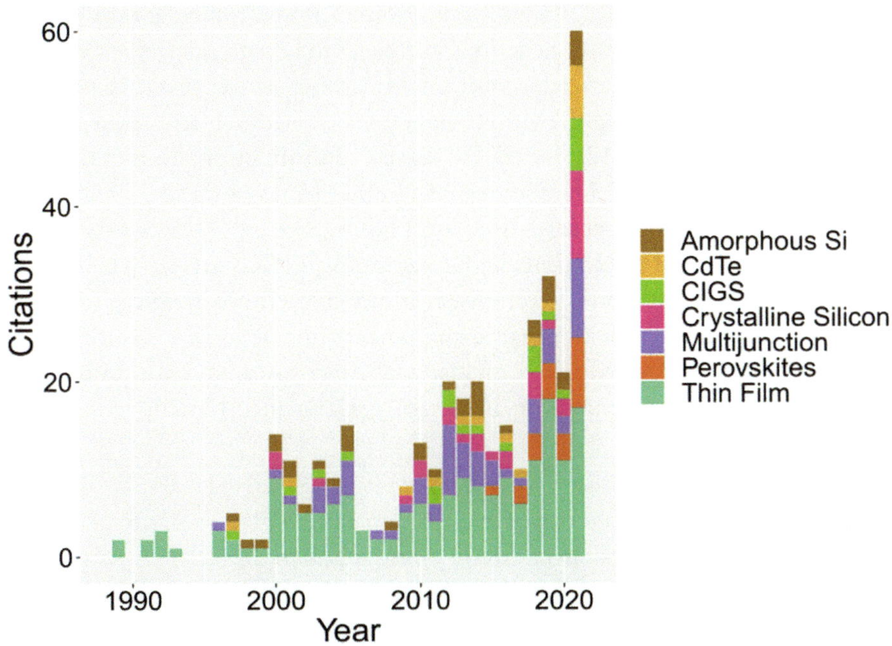

Figure 9.2 Bars show the number of published results plotted by year from running the following query in Scopus: (TITLE-ABS-KEY (solar AND cells) AND TITLE-ABS-KEY (space) AND ALL (satellite OR low AND earth AND orbit OR greater AND earth AND orbit OR am0 OR extraterrestrial OR "proton irradiation")). The results were then refined by the subtopics shown in the legend. *Source: Created using statistics from the literature by the authors.*

unique stressors encountered across space environments. All thin-film absorbers utilize a certain selection of substrates, interlayers, and encapsulation/window layers in their composition; this chapter addresses environmental effects on these components, to provide a general guide to materials choices as new absorbers develop. Promisingly, some thin-film materials have been shown to have increased radiation resistance compared to bulk crystals [3]. Recent data from TFSC payloads show the promise of these materials in powering future missions [6].

9.2 Materials, devices, and impact of the space environment

Photovoltaic arrays have and will continue to play a key role in the generation of power in space. The baseline photovoltaic arrays in space utilize crystalline solar cells [17–19]. While crystalline solar cells have continued to evolve, missions requiring very high specific power and small launch stowed volume may not be feasible using crystalline solar cells in the future. TFSCs built on lightweight substrates may be an enabling technology for future space missions by significantly increasing specific power while reducing launch stowed volume [20–22]. As we have noted, a longstanding group of inorganic TFSC technologies are amorphous silicon, $CuInSe_2$, CdTe, and related materials. While these technologies have experienced significant advances over the years, it may take many years for them to reach the high AM0 efficiency required for many space applications [17–22]. Meanwhile, group 12–16 (II-IV)-based multijunction, thin-film, solar cells have demonstrated the potential to achieve AM0 efficiencies of up to 25% and beyond.

There are a variety of architectures used for TFSCs [23–25], each of which comprise multiple layers with specific functions and constraints. TFSCs are grown on a support material with the devices grown layer-by-layer using methods ranging from thermal evaporation to atmospheric metal organic chemical vapor deposition. In this section of the chapter, we provide an overview of materials, processes, and the impact of both processing and the space environment on device performance [23,24].

9.2.1 Device structures

Devices are deposited on a support material, which can be glass, polymer, or metal foil. Foils are advantageous due to their nature of being lightweight (enabling higher specific powers), flexible, and can be used in roll-to-roll printing, where long or continuous rolls of flexible material are coated, patterned, and laminated by many subsequent steps to

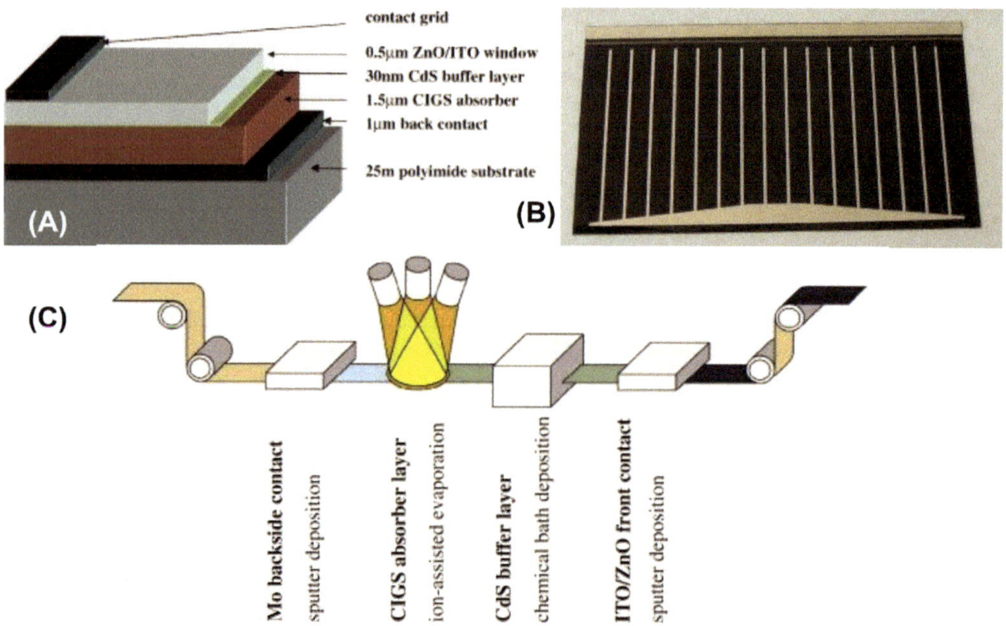

Figure 9.3 (A) Device diagram. (B) Picture of a flexible CIGS TFSC on polyimide substrate (size: $4 \times 7\ cm^2$). (C) Schematic drawing of a continuous roll-to-roll fabrication process for the fabrication of flexible TFSC. *Reproduced with permission from Ref. [12], copyright (2006) Elsevier.*

form a finished product, which lessens production cost (Fig. 9.3) [12,25]. Transparent conducting oxides are used as top contacts, and metals are typically used as back contacts. Interlayers are deposited between the absorber and the contacts and serve many functions. They can block and/or transport selective charges (acting as charge transport or charge blocking layers (CTLs and CBLs, respectively)), can serve as very thin insulating layers to increase shunt resistance, and can serve as the *p*/*n* layers for intrinsic absorbers for *p-i-n* or *n-i-p* configured devices. From a materials perspective, interlayers can also impact the morphology/adhesion of the adjacent layer and the overall device stability. Interlayers are often developed based on combinations of practicality, availability, and resulting device efficiency, but their importance on long-term performance and durability of thin-film devices cannot be overstated [26].

During every step of growth in a thin-film PV device, deposition temperature can affect the previously deposited layers and substrate. Interdiffusion of ions and chemical species between layers is affected by such processing conditions and can have a beneficial or deleterious effect on the overall device performance and stability. Thus, the overall device structure (i.e., whether a superstrate or substrate configuration is used for fabrication) depends not only on individual films and their specific properties, but also on the processing conditions and their effects on the other layers, in the context of the device as a whole. In terms of temperature-sensitive substrates, the specific material constraints set the processing boundaries of the entire device. For these reasons, thin films that require high-temperature processing to reach high efficiencies are at a disadvantage for use in space applications.

9.2.2 Device components

This subsection details various device components and their functions. Substrates and superstrates provide the physical foundation of the devices. Conductive materials—including TCOs and metals—serve as contact layers, guiding charges out of the device. Charge transport layers improve cell performance by enhancing the overall cell electrical field and engineering the direction of current within the cell. All TFSCs are encapsulated for real-world applications, particularly those intended for space. Issues related to operation in the space environment are introduced when most relevant to the device feature(s) being considered.

9.2.2.1 Substrates and superstrates: introduction to mass specific power

When the support material forms the bottom of the TFSC, or the end which faces away from incident sunlight, it is called a substrate. Substrates can be made from a variety of materials, and they are not required to be transparent since they are not involved in the optical path. Conversely, when the base layer forms the top of the TFSC, which faces the Sun, it is referred to as a superstrate. For superstrate devices, the support must be transparent, and is often glass or a polymer followed by a transparent conducting oxide (TCO) layer that functions as the top contact. Historically, rigid base layers have been frequently used, as they are easy to work with; these include materials such as soda-lime glass, borosilicate glass, or quartz for superstrates and substrates, among many other opaque options for substrates [25] (Fig. 9.4).

A significant consideration for TFSCs in space is mass specific power (MSP), or the power generated by a device divided by its weight (kW/kg). According to information available on SpaceX's Falcon 9, which is regularly used to access the international space station (ISS), the current launch costs are approximately \$2700/kg (~\$1000/lb) USD [27]. It is clear, then, that minimizing the weight of the cells themselves becomes a crucial factor for the realization of TFSC use in space. Because of these associated costs, priority has been given to developing lightweight, flexible substrates for solar cells, favorable over the heavier rigid base layers such as glass. Flexible substrates can also benefit from roll-to-roll production techniques. Flexible LEDs are an example of the utilization of roll-to-roll processing [28].

Currently, TFSCs have seen success on flexible substrates made from polymers, metal foils, and flexible ceramics and glasses. For polymers, common choices include polyimide (PI), polyethylene terephthalate (PET), polyethylene naphthalate (PEN), and fluorinated ethylene propylene (FEP) [25]. Polymer films are advantageous in that they may be used in either substrate or superstrate configurations, are lightweight, and

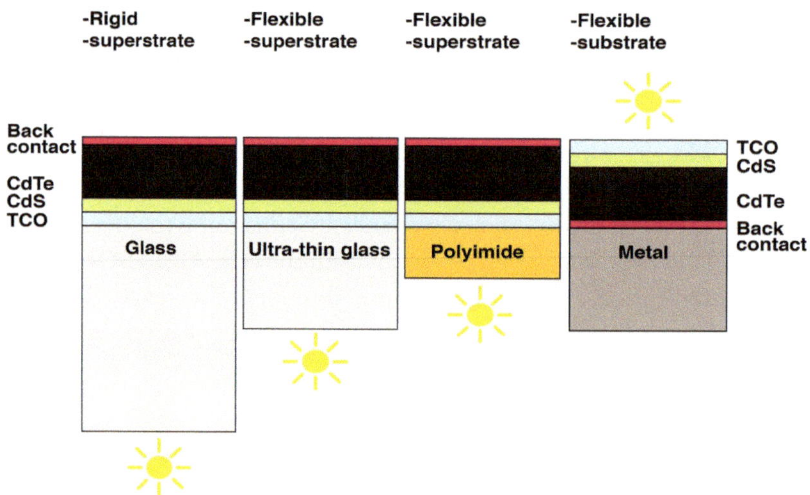

Figure 9.4 Schematic of superstrate and substrate CdTe devices (the substrates' width is not proportional but indicative of the differences between the devices). Both in substrate and superstrate configurations, light enters through wide-bandgap window material (CdS). In the substrate structure, opaque substrates are used; CdS need not undergo high-temperature processing. *Reproduced with permission from Ref. [25], copyright (2020) Elsevier.*

extremely flexible. Perhaps the greatest benefit of polymer films is their insulating nature and their lack of metallic elements. The insulating properties allow polymers to be used in monolithic integrated circuitry, which lends to mass production capability, and the lack of metallic elements prevents unwanted impurities such as iron and nickel from diffusing into the TFSC, which can degrade the solar cell performance [29]. Polymer films have drawbacks, however; they are unstable during high-temperature processing, with a high temperature limitation of <500°C for PI, which is problematic when thin films such as CIGS require processing temperatures >550°C to achieve the large grains necessary for high efficiency [30]. Furthermore, polymers are based on carbon chains, which are susceptible to degradation under UV irradiation and atomic oxygen exposure, both of which come into play in space.

Metal foils are only suitable as a substrate due to their opaque nature, but have properties that remedy many of the drawbacks that polymers face. Their light weight and flexibility allow for flexible roll-to-roll manufacture, their low surface roughness is highly compatible with the hetero-structural growth necessary for TFSCs, and as metals, they are compatible with high processing temperatures. Stainless steel (SS), one of the most common metal foils employed, is problematic on its own as a substrate material, for reasons including high conductivity and a propensity to diffuse contaminants into the absorber layer at high temperatures. To counter this, an electrically insulating diffusion barrier can be deposited on top of the SS before the absorber; common choices include Al_2O_3, Si_3N_4, TiN, and NiP [31]. For CIGS devices, it is common to instead use an enamel coating on the SS, typically containing alkali sources such as sodium, potassium, cesium, and rubidium, to dope the CIGS layer during synthesis for increased electrical performance [32,33]. These diffusion barriers must be of a minimum thickness to avoid potential mechanical defects such as pores or pinholes, and they can require thicknesses ranging into hundreds of microns. Assuming the diffusion barrier encapsulates the foil on both sides to prevent oxidation from atomic oxygen (AO) fluence and is sufficiently grounded, a metal foil may be suitable for TFSC substrate use in space applications. However, this addition of a barrier layer also adds another cost element to the process and must be considered when engineering the TFSC [34].

Flexible glass is the last main competitor in TFSC substrate/superstrate technology. Flexible glass boasts many of the same advantages as metal films, having low surface roughness, high mechanical strength, and high temperature tolerance. Being transparent, flexible glasses are suitable in both substrate and superstrate configurations. Additionally, glasses are inherently insulating and do not suffer from pinholes and other porous defects that are seen in the diffusion barriers necessary in metal foils. Recently, flexible glass was used in a superstrate configuration with CdTe on a successful flight in LEO for 3 years on the AlSat-1N CubeSat mission, demonstrating that this material is suitable for space applications [6].

Until recently, flexible glasses did not have the mechanical properties to be compatible with roll-to-roll processing (Fig. 9.3), limiting their utility in the expanding world of TFSCs. However, Willow Glass, developed by Corning, boasts excellent bend stress per radius figures, allowing flexible glasses of 100- and 200-micron thickness to be utilized in roll-to-roll manufacturing. Furthermore, Willow Glass is a borosilicate, which will not darken or discolor when exposed to cosmic irradiation, making it a prime candidate for TFSCs in space [35]. While still restricted in the extent to which they can bend in comparison to other substrate options, flexible glasses have become a contender in the realm of TFSCs and hold promise for future applications. Table 9.2 summarizes recent reports of the performance of high-efficiency CIS and CdTe on flexible solar cell devices [25,32,33,36–42].

9.2.2.2 Transparent conducting oxides: Impact of a radiation environment

TCOs are broadly used as transparent electrodes in thin-film PV [43]. These oxides are complex mesostructured semiconductors with complicated defect centers that impact their electrical properties. Several metal oxides have the wide optical bandgap and degenerate doping necessary for optical transparency and electrical conductivity; indium tin oxide (ITO) [44], fluorine-doped tin oxide (FTO) [45], and aluminum-doped zinc oxide (AZO) [46] are most commonly used in PV. The selection of a TCO for a PV device depends on several factors; in addition to high optical transparency and conductivity of the bulk material, band alignment and surface energy compatibility with adjacent layers must also be considered. These latter properties are highly surface sensitive. Moreover, the TCO must be stable within a device (i.e., to long-term exposure to adjacent layers) and under performance stressors. ITO is the workhorse TCO used in

Table 9.2 Representative CIGS and CdTe devices on flexible substrates (or superstrates[a]); efficiencies on glass substrates are included for comparison.

Material	Substrate	Eff. (%)	V_{oc} (mV)	J_{sc} (mA/cm^2)	FF (%)	Year	Alkali metals/buffer layer	Refs.
CIGS	Glass	22.6	741	37.8	80.6	2016	KF + RbF	[32]
		22.9	746	38.5	79.7	2019	Cs	[33]
	Polyimide	20.4	736	35.1	78.9	2013	KF	[36]
		20.8	734	36.7	77.2	2019	RbF	[37]
	Stainless steel	17.5	601	40	72.5	2020	–	[38]
		18.0	692	34.5	75.5	2018	RbF + Ni/Cr	[39]
CdTe	Flex glass[a]	16.4	831	25.5	77.4	2015	–	[40]
	Polyimide[a]	13.6	846	22.3	73.4	2012	–	[41]
	Mo foil	11.5	821	21.8	63.9	2013	–	[42]

Adapted from Ref. [25].

development of new optoelectronic devices, due to excellent interlayer stability and above average electrical conductivity.

However, scarcity and difficulty in accessing indium require alternative TCOs to be identified and developed. In this vein, AZO is a viable alternative with excellent bulk properties, composed of earth-abundant elements and accessible by low-temperature (LT) processing. The drawback of this material is that it suffers from stability issues, in particular under exposure to moisture and acidic materials [44–46].

Magnesium-doped ZnO (MZO) films have been shown to enhance the performance of thin-film CdTe solar cells when used as a transparent buffer layer [47]. A study by Bittau et al. tuned the bandgap of MZO buffer layers for CdTe solar cells by increasing the substrate temperature during deposition by radio-frequency magnetron sputtering (RF-MS). Table 9.3 provides a summary of the electrical properties of the MZO/TCO/glass stack compared with FTO/glass. Devices incorporating an optimized MZO buffer layer deposited at 300°C with a bandgap of 3.70 eV yielded a mean efficiency of 12.5% and a highest efficiency of 13.3% under AM1.5 illumination (Fig. 9.5) [47].

A recent study by Barbé et al. measured a record power conversion efficiency of 13.6% (AM0) for PSCs based on AZO. Further, it has been demonstrated that PSCs can withstand proton irradiation up to 10^{13} protons cm^{-2} without significant loss in efficiency [48]. Aluminum-doped ZnO-based TCOs have recently been used in CdTe devices that demonstrated excellent stability during a 3-year mission in LEO [6], described further below. Finally, we must emphasize that there are numerous TCOs that have yet to be explored thoroughly for potential use in TFSCs. The reader is encouraged to consult several recent excellent literature reviews [49–51].

9.2.2.3 Charge transport layers for improved solar cells

As previously stated, a primary component of many TFSCs are the interlayers between the contacts and the absorbers, including CTLs. These can be deposited either on the

Table 9.3 Summary of representative TCO properties.

Material[a]	FTO (SLG)	ITO (SLG)	AZO (SLG)	AZO (BSG)	ITiO (BSG)
Thickness (nm)	450	250	900	700	230
R_{sheet} (Ω/ϒ)	10	4	10	10	8
Carrier density × 10^{20} (cm^{-3})	5.6	18.0	3.7	3.6	3.9
Mobility (cm^2/V·s)	25	34	19	26	89
Resistivity × 10^{-4} (Ω cm)	4.4	1.0	9.0	6.75	1.8

[a]Transparent conducting oxides (TCOs) have been compared to the fluorine-doped tin oxide (FTO) coated glass substrates (TEC10, Pilkington NSG); 4-mm-thick soda-lime glass (SLG) and 1-mm-thick boro-aluminosilicate glass (BSG); magnesium-doped zinc oxide (MZO), aluminum-doped zinc oxide (AZO), titanium-doped indium oxide (ITiO), tin-doped indium oxide (ITO).

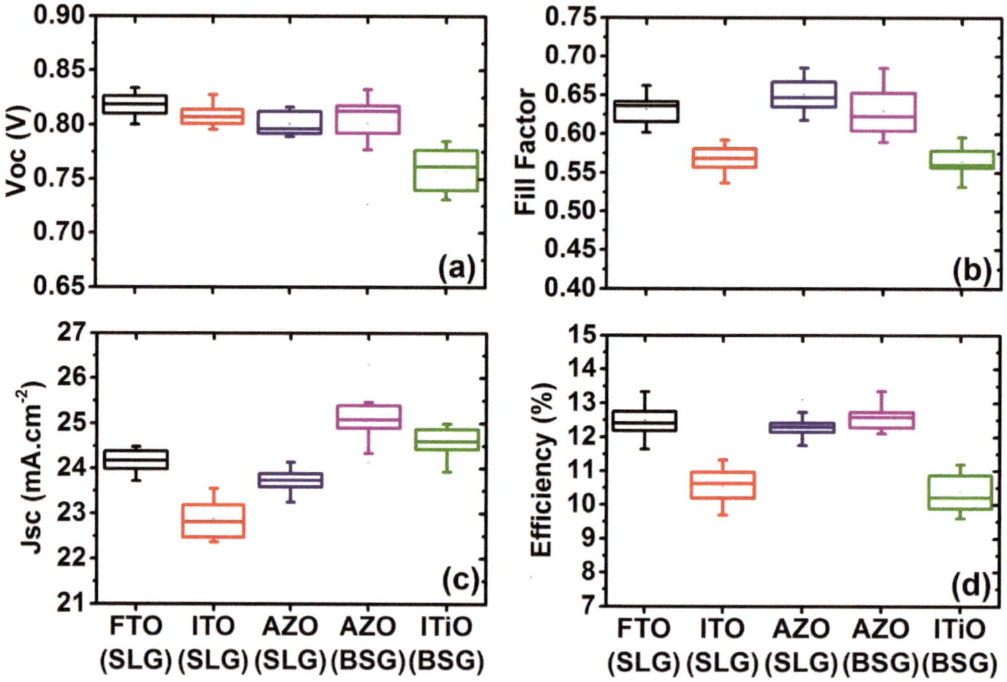

Figure 9.5 Box plots of efficiency, V_{oc}, fill factor, and J_{sc} of CdTe devices incorporating different TCO materials, using a magnesium-doped zinc oxide (MZO) buffer layer deposited at 300°C. Transparent conducting oxides (TCOs) have been compared to the fluorine-doped tin oxide (FTO) coated glass substrates; 4-mm-thick soda-lime glass (SLG) and 1-mm-thick boro-aluminosilicate glass (BSG); aluminum-doped zinc oxide (AZO), titanium-doped indium oxide (ITiO), tin-doped indium oxide (ITO). *Reproduced from Ref. [47], article is distributed under the terms of the Creative Commons (CC-BY 4.0) license (http://creativecommons.org/licenses/by/4.0/), which permits unrestricted use.*

absorber or on the electrode before the absorber deposition, or in both locations, depending on the device. While not used in all TFSC configurations, CTLs improve cell performance by enhancing the overall cell electrical field and engineering the direction of current within the cell. Conduction electrons, ejected from the absorbing layer(s) by incident sunlight, drift in a specified direction under the added influence of the CTLs, increasing the overall open circuit voltage of the TFSC. CTLs may also act as chemical/physical barriers, separating layers within the TFSC, or inhibiting the ingress of problematic species from the ambient such as oxygen and water vapor. CTLs are commonly denoted by their majority carrier type, namely as *n*- or *p*-type materials that facilitate electron or hole conduction, respectively. A material deemed suitable as a CTL in a TFSC is typically selected by comparing its band structure to the adjacent absorbing material. Both organic and inorganic CTLs are used in TFSCs, and each are uniquely advantageous [52].

While crucial for selecting a suitable set of CTLs, band alignment is one of multiple considerations that need to be taken. Chemical inertness, or a low reactivity with neighboring layers within the TFSC during operation, is also important. For example, ZnO tends to absorb oxygen and H_2O from the atmosphere, which creates large defect structures and leads to recombination sites within ZnO, reducing the lifetime of active carriers within the TFSC and diminishing performance [53]. Since these problematic species may also be absorbed during fabrication of ZnO via chemical/solution methods, consideration must be given when selecting a processing method for the specified TFSC configuration. Organic CTLs can have similar pitfalls; organic options such as PEDOT:PSS and Spiro-MeOTAD are favored as HTLs due to their contributions to outstanding electrical performance in the TFSC, but they quickly deteriorate in conditions involving oxygen, H_2O, and high temperatures. Both materials also have undesirable reactivity: PEDOT:PSS solutions are acidic, shown to corrode both AZO and the more stable ITO during deposition [54,55]. Spiro-O-Me-TAD has stability issues due to ionic dopants and is difficult to deposit conformally over large areas [56]. Further, from a materials perspective, organic materials such as spiro-O-Me-TAD [57] and the common electron transfer layer (ETL) PCBM are 100%–800% more expensive than their metal oxide counterparts [58].

In the scope of TFSCs in space, organic CTLs such as PEDOT, PTAA, and PCBM have been shown to be less successful than inorganic options. Organic materials are highly susceptible to deterioration under stressors such as AO, extreme thermal cycling, and UV irradiation, all of which are important factors in LEO [59,60]. Therefore, inorganic CTLs such as metal oxides are preferable in this arena, as they are more stable under these extremes. Al-doped ZnO was recently used in conjunction with CdTe in a successful flight in LEO for 3 years on the AlSat-1N CubeSat mission, where negligible changes in series resistance postflight were indicative that the CTL remained stable over the duration of the mission [6]. Zinc oxide has also proven successful in lab proton irradiation studies [61]; see discussion below (Section 9.2.3.4). In general, metal oxides selected for TFSCs tend to be robust against electron and proton irradiation, and being in an oxidized state prevents most interactions with AO. UV irradiation can be problematic for more unstable metal oxides, namely TiO_2, where oxygen vacancies are created and act as charge trapping sites. This is remedied by replacement with a less photoactive CTL, such as $BaSnO_3$ or SnO_2 [59]. Fig. 9.6 displays the deposition and optical transmittance of six typical metal oxides used as a CTL [62] via a LT solution-based process.

9.2.2.4 Encapsulants, atomic oxygen issues, and International Space Station experiments

All TFSCs must be encapsulated for terrestrial or space applications. Water and oxygen are large contributors to terrestrial TFSC degradation, as they readily react with many

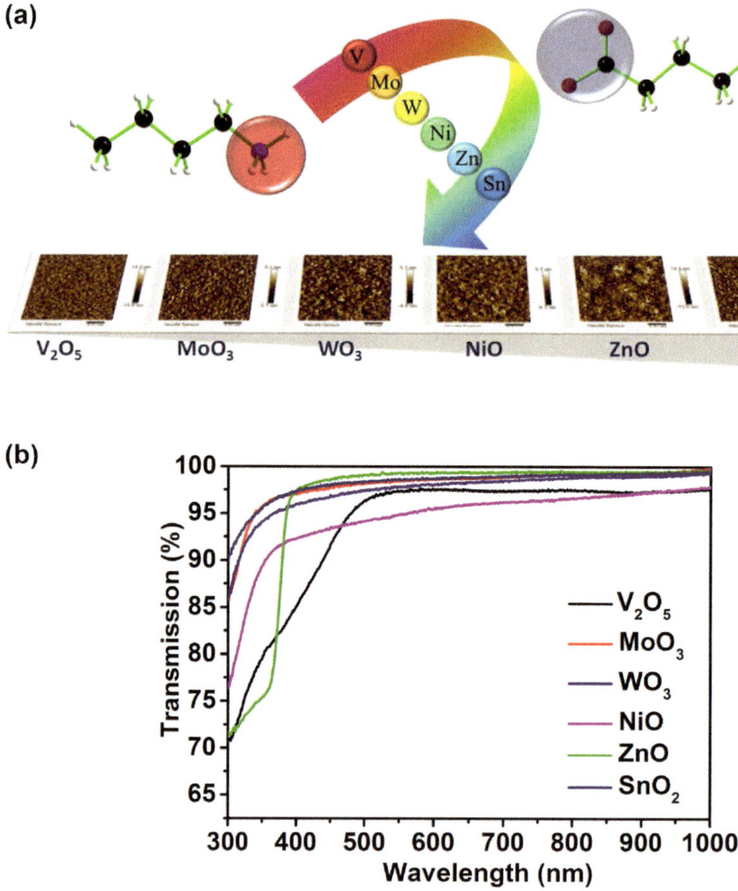

Figure 9.6 (A) Schematic representation of the deposition of six different metal oxide charge transport layers (CTLs), including atomic force microscopy images of the CTLs films, from left to right: V_2O_5, MoO_3, WO_3, NiO, ZnO, and SnO_2. (B) The optical transmittance spectra of six metal oxide CTLs. *Reproduced with permission from Ref. [62], copyright (2022) Elsevier.*

components of the cell. To inhibit this permeation, encapsulation techniques are used, where materials are chosen to cover the TFSC that have low permeability, high chemical inertness, high mechanical strength, and sufficient transparency to minimally impact incident light to the TFSC [63,64]. Two common methods are cover encapsulation and film encapsulation. The former involves the use of adhesive and some form of cover placed on top of the cell, such as a glass slide sealed with silica or epoxy. The latter—film encapsulation—utilizes a deposition process to coat a PSC in a thin film to provide an impervious barrier to the cell from all angles, typically with a metal oxide such as Al_2O_3 or SiO_2 [63]. Polymer encapsulation has also been used for devices on flexible plastic substrates, using materials such as DuPont Tedlar polyvinyl fluoride films, DuPont

Teflon ETFE, or Teflon FEP films (fluoropolymer Tefzel) for back and front sheet/cover encapsulation, glued together with resin encapsulant. The use of these polymer materials tends to be expensive, however, and most polymer materials are not resistant to degradation from UV light in the long term [25].

From the perspective of TFSC for use in space, specifically in LEO where most man-made satellites are located, water vapor becomes a negligible concern. Oxygen, however, continues to be problematic, even more so than in terrestrial applications. AO reacts with all but a small range of inorganic materials, with an extremely high incident particle energy of ~4.5 eV when colliding with fast-moving orbiting objects such as satellites [60,65]. In this environment, polymers and other organics are at a severe disadvantage; most polymers, being carbon-based, decompose when in contact with AO. In fact, the polyimide Kapton H, one of the most chemically and thermally resilient thermoplastics available, is the most characterized standard for measuring AO erosion yields in LEO [60].

The Materials International Space Station Experiment (MISSE) project, first launched in 2001, involves a series of spaceflight missions with experiments flown on the exterior of the ISS to test the performance and durability of materials and devices exposed to the LEO space environment [66]. MISSE-5 was launched on STS-114 and deployed on August 3, 2005, in a zenith/nadir flight orientation; the single passive experiment container (PEC) (Fig. 9.7A) was retrieved on September 15, 2006, after 1.12 years of space exposure and returned as part of the STS-115 mission [66,67]. MISSE-5 was a collaboration between NASA Langley Research Center, NASA Glenn Research Center, The Ohio State University, the Naval Research Laboratory, and the US Naval Academy and consisted of three experiments: PCSat-2, Forward Technology Solar Cell Experiment (FTSCE) (Fig. 9.7C) and the Thin Films Materials Experiment. Fig. 9.7C illustrates key solar cell devices tested on MISSE 5 [67,68], including PowerSphere, a simple prototype integrated power source (Fig. 9.7D) [68]. We will delve more deeply into this novel technology below (Section 9.3.3). As the primary result of the TFSC experiments on MISSE 5 related to the impact of the space environment on the blanket materials [60,66], impact on cell performance was not actively pursued. The reader is directed to the next subsection for results on simulated (Sections 9.2.3.4 and 9.2.3.5) and actual space environments (Section 9.2.3.4) on TFSCs.

For space applications, encapsulation with materials such as glass, graphene, or transparent oxides such as Al_2O_3 or SiO_2 is most advantageous; metal oxides do not interact with AO, and certain oxides are utilized as UV filters. Using encapsulants such as these in the space environment actually provides multiple benefits. All materials launched into space must comply with AIAA-S111A outgassing standards [69]; encapsulants ensure that any volatile byproducts of TFSC decomposition are contained within the cell until decommissioned. Additionally, the encapsulant may also serve as an effective shield for the TFSC against the many sources of cosmic irradiation in space. Some transparent

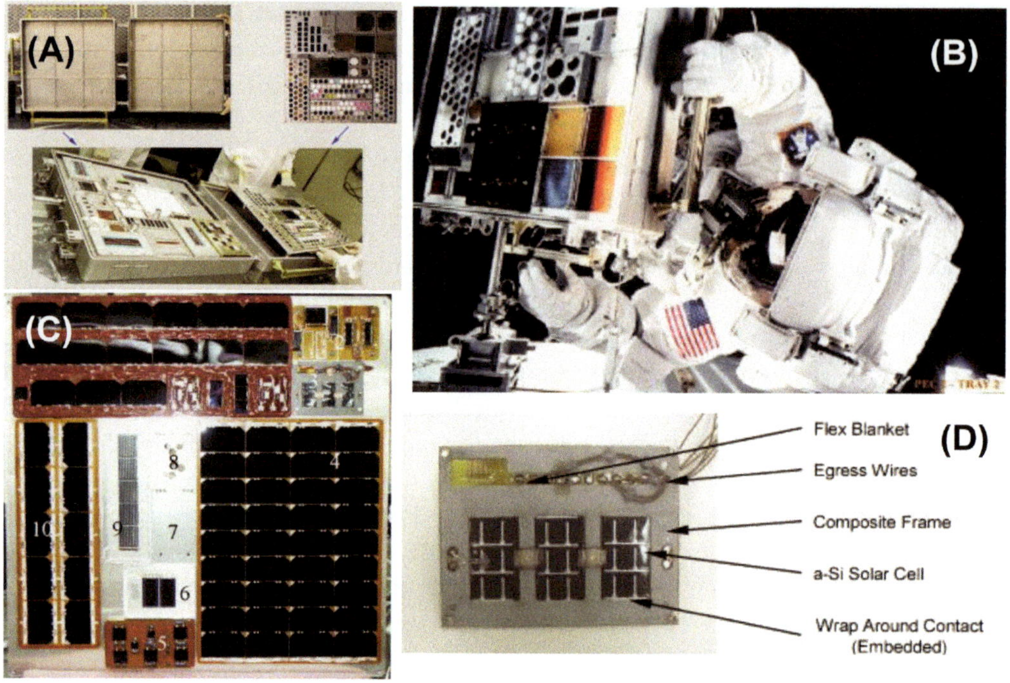

Figure 9.7 (A) Photograph of a passive experiment container (PEC); the PEC was built by NASA and was designed to hold experiments mounted on trays as shown and is closed to protect the experimental samples during transport on the Shuttle and deployment on the International Space Station (ISS). Once deployed, the PEC is opened to expose the experiments to the space environment. At the end of the mission, the PEC is closed and returned to Earth. (B) Photograph of an astronaut deploying the second PEC onto the exterior of the ISS. (C) Photograph of the FTSCE experiment deck placed into the MISSE 5 PEC; samples 3 (amorphous Si on Kapton), 6 (amorphous Si on stainless steel), and 9 (CIGS cells on stainless steel) are blanket-level tests of interconnected thin-film solar cells (note that the contamination monitor (7) and sun sensors (8) are not in this picture). (D) Detailed view of amorphous silicon Powersphere MISSE 5 flight experiment; expanded view of sample 3, second from top in upper right-hand corner of (C). *(A–C) Reproduced with permission from Ref. [67], copyright (2005) Elsevier; (D) from Ref. [68], copyright (2005) Elsevier.*

oxides, such as ZrO_2, are particularly radiation-dense, and as an encapsulant may aid in proton and electron blocking. This area of research is currently being investigated for TFSCs and will see significant development in the coming years as energy production in space continues to grow [70].

9.2.3 Thin-film solar cell materials for space applications

This subsection covers the three main types of inorganic TFSC materials that have been considered for space applications, and a general discussion of studies of their radiation tolerance. The first material is hydrogenated amorphous silicon (a-Si:H), which has been studied for terrestrial and extraterrestrial solar cell applications for decades [4,17,19,71]. Unlike c-Si devices, where the Si has a well-ordered crystalline lattice over the bulk of the materials [72], a-Si:H does not exhibit long-range order and is thus a much more defective material. The other two TFSC absorbers (CdTe, CIGS) are made of chalcogenides, materials with one or more (typically) heavier-than-oxygen chalcogen (e.g., S, Se, Te) making up a substantial constituent. Extensive research has been conducted into terrestrial TFSCs based on CdTe [23,73] and other chalcogenide materials, in particular chalcopyrite (I-III-VI) and kesterite (I_2-II-IV-VI_4) absorbers [23] with mixed success in terms of efficiency, stability, and ramping up actual production.

Accelerated aging studies of PV devices attempt to reproduce performance loss from a variety of stressors in a significantly reduced period of time. Accelerated aging studies for space applications often incorporate exposure to protons and electrons, and in-depth studies will also include alpha particles and heavier ions. The rates of degradation are typically shown as a function of the differential flux spectrum and the total ionizing dose. These conditions are meant to anticipate the stressors present in space applications; the spacecraft location will of course be key to the stressor type and dose. Radiation tolerance of different TFSC materials is further discussed in this and the next section (Section 9.3). A related issue—spacecraft charging—is discussed in-depth in Chapter 2 of this book [74].

9.2.3.1 Thin-film silicon devices: Impact of Staebler-Wronski effect

Unlike c-Si, in a-Si, there is no long-range crystalline order, and not all the atoms are fourfold coordinated [23,25]. The unused valence orbitals are dangling bonds, which act as defect sites. Thus, the material is passivated with H, resulting in a-Si:H, which has a low enough defect density to make solar cells. The change in coordination of the amorphous material results in very different absorption properties; a-Si:H is a direct bandgap semiconductor, with a bandgap of 1.7 eV. Due to its defective nature, it takes orders of magnitude higher dopant levels to make the material *p*- or *n*-type, on the order of several percent. The majority of the dopants are not electrically active, which makes *p*-*n* junction devices unsuitable. Instead, a-Si:H devices are made in a *p-i-n* or *n-i-p* configuration: the intrinsic (undoped) layer acts as the absorber, creating the electron-hole pairs that can be swept away in the electric field created by the adjacent *p*- and *n*-type layers; Fig. 9.8 illustrates three types of amorphous silicon (a-Si) TFSC device diagrams and efficiency progress over the past 50 years [23].

Although inexpensive and easy to deposit, a-Si:H efficiencies are quite low; the record single junction cell has an efficiency of 10.2% [75]. The device performance degrades under operation conditions (i.e., under illumination and current injection), in what is known as the Staebler-Wronski effect [76]. Thicker films experience enhanced degradation. To mitigate this, various improvement strategies have been utilized such as limiting layer thickness, fabricating multijunction cells where many layers in series are

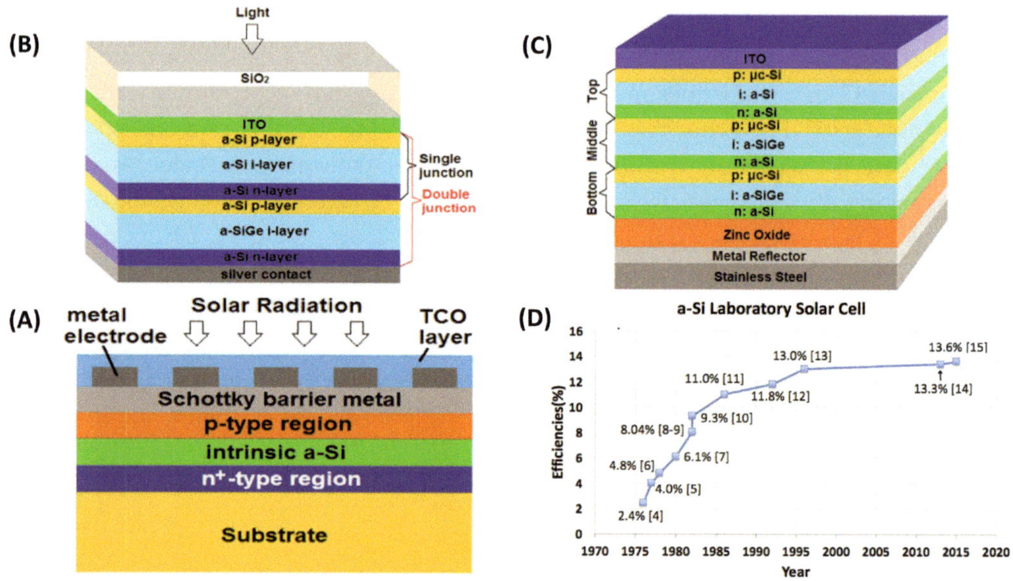

Figure 9.8 Device diagrams of amorphous silicon (a-Si) thin-film solar cells and efficiency progress over the past 50 years; subfigures are clockwise from lower left. (Lower left (A)) Layout of the Schottky barrier a-Si solar cell. An illustration of a Schottky barrier a-Si solar cell with a highly doped *p*-type region adjacent to the Schottky barrier high work function metal. (Upper left(B)) Layout of a double junction a-Si:H/a-SiGe:H solar cell. (Upper right(C)) Layout of a triple-junction a-Si:H/a-SiGe/a-SiGe solar cell. (Lower right(D)) A graph of the improvements in efficiency of laboratory a-Si solar cell. Reference numbers after efficiencies correspond to references from the literature source. *Reproduced with permission from Ref. [23], copyright (2017) Elsevier.*

used to absorb different portions of the solar spectrum, and replacing amorphous Si:H with nano- or microcrystalline silicon, which is more resistant to degradation. Despite these advances, the record efficiencies for double- and triple-junction cells with amorphous and nanocrystalline absorbers are only 12.7% [77] and 14.0% [78], respectively.

For space applications, degradation studies measured a high radiation tolerance relative to c-Si; proton-induced degradation is considerably less for a-Si:H and nc-Si:H cells compared with c-Si devices [79]; this was attributed to the thickness of the absorber layer [80]. Unfortunately, the overall device efficiency remained too low, and Staebler-Wronski degradation was too pronounced to make this material suitable as a primary power source [76]. However, as discussed below (Section 9.3.1), this material may be suitable for power-integrated devices with specialized applications.

9.2.3.2 Legacy work at NASA GRC 1990–2005: Use of single-source precursors

Fabrication of a TFSC is only the first step in the construction of a solar energy conversion system for use in space. Before TFSC or device technology can be fully utilized in space, interconnections, array structures, support structures, and deployment techniques need to be developed that take full advantage of the lightweight, flexible nature of TFSCs. From 1990 to 2005, NASA GRC had a robust in-house and extramural TFSC and electronic materials program. Table 9.4 provides an overview of these efforts, including references that provide more details of the research results and technology applications [10,11,81–120]. This effort was carried out in cooperation with industry and academia, funded through a variety of grant and contract programs. The foundation of the in-house program involved a novel approach to materials fabrication via a single-source precursor (SSP) approach utilizing nonvacuum deposition, particularly atmosphere assisted CVD (AACVD) [82–85,88–96]. The target materials included gallium sulfide for GaAs-based devices [82–85] via AACVD and a variety of I-III-VI_2 (I = Cu, Ag; III = Ga or In; VI = S or Se) materials via AACVD [87–96] or chemical processing [86,97,98].

Other collaborative research efforts included TFSC device fabrication on metal [101–106] and polymer films [101,107,108], including organic-based devices [11,109,110], TFSC power systems and mission studies [20–22,111,114,116], and spin-off efforts such as integrated power sources [112–114,121], and technology transfer [84,85,115,118–120,122]. Several quite recent book chapters review the commercialization of SSPs and processing [121] and development of integrated power source technology and applications [122]. Three example efforts—SSPs and technology transfer, modeling-driven design of dual-junction thin-film devices (Section 9.2.3.3), and integrated power sources—are summarized below (Section 9.3.3) to illustrate the multiple approaches employed and outcomes that can be transferred to other applications. Due to the significant amount of information summarized in Table 9.4, we will not delve more deeply into other topics related to the NASA-funded and GRC effort in this chapter. The reader is encouraged to consult the referenced publications for further information for topics of interest to them.

Fig. 9.9A shows a generic SSP molecule that has numerous sites that can be modified to tailor properties for optimized processing and target material [90,91,94]; production of $CuInS_2$ from an SSP is illustrated by Fig. 9.9B [86–88]. A diagram of a standard AACVD reactor is shown in Fig. 9.9C [93,94,96]. Technology transfer of this reactor design, along with a patented process developed at NASA GRC for the production of multiwall carbon nanotube (MWCNT) [117,118] served as the foundation of Nanotech Innovations LLC, founded in 2008 to commercialize both the reactor design and process (Fig. 9.9D) [119–121].

9.2.3.3 Cadmium telluride dual-junction cells: Simulation and device fabrication

Morel and coworkers at the University of South Florida proposed a four-terminal tandem as an effective means of achieving high efficiency with thin-film compound semiconductor-based absorbers. The proposed structure is shown on the left side of

Table 9.4 Brief overview and summary of efforts on TFSC materials and spin-off technologies: In-house efforts at NASA GRC and research collaborators (academic and industrial), funded (wholly or partly) and/or managed by NASA GRC.

Technology	Material(s)[a]	Details[b]	Partner(s)[c]	Refs.
Single-source precursors for photovoltaics and electronics	- c-GaS/GaAs - I-III-VI_2	- Novel compounds to fabricate thin-film chalcogenide materials. - Primary method was AACVD but later evolved into chemical fabrication of nanomaterials	- NASA GRC - Harvard Univ. - WUStL - MI state Univ. - OAI - R.I.T. - Idaho EPSCoR	[81–98]
Multijunction II-VI solar cells	CIGS, CdSe, ZnSe, CZT	- AMPS modeling - PVD processing	- Univ. Of South Florida (USF)	[99–101]
CIGS on metal foils	- Cu(In,Ga)Se_2 - Metal foil substrates	- Vacuum PVD methods - Next step: selenization or sulfurization	- Florida solar energy center - DayStar technology - ISET - NASA GRC	[101–106]
TFSC devices on polymer substrates	- CIGS, II-VI - Polymer(s)	Exploratory survey of processing methods and device technologies to enable TFSC on polymers	- USF - Industry - NASA GRC	[101,107,108]
TFSC based on organics and polymer substrates	- Polymers - DSSC - Organics	Exploratory survey of processing methods and device technologies to enable hybrid or OPV	- Norfolk state U - Industry (SBIR) - NASA GRC	[11,109,110]
Integrated power sources	- MIM (III-V) solar cell - LIB	Inspired by analog to dual-junction TFSC devices. IPS demonstrated in space on Starshine 3	- NASA GRC - R.I.T. - NRL - Starshine 3 team	[111–115]

(*Continued*)

Table 9.4 Brief overview and summary of efforts on TFSC materials and spin-off technologies: In-house efforts at NASA GRC and research collaborators (academic and industrial), funded (wholly or partly) and/or managed by NASA GRC.—cont'd

Technology	Material(s)[a]	Details[b]	Partner(s)[c]	Refs.
AACVD reactor	MWCNT	NASA mission study and technology transferred to industry	- NASA GRC - NTI LLC	[116–120]

[a]Cubic-gallium sulfide/gallium arsenide; I-III-VI_2 is a generic term for a Group 11-13-16_2 compound (i.e., CIGS); CZT = cadmium zinc telluride; CIGS = $Cu(In,Ga)Se_2$; DSSC = dye-sensitized solar cell; MIM (III-V) = 1 cm^2 GaAs 7 cell micro-array integrated module; LIB = lithium ion battery; MWCNT = multiwall carbon nanotube.
[b]See discussion in text.
[c]WUStL = Washington University, St. Louis; EPSCoR = Established Program to Stimulate Competitive Research partnership among Idaho universities; Industry (SBIR) = consortium of industrial partners funded by internal and/or small business innovative research funds; NRL = US Naval Research Laboratories; Starshine 3 team = consortium of industrial, academic, and government laboratories.

Fig. 9.10 and has been discussed in some detail previously [99,100]. The bottom cell is assumed to be CIGS with a bandgap of 1.0 eV. At the time of this work, efficiencies of 16%–19% have been reported for CIGS devices with bandgaps in the range 1.0–1.2 eV [101], so the bottom cell is assumed to be available. With a 1.0 eV bottom cell, the ideal top cell has a bandgap in the 1.7 eV range. Both CdSe with a bandgap of 1.7 eV and $Cd_{1-x}Zn_xTe$ (CZT) with a tunable bandgap in the 1.45–2.26 eV range are viable candidates for the top cell. Using AMPS-1D simulations, the University of South Florida (USF) group showed that efficiencies of 30% (25% AM0) could be attained for CdSe (or CZT with a bandgap of 1.7 eV) with ideal properties [100]. Potential AM0 efficiencies in this range makes this device a candidate for development for space applications. As can be seen in Fig. 9.10 (right), the projected AM0 efficiency of 20.1% is a combination of 13.8% for the top cell and 6.3% for the underlying CIGS cell. The top cell was altered by adding small amounts of zinc to the CdTe to produce a wider bandgap CZT with electronic properties approximately those of CdSe.

Numeric simulation is a valuable tool in studying solar cells, particularly for novel new devices and structures. In 1997, Prof. Stephen Fonash and colleagues at Penn State developed a one-dimensional analysis of microelectronics and photonic structures (AMPS-1D) simulation tool. The underlying physics in AMPS-1D is quite general and rigorous for the analysis of a variety of design structures and operating environments. However, this software is not compatible with current operating systems; AMPS-1D only runs on older versions of Windows, requiring a virtual machine to run on current platforms. In 2010, a derivative of AMPS-1D was developed by Prof. Angus Rockett and Dr. Yiming Liu at the University of Illinois. Since this derivative was developed using a cross-platform graphical user interface library (wxWidgets), it was

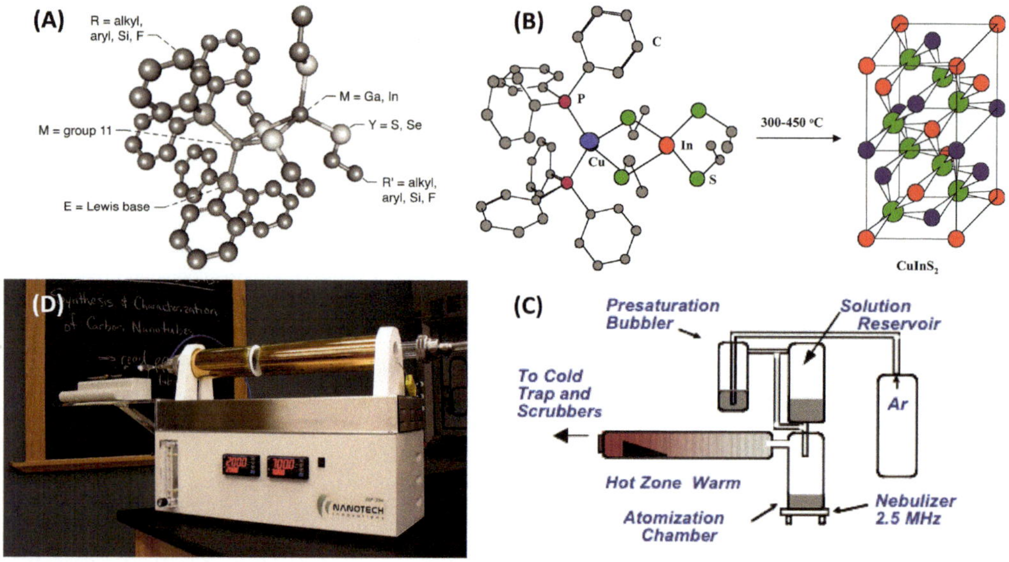

Figure 9.9 Overview of NASA GRC single-source precursor (SSP) technology transfer (clockwise from upper left): (A) schematic demonstrating the versatility of the $[\{ER_3\}_2Cu(QR')_2M(QR')_2]$ (Q = S or Se) architecture. Note that this molecular structure is quite "tunable" and affords the ability to substitute for any of the ternary (Group 11, Group 13, or Group 16) elements in the I-III-VI_2 formula. (B) Schematic of the decomposition of an SSP ($(PPh_3)_2CuIn(SEt)_4$) to produce $CuInS_2$. (C) NASA GRC group-developed atmospheric pressure horizontal hot-wall reactor (AACVD). (D) Nanotech Innovations' portable, bench-top AACVD system, called the SSP-354, developed from the laboratory prototype pictured in (C); it can make research-scale quantities of high-quality MWCNT within a few hours. *(A)–(C) Courtesy NASA; (D) https://www.nanotech-innovations.com/reactor.html, accessed March 4, 2022 (Courtesy Nanotech Innovations LLC).*

named wxAMPS. A recent publication provides an overview of the various simulation methods, their frequency of use, summarizes the basics of AMPS-1D methods, describes issues encountered when attempting to use wxAMPS software, and compares it with AMPS-1D [123]. From 1997 to 2019, the use of AMPS-1D (12%) and wxAMPS (4%) constitute ~1/6 of published studies [123].

Results of the AMPS modeling were used to focus the research of USF on developing high-efficiency, high-bandgap top cells of CdSe and CdZnTe. Several key technologies are required to achieve high specific power thin-film solar arrays for space. To meet the target of a >20% AM0 efficiency II-VI-based thin-film tandem structure on a flexible substrate, several processing steps must be mastered. These include the fabrication of top and bottom cells on thin flexible metal foils or polymers substrates (see Table 9.5). The top cell must be a high-efficiency, wide-bandgap cell, such as $CuInGaS_2$, CdSe, or CdZnTe. To produce a monolithic device, a junction diode will be required.

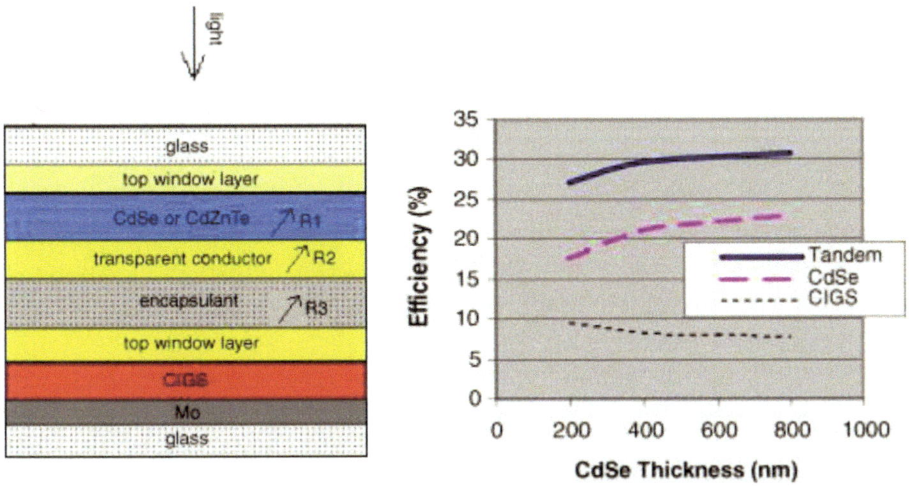

Figure 9.10 (Left) Schematic of a four-terminal tandem device with CIGS as the bottom cell and CdSe or CZT as the top cell. R1, R2, and R3 are additional reflection losses to those experienced by a stand-alone cell. (Right) AM1.5 simulated efficiency versus top cell thickness for a CIGS bottom cell of 15% stand-alone efficiency and an idealized CdSe or 1.7-eV CZT top cell. The estimated AM0 efficiency of this device is ~25%. *Reproduced with permission from Ref. [100], copyright (2005) Elsevier.*

Table 9.5 Solar cell performance characteristics for CZT-based devices.

E_G [eV]	V_{oc} [mV]	J_{sc} [mA/cm^2]	Structure[a]	Heat treatment
1.77	200–260	3.0–6.3	CZT/CdS	H_2/400°C
1.73	670–790	1.0–1.5	ZnTe/CZT/CdS	H_2/400°C
1.65	200	4.4–9.5	CZT/SnO_2	H_2/400°C
1.72	180–220	6.7–8.5	CZT/SnO_2	None
1.67	570–600	1.0	CZT/ZnSe	He/400°C
1.71	360–500	0.2	CZT/ZnSe	None
1.65	200–280	8.0	CZT/CdO	None

[a]CZT = cadmium zinc telluride ($Cd_xZn_{1-x}Te$); SnO_2 = tin (IV) oxide; CdO = cadmium oxide.
Reproduced with permission from Ref. [100], copyright (2005) Elsevier.

Multijunction cells involving TFSC materials and perovskites or other materials are also feasible and the subject of other chapters in this book [15,16].

9.2.3.4 Cadmium telluride: radiation hardness studies

From a terrestrial perspective, CdTe solar cells are arguably the most successful thin-film PV technology. First Solar, a CdTe manufacturer based in Toledo, Ohio, USA, is the only thin-film PV company to regularly appear on the annual top 10 list of solar manufacturers [124]. As an absorber layer, CdTe is a strongly absorbing, thermodynamically stable, direct bandgap material, with a near-ideal 1.44-eV bandgap. Four decades of research have resulted in the evolution of the device from a 4.1% CdTe/CdS heterojunction [125] to a >22% efficient device (19% large area) produced by First Solar [73], enabled by a graded $CdTe/CdSe_xTe_{1-x}$ junction, and a high transparency MZO buffer layer [47]. This performance is 35% higher than the flex glass superstrate device [40] listed in Table 9.2.

Electron and proton irradiation studies of CdTe solar cells show that these devices are stable under high-energy particle irradiation. Romeo et al. observed degradation of CdTe/CdS heterojunction devices at particle fluences two orders of magnitude higher than those used to test monocrystalline Si or III-V devices [73]. Mechanistically, the high-energy particles generated recombination centers, and devices showed performance recovery that predicted little or no damage in space applications [13].

Lamb et al. investigated CdTe in a simulated space radiation environment using superstrate CdTe devices directly deposited onto cerium-doped cover glass. Using the cover glass as the support structure reduced mass and cost by eliminating the need for an additional substrate or superstrate [126]. Their results demonstrated that CdTe had a superior radiation hardness to protons compared with conventional MJ III-V solar cells. Specifically, fluences of 1×10^{12} cm^{-2}, 1×10^{13} cm^{-2}, and 1×10^{14} cm^{-2} resulted in decreases of the relative device efficiency by 5%, 18%, and 96%, respectively. The lowest dose (1×10^{12} cm^{-2}) is estimated to represent a 20-year geostationary Earth orbit (GEO) mission. Further, annealing of the highest dose sample (100°C for 168 h in an inert environment) returned the device to 73% of its initial performance. The proposed mechanism of performance degradation is the formation of interstitial hydrogen from proton irradiation, resulting in a shallow donor level and significantly decreased current. In the space environment, exposure to high temperatures would provide conditions

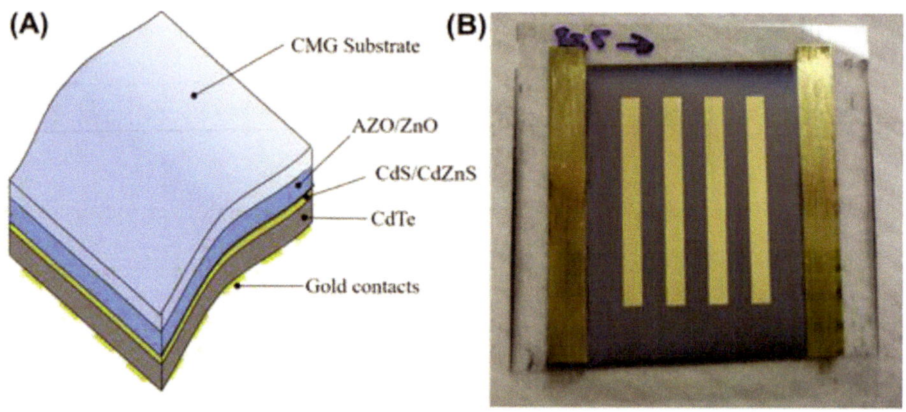

Figure 9.11 Photographs of the CdTe thin-film solar cell(s) tested during the AlSat-1N (3U) CubeSat mission: (A) The CdTe device architecture on 100-micron cover glass. (B) A CdTe device deposited on to 60 × 60-mm 100-micron-thick cover glass; 4 × 1 cm^2 cells defined by gold contacts to the CdTe with two common contacts to the transparent conducting electrode. *Reproduced from Ref. [6]. This open access article is distributed under the terms of the Creative Commons (CC-BY 4.0) license (http://creativecommons.org/licenses/by/4.0/), which permits unrestricted use.*

similar to those in the annealing experiment, allowing performance recovery. In 2021, Lamb et al. published the first IV curve data from CdTe solar cells in space [6]. Fig. 9.11 illustrates the schematic of the device architecture and includes a photograph of the device; a more detailed discussion of the CubeSat experiment can be found below (Section 9.3.2) [6].

9.2.3.5 $CuIn_{1-x}Ga_xSe_2$ (CIGS): Radiation and thermal management

$CuIn_{1-x}Ga_xSe_2$ (CIGS) is another widely studied chalcogenide thin-film absorber. The quaternary material is commonly deposited via vacuum co-evaporation of all constituents on a substrate heated to 400−600°C [25,30,33]; alternate deposition methods will typically include one or more steps at higher temperature to optimize device performance [106,112]. Table 9.1 provides examples of CIGS devices on glass that held terrestrial efficiency records for a polycrystalline thin-film absorber that approached 23% (AM1.5) efficiency [32,33]. Solar Frontier, a leading Japanese CIS company, in partnership with Japan's National Research and Development Agency's New Energy and Industrial Technology Development Organization, announced a new record of 23.35% in 2019 [127], an increase of 0.4% over their previous record [33].

With TFSCs, it is commonly said that "a film is not a device, and a device is not a film." As discussed above, devices comprise graded layers of amorphous and/or polycrystalline materials, ranging in thickness from nanometers to micrometers. During hot phases of growth, and under illumination, heat, and current, diffusion between layers changes the composition and properties of the films, leading to changes in performance over time. The parameter changes in the deposition process induced by the requirements in space applications initially resulted in significantly lower efficiencies. For example, a research program, investigating CIGS deposited on titanium foil, observed a delamination from the substrate at the interconnects when subject to the thermal cycling that might be encountered in space [128]. This is consistent with the lower efficiency of CIGS devices on stainless-steel foils, as shown in Table 9.2.

A recent study by Sellers et al. examined the impact of low intensity-low temperature (LILT) conditions near Mars and the outer planets Jupiter and Saturn on commercial CIGS devices [129] in two states of operation, *relaxed* and *metastable* states (Table 9.6),

Table 9.6 Intensity, temperature, relaxed, and metastable V_{oc} and FF values for CIGS solar cells in the LILT conditions of each planet. Temperatures are the equilibrium flat plate temperature of a solar array.

			Relaxed		Metastable	
Planet	**Intensity (AM0 Suns)**	T_{eq} **(K)**	V_{oc} **(Volts)**	**FF**	V_{oc} **(Volts)**	**FF**
Saturn	0.011	100	0.73	79.1	0.76	80.2
Jupiter	0.037	135	0.74	80	0.77	80.3
Mars	0.430	263	0.67	76.1	0.68	76.7

Reproduced with permission from Ref. [129].

under different conditions due to variations in heating, bias, and/or illumination [129,130]. A solar cell in the "dark" and at temperatures around 330 K for extended periods is described as being in the relaxed state; the metastable state is achieved by illuminating the cell, typically for 1 h with 1 Sun power (AM0 or AM1.5) at room temperature [131].

Concentrated *J-V* measurements of TFSC samples were taken for relaxed and metastable configurations ranging from the light intensity of Saturn up to 10 Sun AM0; samples at an approximate Earth temperature (300 K) were also included for comparison. The V_{oc} and power conversion efficiency (PCE) values extracted from the CPV measurements (Fig. 9.12) enabled valuable insights concerning the respective energy

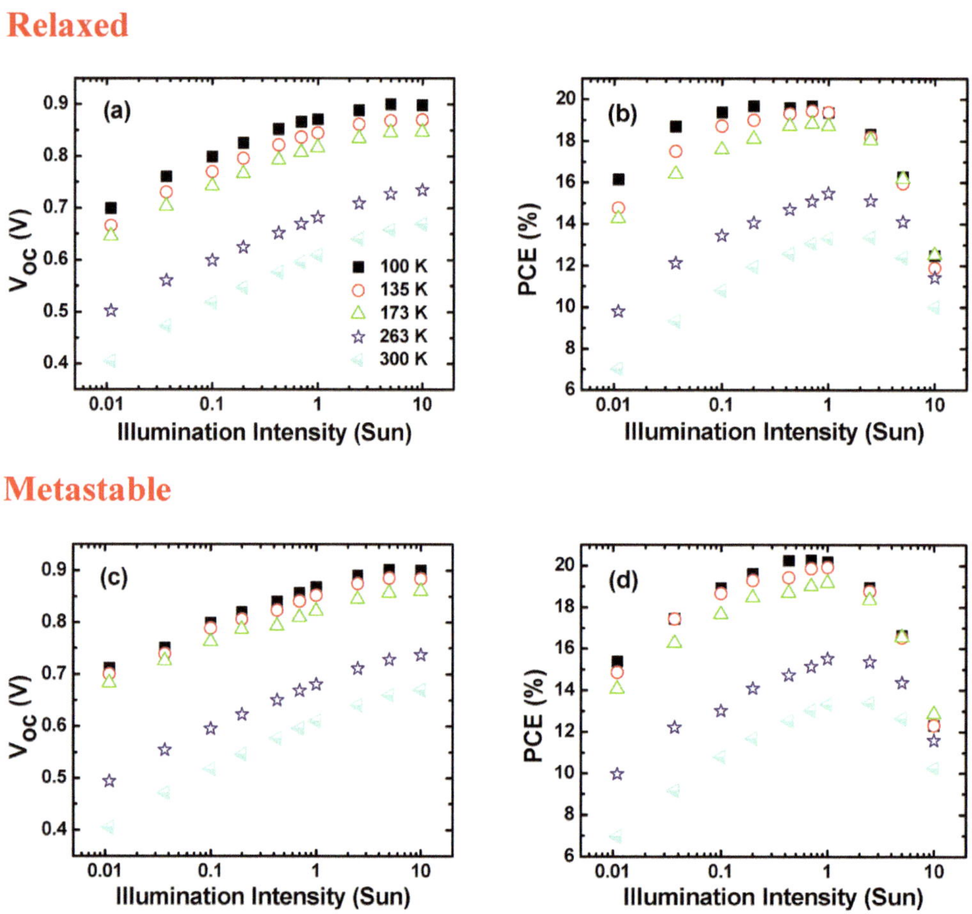

Figure 9.12 V_{oc} and power conversion efficiency (PCE) values extracted from CPV measurements (see text). Subfigures (A) and (B) in the top row show the results of the relaxed state and the bottom row (C) and (D) shows the results of the metastable state, respectively. *Reproduced with permission from Ref. [129], copyright (2020) Elsevier.*

positions of defects and bands. The increase and then saturation of V_{oc} is noticeable for both the relaxed (Fig. 9.12A) and metastable (Fig. 9.12C) states. The magnitude of V_{oc} increases at light intensities from 0.011 to 2 Suns for every temperature. The V_{oc} increases with increasing illumination intensity, approaching the CIGS bandgap; this is evidenced by the plateau of V_{oc} above 2 Suns. Furthermore, the correlation of increasing V_{oc} and excitation intensity is likely due to the gradual occupation and saturation of trap levels both within the bulk absorber and importantly at the CIGS/CdS interface. This results in a reduction in recombination losses and the increase of V_{oc} [132]. The maximum efficiency achieved in the metastable state is marginally higher than that of the relaxed state; this is demonstrated in Fig. 9.12B and D, respectively. The PCE increases due to an increase in both V_{oc} and J_{sc} up to 1 Sun; above 1 Sun, cell performance noticeably decreases.

Unencapsulated CIGS cells were irradiated with high-energy (1.5 MeV) protons at fluences of 1×10^{11}, 5×10^{11}, 1×10^{12}, and 1×10^{13} (H^+/cm^2) (for samples C, D, E, and F, respectively) to investigate the solar cell regions most susceptible to damage. Fig. 9.13 compares the light *J-V* measurements for the reference and irradiated CIGS

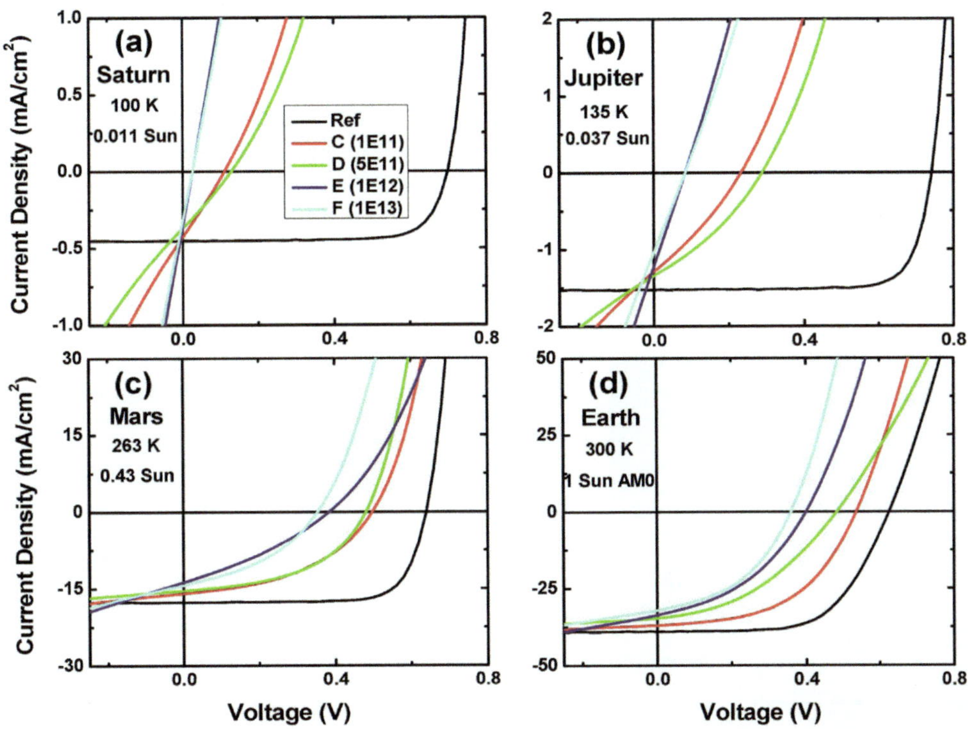

Figure 9.13 Light *J-V* of reference and irradiated samples under various conditions. Samples C, D, E, and F are irradiated with 1.5-MeV protons and fluences of 1×10^{11}, 5×10^{11}, 1×10^{12}, and 1×10^{13} (H^+/cm^2), respectively. *Reproduced with permission from Ref. [129], copyright (2020) Elsevier.*

solar cells under LILT conditions related to the planets of interest. Irradiation leads to an increase in subgap defects and subsequent band tailing; this results in degraded photovoltaic performance. Increased irradiation fluence is correlated with *J-V* reduction in all instances. Although the irradiation damage effects systematically reduce the J_{sc}, the effect on V_{oc} is considerably more significant. Interestingly, at higher temperatures and illuminations such as those simulating Mars (Fig. 9.13C) and Earth (Fig. 9.13D), the impacts of proton irradiation are diminished, yet still dominated by losses in V_{oc}.

The results illustrated in Fig. 9.12 are consistent with illumination saturating defect states at the interface, lowering the barrier to carrier transport and reducing the corresponding resistance. At temperatures above 200 K, there is no difference between the performance of cells in the relaxed or metastable states. The metastability of the CIGS results in a barrier to minority carrier extraction. The presence of this barrier does not affect the performance in LILT conditions as CIGS solar cells operate with a higher PCE. Proton irradiation of the CIGS reduces the performance (Fig. 9.13) via defects. These results suggest while CIGS performs well in LILT conditions, appropriate encapsulation would be required for practical applications in higher radiation environments such as the moons of Jupiter [5].

The Gossamer solar power array (GoSolAr) mission [133] by the German Aerospace Center (DLR) aims to demonstrate the reliability of TFSC technologies in powering spacecraft with applications that require high power such as space tugs, solar electric propulsion (SEP), etc. [134,135]. A recent study by Banik et al. considered a potential solution to thermal issues confronting a TFSC (i.e., CIGS cell) operating in space [136]. The cosmic background is essentially a blackbody at 3 K; it will behave as a heat sink for any solar cell capable of radiating heat in the IR. Due to higher solar insolation and absence of a medium for convection, TFSCs will be operating at higher temperatures in space than on Earth (Fig. 9.14A). A straightforward solution would be coating

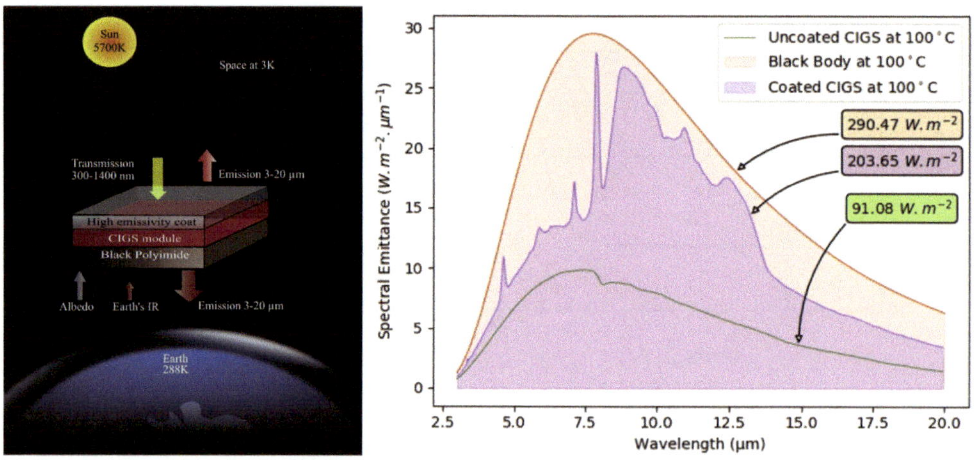

Figure 9.14 (Left) Schematic of light and heat flow into the PV panel in orbit. (Right) Total integral emittance comparison of a blackbody, coated and uncoated module at 100°C. *Reproduced with permission from Ref. [136], copyright (2020 Elsevier.*

technology that combines high transparency in the UV—VIS region of the electromagnetic spectrum, for solar cell efficient operation, with high broad band emissivity in the mid-range infrared (MIR), to facilitate radiative cooling. Achieving a high emissivity space-proof coating would therefore be a significant step toward making flexible solar cells for space applications a reality.

Due to the low PCE of CIGS, ~15% of the absorbed AM0 spectra would be converted to electricity while the remaining energy would be converted to heat. Since the only way to shed this excess heat from such thin-film modules in space is radiative emission, an analysis is required to see what the integral emissive power gain from the coating is. A very broad band emissivity in the 3—20 μm range with high transmissivity in the 0.1—1.4 μm range is most desirable for a coating for the best radiative cooling output from a solar cell. Optical properties of silicon oxycarbonitride (SiCNO) deposited by dip coating and electrical impact of the coating on solar cells were investigated in detail [136]. Commercially available organopolysilazane was used as a precursor to produce SiCNO films on CIGS cells on polyimide substrates; the highest emissivity of 0.72 was achieved with a single layer (~3.2 μm) coating.

Assuming the modules are at 100°C in orbit, as discussed above, we can compare the thermal emittance capability of our coated sample to a noncoated one, as shown in Fig. 9.14B. A theoretical black body at 100°C can emit an integral spectral emittance of 290.47 W/m^2 in the range of 3—20 μm wavelength. The 3.2-μm SiCNO-coated system emitted 203.65 W/m^2 in comparison to an uncoated sample that could only emit 91.08 W/m^2, enhancing emissive power by 123%. Since the resonance peaks of the coating align well with the blackbody peak, it enables the stack to emit radiation with the highest intensity at orbit temperatures.

In summary, the DLR group demonstrated exceptional optical characteristics of a single layer ~3.2 μm high-ε coating on CIGS solar cells; an emissivity of 0.72 was achieved with a polymer-derived SiCNO coating cured under low temperatures. Electrical characterization of coated cells showed that the coating as well as the dip coating and postprocessing steps had a negligible effect on the electrical performance of the solar cells. With already proven flexibility, as well as excellent thermal and electrical behavior, polysilazane-derived coatings may be considered a suitable alternative to cover glasses used on solar cells for space applications. The findings of this investigation complement those of the earlier studies with engineered polysilazane precursor materials [137].

9.3 Thin-film solar cells in space: Past, present, and future

Space exploration missions have historically been largely dependent on classic crystalline technology [17—19,138]. The cost, weight, and structural considerations are motivation for research and development of lighter, flexible thin-film devices [20—22].

As discussed above, many materials—organic, inorganic, and hybrid—have been explored. Radiation testing results are somewhat promising; however, high deposition temperatures require very stable flexible substrates. The recent comprehensive review referenced earlier in this chapter provided advice on appropriate photovoltaic technologies for specific solar system destinations, including optimal materials, device technologies, and key environmental challenges to overcome for future mission concepts and needs [5]. One major finding is that future missions to explore the outer solar system require high-power photovoltaic systems capable of functioning efficiently in LILT conditions and high radiation environments [5]; see the detailed summary above of the study by Sellers and coworkers [129]. Finally, this section details key environmental and spacecraft issues and goes into further details of past and present missions. It concludes with some proposed technologies relevant to TFSCs, as well as future space exploration challenges, technologies, and opportunities.

9.3.1 Integrated power systems: Discrete devices and device-level integration

In the course of the investigation of dual-junction solar cell technology [21,101], the team at GRC noted that an integrated TFSC-battery had a similar structure (Fig. 9.15A) [111,114]. At the annual Small Satellite Conference at Utah State in 1999, we became acquainted with the team from Starshine 3 (Fig. 9.15C), the third in a series of small satellite launches intended to engage students and amateur radio enthusiasts [113] and joined the team to include two NASA GRC power experiments (Fig. 9.15D) [111−114]. The primary mission of Starshine 3 was to measure atmospheric density as a function of altitude. Starshine 3 was essentially a passive satellite; electrical power was not needed to assist orbital tracking.

When Starshine 3 launched in September 2001, a month later than originally scheduled, it had been 60 days since the integrated micropower system (IMPS) units were fully charged; thus, they had only one-third of their initial charge left [112,121]. Any charging for four of the five units would have to come from light reflected off of the surface of the Earth (albedo). Furthermore, when illuminated by only the Earth's albedo, the operating temperature of the IMPS sank as low as −18°C; only 2°C above the manufacturer's stated batteries operating limits [112,114]. During 12 days of monitoring the recharging of batteries by the solar cell, the batteries were maintained at a voltage between 2.5 and 2.9 V [112,121]. This first—and thus far only—demonstration of an IMPS made Starshine 3 an excellent platform to test new power technologies without the burden of mission success depending on the power system. Traditionally, power systems have been one of the most conservatively designed systems on a spacecraft. It is therefore especially difficult to introduce new technologies into the power system design due to the necessity that any new technology must be flight-proven before it can be flown. This reality of space technology development can greatly reduce the pace of innovation.

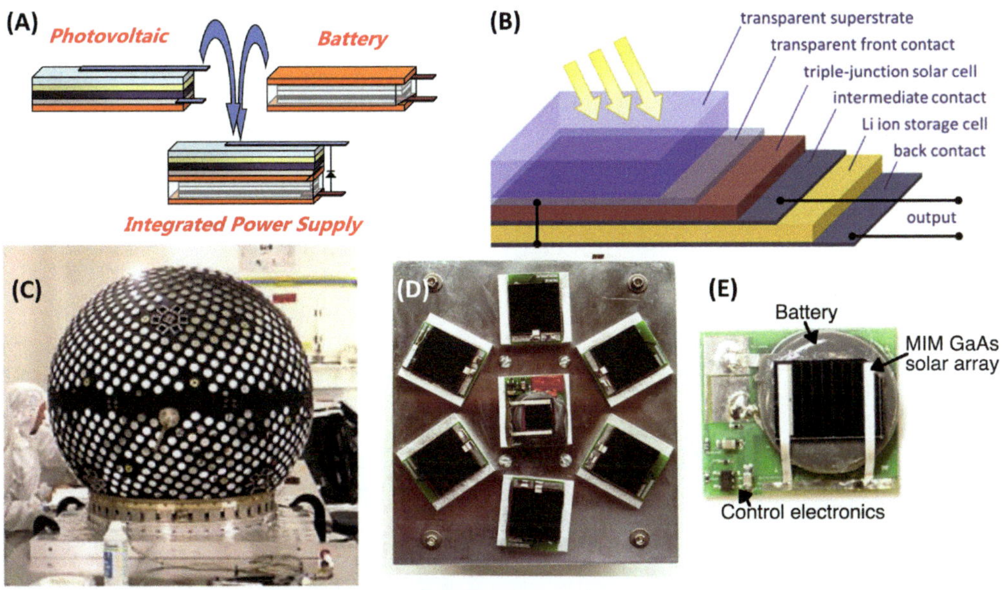

Figure 9.15 (A) Schematic of integrated power system that combines photovoltaic power generation with lithium ion battery storage in a single autonomous device. Provides continuous power under varying or cyclic illumination. For small spacecraft, this technology could eliminate design constraints imposed by integrating all power requirements on a single, centralized electrical bus. (B) Schematic of a monolithically integrated PV-battery cell-to-cell concept reported in the literature. (C) Picture of technicians assembling the Starshine 3 small satellite. (D) Expanded view of constellation (one of five on satellite) of a seven-junction, 1 cm^2 monolithically interconnected GaAs module (MIM) surrounding an integrated (micro-)power device. (E) Detail of an IPS showing a single MIM GaAs solar array, commercial Li-ion battery, and control electronics. *(A), (C)–(E) Courtesy NASA;* (B). *Reproduced with permission from Ref. [140], copyright (2016) Elsevier.*

The IMPS device that was tested on Starshine 3 was inspired by an integrated dual-junction solar cell device, but it was actually a discrete device requiring control electronics (Fig. 9.15E) [121]. An example of a monolithically integrated TFSC-battery power device with a three-electrode configuration reported in 2016 [139] is shown in Fig. 9.15B [140]. A triple-junction thin-film silicon solar cell was deposited on a glass substrate with a suitable transparent front contact with a battery deposited directly on the back contact of the solar cell. Further details [140] of the films that comprise the solar cell, Li-ion battery, and contacts are summarized in a previous review [121]. The measured TFSC efficiency of 12% resulted in an overall system efficiency of 8.5%. This system efficiency is competitive for an integrated power system and one of the highest measured for a TFSC device [121,141].

Table 9.7 provides a summary of examples from the literature that primarily describe integrated devices involving TFSCs [142–148]. Several other examples are included that involve alternative device applications of CIGS [142] or hybrid-organic devices [148] to

Table 9.7 Examples of integrated power technologies including thin-film solar cells (or materials) with aerospace or commercial applications.

Application[a]	Materials/power details[b]	Performance metrics and/or further comments[c]	Refs.
Sun angle detector	CIGS on soda-lime glass with ultra-thin Mo back contact; planar or nonplanar designs feasible	FSI $\eta = 9\%$ optimal $\lambda = 500$ nm; BSI $\eta = 1\%$ optimal $\lambda = 800-900$ nm (dependent on CIGS thickness) CIGS thickness = 700–750 nm optimal; degrades >750 nm	[142]
Solar lamp	Flexible polymer solar cells; Li polymer battery; white LED lamp	$12.5 \times 8.8 \times 2.4$ cm; mass = 50 g	[143]
Temperature sensor	Thin-film Li battery (12 μAh); two 1-mm^2 solar cells (η 5.48%); power management (dc/dc converter) and sensor system	System consumes 7.7 μW; power stored in battery can power sensor for ~5 years	[144]
Wearable textile solar batteries	Six polymer solar cells in series connected to textile battery: $Li_4Ti_5O_{12}$ Anode/$LiFePO_4$ cathode	Supplied LED load consumed 4.2 mW	[145]
Wearable health monitoring device (pulse oximeter)	a-Si solar cell (~230 mW); graphite anode and $LiCoO_2$ cathode, capacity = 47.5 mAh	With appropriate load duty cycle, average load current can be matched to solar module current; battery maintains constant state of charge	[146]
Integrated CL-based lab-on-chip device to detect life markers in extraterrestrial environments	Lab-on-chip device that integrates on a single glass substrate the microfluidic network and array of thin-film a-Si:H photosensors	Detects a range of biomolecules in a liquid aqueous sample extracted from soil, rock, or ice; limits of detection in the nanomolar range	[147]
Wireless portable lightweight self-charging power packs	Perovskite-organic tandem TFSCs integrated with solid-state asymmetric supercapacitors with PANI:$MnSe_2$ composite positive electrode	An overall efficiency of 12.4% solution-processed wireless portable lightweight self-charging power pack; PCE = 17% and energy storage $\eta = 72\%$	[148]

[a]CL = chemiluminescence.

[b]CIGS = copper indium gallium selenide; $Li_4Ti_5O_{12}$ = lithium titanium oxide; $LiFePO_4$ = lithium iron phosphide; $LiCoO_2$ = lithium cobalt oxide; PANI:$MnSe_2$ = polyaniline/manganese selenide.

[c]FSI = front side illumination; BSI = back side illumination; LED = light emitting diode; PCE = power conversion efficiency.

Adapted from Ref. [121] with added information from the literature.

provide context. The reader is directed to the recently published review chapter and the referenced literature for more in-depth discussions of technical and practical aspects of integrated devices for aerospace, commercial, or dual-use applications (Table 9.7).

9.3.2 Small satellite technology experimental platforms

As previously discussed, Lamb et al. published the first IV curve data from CdTe solar cells in space; they detailed the performance of CdTe devices during 3 years in LEO onboard the AlSat-1N (3U) CubeSat, from 2016 to 2019 [6]. Initial data for CdTe devices deployed to space support the predictions made from the laboratory studies [126]. Similar to the laboratory proton irradiation studies above [126], TFSCs were deposited directly onto cerium-doped aluminosilicate glass in the following sequence: AZO as the TCO, a ZnO buffer layer, *n*-type CdZnS and arsenic-doped CdTe p-type layers, followed by a chlorine heat treatment to passivate grain boundaries, and evaporation of the Au back contact. Notably, there were no signs of delamination despite the thermal fluctuations between −3.8 and 51°C. Photographs of the TFSC experiment of the AlSat-1N (3U) CubeSat mission are included in Fig. 9.16 [6].

Current-voltage (*I*-*V*) measurements were collected over 17,000 orbits, during which the cells retained stable short circuit current and series resistance measurements, consistent with a stable front contact (AZO for these devices). The V_{oc} values actually increased over time, which was attributed to lower overall cell temperatures in space and light soaking effects (Fig. 9.17). Degradation of the fill factor (FF) values was observed

Table 9.8 The comparison of practical metrics of three generations of photovoltaic technologies.

	First generation	**Second generation**		**Third generation**	
	Crystalline Si	**Amorphous Si**	**CdTe**	**CIGS**	**GaAs**
Cell efficiency (AM1.5)	26.7% (medium)	14.6% (low)	22.1% (medium)	23.35% (medium)	28.8% (high)
Module efficiency	24.4% (high)	12.3% (low)	18.6% (medium)	19.2% (medium)	25.1% (high)
Flexibility	×	√	√	√	√
Stability	√	×	√	√	√
Price	Low	Medium	Medium	Medium	High
Power to weight	Low	Low	Medium	High	High
Market share	High	Low	Medium	Medium	Low
Toxicity	Low	Low	High	Medium	High
Heuristic figure of merit:	Medium	Low	Medium	High	High
Space applications					

Adapted from Ref. [167].

Figure 9.16 Photographs of the thin-film solar cell payload (TFSC) experiment of the AlSat-1N (3U) CubeSat mission. (A) External PCB of the encapsulated TFSC payload. Gold contacts for the four cells and two common bus bars can be seen on the top and bottom of the glass. These are connected to the PCB electronic circuit via indium/tin solder and gold wires. (B) Internal PCB that, when commanded automatically, measures the four cells and the LM35 temperature sensor. (C) Assembled 3U CubeSat showing the face with the TFSC payload experiment attached alongside the conventional multijunction solar cells that provide power to the spacecraft. *Reproduced from Ref. [6], this article is distributed under the terms of the Creative Commons (CC-BY 4.0) license (http://creativecommons.org/licenses/by/4.0/), which permits unrestricted use.*

due to a decrease in shunt resistance. The authors conjectured that the FF degradation was a result of diffusion of gold from the back contacts through the device, which could result in deep trap states in the CdTe absorber and micro-shunts between the contacts [6].

To summarize, laboratory [126] and GEO [6] data have shown that CdTe devices are suitable for GEO space applications. Notably, the device design incorporated materials unique to space applications. Given the current world record conversion efficiency for First Solar's thin-film CdTe cell of 22.1% (AM1.5) [73], it seems reasonable to target a CdTe solar cell for space applications that is radiation and thermally stable with 20% AM0 efficiency, a specific power of >1.5 kW/kg, and a significantly lower production cost than state-of-the-art III-V multijunction technology.

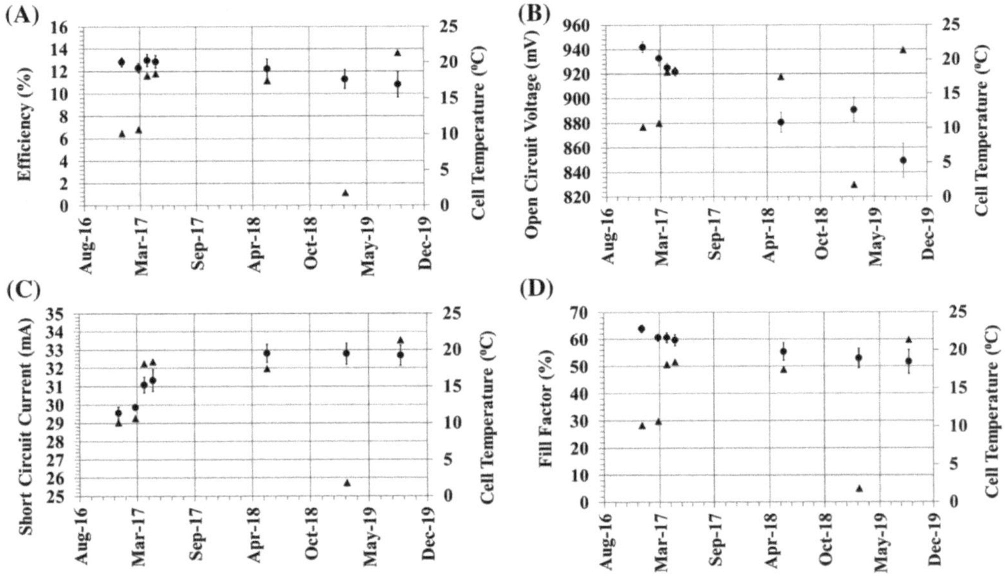

Figure 9.17 The mean (circles) and standard deviation (lines) *I-V* parameters for surveys receiving solar flux above 115 mW/cm^2 over the duration of the mission. Cell temperature (triangles) for each survey is shown on the secondary *y*-axis. (A) Mean and standard deviation efficiency (%) for the four cells. (B) Mean and standard deviation V_{oc} (mV) for the four cells. (C) Intensity corrected mean and standard deviation *Isc* (mA) for the four cells. (D) Mean and standard deviation fill factor (%) for the four cells. *Reproduced from Ref. [6], distributed under the terms of the Creative Commons (CC-BY 4.0) license (http://creativecommons.org/licenses/by/4.0/), which permits unrestricted use.*

As mentioned above (Section 9.2.3.5), DLR's concept for a GoSolAr using thin-film photovoltaics on the small satellite technology experimental platform (S^{2}TEP), a scalable microsatellite class (10–100 kg) bus, is currently under development (Fig. 9.18) [133]. The current focus of the GoSolAr program is to employ CIGS TFSCs on polyamide substrates. Currently, assessment of available CIGS PV technologies is ongoing; two commercial manufacturers, one located in Europe, the other in the United States, are leading supplier candidates. Both have efficiencies around 10%, and both manufacturers have good prospects for efficiency increase in the standard production process in the near future. While use of thin GaAs technology is an alternative option, CIS on polymer has several inherent advantages. They are truly flexible down to a 25-mm roll radius (Fig. 9.18A), have a high MSP (>0.75 kW/kg), and relatively low cost ($25 USD/W).

The study provides an overview of the concept: a large, lightweight, deployable Gossamer PV array based on thin-film PV, DLR's coilable carbon fiber reinforced plastic booms (Fig. 9.18B), and a two-dimensional array deployment (Fig. 9.18D). Together with the PV selection, the investigation of concepts to keep the PV within operating temperature ranges is ongoing, including layering the blanket or using high-emitting

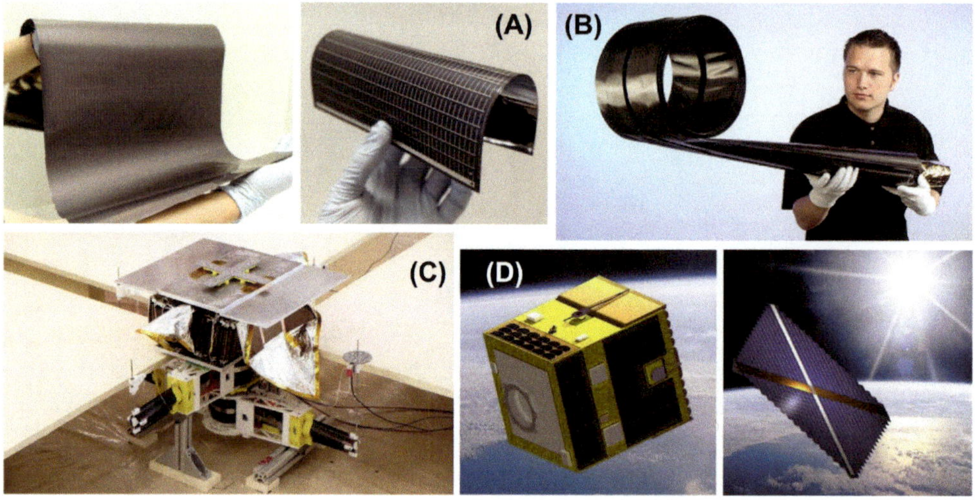

Figure 9.18 German Aerospace Center's (DLR's) concept for a Gossamer solar power array (GoSolAr) using thin-film photovoltaics on the small satellite technology experimental platform (S^2TEP), a scalable microsatellite class (10–100 kg) bus, currently under development at DLR. (A) Flexible CIGS modules. Left: Flisom. Right: Ascent Solar. (B) DLR's tubular, reelable boom in partially deployed state. (C) Deployment test breadboard setup (two boom deployment units with two booms each and one PV blanket stowed in storage box). (D) Artist's view of GoSolAr on S^2TEP (left) and deployed GoSolAr demonstrator array (right). *Reproduced from Ref. [133]. This open access article is distributed under the terms of the Creative Commons (CC-BY 4.0) license (http://creativecommons.org/licenses/by/4.0/), which permits unrestricted use.*

coatings for thermal management [136]. One of the main technical issues is a proper contact between PV and harness, and in between the harness itself, which is currently investigated with high priority. First deployment tests with a demonstrator consisting of a PV blanket dummy and two boom deployment units with two booms each (Fig. 9.18C) were successfully performed.

The long-term goal of DLR's effort is to demonstrate several key concepts and technologies: the applicability of the deployable Gossamer thin-film PV array system, significantly higher MSP ratios compared to conventional PV array technologies, suitability of the thin-film PV for space applications, and potential of the scalability of the main technologies and their combined use as a large system. When considering the increasing efficiencies of the CIGS (and related thin-film) PV systems combined with Gossamer structures, there is clear potential of achieving a very high specific power value (1 kW/kg and beyond), exceeding that of conventional PV systems. Furthermore, the CIGS PV appears to be more radiation resistant and has already surpassed 23% efficiency in laboratories. Such efficiencies are expected to be achieved in the near future in a standard manufacturing process. However, flexible, thin-film GaAs cells are also subject to consideration within GoSolAr. With this prospect, DLR's research has the goal to develop a Gossamer Solar Array (GoSolAr) to provide power for one or more missions described below (Section 9.3.3).

9.3.3 Criticality of mass specific power: Enabling future exploration missions

We have noted that specific power, the ratio of power generated to the device's weight (W/kg), is a bottleneck for TFSCs in space, since every gram of materials used in the solar cell fabrication must be transported into orbit. The associated costs for this have decreased in recent years ($2700/kg USD), but they still remain a large consideration. Because of this limiting factor, TFSCs are favorable for space energy generation when compared to heavy rigid arrays that have been used in the past. The requirement for high specific power is also coupled with the need for stowability, or the power generated per unit volume (W/m^3), and higher power density (W/m^2), where TFSCs also are at an advantage.

Several issues to be addressed include the level of specific power needed for TFSCs to be competitive in space and the amount of power required for a satellite to operate. For example, the ISS reportedly uses 75–90 kW of power, supplied by a large area of Si solar arrays (2500 m^2) divided between eight wings, each measuring 35 × 12 m. Each wing is capable of generating up to 15–20 k W of energy, or 120–160 kW of total electricity for the station; this is approximately 35 W/m^2 [149]. To put this in perspective, this is enough energy to power more than 40 homes on Earth. These arrays weigh in at ~1090 kg each, providing an estimated ~16 W/kg per wing [150].

Recently, lighter, smaller area arrays were deployed to the ISS to "replace," or more accurately supplement, two of the oldest solar cell modules, which were originally installed in 2000 and 2006. These newer roll-out solar arrays for the ISS, or ROSAs, are reported to have a similar power output as the modules they are replacing, but they weigh in at just 690 kg each [151,152]. Assuming identical power generation, this puts their specific power at 25 W/kg, a vast improvement over legacy arrays of the ISS. Additionally, these new arrays are a fraction of the area of the original ISS arrays, measuring 19 × 6 m. These ROSAs are projected to reduce solar array mass by 33% and stowed volume by 75% compared to the legacy arrays [153]. The ROSA arrays make use of a flexible germanium substrate and cover glass assembly, allowing them to be stored in compact rolled sheets, deploying into a flat solar array, which is then mounted in between the legacy solar arrays on the ISS. Additionally, the ROSA solar panels make use of triple-junction III-V materials $GaInP_2/GaAs/Ge$ and InGaP/InGaAs/Ge, setting a precedent for TFSCs in space [19].

To summarize, any improvements on current technologic standards for TFSCs in space must be competitive with this benchmark of 25 W/kg. Panel-level specific power of over 100 W/kg has been reached for conventional arrays, with developed pathways toward higher specific power levels (150–500 W/kg) [19]. One of the promises of TFSCs is the potential to have lightweight devices deposited on lightweight substrates

[19–22,101,106]. Currently, efficient TFSCs are able to reach higher specific power values, but these numbers typically do not account for the additional weights associated with materials required for assembly of the solar arrays onto supporting structures [21,154]. A somewhat related application involves high-altitude pseudo-satellites [155]; this generic term encompasses both unmanned (solar powered) air vehicles (or aircraft) [156] and high-altitude airships [157].

Possible space applications that will require high (mass specific) power supply are missions involving drones that fly over planetary (i.e., Mars Ingenuity [158]) or outer planetary moon surfaces [159], electric propulsion [160] for interplanetary missions, such as SEP for space tugs [134], or missions to the outer planets [135] (Fig. 9.19). An application in the (distant) future for lightweight power generation [18–22] could

Figure 9.19 (A) Artist impression of an Earth-Mars cargo ship approaching Mars. (B) Artist's concept of the solar power sail probe oversize kite-craft for exploration and astronautics in the outer solar system (OKEANOS) planned by the Japan Aerospace Exploration Agency (JAXA). OKEANOS realizes kW-class solar power generation at 5 AU Sun distance using a 40 × 40 m solar power sail; the transit to Jupiter would utilize electric delta-V Earth gravity assist (EDVEGA); EDVEGA is a powerful method for orbital velocity leverage. *Reproduced with permission from: (A) Ref. [134]; (B) Ref. [135], copyright (2021) Elsevier.*

Figure 9.20 An artist's concept shows the deployment of solar arrays to provide power for a lunar base that could eventually grow to a settlement. Immediately after landing, two large flexible solar arrays are rolled out "window shade" fashion from the lander. One panel is tilted toward the eastern horizon, while the other faces west. This allows the array to produce power during the entire 2-week stay without tracking the Sun. These arrays provide power while the astronauts scout the area and prepare the base. Later, the astronauts will set up a larger, Sun-tracking array to provide higher power levels. For exploration, surveying, and lunar prospecting, the astronauts can drive around in a battery-powered moon buggy. The buggy will have its own small solar array, which tops off the battery charge whenever the rover is parked. *Courtesy: NASA.*

include support of extraterrestrial infrastructures or human settlements, built using local resources [161–166] (Fig. 9.20).

9.4 Conclusions

The field of TFSCs is constantly evolving, with researchers exploring new materials and device concepts; see Table 9.8 for a summary of practical metrics of three generations of photovoltaic technologies [167]. In this chapter we have detailed key aspects of materials processing, device structures, and solar cell performance test results that evidence a clear forward momentum toward devices that have potential use for space applications. At the same time, other technologies such as thin III-V cells and hybrid inorganic/organic devices involving perovskites have proven to be attractive alternatives. From the outset, we have clearly outlined the numerous challenges involved with operating in space or in the vicinity of off-world locales. Inorganic thin-film materials have demonstrated the ability to perform under adverse conditions.

It is a truism that most technologies will not succeed or will be superseded in the long run. However, many of the lessons learned in the course of the research discussed in this chapter have found applications well beyond those for which they were developed. The development of robust and effective TCOs for use in a-SiH solar cells resulted in materials and methods that are broadly used across technologies. Interlayers for organic photovoltaics (OPV) such as LiF and MoO_x are incorporated into both perovskite and advanced Si architectures. Indeed, the architectures of the first perovskite devices were initially based on structures that existed for both dye-sensitized solar cells and planar OPV. We have seen other examples of technology transfer or lessons learned involving processing (NTI commercial reactor), single-source materials precursors, and integrated power technologies that are classic examples of NASA "spin-offs." We encourage the reader to apply the lessons learned in this chapter to solve some of their own unique power problems.

References

[1] J. Poortmans, V. Arkhipov (Eds.), Thin Film Solar Cells: Fabrication, Characterization and Applications, John Wiley & Sons, Ltd., 2006, https://doi.org/10.1002/0470091282.
[2] A. Smets, K. Jäger, O. Isabella, R. van Swaaij, M. Zeman, Solar Energy: The Physics and Engineering of Photovoltaic Conversion, Technologies and Systems, UIT Cambridge Ltd., 2016.
[3] J.R. Woodyard, G.A. Landis, Radiation resistance of thin-film solar cells for space photovoltaic power, Sol. Cells (1991) 297–329, https://doi.org/10.1016/0379-6787(91)90103-V.
[4] S.G. Bailey, D.J. Flood, Space photovoltaics, Prog. Photovolt. Res. Appl. 6 (1998) 1–14.
[5] A. Bermudez-Garcia, P. Voarino, Olivier Raccurt, Environments, needs and opportunities for future space photovoltaic power generation: a review, Appl. Energy 290 (2021) 116757.
[6] D.A. Lamb, S.J.C. Irvine, M.A. Baker, C.I. Underwood, S. Mardhani, Thin film cadmium telluride solar cells on ultra-thin glass in low earth orbit—3 years of performance data on the AlSat-1N CubeSat mission, Prog. Photovolt. Res. Appl. 29 (2021) 1000–1007. Available from: https://doi.org/10.1002/pip.3423.
[7] B.A. Banks, S.K. Miller, K.K. de Groh, D.L. Waters, Lessons learned from atomic oxygen interaction with spacecraft materials in low earth orbit, AIP Conf. Proc. 1087 (2009) 312–328. Available from: https://doi.org/10.1063/1.3076845.
[8] C.A. Gueymard, D. Myers, K. Emery, Proposed reference irradiance spectra for solar energy systems testing, Sol. Energy 73 (6) (2003) 443–467.
[9] J.J. Hanak, J. Kaschmitter, The status of lightweight photovoltaic space array technology based on amorphous silicon solar cells, in: Space Photovoltaic Research and Technology, 1991: Proceedings of a Conference Held at NASA Lewis Research Center, Cleveland, Ohio, May 7-9, 1991, CP-3121, for further information see: https://www.osti.gov/biblio/5709019.
[10] R.P. Raffaelle, S.L. Castro, A.F. Hepp, S.G. Bailey, Quantum dot solar cells, Prog. Photovoltaics Res. Appl. 10 (6) (2002) 433–439.
[11] S.G. Bailey, J.D. Harris, A.F. Hepp, E.J. Anglin, R.P. Raffaelle, H.R. Clark, S.T.P. Gardner, S.S. Sun, Thin-film Organic-Based Solar Cells for Space Power, in: 37th Intersociety Energy Conversion Engineering Conference, 2002, pp. 191–196, https://doi.org/10.1109/IECEC.2002.1392006.
[12] K. Otte, L. Makhova, A. Braun, I. Konovalov, Flexible Cu(In,Ga)Se_2 thin-film solar cells for space application, Thin Solid Films 511–512 (2006) 613–622. Available from: https://doi.org/10.1016/j.tsf.2005.11.068.
[13] A. Romeo, D.L. Batzner, H. Zogg, A.N. Tiwari, Potential of CdTe thin film solar cells for space applications, in: Proceedings of the 17th European Photovoltaic Conference and Exhibition 22 - 26 October 2001, Munich, Germany, pp. 2183-2186.
[14] A.K. Jena, A. Kulkarni, T. Miyasaka, Halide perovskite photovoltaics: background, status, and future prospects, Chem. Rev. 119 (5) (2019) 3036–3103. Available from: https://doi.org/10.1021/acs.chemrev.8b00539.

[15] L. McMillon-Brown, T.J. Peshek, Perovskite solar cells: background and prospects for space power applications, in: S.G. Bailey, A.F. Hepp, D.C. Ferguson, R.P. Raffaelle, S.M. Durbin (Eds.), Photovoltaics for Space: Key Issues, Missions and Alternative Technologies, Elsevier, 2022, pp. 129–156.

[16] B.K. Durant, I.R. Sellers, B. Rout, Perovskite solar cells on the horizon for space power systems, in: S.G. Bailey, A.F. Hepp, D.C. Ferguson, R.P. Raffaelle, S.M. Durbin (Eds.), Photovoltaics for Space: Key Issues, Missions and Alternative Technologies, Elsevier, 2022, pp. 175–195.

[17] A.F. Hepp, S.G. Bailey, R.P. Raffaelle, Inorganic photovoltaic materials and devices: past, present, and future, in: S.-S. Sun, N.S. Sariciftci (Eds.), Organic Photovoltaics: Mechanisms, Materials and Devices, CRC Press, Boca Raton, FL, USA, 2005, pp. 19–36.

[18] S. Bailey, R. Raffaelle, Chapter II-4-B - operation of solar cells in a space environment, in: S. Kalogirou (Ed.), McEvoy's Handbook of Photovoltaics (Third Edition) Fundamentals and Applications, Academic Press, 2017, pp. 987–1003.

[19] R.P. Raffaelle, An introduction to space photovoltaics: technologies, issues, and missions, in: S.G. Bailey, A.F. Hepp, D.C. Ferguson, R.P. Raffaelle, S.M. Durbin (Eds.), Photovoltaics for Space: Key Issues, Missions and Alternative Technologies, Elsevier, 2022, pp. 3–27.

[20] G.A. Landis, A.F. Hepp, Applications of thin-film photovoltaics for space, in: Proceedings of the 26th Intersociety Energy Conversion Engineering Conference, 2, August 4–9, 1991, pp. 256–261.

[21] D.J. Hoffman, T.W. Kerslake, A.F. Hepp, M.K. Jacobs, D. Ponnusamy, Thin-film photovoltaic solar array parametric assessment, in: Proceedings of the 35th Intersociety Energy Conversion Engineering Conference, 1, July 24–28, 2000, pp. 670–680. AIAA-2000-2919, also: NASA/TM-2000-210342.

[22] S.G. Bailey, A.F. Hepp, R.P. Raffaelle, Thin film photovoltaics for space applications, in: Proceedings of the 36th Intersociety Energy Conversion Engineering Conference, 1, July 29–August 2, 2001, pp. 235–238.

[23] T.D. Lee, A.E. Ubong, A review of thin film solar cell technologies and challenges, Renew. Sustain. Energy Rev. 70 (April 2017) 1286–1297. Available from: https://doi.org/10.1016/j.rser.2016.12.028.

[24] G. Han, S. Zhang, P.P. Boix, L.H. Wong, L. Sun, S.-Y. Lien, Towards high efficiency thin film solar cells, Prog. Mater. Sci. 87 (June 2017) 246–291. Available from: https://doi.org/10.1016/j.pmatsci.2017.02.003.

[25] J. Ramanujam, D.M. Bishop, T.K. Todorov, O. Gunawan, J. Rath, R. Nekovei, E. Artegiani, A. Romeo, Flexible CIGS, CdTe and a-Si:H based thin film solar cells: a review, Prog. Mater. Sci. 110 (2020) 100619.

[26] P.-K. Kung, M.-H. Li, P.-Y. Lin, Y.-H. Chiang, C.-R. Chan, T.-F. Guo, P. Chen, A review of inorganic hole transport materials for perovskite solar cells, Adv. Mater. Interfaces 5 (22) (2018) 1800882, https://doi.org/10.1002/admi.201800882.

[27] https://www.spacex.com/media/Capabilities&Services.pdf. (Accessed 16 March 2022).

[28] K. Keränen, P. Korhonen, J. Rekilä, O. Tapaninen, T. Happonen, P. Makkonen, K. Rönkä, Roll-to-roll printed and assembled large area LED lighting element, Int. J. Adv. Manuf. Technol. 81 (1) (2015) 529–536, https://doi.org/10.1007/s00170-015-7244-6.

[29] K. Kim, J. Kim, M.G. Gang, S.-H. Kim, S. Song, Y. Cho, D. Shin, et al., A simple and robust route toward flexible CIGS photovoltaic devices on polymer substrates: atomic level microstructural analysis and local opto-electronic investigation, Sol. Energy Mater. Sol. Cells 195 (June 2019) 280–290, https://doi.org/10.1016/j.solmat.2019.03.008.

[30] J. Gwak, M. Lee, J.H. Yun, S. Ahn, A. Cho, S. Ahn, J.P. Hulme, K.S. Shin, K.H. Yoon, Selenium flux effect on $Cu(In,Ga)Se_2$ thin films grown by a 3-stage Co-evaporation process, Isr. J. Chem. 55 (10) (2015) 1115–1122, https://doi.org/10.1002/ijch.201500011.

[31] L. Guo, M. Mason, M. Hopstaken, A. Kellock, L.T. Romankiw, H. Deligianni, A diffusion barrier for flexible thin film photovoltaics, J. Electrochem. Soc. 160 (3) (January 2013) D102, https://doi.org/10.1149/2.045303jes.

[32] P. Jackson, R. Wuerz, D. Hariskos, E. Lotter, W. Witte, M. Powalla, Effects of heavy alkali elements in $Cu(In,Ga)Se_2$ solar cells with efficiencies up to 22.6%, Phys. Status Solidi Rapid Res. Lett. 10 (8) (2016) 583–586, https://doi.org/10.1002/pssr.201600199.

[33] T. Kato, J.-L. Wu, Y. Hirai, H. Sugimoto, V. Bermudez, Record efficiency for thin-film polycrystalline solar cells up to 22.9% achieved by Cs-treated $Cu(In,Ga)(Se,S)_2$, IEEE J. Photovolt. 9 (1) (2019) 325–330, https://doi.org/10.1109/JPHOTOV.2018.2882206.

[34] R. Wuerz, A. Eicke, F. Kessler, S. Paetel, S. Efimenko, C. Schlegel, CIGS thin-film solar cells and modules on enamelled steel substrates, Sol. Energy Mater. Sol. Cells 100 (2012) 132–137, https://doi.org/10.1016/j.solmat.2012.01.004.

[35] Corning Willow Glass. https://www.corning.com/worldwide/en/innovation/corning-emerging-innovations/corning-willow-glass.html. (Accessed 29 January 2022).

[36] A. Chirilă, P. Reinhard, F. Pianezzi, P. Bloesch, A.R. Uhl, C. Fella, L. Kranz, et al., Potassium-induced surface modification of $Cu(In,Ga)Se_2$ thin films for high-efficiency solar cells, Nat. Mater. 12 (2013) 1107–1111.

[37] R. Carron, S. Nishiwaki, T. Feurer, R. Hertwig, E. Avancini, J. Löckinger, S.-C. Yang, S. Buecheler, A.N. Tiwari, Advanced alkali treatments for high-efficiency $Cu(In,Ga)Se_2$ solar cells on flexible substrates, Adv. Energy Mater. 9 (24) (June 2019) 1900408.

[38] Y. Kawano, J. Chantana, T. Nishimura, A. Mavlonov, T. Minemoto, Manipulation of [Ga]/([Ga] + [In]) profile in 1.4-μm-thick $Cu(In,Ga)Se_2$ thin film on flexible stainless steel substrate for enhancing short-circuit current density and conversion efficiency of its solar cell, Sol. Energy 204 (July 2020) 231–237. Available from: https://doi.org/10.1016/j.solener.2020.04.069.

[39] L. Zortea, S. Nishiwaki, T.P. Weiss, S. Haass, J. Perrenoud, L. Greuter, T. Feurer, G. Palaniswamy, S. Buecheler, A.N. Tiwari, $Cu(In,Ga)Se_2$ solar cells on low cost mild steel substrates, Sol. Energy 175 (2018) 25–30, https://doi.org/10.1016/j.solener.2017.12.057.

[40] H.P. Mahabaduge, W.L. Rance, J.M. Burst, M.O. Reese, D.M. Meysing, C.A. Wolden, J. Li, et al., High-efficiency, flexible CdTe solar cells on ultra-thin glass substrates, Appl. Phys. Lett. 106 (13) (March 2015) 133501.

[41] J.C. Perrenoud, Low Temperature Grown CdTe Thin Film Solar Cells for the Application on Flexible Substrates, Ph.D. Thesis, ETH Zürich, 2012, https://doi.org/10.3929/ethz-a-007339371. CH, No. 20460.

[42] L. Kranz, C. Gretener, J. Perrenoud, R. Schmitt, F. Pianezzi, F. La Mattina, P. Blösch, et al., Doping of polycrystalline CdTe for high-efficiency solar cells on flexible metal foil, Nature Commun 4 (2013) 2306. Available from: https://doi.org/10.1038/ncomms3306.

[43] S. Calnan, A.N. Tiwari, High mobility transparent conducting oxides for thin film solar cells, Thin Solid Films 518 (7) (January 2010) 1839–1849.

[44] R.G. Gordon, Criteria for choosing transparent conductors, MRS Bull. 25 (8) (2011) 52–57, https://doi.org/10.1557/mrs2000.151.

[45] S.C. Dixon, D.O. Scanlon, C.J. Carmalt, I.P. Parkin, N-Type doped transparent conducting binary oxides: an overview, J. Mater. Chem. C 4 (29) (2016) 6946–6961. Available from: https://doi.org/10.1039/C6TC01881E.

[46] B. Dou, E.M. Miller, J.A. Christians, E.M. Sanehira, T.R. Klein, F.S. Barnes, S.E. Shaheen, et al., High-performance flexible perovskite solar cells on ultrathin glass: implications of the TCO, J. Phys. Chem. Lett. 8 (19) (2017) 4960–4966, https://doi.org/10.1021/acs.jpclett.7b02128.

[47] F. Bittau, C. Potamialis, M. Togay, A. Abbas, P.J.M. Isherwood, J.W. Bowers, J.M. Walls, Analysis and optimisation of the glass/TCO/MZO stack for thin film CdTe solar cells, Sol. Energy Mater. Sol. Cells 187 (December 2018) 15–22.

[48] J. Barbé, D. Hughes, Z. Wei, A. Pockett, H.K.H. Lee, K.C. Heasman, M.J. Carnie, T.M. Watson, W.C. Tsoi, Radiation hardness of perovskite solar cells based on aluminum-doped zinc oxide electrode under proton irradiation, RRL Solar 3 (12) (December 2019) 1900219. Available from: https://onlinelibrary.wiley.com/doi/10.1002/solr.201900219.

[49] Y. Zhang, S.-W. Ng, X. Lu, Z. Zheng, Solution-processed transparent electrodes for emerging thin-film solar cells, Chem. Rev. 120 (4) (2020) 2049–2122, https://doi.org/10.1021/acs.chemrev.9b00483 (January 2020).

[50] J. Willis, D.O. Scanlon, J. Mater. Chem. C 9 (36) (2021) 11995–12009.

[51] M. Singh, S. Rana, A.K. Singh, Advanced nanomaterials utilized as top transparent electrodes in semi-transparent photovoltaic, Colloid Interface Sci. Commun 46 (January 2022) 100563. Available from: https://doi.org/10.1016/j.colcom.2021.100563.

[52] Z. Song, S.C. Watthage, A.B. Phillips, M.J. Heben, Pathways toward high performance perovskite solar cells: review of recent advances in organo-metal halide perovskites for photovoltaic applications, J. Photon. Energy 6 (2) (April—June 2016) 022001. Available from: https://doi.org/10.1117/1.JPE.6.022001.

[53] P. Zhang, J. Wu, T. Zhang, Y. Wang, D. Liu, H. Chen, L. Ji, et al., Perovskite solar cells with ZnO electron-transporting materials, Adv. Mater. 30 (3) (2018) 1703737. Available from: https://doi.org/10.1002/adma.201703737.

[54] M.P. De Jong, L.J. Van Ijzendoorn, M.J.A. De Voigt, Stability of the interface between indium-tin-oxide and poly (3, 4-ethylenedioxythiophene)/poly (styrenesulfonate) in polymer light-emitting diodes, Appl. Phys. Lett. 77 (14) (2000) 2255—2257.

[55] M. Jørgensen, K. Norrman, F.C. Krebs, Stability/degradation of polymer solar cells, Sol. Energy Mater. Sol. Cells 92 (7) (2008) 686—714. Available from: https://doi.org/10.1016/j.solmat.2008.01.005.

[56] Y. Hou, X. Du, S. Scheiner, D.P. McMeekin, Z. Wang, N. Li, M.S. Killian, et al., A generic interface to reduce the efficiency-stability-cost gap of perovskite solar cells, Science 358 (2017) 1192—1197.

[57] Z. Hawash, L.K. Ono, Y. Qi, Recent advances in spiro-MeOTAD hole transport material and its applications in organic—inorganic halide perovskite solar cells, Adv. Mater. Interfaces 5 (2018) 1700623.

[58] Z. Li, Y. Zhao, X. Wang, Y. Sun, Z. Zhao, Y. Li, H. Zhou, Q. Chen, Cost analysis of perovskite tandem photovoltaics, Joule 2 (2018) 1—14, https://doi.org/10.1016/j.joule.2018.05.001.

[59] Y. Tu, J. Wu, G. Xu, X. Yang, R. Cai, Q. Gong, R. Zhu, W. Huang, Perovskite solar cells for space applications: progress and challenges, Adv. Mater. 33 (21) (2021) 2006545. Available from: https://doi.org/10.1002/adma.202006545.

[60] K.K. de Groh, B.A. Banks, Atomic Oxygen Erosion Data from the MISSE 2-8 Missions, NASA/TM—2019-219982, May 2019. Available from: https://ntrs.nasa.gov/citations/20190025445.

[61] F. Lang, M. Jost, K. Frohna, E. Köhnen, A. Al-Ashouri, A.R. Bowman, T. Bertram, et al., Proton radiation hardness of perovskite tandem photovoltaics, Joule 4 (May 2020) 1054—1069. Available from: https://doi.org/10.1016/j.joule.2020.03.006.

[62] Y. Wang, Y. Liu, J. Tong, X. Shia, L. Huang, Z. Xiao, G. Wan, D. Pan, A general and facile solution approach for deposition of high-quality metal oxide charge transport layers, Sol. Energy Mater. Sol. Cells 236 (March 2022) 111511. Available from: https://doi.org/10.1016/j.solmat.2021.111511.

[63] Q. Lu, Z. Yang, X. Meng, Y. Yue, M.A. Ahmad, W. Zhang, S. Zhang, Y. Zhang, Z. Liu, W. Chen, A review on encapsulation technology from organic light emitting diodes to organic and perovskite solar cells, Adv. Funct. Mater. 31 (23) (2021) 2100151. Available from: https://doi.org/10.1002/adfm.202100151.

[64] A. Uddin, M.B. Upama, H. Yi, L. Duan, Encapsulation of organic and perovskite solar cells: a review, Coatings 9 (2) (February 2019) 2, https://doi.org/10.3390/coatings9020065.

[65] B.A. Banks, S.K.R. Miller, K.K. deGroh, R. Demko, Atomic Oxygen Effects on Spacecraft Materials, NASA/TM-2003-212484, June 2003. Available from: https://ntrs.nasa.gov/citations/20030062195.

[66] https://www1.grc.nasa.gov/space/iss-research/misse/. (Accessed 1 March 2022).

[67] R.J. Walters, J.C. Garner, S.N. Lam, J.A. Vasquez, W.R. Braun, R.E. Ruth, S.R. Messenger, et al., Materials on the International Space Station—forward technology solar cell experiment, Mater. Sci. Eng. B 116 (2005) 257—263.

[68] E.J. Simburger, J.H. Matsumoto, T.W. Giants, A. Garcia III, S. Liu, S.P. Rawal, A.R. Perry, et al., Development of a thin film solar cell interconnect for the PowerSphere concept, Mater. Sci. Eng. B 116 (2005) 321—325.

[69] Standard: Qualification and Quality Requirements for Space Solar Cells (AIAA S-111A), American Institute of Aeronautics and Astronautics, Inc., Washington, DC, 2014. Available from: https://doi.org/10.2514/4.102806.

[70] R.H. Ahangharnejhad, Z. Song, T. Mariam, J.J. Gardner, G.K. Liyanage, Z.S. Almutawah, B.M.M. Anwar, et al., Protecting perovskite solar cells against moisture-induced degradation with sputtered inorganic barrier layers, ACS Appl. Energy Mater. 4 (8) (2021) 7571–7578. Available from: https://doi.org/10.1021/acsaem.1c00816.

[71] A. Banerjee, et al., Advances in cell efficiency of a-Si:H and nc-Si:H-based multijunction solar cells for space and near-space applications, in: Proc 35th IEEE Photovoltaic Specialist Conference, 2010, pp. 2651–2655.

[72] C. Battaglia, A. Cuevas, S. De Wolf, High-efficiency crystalline silicon solar cells: status and perspectives, Energy Environ. Sci. 9 (5) (May 2016) 1552–1576.

[73] A. Romeo, E. Artegiani, CdTe-based thin film solar cells: past, present and future, Energies 14 (6) (2021) 1684. Available from: https://doi.org/10.3390/en14061684.

[74] D.C. Ferguson, D.P. Engelhart, R.C. Hoffmann, V.J. Murray, E.A. Plis, Space solar arrays and spacecraft charging, in: S.G. Bailey, A.F. Hepp, D.C. Ferguson, R.P. Raffaelle, S.M. Durbin (Eds.), Photovoltaics for Space: Key Issues, Missions and Alternative Technologies, Elsevier, 2022, pp. 29–50.

[75] T. Matsui, A. Bidiville, K. Maejima, H. Sai, T. Koida, T. Suezaki, M. Matsumoto, K. Saito, I. Yoshida, M. Kondo, High-efficiency amorphous silicon solar cells: impact of deposition rate on metastability, Appl. Phys. Lett. 106 (52) (February 2015) 053901.

[76] H. Fritzsche, Development in understanding and controlling the Staebler-Wronski effect in a-Si:H, Annu. Rev. Mater. Sci. 31 (2001) 47–79.

[77] T. Matsui, K. Maejima, A. Bidiville, H. Sai, T. Koida, T. Suezaki, M. Matsumoto, K. Saito, I. Yoshida, M. Kondo, High-efficiency thin-film silicon solar cells realized by integrating stable a-Si:H absorbers into improved device design, Jpn. J. Appl. Phys. 54 (8S1) 08KB10.

[78] H. Sai, T. Matsui, K. Matsubara, Stabilized 14.0%-efficient triple-junction thin-film silicon solar cell, Appl. Phys. Lett. 109 (18) (2016) 183506. Available from: https://doi.org/10.1063/1.49669986.

[79] S.-I. Sato, H. Sai, T. Ohshima, M. Imaizumi, K. Shimazaki, M. Kondo, Temperature influence on performance degradation of hydrogenated amorphous silicon solar cells irradiated with protons, Prog. Photovolt. Res. Appl. 21 (2013) 1499–1506.

[80] A.D. Verkerk, J.K. Rath, R.E.I. Schropp, Degradation of thin film nanocrystalline silicon solar cells with 1 MeV protons, Energy Proc. 2 (1) (2010) 221–226. Available from: https://doi.org/10.1016/j.egypro.2010.07.032.

[81] A.N. Maclnnes, M.B. Power, A.R. Barron, Chemical vapor deposition of gallium sulfide: phase control by molecular design, Chem. Mater 5 (9) (September 1993) 1344–1351.

[82] A.N. MacInnes, M.B. Power, A.R. Barron, P.P. Jenkins, A.F. Hepp, Enhancement of photoluminescence intensity of GaAs with cubic GaS chemical vapor deposited using a structurally designed single-source precursor, Appl. Phys. Lett. 62 (7) (February 1993) 711–713.

[83] M. Tabib-Azar, S. Kang, A.N. MacInnes, M.B. Power, A.R. Barron, P.P. Jenkins, A.F. Hepp, Electronic passivation of n- and p-type GaAs using chemical vapor deposited GaS, Appl. Phys. Lett. 63 (5) (August 1993) 625–627.

[84] A.R. Barron, P.P. Jenkins, A.N. MacInnes, A.F. Hepp, Metal, passivating layer, semiconductor, field-effect transistor, June 2, 1998. United States Patent 5,760,462.

[85] A.R. Barron, P.P. Jenkins, A.N. MacInnes, A.F. Hepp, Minority carrier device comprising a passivating layer including a group 13 element and a chalcogenide component, December 28, 1999. United States Patent 6,008,525.

[86] W. Hirpo, S. Dhingra, A.C. Sutorik, M.G. Kanatzidis, Synthesis of mixed copper-indium chalcogenolates. Single-source precursors for the photovoltaic materials $CuInQ_2$ (Q = S, Se), J. Am. Chem. Soc. 115 (4) (February 1993) 1597–1599.

[87] J.A. Hollingsworth, A.F. Hepp, W.E. Buhro, Spray CVD of copper indium sulfide films: control of microstructure and crystallographic orientation, Chem. Vap. Depos. 5 (3) (June 1999) 105–108.

[88] K.K. Banger, J. Cowen, A.F. Hepp, Synthesis and characterization of the first liquid single-source precursors for the deposition of ternary chalcopyrite ($CuInS_2$) thin film materials, Chem. Mater. 13 (11) (November 2001) 3827–3829.

[89] K.K. Banger, J.A. Hollingsworth, J.D. Harris, J. Cowen, W.E. Buhro, A.F. Hepp, Ternary single-source precursors for polycrystalline thin-film solar cells, Appl. Organomet. Chem. 16 (November 2002) 617–627.

[90] K.K. Banger, J.A. Hollingsworth, M.H.-C. Jin, J.D. Harris, J.E. Cowen, E.W. Bohannan, J.A. Switzer, W.E. Buhro, A.F. Hepp, Single source precursors for fabrication of I-III-VI_2 thin-film solar cells via spray CVD, Thin Solid Films 431–432 (May 2003) 63–67.

[91] K.K. Banger, M.H.-C. Jin, J.D. Harris, P.E. Fanwick, A.F. Hepp, A new facile route for the preparation of single-source precursors for bulk, thin-film, and nanocrystallite I-III-VI semiconductors, Inorg. Chem. 42 (24) (December 2003) 7713–7715.

[92] J.E. Lau, J.D. Harris, D.G. Hehemann, O. Khan, N.V. Duffy, M.D. Hoops, A.F. Hepp, Spray chemical vapor deposition of tris(bis(phenylmethyl)carbamo-dithioato-S,S') indium (III) for $CuInS_2$ polycrystalline thin-film photovoltaic devices, Mater. Sci. Eng. B 116 (3) (February 2005) 381–389.

[93] C.V. Kelly, M.H.-C. Jin, K.K. Banger, J.S. McNatt, J.E. Dickman, A.F. Hepp, Parametric study on non-vacuum chemical vapor deposition of $CuInS_2$ from a single-source precursor, Mater. Sci. Eng. B 116 (3) (February 2005) 403–408.

[94] K.K. Banger, A.F. Hepp, J.D. Harris, M.H. Jin, S.L. Castro, Single-source precursors for ternary chalcopyrite materials, and methods of making and using the same, January 31, 2006. United States Patent 6,992,202.

[95] R.G. Rodriguez, D.J.V. Pulsipher, L.D. Lau, E. Shurda, J.J. Pak, M.H. Jin, K.K. Banger, A.F. Hepp, Pulsed-spray radiofrequency PECVD of $CuInS_2$ thin films, Plasma Chem. Plasma Proc. 26 (2) (April 2006) 137–148.

[96] A.F. Hepp, K.K. Banger, M.H.-C. Jin, J.D. Harris, J.S. McNatt, J.E. Dickman, Spray Chemical Vapor Deposition of Single-Source Precursors for I-III-VI_2 Thin-Film Materials, NASA/TP-2008-214937, November 2008, 32 pp.

[97] S.L. Castro, S.G. Bailey, R.P. Raffaelle, K.K. Banger, A.F. Hepp, Nanocrystalline chalcopyrite materials ($CuInS_2$ and $CuInSe_2$) via low-temperature pyrolysis of molecular single-source precursors, Chem. Mater. 15 (2003) 3142–3147.

[98] S.L. Castro, S.G. Bailey, R.P. Raffaelle, K.K. Banger, A.F. Hepp, Synthesis and characterization of colloidal $CuInS_2$ nanoparticles from a molecular single-source precursor, J. Phys. Chem. B 108 (2004) 12,429–12,435.

[99] P. Mahawela, S. Jeedigunta, C.S. Ferekides, D.L. Morel, Analysis of carrier transport phenomena in high band gap II–VI-based MIS photovoltaic devices, Thin Solid Films 431–432 (May 2003) 461–465.

[100] P. Mahawela, G. Sivaraman, S. Jeedigunta, J. Gaduputi, M. Ramalingam, S. Subramanian, S. Vakkalanka, C.S. Ferekides, D.L. Morel, II–VI compounds as the top absorbers in tandem solar cell structures, Mater. Sci. Eng. B 116 (2005) 283–291.

[101] J.E. Dickman, A.F. Hepp, D.L. Morel, C.S. Ferekides, J.R. Tuttle, D.J. Hoffmann, N.G. Dhere, Utility of thin film solar cells on flexible substrates for space power, in: Proceedings of the 1st International Energy Conversion Engineering Conference, August 2003. Portsmouth, VA, AIAA 2003-5922, 10 pp.

[102] N.G. Dhere, S.R. Ghongadi, M.B. Pandit, A.H. Jahagirdar, D. Scheiman, CIGS2 thin-film solar cells on flexible foils for space power, Prog. Photovoltaics Res. Appl. 10 (6) (September 2002) 407–416.

[103] N.G. Dhere, V.S. Gade, A.A. Kadam, A.H. Jahagirdar, S.S. Kulkarni, S.M. Bet, Development of CIGS2 thin film solar cells, Mater. Sci. Eng., B 116 (3) (February 2005) 303–309.

[104] B.M. Basol, V.K. Kapur, A. Halani, C. Leidholm, Copper indium diselenide thin film solar cells fabricated on flexible foil substrates, Sol. Energy Mater. Sol. Cells 29 (2) (March 1993) 163–173.

[105] J.A. AbuShama, J. Wax, T. Berens, J. Tuttle, Progress toward improved device performance in large-area $Cu(In,Ga)Se_2$ thin film solar cells, in: 2006 IEEE 4th World Conference on Photovoltaic Energy

Conference, 2006, pp. 487–490, https://doi.org/10.1109/WCPEC.2006.279497. Available from: https://ieeexplore.ieee.org/document/4059670.

[106] R.P. Raffaelle, A.F. Hepp, J.E. Dickman, D.J. Hoffman, N. Dhere, J.R. Tuttle, M.H.-C. Jin, C.V. Kelly, Thin-film solar cells on metal foil substrates for space power, in: Proceedings of 2nd International Energy Conversion Engineering Conference, Providence, RI, AIAA-2004-5735, August 2004, 18 pp.

[107] B.M. Basol, V.K. Kapur, C.R. Leidholm, A. Halani, K. Gledhill, Flexible and light weight copper indium diselenide solar cells on polyimide substrates, Sol. Energy Mater. Sol. Cells 43 (1) (August 1996) 93–98.

[108] A.F. Hepp, J.S. McNatt, J.E. Dickman, D.L. Morel, C.S. Ferekides, M.H.-C. Jin, K.K. Banger, N. Orbey, M. Cushman, R.W. Birkmire, W.S. Shafarman, Thin-film solar cells on polymer substrates for space power, in: Proceedings of 2nd International Energy Conversion Engineering Conference, Providence, RI, AIAA-2004-5537, August 2004, p. 18 pp.

[109] A.F. Hepp, J.S. McNatt, S.G. Bailey, J.E. Dickman, R.P. Raffaelle, B.J. Landi, A. Anctil, et al., Ultra-lightweight Hybrid Thin-Film Solar Cells: A Survey of Enabling Technologies for Space Power Applications, AIAA-2007-4721; Re-print Published as NASA/TM-2007-214955, August 2007, 22 pp.

[110] A.F. Hepp, J.S. McNatt, S.G. Bailey, R.P. Raffaelle, B.J. Landi, S.-S. Sun, C.E. Bonner, K.K. Banger, D. Rauh, Ultra-lightweight space power from hybrid thin-film solar cells, IEEE Aerosp. Electron Syst. 23 (9) (September 2008) 31–41.

[111] R.P. Raffaelle, A.F. Hepp, G.A. Landis, D.J. Hoffman, Mission applicability assessment of integrated power components and systems, Prog. Photovoltaics Res. Appl. 10 (6) (August 2002) 391–397.

[112] P. Jenkins, D. Scheiman, D. Wilt, R. Raffaelle, R. Button, T. Kerslake, M. Batchelder, D. Lefevre, R.G. Moore, Results from the advance power technology experiment on the Starshine 3 satellite, in: Proceedings of the 16th Annual/USU Conference on Small Satellites, Logan, UT, USA, August 2002 paper SSC02-X-3. Available from: https://digitalcommons.usu.edu/smallsat/. (Accessed 15 February 2022).

[113] http://www.azinet.com/starshine/, February 2003. (Accessed 15 February 2022).

[114] D.M. Wilt, A.F. Hepp, M. Moran, P.P. Jenkins, D.A. Scheiman, R.P. Raffaelle, Integrated micro-power system (IMPS) development at NASA Glenn Research Center, in: Electrochemical Society Proceedings 2002-2025, 2003, pp. 194–202.

[115] E.J. Simburger, J.H. Matsumoto, P.A. Gierow, A.F. Hepp, Integrated thin film battery and circuit module, May 16, 2006. U.S. Patent 7,045,246.

[116] R.P. Raffaelle, B.J. Landi, J.D. Harris, S.G. Bailey, A.F. Hepp, Carbon nanotubes for power applications, Mater. Sci. Eng. B 116 (3) (February 2005) 233–243.

[117] J.D. Harris, R.P. Raffaelle, T. Gennett, B.J. Landi, A.F. Hepp, Growth of multi-walled carbon nanotubes by injection CVD using cyclopentadienyliron dicarbonyl dimer and cyclooctatetraene iron tricarbonyl, Mater. Sci. Eng. B 116 (2005) 369–374.

[118] A.F. Hepp, J.D. Harris, Process and apparatus utilizing mixed ligand organometallic catalysts for in situ growth of high purity, low defect density carbon nanotubes, July 27, 2010. United States Patent 7,763,230 B2.

[119] D.M. Flood, D.J. Flood, J.D. Harris, A.F. Hepp, in: B. Schwerin (Ed.), Nanotube production devices expand research capabilities, Spinoff 2011, March 2012, pp. 148–149.

[120] https://www.nanotech-innovations.com/. (Accessed 15 February 2022).

[121] A.F. Hepp, P.N. Kumta, O.I. Velikokhatnyi, R.P. Raffaelle, Batteries for integrated power and CubeSats: recent developments and future prospects, in: P.N. Kumta, A.F. Hepp, M.K. Datta, O.I. Velikokhatnyi (Eds.), Silicon Anode Systems for Lithium-Ion Batteries, Elsevier Inc., Cambridge, MA, USA, 2022, pp. 457–508, Available from, https://doi.org/10.1016/B978-0-12-819660-1.00011-6.

[122] A.F. Hepp, J.D. Harris, A.W. Apblett, A.R. Barron, Commercialization of single-source precursors: applications, intellectual property, and technology transfer, in: A.W. Apblett, A.R. Barron, A.F. Hepp (Eds.), Nanomaterials via Single-Source Precursors: Synthesis, Processing and

Applications, Elsevier Inc., Cambridge, MA, USA, 2022, pp. 563–601. Available from: https://doi.org/10.1016/B978-0-12-820340-8.00014-9.

[123] J. Smucker, J. Gong, A comparative study on the band diagrams and efficiencies of silicon and perovskite solar cells using wxAMPS and AMPS-1D, Sol. Energy 228 (November 2021) 187–199.

[124] U.S. Solar Panel Manufacturers. https://www.solarpowerworldonline.com/u-s-solar-panel-manufacturers/. (Accessed 30 January 2022).

[125] D. Bonnet, H. Rabenhorst, New results on the development of a thin film *p*-CdTe–*n*-CdS heterojunction solar cell, in: Proceedings of the 9th Photovoltaic Specialists Conference, Silver Spring, MD, USA, 1972, pp. 129–131.

[126] D.A. Lamb, C.I. Underwood, V. Barrioz, R. Gwilliam, J. Hall, M.A. Baker, S.J. Irvine, Proton irradiation of CdTe thin film photovoltaics deposited on cerium-doped space glass, Prog. Photovoltaics Res. Appl. 25 (12) (2017) 1059–1067.

[127] https://www.solar-frontier.com/eng/news/2019/0117_press.html. (Accessed 1 March 2022).

[128] R. Zwanenburg, M. Kroon, Requirements for thin film solar arrays and the development status of a new solar cell blanket concept, in: European Space Agency, (Special Publication) ESA SP Issue 589, Seventh European Space Power Conference, 9 May - 13 May 2005, 2005, pp. 299–304.

[129] H. Afshari, B.K. Durant, C.R. Brown, K. Hossain, D. Poplavskyy, B. Rout, I.R. Sellers, The role of metastability and concentration on the performance of CIGS solar cells under Low-Intensity-Low-Temperature conditions, Sol. Energy Mater. Sol. Cells 212 (August 2020) 110571.

[130] C. Morioka, K. Shimazaki, S. Kawakita, M. Imaizumi, H. Yamaguchi, T. Takamoto, S.-i. Sato, T. Ohshima, Y. Nakamura, K. Hirako, M. Takahashi, First flight demonstration of film-laminated InGaP/GaAs and CIGS thin-film solar cells by JAXA's small satellite in LEO, Prog. Photovoltaics Res. Appl. 19 (2011) 825–833, https://doi.org/10.1002/pip.1046.

[131] C.R. Brown, V.R. Whiteside, D. Poplavskyy, K. Hossain, M.S. Dhoubhadel, I.R. Sellers, Flexible Cu(In,Ga)Se_2 solar cells for outer planetary missions: Investigation under low-intensity low-temperature conditions, IEEE J. Photovoltaics 9 (2019) 552–558, https://doi.org/10.1109/JPHOTOV.2018.2889179.

[132] S.B. Zhang, S.-H. Wei, A. Zunger, H. Katayama-Yoshida, Defect physics of the $CuInSe_2$ chalcopyrite semiconductor, Phys. Rev. B 57 (1998) 9642–9656. Available from, https://doi.org/10.1103/Phys-RevB.57.9642.

[133] T. Sproewitz, U. Banik, J.-T. Grundmann, F. Haack, M. Hillebrandt, H. Martens, S. Meyer, et al., Concept for a Gossamer solar power array using thin-film photovoltaics, CEAS Space J 12 (2020) 125–135. Available from: https://doi.org/10.1007/s12567-019-00276-6.

[134] F. Chang Díaz, J. Carr, L. Johnson, W. Johnson, G. Genta, P.F. Maffione, Solar electric propulsion for human Mars missions, Acta Astronaut. 160 (July 2019) 183–194.

[135] Y. Takao, O. Mori, M. Matsushita, A.K. Sugihara, Solar electric propulsion by a solar power sail for small spacecraft missions to the outer solar system, Acta Astronaut. 181 (April 2021) 362–376.

[136] U. Banik, K. Sasaki, N. Reininghaus, K. Gehrke, M. Vehse, M. Sznajder, T. Sproewitz, C. Agert, Enhancing passive radiative cooling properties of flexible CIGS solar cells for space applications using single layer silicon oxycarbonitride films, Sol. Energy Mater. Sol. Cells 209 (2020) 110456. Available from: https://doi.org/10.1016/j.solmat.2020.110456.

[137] M. Pscherer, M. Günthner, C.A. Kaufmann, A. Rahm, G. Motz, Thin-film silazane/alumina high emissivity double layer coatings for flexible Cu(In,Ga)Se_2 solar cells, Sol. Energy Mater. Sol. Cells 132 (January 2015) 296–302, https://doi.org/10.1016/j.solmat.2014.09.015.

[138] C.R. Mercer, Solar array designs for deep-space science missions, in: S.G. Bailey, A.F. Hepp, D.C. Ferguson, R.P. Raffaelle, S.M. Durbin (Eds.), Photovoltaics for Space: Key Issues, Missions and Alternative Technologies, Elsevier, 2022, pp. 349–378.

[139] A. Gurung, Q. Qiao, Solar charging batteries: advances, challenges, and opportunities, Joule 2 (7) (July 2018) 1217–1230.

[140] S.N. Agbo, T. Merdzhanova, S. Yu, H. Tempel, H. Kungl, R.-A. Eichel, U. Rau, O. Astakhov, Development towards cell-to-cell monolithic integration of a thin-film solar cell and lithium-ion accumulator, J. Power Sources 327 (September 2016) 340–344.

[141] V. Vega-Garita, L. Ramirez-Elizondo, N. Narayan, P. Bauer, Integrating a photovoltaic storage system in one device: a critical review, Prog. Photovolt. 27 (4) (2019) 346–370.
[142] T. Böhnke, M. Edoff, Copper indium gallium diselenide thin films for sun angle detectors in space applications, Thin Solid Films 517 (6) (January 2009) 2063–2068.
[143] F.C. Krebs, T.D. Nielsen, J. Fyenbo, M. Wadstrom, M.S. Pedersen, Manufacture, integration and demonstration of polymer solar cells in a lamp for the "Lighting Africa" initiative, Energy Environ. Sci. 3 (5) (May 2010) 512–525.
[144] M. Fojtik, D. Kim, G. Chen, Y.-S. Lin, D. Fick, J. Park, M. Seok, M.-T. Chen, Z. Foo, D. Blaauw, D. Sylvester, A millimeter-scale energy-autonomous sensor system with stacked battery and solar cells, IEEE J. Solid State Circ. 48 (3) (March 2013) 801–813, https://doi.org/10.1109/JSSC.2012.2233352.
[145] Y.H. Lee, J.S. Kim, J. Noh, I. Lee, H.J. Kim, S. Choi, J. Seo, S. Jeon, T.S. Kim, J.Y. Lee, J.W. Choi, Wearable textile battery rechargeable by solar energy, Nano Lett. 13 (11) (November 2013) 5753–5761, https://doi.org/10.1021/nl403860k.
[146] A.E. Ostfeld, A.M. Gaikwad, Y. Khan, A.C. Arias, High-performance flexible energy storage and harvesting system for wearable electronics, Sci. Rep. 6 (May 2016) 26122.
[147] A. Nascetti, M. Mirasoli, E. Marchegiani, M. Zangheri, F. Costantini, A. Porchetta, L. Iannascoli, Integrated chemiluminescence-based lab-on-chip for detection of life markers in extraterrestrial environments, Biosens. Bioelectron. 123 (January 2019) 195–203.
[148] T. Zhu, Y. Yang, Y. Liu, R. Lopez-Hallman, Z. Ma, L. Liu, X. Gong, Wireless portable light-weight self-charging power packs by perovskite-organic tandem solar cells integrated with solid-state asymmetric supercapacitors, Nano Energy 78 (December 2020) 105397.
[149] https://www.nasa.gov/mission_pages/station/structure/elements/solar_arrays-about.html. (Accessed 2 March 2022).
[150] https://www.nasa.gov/mission_pages/station/behindscenes/truss_segment.html. (Accessed 2 March 2022).
[151] https://www.nasa.gov/feature/new-solar-arrays-to-power-nasa-s-international-space-station-research. (Accessed 2 March 2022).
[152] https://spaceflightnow.com/2021/06/02/new-solar-arrays-ready-to-upgrade-international-space-stations-power-grid/. (Accessed 2 March 2022).
[153] M.K. Chamberlain, S.H. Kiefer, M. LaPointe, Pete LaCorte, on-orbit flight testing of the roll-out solar array, Acta Astronaut. 179 (February 2021) 407–414.
[154] K. Reed, H.J. Willenberg, Early commercial demonstration of space solar power using ultra-lightweight arrays, Acta Astronaut. 65 (9–10) (November–December 2009) 1250–1260.
[155] J. Gonzalo, D. López, D. Domínguez, A. García, A. Escapa, On the capabilities and limitations of high altitude pseudo-satellites, Prog. Aero. Sci. 98 (April 2018) 37–56.
[156] X. Zhu, Z. Guo, Z. Hou, Solar-powered airplanes: A historical perspective and future challenges, Prog. Aero. Sci. 71 (November 2014) 36–53.
[157] Y. Xu, W. Zhu, J. Li, L. Zhang, Improvement of endurance performance for high-altitude solar-powered airships: a review, Acta Astronaut. 167 (February 2020) 245–259.
[158] https://mars.nasa.gov/technology/helicopter/. (Accessed 6 March 2022).
[159] M. Hassanalian, D. Rice, A. Abdelkefi, Evolution of space drones for planetary exploration: A review, Prog. Aero. Sci. 97 (February 2018) 61–105.
[160] K. Holste, P. Dietz, S. Scharmann, K. Keil, T. Henning, D. Zschätzsch, M. Reitemeyer, et al., Ion thrusters for electric propulsion: Scientific issues developing a niche technology into a game changer, Rev. Sci. Instrum. 91 (6) (2020) 061101.
[161] G.A. Landis, Materials refining on the moon, Acta Astronaut. 60 (10–11) (May–June 2007) 906–915. Accessible at: https://doi.org/10.1016/j.actaastro.2006.11.004.
[162] M.Z. Naser, Space-native construction materials for earth-independent and sustainable infrastructure, Acta Astronaut. 155 (February 2019) 264–273.
[163] S.O. Starr, A.C. Muscatello, Mars in situ resource utilization: A review, Planet. Space Sci. 182 (2020) 104824, https://doi.org/10.1016/j.pss.2019.104824.

[164] N. Kalapodis, G. Kampas, O.-J. Ktenidou, A review towards the design of extraterrestrial structures: from regolith to human outposts, Acta Astronaut. 175 (October 2020) 540–569.
[165] S.K. Borowski, S.W. Ryan, D.R. McCurdy, B.G. Sauls, Key technologies, systems, and infrastructure enabling the commercialization and human settlement of low Earth orbit, the Moon, and cislunar space, New Space 9 (2) (2021) 104–122.
[166] T.E. Girish, T.N. Satrhyan, V.M. Kumar, Technological relevance and photovoltaic applications of high-quality silica deposits on Mars, in: S.G. Bailey, A.F. Hepp, D.C. Ferguson, R.P. Raffaelle, S.M. Durbin (Eds.), Photovoltaics for Space: Key Issues, Missions and Alternative Technologies, Elsevier, 2022, pp. 433–441.
[167] S. Shi, L. Yao, P. Ma, Y. Jiao, X. Zheng, D. Ning, M. Chen, F. Sui, H. Liu, C. Yang, W. Li, Recent progress in the high-temperature-resistant PI substrate with low CTE for CIGS thin-film solar cells, Mater. Today Ener. (June 2021) 100640.

CHAPTER TEN

Inverted lattice-matched GaInP/GaAs/GaInNAsSb triple-junction solar cells: Epitaxial lift-off thin-film devices and potential space applications

Naoya Miyashita[1,2] and Yoshitaka Okada[1]
[1]Research Center for Advanced Science and Technology (RCAST), The University of Tokyo, Tokyo, Japan
[2]Department of Engineering Science, The University of Electro-Communications, Chofu, Japan

10.1 Introduction

Modern high-efficiency III-V multijunction solar cells (MJSCs) are ideal for a variety of space applications due to their higher efficiency and potential radiation damage resistance [1,2]. A recent in-depth introduction to III-V solar cell materials technologies and space applications is provided by a chapter by Chiu in the first section of this book [3]. Concentrator photovoltaic(s) (CPV) technology is regarded as one of the important applications for III-V MJSCs. However, although the cell itself is capable of high efficiency, fabrication is significantly expensive. Consequently, integration with concentrator optics enables a reduction in the total implementation cost while simultaneously improving the efficiency by operating under concentration. The solar cell device is usually placed near the focal point (i.e., loosely focused point) in the CPV configuration, where the incident light comes from normal direction to the cell surface. A chapter by O'Neill in the final section of this book addresses space exploration missions enabled by space photovoltaic concentrators utilizing ultralight Fresnel lenses [4]. In general, CPV systems need to track the Sun to maximize daily power generation, and they need heavy tracker frames. Conventionally, CPV is considered mainly for large-scale solar power plants. In recent years, new types of CPV applications including a nontracking-type CPV for automobiles [5] and space exploration have been proposed and/or explored [2,4].

Meanwhile, nonconcentrator MJSCs have advantages in some applications because they need neither concentrator optics nor mechanical trackers. In terms of efficiency records, nonconcentrator efficiencies have also been improved along with CPV efficiencies [6]. These improvements are due to the progress in integration technologies of lattice-mismatched materials, such as compositionally-graded buffer layers or wafer bonding. Use of such options allows monolithic stacking of single crystal lattice-mismatched subcells while minimizing degradation due to misfit dislocations. So far,

Photovoltaics for Space
ISBN 978-0-12-823300-9, https://doi.org/10.1016/B978-0-12-823300-9.00013-3

nonconcentrated efficiencies of 37.9% for inverted metamorphic (IMM) three-junction (3J) and 39.2% for IMM six-junction cells [7] have been reported.

On the other hand, MJSCs consisting of lattice-matched (LM) subcells such as GaInP/GaAs/Ge have the advantage of not requiring compositionally-graded buffer layers or a wafer bonding process. In terms of this kind of Ge-based 3JSCs, which is commercially available, the growth method is well-established and therefore regarded as an industrial standard. However, there is a large mismatch in the photocurrent among the subcells; specifically, the Ge cell generates a current nearly double that of the other subcells. This issue is the primary contributor to a limiting efficiency of 41.6% [8]. To achieve increased efficiency, 1.0-eV subcells that can be inserted between the GaAs (1.4 eV) and Ge (0.66 eV) subcells to make a 4J structure have been considered.

Most conventional 1.0-eV materials (e.g., GaInAs) cannot satisfy the constraint of lattice matching to the Ge bottom cell except in the case of dilute nitrides, for example GaInNAs, GaNAsSb, and GaInNAsSb. With changing the N, In, and Sb compositions in the GaInNAs(Sb) or GaNAsSb, the bandgap energy can be tuned to below ~1.4 eV (i.e., the bandgap of GaAs) while maintaining an LM condition. Using GaInNAs as a bottom cell for a 3JSC, in 2012 Solar Junction demonstrated a 44.0% concentrator efficiency [9].

In addition, LM MJSCs are compatible with epitaxial lift-off (ELO), which enables the device layers to be released from the substrate [10—15]. Because the separated epilayer itself is fragile and difficult to be handled through subsequent processes, it requires schemes for mechanical support. Furthermore, due to lattice strain, the separated layer may suffer from risk of breaking by the freed stress. Consequently, solar cell structures consisting of LM materials are considered to be a preferred choice for fabrication of ELO thin-film devices. On the other hand, given the importance of high mass specific power [1,2], space applications may prove to be suitable for ELO III-V devices [14].

Post-ELO, the released substrates can be reused for subsequent growth that leads to an expected reduction of the total cell cost. This is one of the strongest motivations to develop ELO-related technologies such as large scale, high speed, multiple release, etc. In addition, the released stand-alone device layer opens novel applications, such as flexible and lightweight solar cells as power sources for portable, stratospheric, and space usages. Furthermore, in III-V ELO thin-film solar cells, high conversion efficiencies have been reported [16,17], which are due in part to efficient utilization of the photon recycling effect. In thin-film solar cells, several approaches for optical management can be applied to enhance the light absorption, e.g., increasing the optical path length (OPL) with a light-scattering back pattern [18,19] or using Fabry-Pérot (FP) cavity resonances [20]. These approaches are particularly effective for multiquantum well [21—24] and quantum dot [25—27] solar cells whose optical thickness is often limited owing to growth constraints.

Recently, single-junction solar cell devices including GaInNAsSb layers have been successfully released from a GaAs substrate by ELO without any damage or cracks [28]. This provides motivation to seriously consider use of dilute nitrides such as GaInNAs(Sb) as a 1.0-eV subcell in LM-MJSCs, since it can be grown lattice matched to the GaAs substrate, while its band gap can be tuned in the range of 0.8−1.4 eV. For this, a method was developed to fabricate MJSCs with a molecular beam epitaxy (MBE)-grown GaInNAs(Sb) subcell. Approach is based on combining the two growth methods of metalorganic chemical vapor deposition (MOCVD) and MBE creating a type of hybrid growth, with the goal being to select the better growth method for each subcell. This is the primary focus of the remainder of this chapter.

10.2 Design and growth of GaInNAs 1.0-eV subcells

One of the optimum combinations of band gap energies (E_g) for a 4JSC that is expected to perform beyond 50% of solar energy conversion efficiency under light concentration uses 1.88 eV/1.42 eV/1.04 eV/0.66 eV where GaAs and Ge are assigned as the second and fourth subcells respectively [29,30]. In terms of realizing this structure in a LM system, the dilute nitride semiconductor alloy GaInNAs has attracted considerable interest [29−37]. It has been shown that introduction of a small amount of N into GaAs can reduce the band gap energy significantly below 1.4 eV. The presence of N also reduces the GaAs lattice constant, which allows lattice matching to GaAs and Ge by incorporating In into GaNAs. Therefore, a GaInNAs subcell enables the 1.0−1.4 eV region of the solar spectrum to be covered in LM 4JSCs. Relevance of this materials technology for space applications has been addressed [36]. Previous studies revealed improved material quality by a supply of Sb during the growth of both GaInNAs quantum wells [38−40] and thin films [41−43]. The Sb acts as a surfactant and yields improvements of the heterointerfaces properties [38], photoluminescence efficiency [39−42], and suppression of nitrogen-related defects [43].

In an MJSC, each subcell is connected in series, so the target current values generated in each subcell should be designed to be identical to obtain optimum solar cell performance. For the 4J (1.88 eV/1.42 eV/1.04 eV/0.66eV) or 3J (1.88 eV/1.42 eV/1.04 .eV) SCs, the output current density is limited by the second 1.42-eV subcell when each subcell fully generates and collects carrier. Then the output current density (i.e., second cell-limited current density, J_{Lim}) becomes approximately 13−14 mA/cm^2 under terrestrial air mass (AM1.5) direct (AM1.5d) or global (AM1.5g) spectra and ~16 mA/cm^2 under space irradiation (AM0), respectively (Fig. 10.1). In the case of AM1.5d or 1.5g, a GaInNAsSb 1.0-eV absorber layer at least 2−2.5 μm thick is required to generate the necessary photocurrent due to the limited solar photon flux in the 1.04−1.42 eV range. On the other hand, assuming AM0 conditions, an absorber with a thickness of

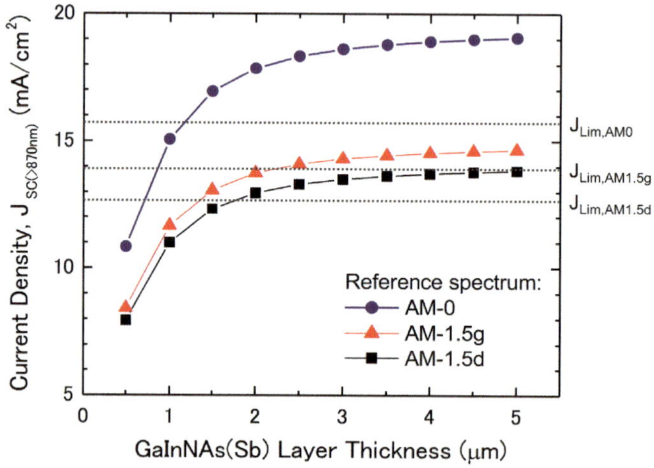

Figure 10.1 Plots of the projected current densities versus thickness of GaInNAs(Sb) 1.0-eV absorber with respect to the reference solar spectra (AM0, AM1.5g, and AM1.5d). Calculations were performed for wavelengths longer than 870 nm.

approximately 1.2 μm can match the $J_{Lim,AM0}$, since there are no atmospheric absorption bands in the 900–990 nm and 1070–1170 nm wavelength ranges.

In terms of the dilute nitride materials, it is well-known that incorporation of N atoms into the host matrix generally leads to degradation of electrical and optical characteristics [29]. As for solar cell materials, poor minority carrier properties (specifically the carrier diffusion length and lifetime) limit the photocurrent collected. Therefore, improvement of material quality is one of the most important issues. Meanwhile, from the aspect of device engineering, it is desireable to consider a p-i-n junction type structure, where photocarriers can be collected through the drift process by the depletion region electric field [35,44,45]. However, the width of the depletion region is not always equal to the thickness of the intrinsic (or undoped) layer (*i*-layer) because the depletion width shrinks with increasing *unintentional* background carrier concentration (BGCC). In this case, the contribution from the field-assisted collection of photocarriers decreases. This issue becomes serious when the thickness of the *i*-GaInNAs(Sb) layer increases. As mentioned earlier, although the 1.0 eV absorber thickness needs to be greater than 2 μm under AM1.5d or 1.5g, it might be difficult to have a depletion width as large as 2 μm.

Fig. 10.2A shows band diagrams of an *n*-GaAs/*i*-GaInNAs(Sb)/*p*-GaAs double heterostructure with GaInNAs(Sb) thicknesses, t_i, of 1.0, 2.0, and 3.0 μm, where we calculated using a doping level of the *i*-layer of $n = 1 \times 10^{15}$ cm^{-3}. Plotted in Fig. 10.2B is the magnitude of the electric field along the *i*-GaInNAs(Sb) region for all three thickness values. For $t_i = 1.0$ μm, the *i*-layer depletes fully, and the photocarriers are expected to be collected via the drift process. However, the $t_i = 2.0$ and 3.0 μm

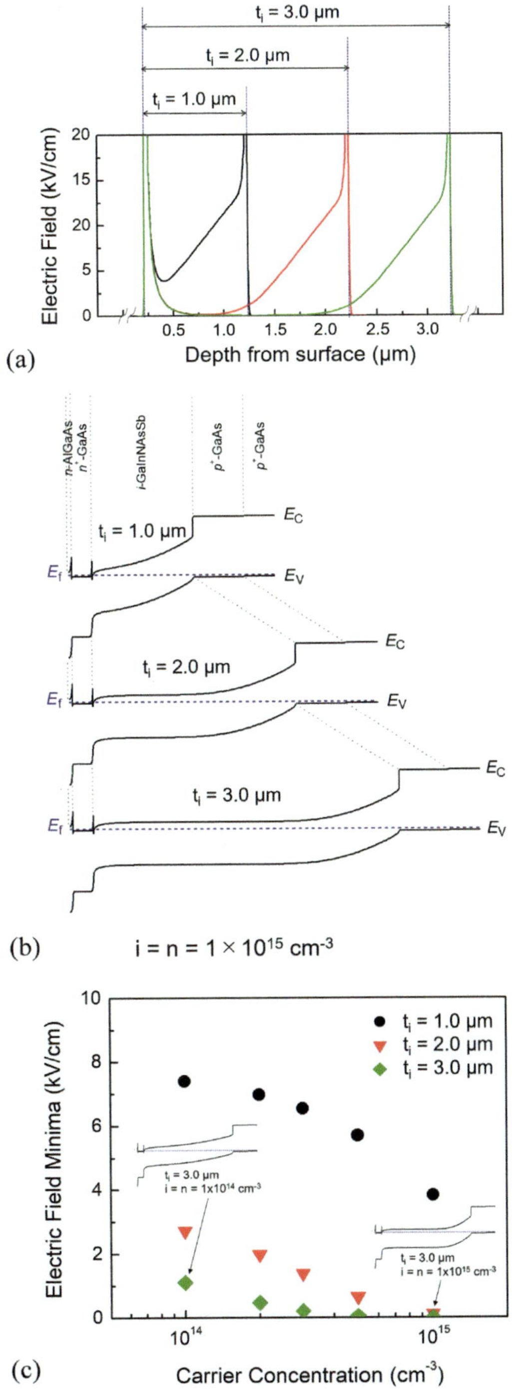

Figure 10.2 (A) Calculated electric field distribution across the GaInNAsSb layers in the n-i-p structure for an electron concentration of $n = 1 \times 10^{15}$ cm^{-3} and (B) corresponding band alignment for $t_i = 1.0$, 2.0, and 3.0 μm-thick devices. (C) Plots of the minima of the electric field magnitude for the $t_i = 1.0$, 2.0, and 3.0 μm devices with carrier concentrations between n $\sim 10^{14}$ and $\sim 1 \times 10^{15}$ cm^{-3}.

structures have a region of nearly zero electric field, where some of the photocarriers are likely to be lost through recombination because of the absence of field assistance. Therefore, for devices with a relatively thick i layer, the BGCC level should be low enough to realize a space charge region that fully extends across the GaInNAsSb layer. For the purpose of estimation, we calculated the electric field distributions across the 1.0 to 3.0 cm-thick GaInNAsSb region in conjunction with lowering the BGCC level from $1 \times 10^{15}\ cm^{-3}$ to $1 \times 10^{14}\ cm^{-3}$; the minima of the electric field magnitude are plotted in Fig. 10.2C. Specifically, we note that the electric field minimum increases from nearly zero at BGCC $= 1 \times 10^{15}\ cm^{-3}$ to 2.7 and 1.1 kV/cm for the $t_i = 2.0$ and 3.0 μm structures, respectively, when the BGCC is decreased to $1 \times 10^{14}\ cm^{-3}$. This suggests that the level of BGCC on the order of low-$10^{14}\ cm^{-3}$ yields a sufficient electric field even for a 3.0 μm-thick GaInNAsSb layer. Therefore, control of BGCC is considered necessary to ensure adequate current collection in GaInNAsSb-based solar cells.

For growth using MOCVD, dilute nitrides are known to suffer from unintentional incorporation of carbon and hydrogen atoms, both originating from the precursors of the source gases in the MOCVD growth. Both impurities can act as unintentional dopants and potentially increase the carrier concentration of the i layer [34,46,47]. In solid source MBE, on the other hand, the dilute nitrides are usually realized with such impurities below detection limit levels [46,47] and thus characterized by a lower level of BGCC [45].

As part of this study, we grew a series of single-junction GaInNAsSb solar cell samples by MBE and specifically investigated the BGCC levels in the undoped layer. The structure is based on the n-i-p cells consisting of an *n*-GaAs/*i*-GaInNAsSb/*p*-GaAs double heterostructure on a p-type GaAs substrate. The GaInNAsSb absorber thickness, t_i, was set at 1.0, 2.0, and 3.0 μm. Subsequently, an *n*-AlGaAs window and an n^+-GaAs contact were grown on top of the junction. These layers were grown at 570°C except for the GaInNAsSb, for which a typical growth temperature was 480–520°C.

To estimate the BGCC, capacitance-voltage (C–V) measurements were conducted at room temperature. Here, the unintentional net dopant concentration, N_d, is regarded as the BGCC. N_d can be deduced from a Mott-Schottky plot, where the dielectric constant of the GaInNAsSb is assumed to be the same as that of the GaAs. From the measured capacitance values (C) as a function of the swept voltage (V), the depletion width (W) can be obtained from Eq. (10.1):

$$W = \varepsilon_0 \varepsilon_S / C \tag{10.1}$$

where ε_0, and ε_S are the vacuum permittivity and the relative dielectric constant of GaInNAsSb, respectively, and the ionized dopant concentration (N_d) is obtained from Eq. (10.2):

$$N_d = \frac{2}{q\varepsilon_0\varepsilon_S\left[d\left(C^{\frac{1}{2}}\right)/dV\right]} \tag{10.2}$$

where q is the electronic charge.

From this analysis, the depletion widths were found to be 0.89, 0.98, and 0.86 μm under short circuit conditions for $t_i = 1.0$, 2.0, and 3.0 μm, respectively. Because both the n-GaAs and p-GaAs layers were highly doped, the depletion region was expanded within the GaInNAsSb layer. An almost constant N_d value of $\sim 1 \times 10^{15}$ cm^{-3} was observed for the set of samples, so it can be reasonably interpreted that the BGCC value was $\sim 1 \times 10^{15}$ cm^{-3} for the as-grown undoped GaInNAsSb layer. As mentioned earlier, with $N_d = 1 \times 10^{15}$ cm^{-3} the i-GaInNAsSb region is not fully depleted for $t_i > 2$ μm.

The light current density-voltage (J-V) and external quantum efficiency (EQE) curves for the three samples are shown in Fig. 10.3, where the extracted values of depletion width and BGCC reasonably explain the observed trends. For the $t_i = 1.0$ μm sample, a fill factor (FF) of 0.72 and peak EQE value of ~70% were obtained (note that no antireflection coating was applied, and the surface reflection loss is estimated to be as high as ~30%). However, for the 2.0 and 3.0 μm samples, the short circuit current density (J_{SC}), open circuit voltage (V_{oc}), and FF are all degraded. This behavior indicates insufficient carrier collection across the GaInNAsSb region having a narrow depletion width for the $t_i = 2.0$ and 3.0 μm samples.

The change of J_{sc} is correlated to both the generation and collection of photocarriers. The spectral properties of EQE were investigated to see the influence of the carrier recombination in the $t_i = 2.0$ and 3.0 μm structures [45]. Compared with the $t_i = 1.0$ μm sample, as shown in Fig. 10.3B, a reduced EQE response is observed for the $t_i = 2.0$ and 3.0 μm samples, especially in the short wavelength region (<900 nm). This is attributed to an increase of recombination loss of photocarriers generated in the n-GaAs emitter layer. For the $t_i = 3.0$ μm structure, the degradation is even larger, and the EQE response reduced in the entire wavelength range.

Under a reverse bias condition, the magnitude of the electric field increases, and the space charge region across the GaInNAsSb expands. The EQE response for the $t_i = 1.0$ μm sample was almost unchanged over an applied bias of 0 to −5 V. However, for the $t_i = 2.0$ and 3.0 μm samples, recovery of EQE was observed, and the responses saturated at approximately −4 V and −9 V, respectively [45]. Under such negative biases, almost all photocarriers are expected to be collected by drifting across the depletion region before recombination. Short wavelength EQE saturated at ~70% for all samples. It is worth noting that an EQE of ~70% is nearly the maximum value due to the reflection loss at the front air-semiconductor interface. Meanwhile, at long wavelengths, saturated EQE values tend to increase with t_i. For example, at 1000 nm, EQE under the present maximum bias conditions is 45.9% (1.0 μm), 62.2% (2.0 μm), and 69.9%

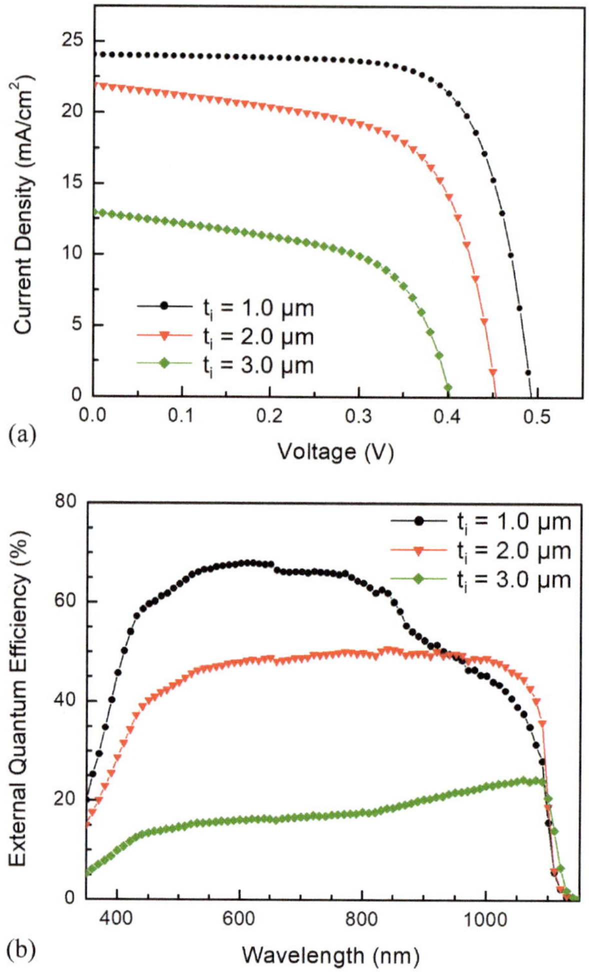

Figure 10.3 (A) Light J-V curves and (B) EQE spectra for the as-grown solar cells for three different GaInNAsSb thicknesses ($t_i = 1.0$, 2.0, and 3.0 μm).

(3.0 μm), respectively, for the three thickness values. This indicates that the amount of photon absorption increases with increasing the thickness of the GaInNAsSb layers. Therefore, if the *i*-GaInNAsSb region is sufficiently depleted to obtain efficient drift-assisted carrier collection, then the desired current output from the GaInNAsSb can be realized.

To improve the GaInNAsSb material quality, we investigated the optimal postanneal condition. Annealing is a common technique for improving the material quality of dilute nitrides [48–54]. It is considered that the annealing causes an annihilation of the crystal defects in the dilute nitrides such as interstitials, antisite defects, point defects, and so on, which results in the improvement of radiative recombination efficiencies. Annealing was performed in a nitrogen atmosphere for 10 min at 650, 700, 750, 800, and 850°C using a rapid thermal annealing (RTA) system. The sample with $t_i = 3.0$ μm was grown by MBE and subsequently cleaved into pieces. This sample contains a rather increased N composition compared to the samples characterized in Fig. 10.3, resulting in a redshift of the band edge as well as the decrease in the overall EQE response due to the effect of N-induced defects. EQE and light J-V curves are plotted in Fig. 10.4. Solar cell parameters (J_{sc}, V_{oc}, FF, and efficiencies) along with N_d values at short circuit (0 V) condition are summarized in Fig. 10.5 and Table 10.1.

It is observed that the values for maximum EQE (EQE_{max}) of 7%–9% and J_{SC} ~0.8–1.0 mA/cm^2 are similar in both the as-grown and the 650°C-annealed samples. These low values suggest inferior collection efficiency of photocarriers. Given that the N_d values at short circuit (0 V) condition are ~4×10^{15} cm^{-3} in these two samples, we speculate that this is responsible in part for the degraded photocarrier collection, since high BGCC values can hinder the electric field-assisted drift process of photocarriers across an i-GaInNAsSb region as thick as 3 μm. When the annealing temperature was raised to 700°C, N_d decreased to 1.6×10^{15} cm^{-3} at 0 V. At the same time, the spread of values in EQE and J-V properties among the devices annealed at 700°C became large (see Fig. 10.5), indicating that the annealing-induced improvement started to appear at this temperature.

Although in Fig. 10.4 we plot the results of one of the low-performing devices annealed at 700°C, a slight increase in the photocurrent appeared with $EQE_{max} = 14\%$ and $J_{SC} = 3.3$ mA/cm^2, indicating that the reduction of BGCC partly contributed to the improvement. On the other hand, more efficient photocarrier collection (J_{sc} >

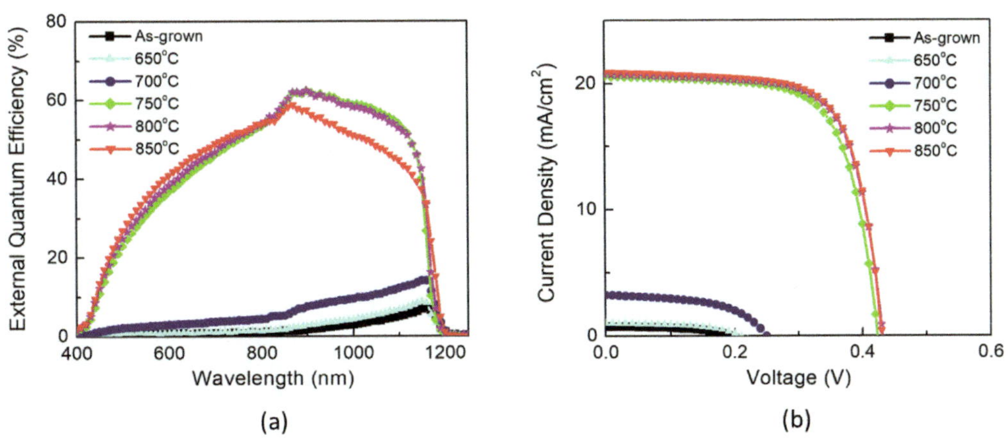

Figure 10.4 (A) EQE spectra and (B) light J-V curves for the as-grown and annealed GaInNAsSb solar cells with $t_i = 3.0$ μm. Note that no antireflection layer was applied for these devices.

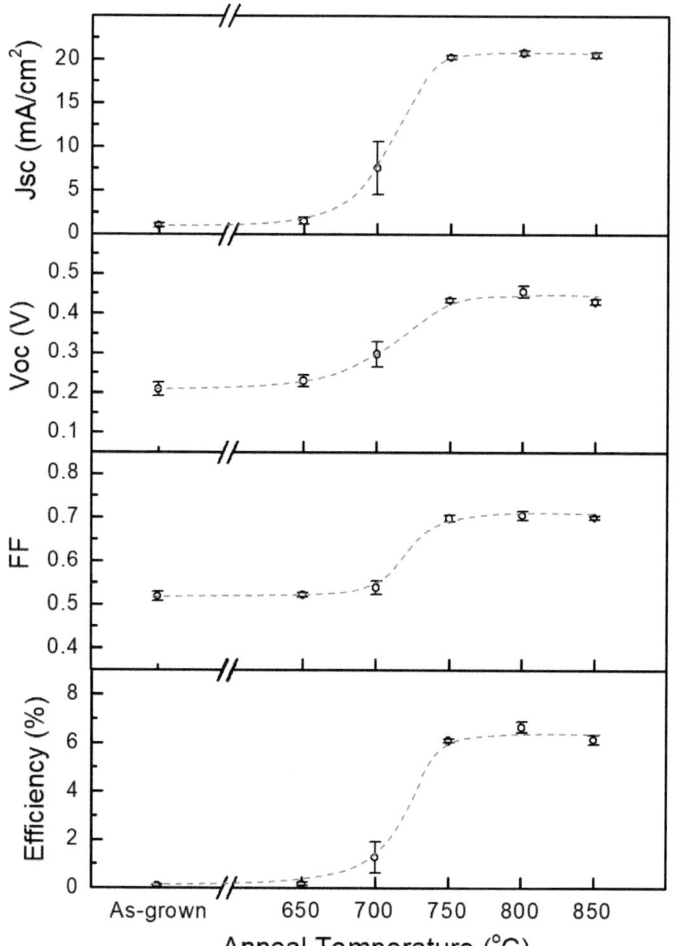

Figure 10.5 Plots of the J_{sc}, V_{oc}, FF, and efficiency of the GaInNAsSb solar cells as a function of annealing temperature. Note that no antireflection layer was applied for these devices. The *dashed lines* are guides for the eye.

Table 10.1 Summary of short circuit current density (J_{sc}), open circuit voltage (V_{oc}), fill factor (FF), and N_D at 0 V of the as-grown and annealed GaInNAsSb solar cells.

Annealing temperature (°C)	J_{sc} (mA/cm^2)	V_{oc} (V)	FF	N_D at 0 V (cm^{-3})
None (as-grown)	0.75	0.19	0.51	4.1×10^{15}
650	1.0	0.21	0.52	3.6×10^{15}
700	3.3	0.25	0.52	1.6×10^{15}
750	20.6	0.42	0.69	7.1×10^{14}
800	20.7	0.43	0.70	7.9×10^{14}
850	20.9	0.43	0.70	4.1×10^{14}

20.0 mA/cm^2) was achieved in the 750, 800, and 850°C-annealed samples, where they exhibited lower short circuit N_d values (7.1×10^{14} cm^{-3}, 7.9×10^{14} cm^{-3}, and 4.1×10^{14} cm^{-3}, respectively). Furthermore, V_{oc} and FF were also improved, achieving values in the range of 0.42−0.43 V and 0.69−0.70, respectively. This suggests that the annealing process caused a characteristic change related to recombination centers along with lowering the BGCC levels.

When the GaInNAsSb samples were annealed above 750°C in nitrogen atmosphere, improvement in the radiative efficiency in low temperature (18 K) photoluminescence (PL) [55] and room temperature electroluminescence [56] was observed. In the time-resolved PL (TRPL) characterization at room temperature, the PL decay time enhanced from 0.055 ns (as-grown) to 0.11 ns (750°C), 0.21 ns (800°C), and 0.10 ns (850°C), respectively [55]. The tendency of the improved decay times with increasing annealing temperature is similar to the improvement observed in the PL intensities, and these behaviors suggest an increase in radiative recombination and/or decreased non-radiative recombination rate.

Further studies focusing on the dynamics of deep levels upon the annealing process were subsequently performed [56] for the GaInNAsSb samples whose cell performances were shown in Fig. 10.4. The deep levels were characterized by admittance spectroscopy based on the capacitance-frequency (C-f) measurement, and the results are summarized in Table 10.2. The detected defects are considered to be electron traps, and their activation energies, E_a, indicate energy levels lying below the conduction band because the i-GaInNAsSb layers exhibited slight n-type conductivity in separate Hall effect measurements for control samples consisting of an undoped GaInNAsSb thin-film layer.

For the as-grown sample, a single peak relevant to a deep level (E1) was observed, as seen in Fig. 10.6, and the intensity of this peak decreased after the anneal at 650°C. At 700°C, two new peaks (E2 and E3) appeared, suggesting some changes in the nature of the energy levels. At 750 and 800°C, E2 and E3 disappeared and no apparent peak was detected. However, for the 850°C-annealed sample, although no peak was found above 180 K, a new peak appeared in the low-temperature region (<140 K).

Table 10.2 Summary of activation energy E_a, capture cross-section σ_T, and density N_T of the defects in the GaInNAsSb solar cells.

Anneal temperature	Peak label	E_a (eV)	σ_T (cm^2)	N_T (cm^{-3})
As-grown	E1	0.23	8.4×10^{-15}	7.0×10^{14}
650°C	E1	0.26	6.0×10^{-15}	4.3×10^{14}
700°C	E2	0.32	5.5×10^{-14}	5.0×10^{13}
	E3	0.48	2.4×10^{-12}	1.5×10^{13}
750°C	Under detection limit	—	—	—
800°C	Under detection limit	—	—	—
850°C	E4	0.06	1.1×10^{-15}	1.2×10^{14}

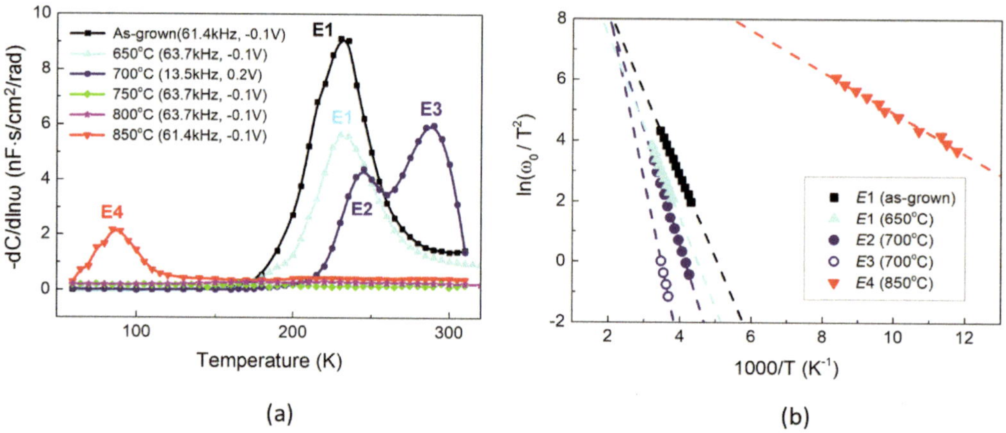

Figure 10.6 (A) Admittance spectra of the as-grown and the annealed GaInNAsSb solar cells and (B) Arrhenius plots of deep levels detected. Frequencies and DC biases indicated were chosen for each cell so that all peaks can be viewed. *Dashed lines* in Fig. 10.6B are linear fits to the data.

According to the analyses, there was observed a similarity in defect properties in the as-grown ($E_a = 0.23$ eV, $\sigma_T = 8.4 \times 10^{-15}$ cm^2) and the 650°C-annealed samples ($E_a = 0.26$ eV, $\sigma_T = 6.0 \times 10^{-15}$ cm^2). These levels can be considered to be of the same origin (labeled as E1). In the 700°C-annealed sample, the defect E1 diminished, and new defects E2 with $E_a = 0.32$ eV, $\sigma_T = 5.5 \times 10^{-14}$ cm^2 and E3 with $E_a = 0.48$ eV, $\sigma_T = 2.4 \times 10^{-12}$ cm^2 appeared. Here, deeper E_a and larger σ_T reveal an entire change of defect dynamics from the defect E1. The difference of the defect characteristics between E1, E2, and E3 was also observed as a peak shift of the differential C-f curves against the change in the DC-bias applied [56]. From the analyses, it was revealed that E1 likely acted as electron traps, whereas E2 was assigned as a bulk-type defect and E3 as an interface-type defect. In the 750 or 800°C-annealed GaInNAsSb, defects were below the detection limit, whereas a new defect E4 appeared after the 850°C annealing. E4 with $E_a = 0.06$ eV is likely attributable to the arsenic vacancy [57], which is supposed to be induced by desorption of arsenic atoms during a high-temperature annealing process. The shallow energy level of E4 suggests that it might act as a dopant.

The dynamics of deep-level defects revealed the relationship between annealing temperature and solar cell properties. Because the E1 level acts as donor with a density as high as $N_T = 7.0 \times 10^{14}$ cm^{-3}, this defect level might partially contribute to the BGCC in the as-grown GaInNAsSb layer. The density of E1 decreased slightly (by $\sim 3 \times 10^{14}$ cm^{-3}) after annealing at 650°C, along with a reduction in BGCC by $\sim 5 \times 10^{14}$ cm^{-3}. However, according to the comparable N_D values in these samples, as well as the similarities in EQE, *J-V*, and admittance spectroscopy, suggest that annealing at 650°C yields only a small impact on the crystal structure. On the other hand, a discrepancy exists between the values of BGCC and N_T in these samples, implying some other origin for free carriers. Based on the clear decrease in the level of N_D and improvement in the solar cell properties of the samples annealed above 700°C, in which E1-related peaks were not observed, suppression of the E1 level is one of the important requirements to enhance collection of photocarriers. However, the new defect levels E2 and E3 that appeared in the 700°C-annealed cell may act as recombination centers and lead to insufficient photocarrier collection. Therefore, suppressing the formation of E2 and E3 defects is as important as suppressing the E1 defects.

In the 750 and 800°C-annealed samples, defect levels are significantly reduced - under the detection limit. Simultaneously, moderate radiative characteristics for these samples suggest that nonradiative recombination centers were effectively suppressed by annealing, which is likely correlated to the reduction of defect levels. Noting that values of EQE, J-V, and N_D for these samples are comparable and indicative of reasonably high-quality devices, we can conclude that annealing at 750–800°C for 10 min is a worthwhile processing step. In the 850°C-annealed sample, although the shallow E4 level exists, absence of the E1, E2, and E3 defects results in EQE and J-V characteristics almost as good as those for 750 and 800°C-annealed samples.

10.3 Hybrid growth approach for multijunction solar cells and epitaxial lift-off devices

There are three distinct ways to integrate a dilute nitride GaInNAsSb cell into an MJSC configuration:

(1) MOCVD

(2) MBE

(3) a hybrid growth by MOCVD and MBE

The MOCVD-grown dilute nitride solar cells usually suffer from a high unintentional BGCC, which leads to degraded carrier collection properties in the dilute nitride layers [58]. This is because the dilute nitrides have poor minority carrier properties, i.e., short diffusion length and lifetime [33,34], so a thick dilute nitride layer with a high BGCC reduces the electric field–assisted carrier collection efficiency. To address this problem, 5J and 6J configurations have been proposed to divide the solar spectrum above the 1.4 eV region into more than three parts to achieve a current matching to the dilute nitride subcell, which generates a small photocurrent [59,60]. On the other hand, MBE growth of 3J solar cells was demonstrated successfully [61,62], and Solar Junction reported a world record efficiency of 44.0% under 942 suns with GaInP/GaAs/GaInNAs 3J cells [9]. The third option is to fabricate MJSCs by a combination of MOCVD and MBE hybrid growth. In applying hybrid growth, it is expected to combine the subcells grown by the best method to date to reach a high efficiency.

Thus far, we have investigated three types of hybrid-grown MJSCs: GaInNAsSb/Ge, GaInP/GaAs/GaInNAsSb, and GaInP/GaAs/GaInNAsSb/Ge, where the GaInNAsSb subcell was grown by MBE and the other cells by MOCVD [63,64]. They were grown in the upright configuration with the normal growth order from the bottom to the top cells. It was observed that the MOCVD process for the growth of the upper GaInP/GaAs subcells in the case of 3J and 4J induces degradation of the carrier collection efficiency in the MBE-grown GaInNAsSb cell, although in the case of bottom GaInNAsSb/Ge 2J without MOCVD overgrowth process, the MBE-GaInNAsSb cell on the MOCVD-Ge cell showed a sound performance.

This degradation has been correlated to the unintentional incorporation of hydrogen originated from the MOCVD precursors, which can lead to an increased BGCC in the GaInNAsSb region [64], together with the generation of additional defects [65]. It has also been found that a subsequent annealing in nitrogen atmosphere decreases the level of this unintentional hydrogen incorporation, which contributed in part to recovery of the carrier collection properties for the 3μm-thick GaInNAsSb single-junction solar cells [64].

However, in the case of an upright GaInP/GaAs/GaInNAsSb hybrid 3J structure, the subsequent nitrogen-annealing process at 800°C did not fully lower the hydrogen concentration. The annealing-induced improvement in the hydrogen-incorporated GaInNAsSb 1J device is partly linked to the out-diffusion of the hydrogen atoms due to the thermal energy, and there can therefore be some difference in the effect of the hydrogen out-diffusion with the existence of upper layers. Owing to this insufficient effect to lower the hydrogen concentration, recovery of the carrier collection efficiency in the GaInNAsSb subcell to something comparable to the case of 1J structure has not yet been achieved.

In addition to these straightforward approaches to upright integration, postgrowth hydrogen incorporation can be avoided by choosing an inverted sequence in the hybrid growth: i.e., first, an inverted GaInP/GaAs n-on-p tandem structure is grown by MOCVD, followed by an inverted n-i-p GaInNAsSb cell by MBE. With this concept, we performed hybrid growth of 3J solar cells in the inverted order (see Fig. 10.7). First, a GaInP etch stop layer was grown on a GaAs substrate. Then, a GaInP top subcell and a GaAs middle subcell in p-on-n configuration (i.e., inverted structure) and two tunnel junctions, made of n^+-GaInP and p^+-AlGaAs, were grown by MOCVD. The base thickness of the GaInP and GaAs subcells was 0.9 and 3.0 μm, respectively. After the MOCVD process, the samples were transferred under air ambient to MBE. The native oxide layer formed on the sample surface during the transfer sequence was removed thermally in the MBE system before overgrowth. The transferred sample was heated to approximately 600°C, and removal of the oxide layer was confirmed by monitoring reflection high-energy electron diffraction. Lastly, a GaInNAsSb dilute nitride bottom cell consisting of p-GaAs/i-GaInNAsSb/n-GaAs double heterostructure was overgrown. The thickness of the undoped GaInNAsSb absorber was 1.0–2.0 μm in this study.

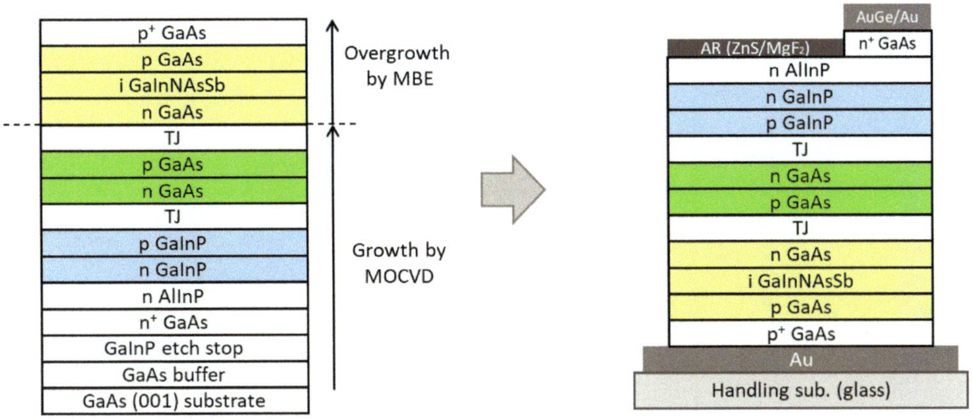

Figure 10.7 Schematics of the hybrid-grown inverted 3JSC and processed device structures.

After completing the growth, the devices were annealed at 750°C or 800°C in nitrogen ambient in an RTA system to improve the carrier collection process in the GaInNAsSb subcell [45,66]. A gold rear contact layer was then deposited on the bottom cell surface, and the cell was flipped and transferred onto a glass handling substrate coated with epoxy. Subsequently, the GaAs substrate and the etch stop layer were removed by $H_3PO_4/H_2O_2/H_2O$ and HCl etching, respectively. Finally, a front metal grid made of AuGe/Au was thermally deposited after photolithography patterning. Each 2 × 2 mm device was separated by mesa etching, and the cell aperture area was 0.032 cm^2.

Fig. 10.8 compares EQE and light J-V curves of three samples without annealing (as-grown) and annealed at either 750 or 800°C for 5 min. For this investigation, the cells contained an undoped 2.0-μm-thick GaInNAsSb 1.0-eV absorber. In the top and middle cells, EQEs reach ~70%, which is comparable to an original 2JSC without overgrowth, indicating no degradation resulted from the hybrid growth and substrate removal processes. On the other hand, in the GaInNAsSb bottom cells, the as-grown devices showed low EQE values of 10%–20% and exhibited a bias dependency, indicating that the carrier collection efficiency of the bottom cell was not optimal. In this case, the GaInNAsSb junction limited the total output current, so the poor light J-V curve reflected the limiting performance of the bottom cell.

After annealing at either 750 or 800°C, the EQEs for the top and middle cells were almost unchanged from the as-grown device. In terms of the bottom cell, the EQE significantly recovered to as high as the top and middle cells. Meanwhile, the specific bias dependence feature in the bottom cell disappeared. As seen in Fig. 10.8B, annealing led to significant improvement in the J-V properties, which is attributed to enhancement of the carrier collection efficiency, or photocurrent, in the GaInNAsSb junction. In the present samples, the current-limiting cell for the 750 and 800°C-annealed devices are the GaInP top and the GaInNAsSb bottom cell, respectively. The V_{oc} also increased after

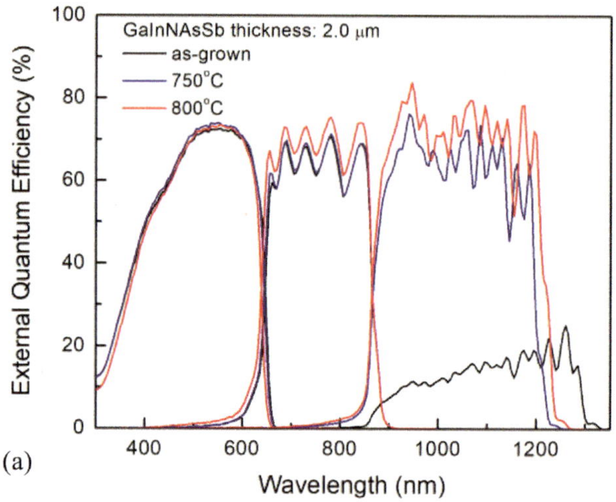

(a)

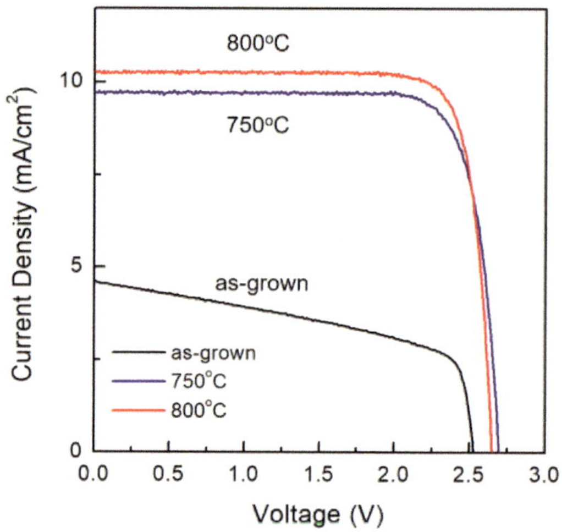

(b)

Figure 10.8 (A) EQE spectra and (B) light J-V curves for the inverted grown 3J solar cells with 2.0-μm GaInNAsSb absorber with different postannealing conditions: nonannealed (as-grown), annealed at 750 and 800°C in the nitrogen ambient for 5 min. No antireflection coating (ARC) was employed for these devices. The light J-V measurement was performed under AM1.5g illumination.

the annealing from 2.53 V (as-grown) to 2.70 (750°C) and 2.65 V (800°C). As a result, it is confirmed that the postannealing effectively improves the carrier collection efficiency of the *n-i-p*-GaInNAsSb junction even in the inverted 3J configuration.

Next, the effect of the GaInNAsSb thickness on the solar cell performance of the inverted 3J samples is investigated. We prepared 3JSCs with a GaInNAsSb thickness of 1.0, 1.5, and 2.0 μm, and the postanneal was carried out at 800°C. For these devices, an

MgF_2/ZnS dual-layer ARC was employed to minimize reflection loss. The inverted configuration has the benefit of easy-to-apply light management structures on the bottom side because there is no structure such as a thick substrate in the case of upright alignment. In the present devices, a gold metal layer served as the rear surface contact. The gold thickness was ~500 nm, which is thick enough to work as an optical reflector, so an increase in the OPL can be expected even with a thinner absorber. Here, the GaInNAsSb absorber layer is not thick enough to fully absorb photons in a single path. If the device is assumed to have a perfect ARC and a perfect reflector with no parasitic absorption, the OPL is doubled. Consequently, the nominal OPL equivalent to 2–4 μm is expected with an absorber thickness of 1–2 μm, and a near saturation level of the photocurrent can be realized even with the absorber thinner than 2 μm, as seen in Fig. 10.9.

In Fig. 10.10, EQE spectra and light J-V curves for the three 3JSCs are plotted. EQE for all subcells reached 80%–90%, which can be attributed to light trapping. The bottom cell EQEs for the three samples are similar, with values as high as ~90% owing to the light-trapping effect. The difference in the EQEs between these devices for the top and middle cell ranges is most likely due to the fluctuation in the fabrication process for the transferred thin-film devices. According to the analyses of the EQE spectra, the GaInP top cell was found to be a current-limiting cell for the three devices under AM1.5g conditions. The low photocurrent in the top cell is mainly attributed to a non-optimized ARC design where reflection loss at around 300–400 nm was not suppressed effectively.

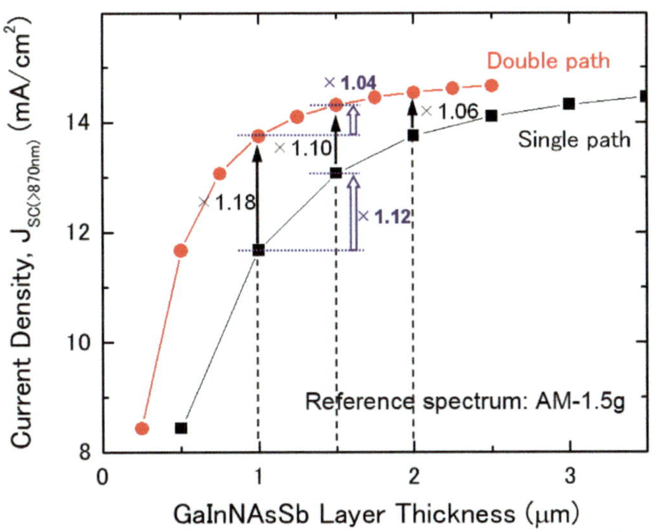

Figure 10.9 Calculated current densities, $J_{SC(>870\ nm)}$ as a function of the GaInNAsSb 1.0-eV absorber thickness under the AM1.5g reference spectrum for the single path and double path situations.

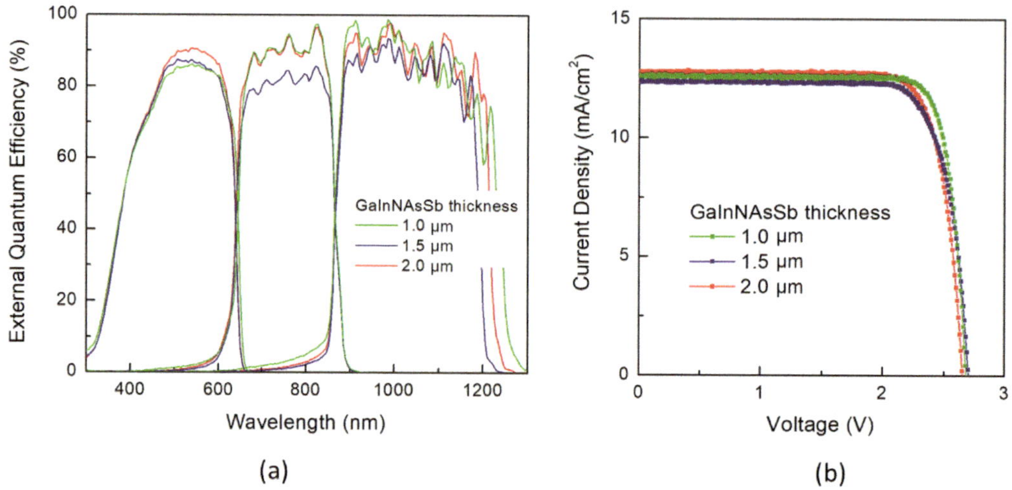

Figure 10.10 (A) EQE spectra and (B) light *J-V* curves for the inverted grown 3J solar cells with GaInNAsSb thickness of 1.0, 1.5, and 2.0 μm, annealed at 800°C. A MgF_2/ZnS ARC was employed for these devices. The light *J-V* measurement was performed under AM1.5g corrected illumination.

The light *J-V* curves were characterized under AM1.5g illumination, and J_{SC} values were limited by the top cell as mentioned earlier. This was also experimentally confirmed by increasing the intensity of the 440-nm LED, which resulted in an increase to J_{SC}, whereas no apparent changes were observed when increasing the intensity of the 780- and 970-nm LEDs. Fill factors of ~80% are ascribed to the top cell limiting feature. Slightly low FFs for 1.5-μm devices, measured under AM1.5g illumination, might be due to the close J_{SC} values for the top and bottom cells. As a result of convoluted *J-V* curves of the subcells near the maximum power point, FF of the total *J-V* curve is affected by the *J-V* feature of the low FF subcell. Typically, the diode ideality factor of the GaInNAsSb was larger than for GaInP and GaAs [67,68], so this kind of competitive situation can yield a poorer FF; see Table 10.3.

The devices have reasonably thick top and middle cell absorbers, where almost all photons above 1.42 eV (i.e. GaAs band gap) were fully absorbed. On the basis of this configuration, the GaInNAsSb bottom subcell exhibited current as high as that of the GaAs middle cell, indicating that it achieves high carrier collection efficiency even with the 2.0-μm *i* layer thickness. On the other hand, a higher FF value was exhibited for the 1.0-μm cell. This is because the thinner the i layer in the *p-i-n* junction, the lower the recombination current component in the depletion region (i.e. *i* layer), which can lead to a diode ideality factor close to unity. As mentioned before, improvement in the FF of the GaInNAsSb junction can result in the increase in the FF of the total *J-V* characteristics. In addition, the subcell J_{SC} of the 1.0-μm GaInNAsSb junction is greater than the other subcells. This can hinder the effect of the poorer FF in the GaInNAsSb junction in the

Table 10.3 Summary of the calculated subcell J_{SC} values, under the AM1.5g reference solar spectrum and solar cell parameters extracted from the light *J-V* measurement for the devices shown in Fig. 10.10.

$t_{GaInNAsSb}$	J_{sc}^{TOP} (mA/cm²)	J_{sc}^{Middle} (mA/cm²)	J_{sc}^{Bottom} (mA/cm²)	Efficiency (%)	J_{sc} (mA/cm²)	V_{oc} (V)	FF
1.0 μm	12.62	14.16	15.08	28.2	12.6	2.68	0.831
1.5 μm	12.39	12.97	12.68	26.3	12.4	2.70	0.787
2.0 μm	12.76	14.27	14.12	27.2	12.8	2.65	0.802

competition among the subcells, and consequently the FF is limited by the top or middle junctions whose FF is better than the GaInNAsSb. Nonconcentrator efficiencies of 26% −28% were achieved for the inverted LM 3JSCs. By optimizing the ARC design and enlarging the cell area size as well as improving the fabrication process of the inverted cells, higher efficiency can be expected. With assuming AM0 conditions, because of the lack of the atmospheric absorption bands (900−990 and 1070−1170 nm), the bottom cell generates more subcell current as high as 18 mA/cm^2 [69], which might be also attractive for space applications.

The inverted 3JSCs are considered to be suitable for applying ELO for cell fabrication, along with the potential for cell cost reduction by reusing the released semiconductor substrates. In the inverted cells, the rear surface of the device structure is exposed at the front side as-grown, so the rear contact metal can be deposited simply onto the surface first. It is followed by ELO to separate the device thin-film layer, and then a front contact metal is formed on the opposite surface. This can lead to a reduction of the number of thin-film transfer processes compared to the upright structure. The separation of the epitaxial layer in the ELO concept is based on the selectivity of the wet etching between a preinserted release layer (or sacrificial layer) and the other neighboring layer materials. One of the most common combinations for ELO in GaAs-based devices is $Al_xGa_{1-x}As$ and hydrofluoric acid (HF) for the release layer and the etchant, respectively, where the etching selectivity of the HF between the $Al_xGa_{1-x}As$ ($x > 0.7$) and the GaAs is on the order of 10^5−10^6 [70].

In Fig. 10.11, we summarize the sequence for the fabrication of the ELO thin-film 3JSCs in this study. The device structures were grown by the hybrid growth technique in inverted order. First an AlAs ELO release layer was grown on the GaAs substrate, followed by a GaInP top cell, a tunnel junction, a GaAs middle cell, and another tunnel junction by MOCVD. Then, a GaInNAsSb dilute nitride bottom cell was overgrown on it by MBE. The bottom cell was consisted of *n*-GaAs/*i*-GaInNAsSb/*p*-GaAs double heterojunction. After the epitaxial growth process, the sample was annealed at 750−800°C in a nitrogen atmosphere, aiming to reduce crystal defects as well as BGCC in the GaInNAsSb layer to improve the carrier collection efficiency [45,65]. For

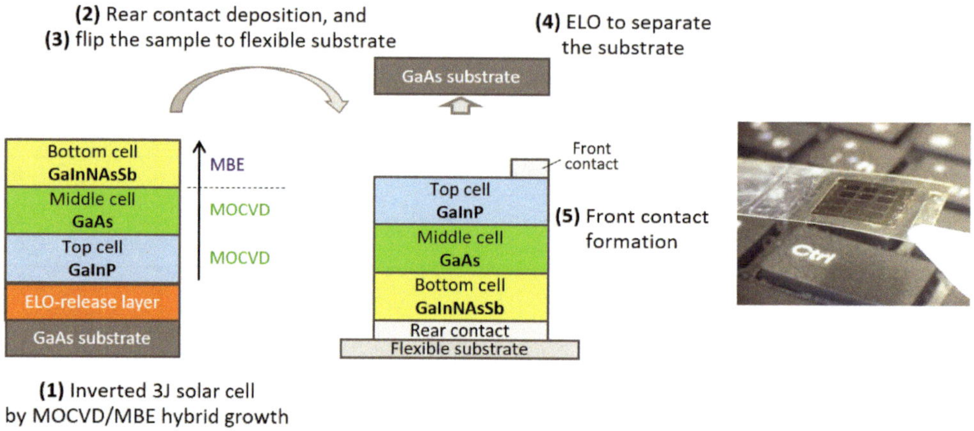

Figure 10.11 Sequence for fabrication of the 3J GaInP/GaAs/GaInNAsSb thin-film solar cells using ELO and the MOCVD/MBE hybrid growth. A picture of an actual device is shown on the right hand side of the figure.

separating the device layer via ELO, a metal layer (Au) was first thermally deposited on the bottom cell surface, and a flexible PET film was attached on it to serve a mechanical support for the thin-film device layer to be separated. The ELO in this study was performed at room temperature with applying few drops of hydrofluoric acid at an edge of the sample chip. Simultaneously, a weight was applied to make the ELO slit open to accelerate the chemical reaction at the etch front. After the device layer was released from its substrate, a front contact grid was patterned, followed by the device separation through mesa etching. No antireflection coating layer was applied to the device.

Fig. 10.12 shows EQE and light J-V characteristics for a GaInP/GaAs/GaInNAsSb inverted LM 3J (ILM-3J) solar cell fabricated through the ELO process. The thickness of the GaInNAsSb bottom cell absorber is 1.0 μm. EQE and J-V for a 2J ELO thin-film cell, which was fabricated without an overgrowth of a GaInNAsSb cell, is also plotted as a reference. Comparing EQEs of the GaInP and GaAs subcells in the 2J and 3J devices, no apparent difference is observed, suggesting that the additional processes related to the bottom cell such as the MBE overgrowth and the postanneal likely induced minor influence on the quality of top and middle subcells.

J-V curves, shown in Fig. 10.12B, indicate that short circuit current densities (J_{SC}) have similar values of 7.40 mA/cm^2 for the 2J cell and 7.22 mA/cm^2 for the 3J cell, and V_{oc} and FF are also comparable at $V_{oc} = 2.27$ V, FF = 0.85 and $V_{oc} = 2.57$ V, FF = 0.82, respectively. J_{sc} is limited by the top cell for both devices. This can be confirmed by adding illumination from LEDs during the light *J-V* measurement. Adding 440-nm light illumination resulted in an increase in the J_{sc}, whereas J_{sc} was almost unchanged by adding illumination from 780- or 970-nm LEDs. On the other hand, in the test of 970-

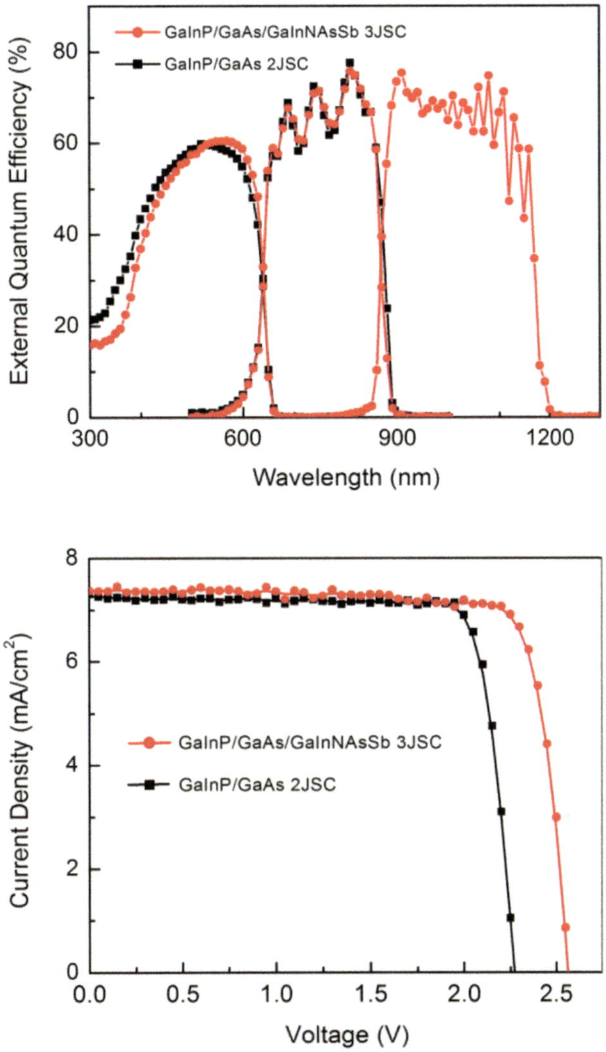

(a)

(b)

Figure 10.12 (A) EQE spectra and (B) light J-V curves for the ELO thin-film solar cells (GaInP/GaAs 2J cell and GaInP/GaAs/GaInNAs 3J cell). The bottom cell absorber thickness of the 3J cell is $t_i = 1.0$ μm. No ARC was applied to these devices.

nm LED addition, improvement of FF was observed. This is a result of the overgeneration of photocarriers in the GaInNAsSb cell, so the effect of relatively low FF on the total *J*-*V* characteristics can be hindered.

Therefore, we investigate the cell structure so that the balance of current generation from each subcell is improved. It is preferable to have an overgeneration of photocarriers in the GaInNAsSb cell. Thus, to increase the carrier generation, first, the thickness of the

GaInNAsSb absorber in the bottom cell was increased from 1.0 to 1.5 μm. EQE spectra and J-V curves for these 3J cells are plotted in Fig. 10.13. Both 3J cells show almost similar characteristics. With a specific focus on the bottom GaInNAsSb subcell, it can be

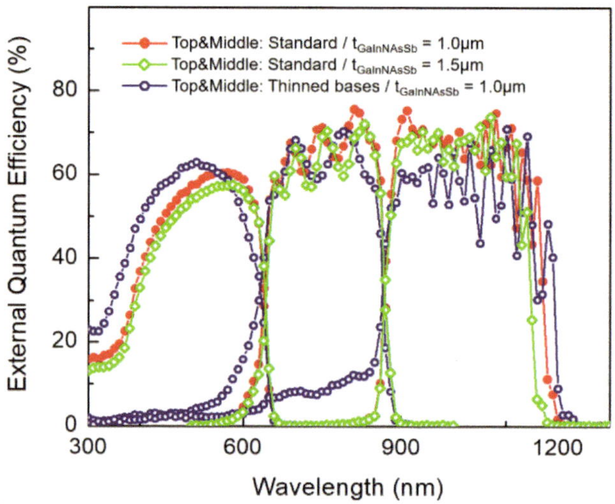

(a)

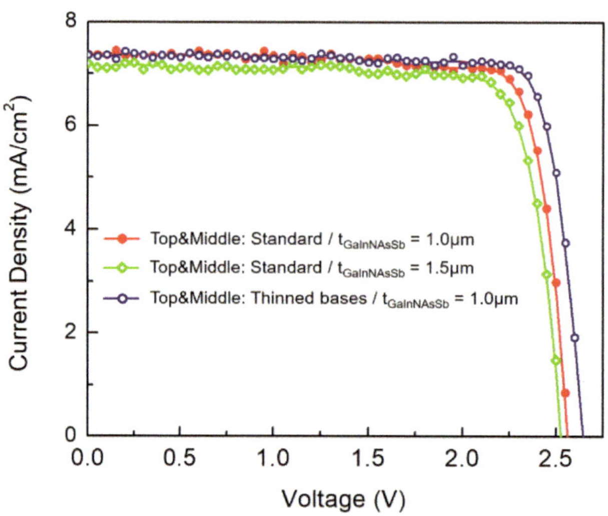

(b)

Figure 10.13 (A) EQE spectra and (B) current density-voltage (J-V) curves for the ELO thin-film 3J solar cells. The GaInNAsSb thickness (t_i) of 1.0 μm for the bottom cell with standard base layers for top and middle cells, t_i of 1.5 μm for the bottom cell with standard base layers for top and middle cells, and t_i of 1.0 μm for the bottom cell with thinned base layers for top and middle cells. The first one is the same data shown in Fig. 10.2. No ARC was applied to these devices.

observed that carrier collection properties are reasonably good for both the 1.0 and 1.5-μm devices, and the EQE values are comparable to each other despite the thickness being changed. We can attribute this to the light-trapping effect by the rear side Au reflector resulting in the increase of the OPL. Theoretically, a difference in the projected J_{SC} between the GaInNAsSb thicknesses of 1.5 and 1.0 μm of $J_{sc}(1.5\ \mu m)/J_{sc}(1.0\ \mu m) = 1.12$ for the single path case (see Fig. 10.9) is expected. However, the ratio approaches unity (1.04) if one assumes the OPL is doubled.

Next, we focused on a new 3J structure, whose top and middle cells have thinner base layers to transmit a part of incident photons and generate more current in the lower junction. A thin GaInP/GaAs inverted 2J cell structure was newly grown by MOCVD, and a GaInNAsSb bottom cell with 1.0-μm-thick absorber was overgrown by MBE. The ELO thin-film 3J device was fabricated in the same manner as described earlier. In Fig. 10.13A, EQE spectra for 3J cells with standard (plotted in red lines) and thinned (blue lines) GaInP/GaAs top 2J structure are compared. In the thinned device, long wavelength sides of the EQE spectra for the InGaP and GaAs subcells are lowered, whereas short-wavelength sides increase for the GaAs and GaInNAsSb cells. As expected, this increase is owing to the carriers generated by the photons transmitting through the upper subcells. As a result, photocurrent under AM1.5g conditions in the GaInNAsSb cell for the thinned device increases to 10.09 mA/cm^2 compared with 9.16 mA/cm^2 for the standard case. This also contributes to the improvement of FF = 0.84 in the thinned device along with an increase in the V_{oc} by 78 mV. We also compared the V_{oc} values for nonovergrown 2J devices (normal and thinned 2J structures), and that for the thinned 2J device was slightly higher by 48 mV. Thus, the improvement of subcell J_{SC} in the bottom GaInNAsSb cell is considered to contribute to ~30 mV of the V_{oc} gain in the thinned 3J cell. A recent study addresses the utility of dilute nitride triple-junction solar cells for space applications [71].

10.4 Conclusion

In this chapter, we detailed recent results on the development of dilute nitride GaInNAsSb as a 1.0-eV absorber material in III-V-based MJSCs. We delved into the improvement of GaInNAsSb to optimize solar cell properties. These efforts included hybrid growth by a combination of metalorganic chemical vapor deposition and MBE for developing MJSCs. ELO processing for LM GaInP/GaAs/GaInNAsSb triple-junction solar cells was explored. As discussed at the outset, high-efficiency, radiation-tolerant solar cell technology, such as III-V MJSC devices, is essential for numerous space applications. To discover new approaches to such potential solutions to the challenges of space photovoltaics, it is imperative to continue to explore new processing technologies such as those detailed in this chapter. These efforts will continue in our laboratories and around the world.

Acknowledgments

We gratefully acknowledge Dr. W. Walukiewicz and K. M. Yu of Lawrence Berkeley National Laboratory, Prof. J-F. Guillemoles of Center National de la Recherche Scientifique (CNRS), Dr. T. Takamoto of SHARP company, Dr. N. Pan and R. Tatavarti of MicroLink Devices, Prof. T. Kita of Kobe University, and Prof. C. Algora of Universidad Politècnica de Madrid for their collaboration. This work is supported in part by the New Energy and Industrial Technology Development Organization (NEDO), and the Ministry of Economy, Trade and Industry (METI), Japan.

References

[1] S.G. Bailey, R.P. Raffaelle, Chapter II-4-B - operation of solar cells in a space environment, in: S.A. Kalogirou (Ed.), McEvoy's Handbook of Photovoltaics Fundamentals and Applications, third ed., Academic Press, New York, NY USA, 2012, pp. 987–1003.

[2] A. Bermudez-Garcia, P. Voarino, O. Raccurt, Environments, needs and opportunities for future space photovoltaic power generation: A review, Appl. Energy 290 (May 15 , 2021) 116757.

[3] P.T. Chiu, Space applications of III-V single- and multi-junction solar cells, in: S.G. Bailey, A.F. Hepp, D.C. Ferguson, R.P. Raffaelle, S.M. Durbin (Eds.), Space Photovoltaics: Materials, Missions and Alternative Technologies, Elsevier, Cambridge, MA USA, 2022, pp. 79–127.

[4] M.J. O'Neill, Space photovoltaic concentrators for outer planet and near-Sun missions using ultralight Fresnel lenses, in: S.G. Bailey, A.F. Hepp, D.C. Ferguson, R.P. Raffaelle, S.M. Durbin (Eds.), Space Photovoltaics: Materials, Missions and Alternative Technologies, Elsevier, Cambridge, MA USA, 2022, pp. 411–432.

[5] D. Sato, K.H. Lee, K. Araki, T. Masuda, M. Yamaguchi, N. Yamada, Design of low-concentration static III-V/Si partial CPV module with 27.3% annual efficiency for car-roof application, Prog. Photovolt. 27 (2019) 501–510, https://doi.org/10.1002/pip.3124.

[6] https://www.nrel.gov/pv/cell-efficiency.html.

[7] M. Green, E. Dunlop, J. Hohl-Ebinger, M. Yoshita, N. Kopidakis, X. Hao, Solar cell efficiency tables (version 59), Prog. Photovolt. 30 (1) (2022) 3–15, https://doi.org/10.1002/pip.3371.

[8] R.R. King, A. Boca, W. Hong, X.-Q. Liu, D. Bhusari, D. Larrabee, K.M. Edmondson, D.C. Law, C.M. Fetzer, S. Mesropian, N.H. Karam, Band-gap-engineered architectures for high-efficiency multijunction concentrator solar cells, in: Proc. Of 24th European Photovoltaic Solar Energy Conference and Exhibition, 2009, https://doi.org/10.4229/24thEUPVSEC2009-1AO.5.2.

[9] M.A. Green, K. Emery, Y. Hishikawa, W. Warta, E.D. Dunlop, Solar cell efficiency tables (version 41), Prog. Photovolt. 21 (2013) 1–11.

[10] M. Konagai, M. Sugimoto, K. Takahashi, High efficiency GaAs thin film solar cells by peeled film technology, J. Cryst. Growth 45 (1978) 277–280.

[11] E. Yablonovitch, D.M. Hwang, T.J. Gmitter, L.T. Florez, J.P. Harbison, Van der Waals bonding of GaAs epitaxial liftoff films onto arbitrary substrates, Appl. Phys. Lett. 56 (1990) 2419–2421.

[12] J.J. Schermer, G.J. Bauhuis, P. Mulder, W.J. Meulemeesters, E. Haverkamp, M.M.A.J. Voncken, P.K. Larsen, High rate epitaxial lift-off of InGaP films from GaAs substrates, Appl. Phys. Lett. 76 (2000) 2131–2133.

[13] B.M. Kayes, H. Nie, R. Twist, S.G. Spruytte, F. Reinhardt, I.C. Kizilyalli, G.S. Higashi, 27.6% conversion efficiency, a new record for single-junction solar cells under 1 sun illumination, in: Proc. IEEE 37th Photovolt. Spec. Conf, 2011, pp. 4–8.

[14] J.J. Schermer, P. Mulder, G.J. Bauhuis, P.K. Larsen, G. Oomen, E. Bongers, Thin-film GaAs epitaxial lift-off solar cells for space applications, Progress in Photovoltaics 13 (7) (November 2005) 587–596.

[15] K. Lee, J.D. Zimmerman, T.W. Hughes, S.R. Forrest, Non-destructive wafer recycling for low-cost thin-film flexible optoelectronics, Adv. Funct. Mater. 24 (2014) 4284–4291.

[16] M. Yamaguchi, Japanese R&D activities of high efficiency III-V compound multi-junction and concentrator solar cells, Energy Procedia 15 (2012) 265–274.

[17] R. Tatavarti, A. Wibowo, G. Martin, F. Tuminello, C. Youtsey, G. Hillier, N. Pan, M.W. Wanlass, M. Romero, InGaP/GaAs/InGaAs inverted metamorphic (IMM) solar cells on 4" epitaxial lifted off

(ELO) wafers, in: Proc. 35th IEEE Photovoltaic Specialists Conference (PVSC), 2010, pp. 2125–2128.

[18] K. Watanabe, B. Kim, T. Inoue, H. Sodabanlu, M. Sugiyama, M. Goto, S. Hayashi, K. Miyano, Y. Nakano, Thin-film InGaAs/GaAsP MQWs solar cell with backside nanoimprinted pattern for light trapping, IEEE J. Photovolt. 4 (2014) 1086–1090.

[19] K. Watanabe, T. Inoue, K. Toprasertpong, A. Delamarre, H. Sodabanlu, J.-F. Guillemoles, M. Sugiyama, Y. Nakano, Enhancement of photocurrent in epitaxial lift-off thin-flim ganlnNAsSb solar cells due to light-confinement structure, in: Proc. 43rd IEEE Photovoltaic Specialists Conference (PVSC), 2016, pp. 1268–1272.

[20] B. Behaghel, R. Tamaki, N. Vandamme, K. Watanabe, C. Dupuis, N. Bardou, H. Sodabanlu, A. Cattoni, Y. Okada, M. Sugiyama, S. Collin, J.F. Guillemoles, Appl. Phys. Lett. 106 (2015) 081107.

[21] K.W.J. Barnham, I. Ballard, J.P. Connolly, N.J. Ekins-Daukes, B.G. Kluftinger, J. Nelson, C. Rohr, Physica E 14 (2002) 27–36.

[22] Y. Okada, N. Shiotsuka, Proceedings 31st IEEE Photovoltaic Specialist Conference (PVSC), 2005, p. 591.

[23] T. Nakata, K. Watanabe, N. Miyashita, H. Sodabanlu, M. Giteau, Y. Nakano, Y. Okada, M. Sugiyama, Thin-film multiple-quantum-well solar cells fabricated by epitaxial lift-off process, Jpn. J. Appl. Phys. 57 (2018) 08RF03.

[24] Y. Okada, N. Kobayashi, N. Sasaki, Proceedings IEEE 4th World Conference on Photovoltaic Energy Conversion (WCPEC), 2006, p. 865.

[25] R. Oshima, T. Hashimoto, H. Shigekawa, Y. Okada, J. Appl. Phys. 100 (2006) 083110.

[26] Q. Xie, A. Madhukar, P. Chen, N.P. Kobayashi, Phys. Rev. Lett. 75 (1995) 2542.

[27] T. Sogabe, Y. Shoji, P. Mulder, J. Schermer, E. Tamayo, Y. Okada, Appl. Phys. Lett. 105 (2014) 073118.

[28] N. Miyashita, B. Behaguel, J.-F. Guillemoles, Y. Okada, Enhancement of photocurrent in epitaxial lift-off thin-film GaInNAsSb solar cells due to light-confinement structure, Appl. Phys. Express 11 (2018) 072301.

[29] D.J. Friedman, J.F. Geisz, S.R. Kurtz, J.M. Olson, 1-eV solar cells with GaInNAs active layer, J. Cryst. Growth 195 (1998) 409–415.

[30] D.J. Friedman, Progress and challenges for next-generation high-efficiency multijunction solar cells, Curr. Opin. Solid State Mater. Sci. 14 (2010) 131–138.

[31] M. Kondow, K. Uomi, A. Niwa, T. Kitatani, S. Watahiki, Y. Yazawa, GaInNAs: Novel material for long-wavelength-range laser diodes with excellent high-temperature performance, Jpn. J. Appl. Phys. 35 (1996) 1273–1275.

[32] F. Dimroth, C. Baur, A.W. Bett, K. Volz, W. Stolz, Comparison of dilute nitride growth on a single- and 8 x 4-inch multiwafer MOVPE system for solar cell applications, J. Cryst. Growth 272 (2004) 726–731.

[33] S.R. Kurtz, A.A. Allerman, E.D. Jones, J.M. Gee, J.J. Banas, B.E. Hammons, InGaAsN solar cells with 1.0 eV band gap, lattice matched to GaAs, Appl. Phys. Lett. 74 (1999) 729–731.

[34] J.F. Geisz, D.J. Friedman, J.M. Olson, S.R. Kurtz, B.M. Keyes, Photocurrent of 1 eV GaInNAs lattice-matched to GaAs, J. Cryst. Growth 195 (1998) 401–408.

[35] N. Miyashita, Y. Shimizu, N. Kobayashi, Y. Okada, M. Yamaguchi, Fabrication of GaInNAs-based solar cells for application to multi-junction tandem solar cells, in: Proceedings of the IEEE 4th World Conference on Photovoltaic Energy Conference, 2006, pp. 869–872.

[36] K. Volz, D. Lackner, I. Németh, B. Kunert, W. Stolz, C. Baur, F. Dimroth, A.W. Bett, Optimization of annealing conditions of (GaIn)(NAs) for solar cell applications, J. Cryst. Growth 310 (7–9) (April 2008) 2222–2228.

[37] A.J. Ptak, R. France, C.-S. Jiang, M.J. Romero, Improved performance of GaInNAs solar cells grown by molecular-beam epitaxy using increased growth rate instead of surfactants, J. Cryst. Growth 311 (2009) 1876–1880.

[38] X. Yang, J.B. Heroux, L.F. Mei, W.I. Wang, InGaAsNSb/GaAs quantum wells for 1.55 μm lasers grown by molecular beam epitaxy, Appl. Phys. Lett. 78 (2001) 4068–4070.

[39] J.S. Harris, The opportunities, successes and challenges for GaInNAsSb, J. Cryst. Growth 278 (2005) 3–17.

[40] J.A. Gupta, G.I. Sproule, X. Wu, Z.R. Wasilewski, Molecular beam epitaxy growth of 1.55 μm GaInNAs(Sb) double quantum wells with bright and narrow photoluminescence, J. Cryst. Growth 291 (2006) 86–93.

[41] N. Miyashita, S. Ichikawa, Y. Okada, Improvement of GaInNAsSb films fabricated by atomic hydrogen-assisted molecular beam epitaxy, J. Cryst. Growth 311 (2009) 3249–3251.

[42] N. Miyashita, N. Ahsan, Y. Okada, Effect of antimony on uniform incorporation of nitrogen atoms in GaInNAs films for solar cell application, Sol. Energy Mater. Sol. Cells 111 (2013) 127–132.

[43] M.M. Islam, N. Miysahita, N. Ahsan, T. Sakurai, K. Akimoto, Y. Okada, Effect of antimony on the deep-level traps in GaInNAsSb thin films, Appl. Phys. Lett. 105 (2014) 112103.

[44] A.J. Ptak, D.J. Friedman, S. Kurtz, J. Kiehl, Proceedings of the 31st IEEE Photovoltaic Specialist Conference, 2005, pp. 603–606.

[45] N. Miyashita, N. Ahsan, Y. Okada, Generation and collection of photocarriers in dilute nitride GaInNAsSb solar cells, Prog. Photovoltaics Res. Appl. 24 (2016) 28–37.

[46] T. Miyamoto, T. Kageyama, S. Makino, D. Schlenker, F. Koyama, K. Iga, CBE and MOCVD growth of GaInNAs, J. Cryst. Growth 209 (2000) 339–344.

[47] A.J. Ptak, S.W. Johnston, S. Kurtz, D.J. Friedman, W.K. Metzger, A comparison of MBE- and MOCVD-grown GaInNAs, J. Cryst. Growth 251 (2003) 392–398.

[48] A. Hierro, J.-M. Ulloa, J.-M. Chauveau, A. Trampert, M.-A. Pinault, E. Toutnié, A. Guzmán, J.L. Sánchez-Rojas, E. Calleja, Annealing effects on the crystal structure of GaInNAs quantum wells with large In and N content grown by molecular beam epitaxy, J. Appl. Phys. 94 (2003) 2319–2324.

[49] M. Kondow, T. Kitatani, In situ annealing of GaInNAs up to 600oC, Jpn. J. Appl. Phys. 40 (2001) 108–109.

[50] S.G. Spruytte, C.W. Coldren, J.S. Harris, W. Wampler, P. Krispin, K. Ploog, M.C. Larson, Incorporation of nitrogen in nitride-arsenides: Origin of improved luminescence efficiency after anneal, J. Appl. Phys. 89 (2001) 4401–4406.

[51] H. Zhao, Y.Q. Xu, H.Q. Ni, S.Y. Zhang, Q. Han, Y. Du, X.H. Yang, R.H. Wu, Z.C. Niu, Post-growth and in situ annealing on GaInNAs(Sb) and their application in 1.55 μm lasers, Semicond. Sci. Technol. 21 (2006) 279–282.

[52] S.R. Bank, H.B. Yuen, H. Bae, M.A. Wistey, J.S. Harris, Overanneaing effects in GaInNAs(Sb) alloys and their importance to laser applications, Appl. Phys. Lett. 88 (2006) 221115.

[53] C.S. Peng, H.F. Liu, J. Konttinen, M. Pessa, Mechanism of photoluminescence blue shift in InGaAsN/GaAs quantum wells during annealing, J. Cryst. Growth 278 (2005) 259–263.

[54] J.-M. Chauveau, A. Trampert, K.H. Ploog, E. Tournie, Nanoscale analysis of the In and N spatial redistributions upon annealing of GaInNAs quantum wells, Appl. Phys. Lett. 84 (2004) 2503–2505.

[55] N. Miyashita, N. Ahsan, Y. Okada, Characterization of 1.0 eV GaInNAsSb solar cells for multi-junction applications and the effect of annealing, in: Proc. Of the 31st European Photovoltaic Solar Energy Conference, 2015, p. 1461.

[56] N. Miyashita, Y. He, N. Ahsan, Y. Okada, Anneal mediated deep-level dynamics in GaInNAsSb dilute nitrides lattice-matched to GaAs, J. Appl. Phys. 126 (2019) 143104.

[57] D. Pons, J.C. Bourgoin, J. Phys. C Solid State Phys. 18 (1985) 3839.

[58] A.J. Ptak, D.J. Friedman, S. Kurtz, R.C. Reedy, Low-acceptor-concentration GaInNAs grown by molecular-beam epitaxy for high-current p-i-n solar cell applications, J. Appl. Phys. 98 (2005) 094501.

[59] R.R. King, P.C. Colter, D.E. Joslin, K.M. Edmondson, D.D. Krut, N.H. Karam, S. Kurtz, High-voltage, low-current GalnP/GalnP/GaAs/GalnNAs/Ge solar cells, in: Proc. 29th IEEE Photovolt. Spec. Conf, 2002, pp. 852–855.

[60] R.R. King, C.M. Fetzer, D.C. Law, K.M. Edmondson, H. Yoon, G.S. Kinsey, D.D. Krut, et al., Advanced III-V multijunction solar cells for space, in: Proc. IEEE 4th World Conf. Photovolt. Energy Convers, 2006, pp. 1757–1762.

[61] M.A. Wiemer, V. Sabnis, H. Yuen, 43.5% efficient lattice matched solar cells, Proc. SPIE 8108, High and Low Concentrator Systems for Solar Electric Applications VI (2011) 810804, https://doi.org/10.1117/12.897769.

[62] A. Aho, R. Isoaho, A. Tukiainen, V. Polojärvi, T. Aho, M. Raappana, M. Guina, Temperature coefficients for GaInP/GaAs/GaInNAsSb solar cells, AIP Conf. Proc. 1679 (2015), https://doi.org/10.1063/1.4931522.

[63] N. Miyashita, N. Ahsan, Y. Okada, Concentrating photovoltaic properties of GaInNAsSb/Ge dual junction tandem solar cell, in: Proc. IEEE 40th Photovolt. Spec. Conf, 2014, pp. 520–523.

[64] N. Miyashita, et al., Incorporation of hydrogen into MBE-grown dilute nitride GaInNAsSb layers in a MOCVD growth ambient, Sol. Energy Mater. Sol. Cells 185 (2018) 359–363.

[65] Y. He, N. Miyashita, Y. Okada, N-H-related deep-level defects in dilute nitride semiconductor GaInNAs for four-junction solar cells, Jpn. J. Appl. Phys. 57 (2018), 08RD11.

[66] N. Miyashita, N. Ahsan, Y. Okada, Improvement of 1.0 eV GaInNAsSb solar cell performance upon optimal annealing, Phys. Status Solidi 214 (2017) 1600586.

[67] J.B. Jackrel, et al., Dilute nitride GaInNAs and GaInNAsSb solar cells by molecular beam epitaxy, J. Appl. Phys. 101 (2007) 114916.

[68] N. Miyashita, N. Ahsan, Y. Okada, Evaluation of concentrator photovoltaic properties of GaInNAsSb solar cells for multijunction solar cell applications, Jpn. J. Appl. Phys. 54 (2015), 08KE06.

[69] N. Miyashita, Y. He, T. Agui, H. Juso, T. Takamoto, Y. Okada, Inverted lattice-matched triple junction solar cells with 1.0 eV GaInNAsSb subcell by MOCVD/MBE hybrid growth, IEEE J. Photovoltaics 9 (2019) 666–672.

[70] M.M.A.J. Voncken, J.J. Schermer, G.J. Bauhuis, P. Mulder, P.K. Larsen, Multiple release layer study of the intrinsic lateral etch rate of the epitaxial lift-off process, Appl. Phys. A 79 (7) (November 2004) 1801–1807.

[71] A. Aho, R. Isoaho, A. Tukiainen, G. Gori, R. Campesato, M. Guina, Dilute nitride triple junction solar cells for space applications: Progress towards highest AM0 efficiency, Progress in Photovoltaics 26 (9) (September 2018) 740–744.

CHAPTER ELEVEN

Summary of the design principles of betavoltaics and space applications

Tariq Rizvi Alam[1], Modeste Tchakoua Tchouaso[2] and Mark Antonio Prelas[3]
[1]Nuclear Environments and System Assessments, Applied Research Associates, Santa Barbara, CA, United States
[2]Physics, North Carolina A&T State University, Greensboro, NC, United States
[3]Electrical Engineering, University of Missouri, Columbia, MO, United States

11.1 Nuclear batteries

A nuclear battery converts radioisotope energy into electrical energy [1,2]. It has an advantage over other types of batteries due to its high energy density. Energy density is the total energy content per unit mass. The energy density of a nuclear battery is about 10^4 times higher than a chemical battery [3]. On the other hand, a nuclear battery has a very low power density compared to other types of batteries. Power density is the rate that it can output the power for a given size. As a result, a nuclear battery cannot compete with a fuel cell or a chemical battery for applications that require high power output. Therefore, the goal of the nuclear battery design is not to replace the chemical battery but to aid chemical batteries such as hybrid batteries and find applications where chemical batteries are not feasible. Thus, the targeted applications for a nuclear battery are mainly miniaturized low power output applications that cannot be fulfilled by chemical batteries. Other advantages of nuclear batteries are their reliability and longevity. A nuclear battery can output power for decades to a hundred years. A schematic Ragone plot [4] for the comparison of a nuclear battery with a fuel cell and a chemical battery is shown in Fig. 11.1.

11.2 Different types of nuclear batteries

The radioisotope energy can be harvested using different mechanisms as shown in Fig. 11.2. A brief discussion of some mechanisms is given below. A fission battery as shown in the schematic in Fig. 11.3 is still only a conceptual design [5]. It is a small reactor that does not require turbines and heat exchangers. Micron-scale fuel is coated as a ceramic on wire meshes that reside in a liquid metal bath that transfers heat to a conventional secondary heat transfer system. The resulting fission products carry the bulk of the kinetic energy. They will deposit most of their energy in the high-Z metal and create a knock-on electrons shower that will be able to pass through the thin

Photovoltaics for Space
ISBN 978-0-12-823300-9, https://doi.org/10.1016/B978-0-12-823300-9.00003-0

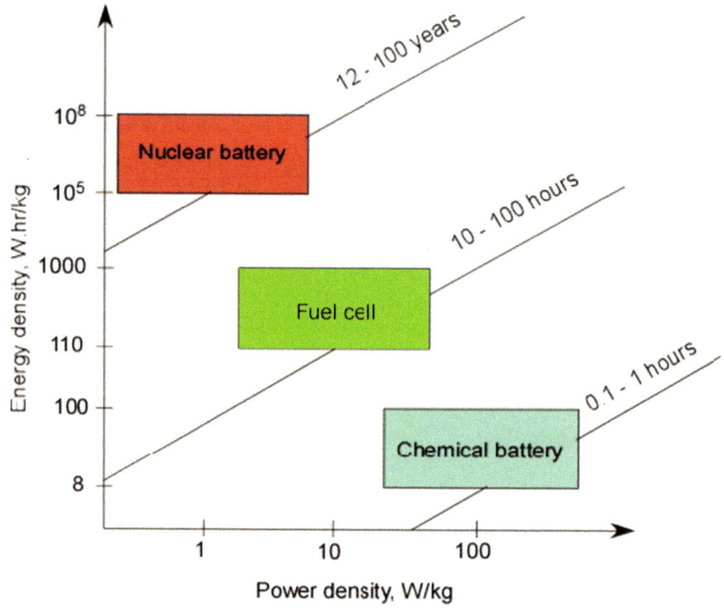

Figure 11.1 A schematic Ragone plot for nuclear batteries, fuel cells, and chemical battery.

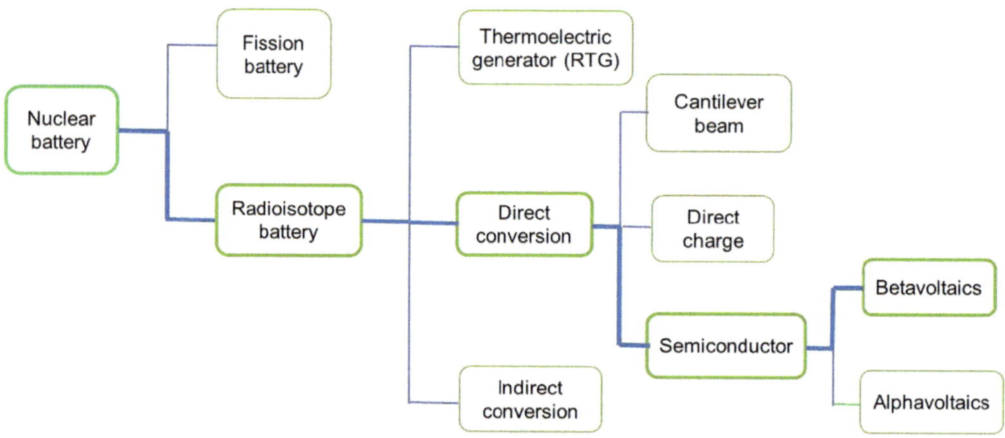

Figure 11.2 Nuclear battery classifications.

insulator layer but will be stopped in the sufficiently thick low-Z metal. The high-Z metals will be positively charged, and the low-Z metals will be negatively charged for the battery.

Thermocouples, as shown in Fig. 11.4, or small Sterling engines are used as thermoelectric devices in a radioisotope thermoelectric generator. A large amount of radioisotopes is used to generate decay heat energy that is harvested to generate electrical

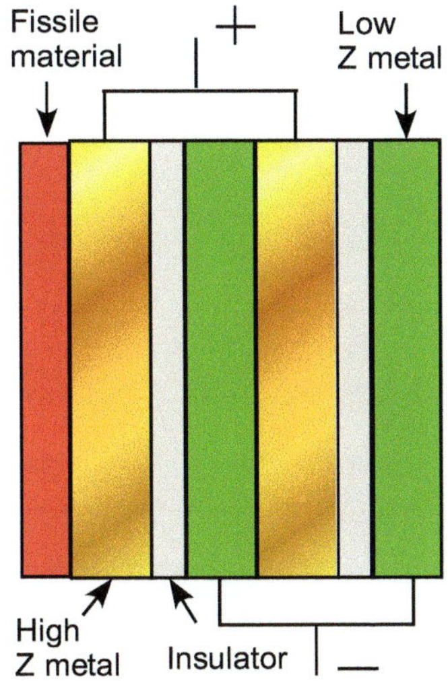

Figure 11.3 Design concept of a fission battery. *Adapted from Ref.* [5].

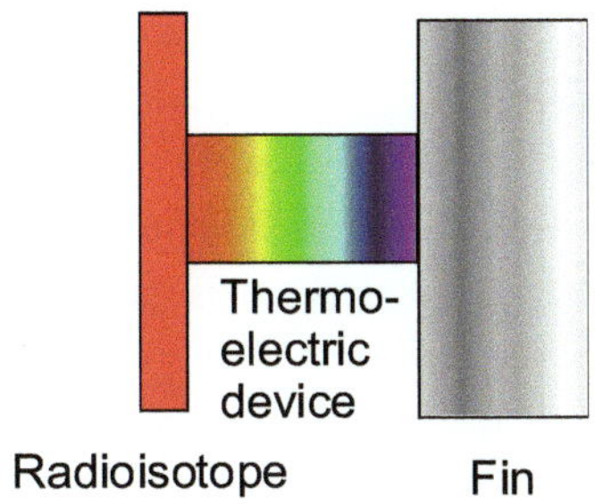

Figure 11.4 Radioisotope thermoelectric generator. *Adapted from Ref.* [6].

energy. This type of battery has applications mainly in space; the efficiency of the battery is about 6% [6].

Fig. 11.5 shows an indirect conversion method to generate electrical energy from the radioisotope energy. Production of photons is the intermediary step of energy conversion in this method. The challenge in this type of battery is the low photon intensity due to opacity of the radioluminescent materials. It mostly uses high-energy alpha particles. Using 300 mCi of Pu-238 with a phosphor screen and aluminum gallium arsenide (AlGaAs) photovoltaics resulted in generating a short circuit current of 14 μA and an open circuit voltage of 2.3 V using five cells with an efficiency of 0.11% [7]. It was used to power a calculator and a wristwatch. However, degradation of the output power was very rapid from damage of the radioluminescent material for this type of battery. The power output drops significantly due to the damage.

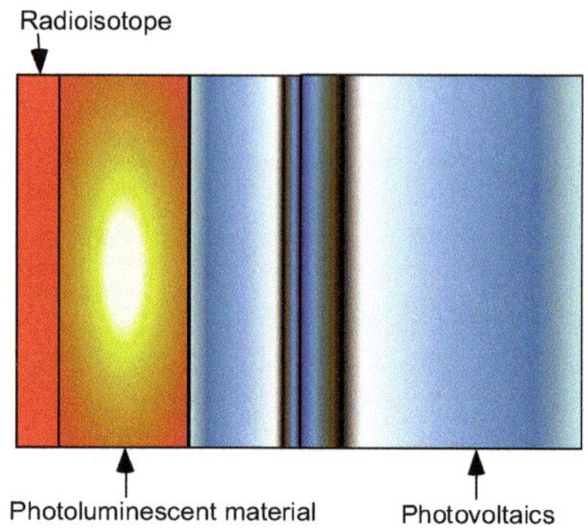

Figure 11.5 Indirect energy conversion method. *Adapted from Ref.* [7].

Fig. 11.6 shows a direct charge battery where the radioisotope and the electrode are separated by vacuum, air, or any other dielectric medium. This type of battery provides a very high open circuit voltage, and the efficiency of the battery is comparatively high. For example, 2.6 Ci of Pm-147 in a vacuum generated an open circuit voltage of 60 kV and a short circuit current of 6 nA with an efficiency of 14% [4]. This type of battery has applications for electrostatic motors.

Fig. 11.7 shows a cantilever method of energy conversion. The electrostatic force developed between the cantilever and the radioisotope pulls the cantilever beam to the radioisotope, and after discharging, it starts to oscillate. This self-reciprocating motion can be used as an electromechanical sensor. For example, 1 mCi of Ni-63 with a Cu cantilever beam (5 cm $\times$ 4 mm $\times$ 60 μm) generated 0.4 pW power with an efficiency of 0.0004% [8].

Alphavoltaics and betavoltaics are direct energy conversion methods using semiconductors, as shown for betavoltaics in Fig. 11.8. It can be used with alpha- or beta-emitting radioisotopes. However, damage of the semiconductor crystal structure is high for alpha particles. This method of energy conversion using beta particles is known

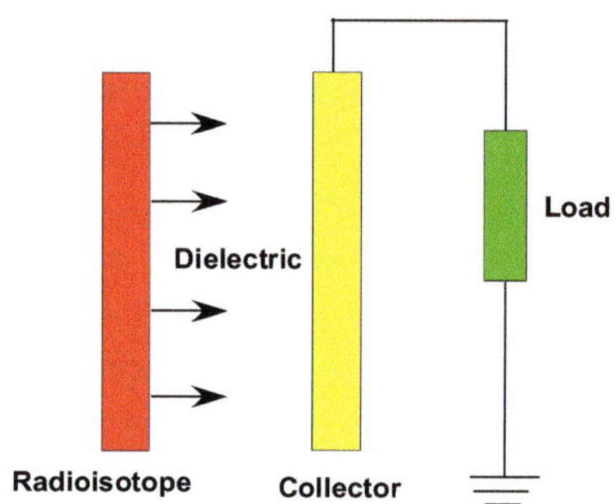

Figure 11.6 Direct charge nuclear battery. *Adapted from Ref.* [4].

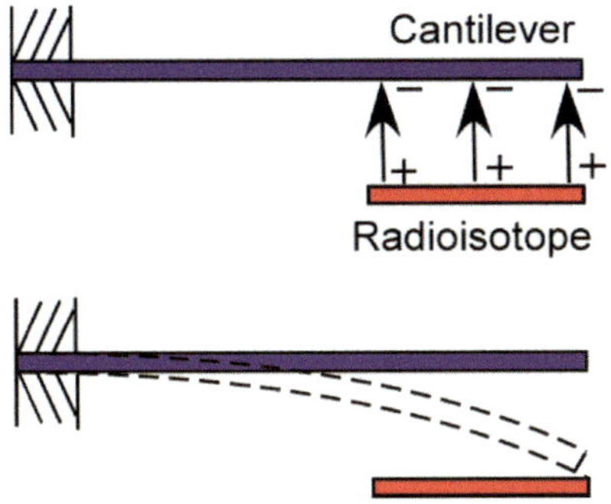

Figure 11.7 Schematic for the cantilever beam method. *Adapted from Ref.* [8].

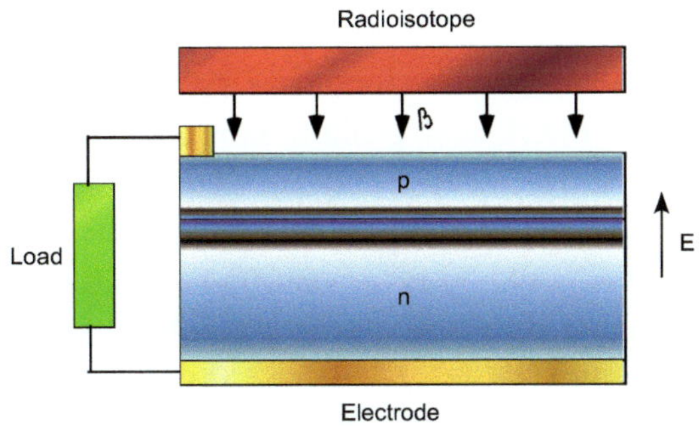

Figure 11.8 A betavoltaic battery design. *Adapted from [1].*

as the betavoltaic battery. It appears in the literature that most betavoltaic battery models overpredict the experimental results [9–13]. Therefore, a better model and design analysis are required. The betavoltaic battery design and analysis are provided in detail in the next few sections. The sections are organized for basics of betavoltaic batteries, analysis for radiation damage in semiconductors, beta emitter selection, and various radioisotopes and semiconductor results. Finally, principles of betavoltaic battery design are drawn, and space applications are discussed.

11.3 Betavoltaic batteries

Betavoltaic batteries fall under the classification of nuclear batteries, which is a large area, as shown in Fig. 11.2. Betavoltaic batteries directly convert the radioactive decay energy of a beta-emitting radioisotope into electrical energy employing a semiconductor method of charge collection. Betavoltaic batteries have the advantage that they can be made into very small, reliable, and durable power sources for micro-scale applications. Betavoltaic batteries can be further divided into different categories based on whether they are a solid or liquid. Solid betavoltaic batteries can have either crystalline or amorphous structures.

Betavoltaic batteries are not new. The concept of radioisotope batteries was first suggested by Mosely in 1913 [14]. A betavoltaic battery design was first investigated by Rappaport at Radio Corporation of America in 1954 [15], and later a radioactive battery using an intrinsic semiconductor was patented in 1956 [16–21]. Alphavoltaic and betavoltaic batteries were described by Pfann and Olsen [19] in 1954, where Pfann observed the lattice damage of semiconductors due to irradiation of high-energy alpha and beta particles [14]. As a result, low-energy betavoltaic batteries became a popular research topic in this area. In the 1970s, betavoltaics made of *p-n* junctions of Si at Donald Douglas Lab were used in clinical trials for pacemakers [22,23]. In early 1970, Olsen [24] suggested that betavoltaic batteries using wide- and indirect-bandgap semiconductors would have higher efficiency. Since 1989, many researchers [21] have worked on different semiconductor materials such as silicon (Si), AlGaAs, amorphous Si (a-Si), porous Si, silicon carbide (SiC), and gallium nitride (GaN). Many of these semiconductor materials were initially developed for solar photovoltaic applications. They found that wide-bandgap semiconductors have advantages of higher open circuit voltage, higher conversion efficiency, and greater radiation resistance.

Development of betavoltaic batteries has drawn additional researchers in recent years due to advancements in nanotechnology. Small size, reliability, and long-lasting durable power sources are required for future generations of electronics. Betavoltaic batteries are very promising sources of power that can fulfill these requirements. They can be miniaturized to the size of a human hair [25]. On the other hand, miniaturization of chemical batteries is restricted by their low energy density [26]. As an alternative, some researchers are working on scaling down power sources from fossil fuels and fuel cells. However, this is difficult because one must replenish the liquid fuel supply while eliminating by-products inside the electronics. It also results in a low energy density even though it is five to ten times better [26] than lithium ion batteries. A betavoltaic battery has an energy density that is 10^2 to 10^4 times [3] higher than that of chemical or fossil fuels. It has a long lifetime potential of several tens of years to several hundreds of years. Betavoltaic batteries are light, tiny, and integrated with the semiconductors to supply on-chip power [3,27] without any performance compromise to the surrounding environment.

Betavoltaic batteries have applications in microelectromechanical systems (MEMS), remote sensors, and implantable medical devices [28] such as pacemakers [23]. Due to their high energy density, long lifetime, and antijamming capabilities, they can also be used for remote applications including powering scientific apparatus in spacecraft, in undersea exploration, in the oil and mining industries, underground, in polar regions, in high mountainous regions, in military equipment, in sensor networks for environmental monitoring, and in bridges with embedded sensors [20,27,29]. Betavoltaic batteries are particularly useful for applications where solar photovoltaics are not a viable power option. They mainly work in low-power electronics that require nano-power sources [30]. Some researchers [23,31] have proposed the idea of hybrid batteries for trickle charging of rechargeable chemical batteries. In other words, a betavoltaic battery can be used in combination with rechargeable chemical batteries and supercapacitors for trickle charging. Trickle charging is effective for electronics that only require intermittent power; such an application is that of radiofrequency identification sensors. Betavoltaic batteries can also provide an on-chip-battery to power next-generation sensors and low-power processors without the need for an external power source.

11.4 Basic operation of betavoltaic batteries

An analysis of betavoltaic batteries is useful to identify design parameters and their optimization to understand the limitations and opportunities for further improvement. Betavoltaic batteries convert the kinetic energy of beta particles into electrical energy, as depicted in Fig. 11.8. However, not all the kinetic energy of the beta particles contributes toward generating electron-hole pairs (EHPs). A large percentage of the energy is lost to the lattice structure of the semiconductors by phonon interaction and through X-ray generation [21,32]. It is important to study the behavior of beta particles in semiconductor materials to optimize battery design. To do so, it is also important to understand how semiconductors behave differently from other materials.

The conductivity of a semiconductor changes with temperature as a semiconductor behaves like an insulator at absolute zero temperature. However, as the temperature increases, it conducts more and more electrons. Thus, it behaves like a metal at higher temperatures. When thermal energy is absorbed by the semiconductor lattice structure, electrons are released by breaking the bonds with neighboring atoms allowing them to move freely in the crystal. In the absence of an electric field, the free electrons may take part in re-bonding again with neighboring atoms after giving off their energy to other electrons and lattice vibration. This unique dynamic process takes place in semiconductors at temperatures higher than absolute zero.

An intrinsic semiconductor refers to a pure semiconductor material with no impurities in the crystal lattice. The process of adding impurity atoms by diffusion is called doping. Doping is a controlled method to create excessive holes and electrons in the semiconductors. When the semiconductor is doped with trivalent atoms such as boron, it creates a hole in the process of bonding with the four surrounding Si atoms. A hole, or vacancy, is the name given to the absence of an electron in the bond. However, holes along with electrons contribute to the total current. Those semiconductors that have excessive holes are called *p*-type semiconductors because holes are similar to a positive charge in that they attract an electron to the vacant bond. However, a *p*-type semiconductor is not really positively charged, and all the atoms are neutral. Similarly, if it is doped with a pentavalent atom such as phosphorous, it will generate excessive electrons in the semiconductor. This type of semiconductor is called an *n*-type semiconductor due to the negative charge of the electrons, which are easily freed from the lattice. An *n*-type semiconductor is also neutrally charged similar to a *p*-type semiconductor.

Fig. 11.9 shows the majority carriers, which are the holes and electrons for a *p*-type and an *n*-type Si semiconductor respectively. Both the majority and the minority carriers for a *p*-type and an *n*-type semiconductor are shown in Fig. 11.10. Holes and electrons always move in the opposite direction. When an electron jumps to a vacant bond, it leaves a vacancy behind. This is how holes in the semiconductor appear to move. When *p*-type and *n*-type semiconductor materials are placed together to form a *p*-*n* junction,

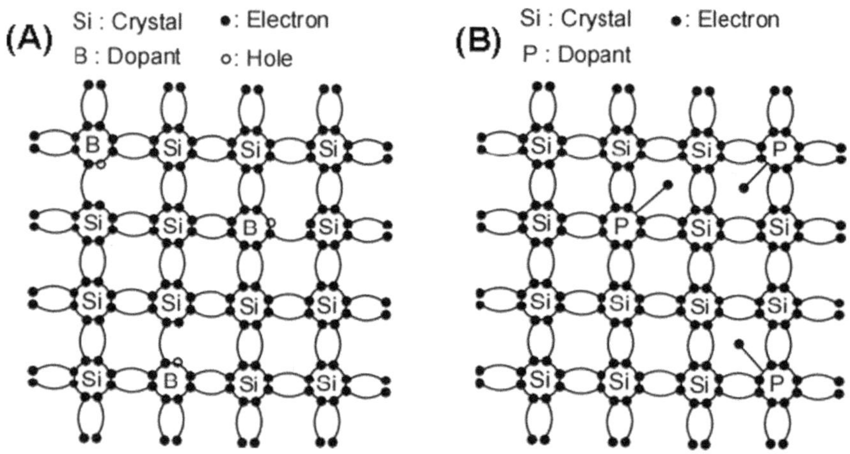

Figure 11.9 Holes or electrons in (A) *p*-type and (B) *n*-type silicon semiconductor lattice. Adapted from [1].

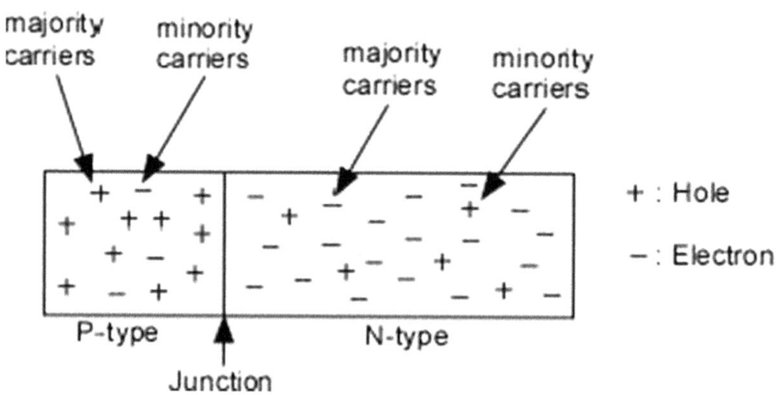

Figure 11.10 Majority and minority carriers for a *p*-type and an *n*-type semiconductor prior to diffusion. Adapted from [1].

the holes and electrons diffuse from regions of higher concentration to lower concentration.

This movement upsets the charge equilibrium of the atoms on either side of the junction and creates a space charge region and resulting electric field. When excess electrons of the donor atoms from the *n*-type region diffuse to the *p*-type region, they are attracted to the vacancies of the acceptor dopant atoms on the *p*-type region. As a result, the donor dopant atoms release electrons that are accepted by the acceptor atoms. By doing so, the mobile charge carriers around the junction become depleted. As a result, a space charge region forms around the junction, which is also called the depletion region. A depletion region of about 1 μm thickness (depending on the dopant concentration and semiconductor material) occurs around the *p-n* junction. The dopant atoms on the *p*-side and the *n*-side of the depletion region become negatively and positively charged ions, respectively. This effect generates a potential, known as the built-in potential, in the depletion region. It acts as a barrier for further diffusion of holes and electrons. Thus, an equilibrium condition is reached by the repulsion forces of the ionized atoms at the edges of the depletion region.

Fig. 11.11 shows the formation of the depletion region for a *p-n* junction after the diffusion of holes and electrons. It deliberately shows that the depletion region is asymmetric around the junction. This varies based on the difference in dopant concentrations inside the *p*-type and *n*-type regions. In betavoltaics, typically the *p*-type region is heavily doped. Therefore, the higher concentration of holes from the *p*-type region diffuses further into the *n*-type region due to the large diffusion gradient. As a result, the depletion region extends more into the *n*-type region in this case. The resulting width of the depletion region is very important for betavoltaic battery design.

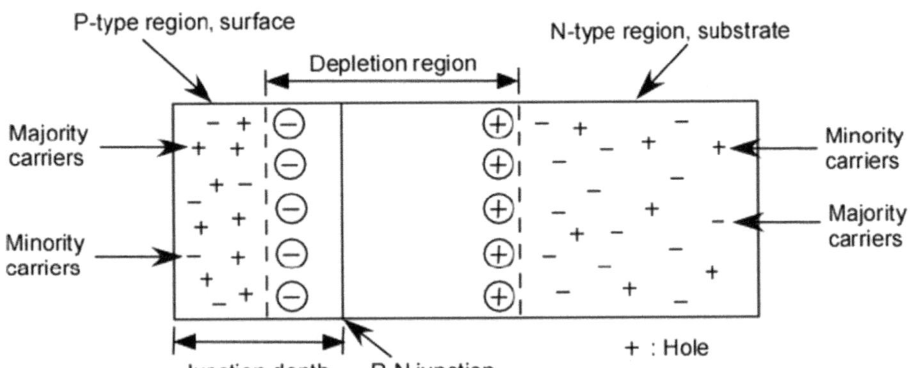

Figure 11.11 Formation of a depletion region for a *p-n* junction semiconductor after diffusion of the carriers.

A key aspect of betavoltaic battery design is to understand how EHPs are generated in the semiconductor. A band diagram explains the generation of EHPs in a simple multistep method [33], as shown in Fig. 11.12. The first step is called the excitation step. In this step, when the electrons in the valence band receive sufficient energy from collisions with the beta particles, they jump to the conduction band. Some electrons may even jump from the deeper bands below the valence band. This process leaves holes in the valence and deeper bands as well as increases the number of electrons in the conduction band. This process takes about 10 ps.

The next step is called the de-excitation step. In this step, electrons move to the bottom of the conduction band and holes move to the top of the valence band. Additional EHPs are generated during the de-excitation process. There is also some energy imparted to the crystal lattice structure due to the conservation of momentum that generates phonons during this process. There are many modes of crystal vibrations available at lower energies than the bandgap energy. This multistep band diagram method explains why the required energy to generate one EHP is higher than the bandgap energy of the semiconductor. On average, the energy to generate one EHP is approximately three times the bandgap energy. For example, the average energy required to generate one EHP in Si is 3.64 eV, which has a bandgap of 1.12 eV. It takes into account the thermalization and phonon losses in the process of EHP generation [34,35].

There are also different types of junctions available for solid betavoltaic batteries such as a Schottky junction or a *p-n* junction. A beta-emitting radioisotope can be placed on either the *p*-type region surface or *n*-type region surface of the semiconductor. A typical solid *p-n* junction betavoltaic battery design is shown in Fig. 11.13 where the radioisotope is placed on the *p*-type region surface. This is because many semiconductors are built up with a thick *n*-type layer on a substrate using chemical vapor deposition. The *p*-type region is then created using diffusion of the dopant atoms, which is typically much

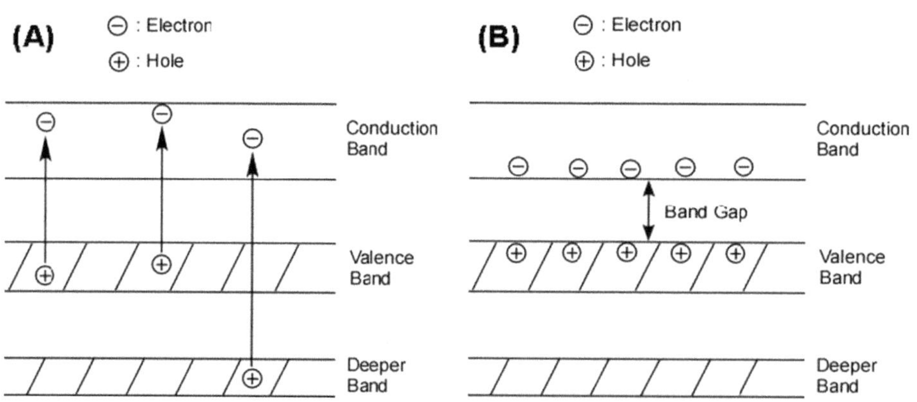

Figure 11.12 Generation of electron-hole pairs by (A) excitation and (B) de-excitation process.

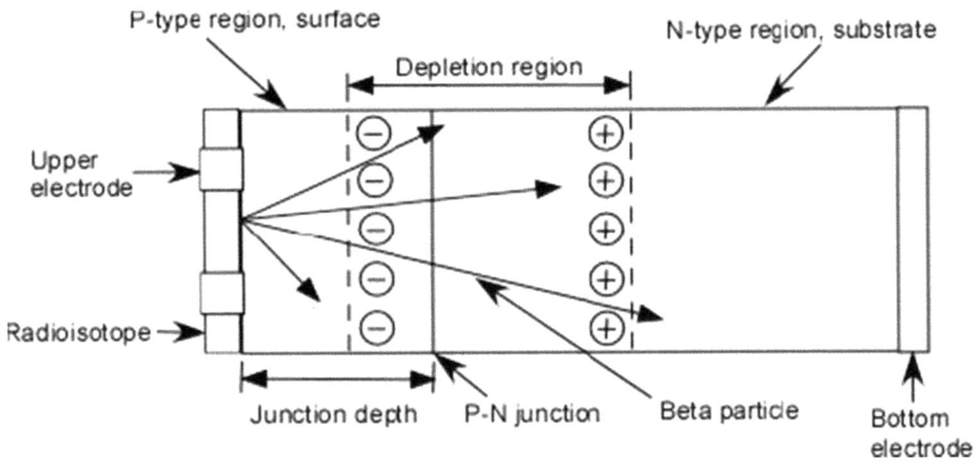

Figure 11.13 Solid *p-n* junction betavoltaic battery design.

thinner than the *n*-type region. Placing the radioisotope on the thinner *p*-type side allows the beta particles to reach the depletion region.

A Schottky diode is built with a Schottky metal and either a *p*-type or *n*-type substrate. The depletion region is formed in between the metal contact and the p-type or n-type region. This type of contact is known as a Schottky contact, and the metal used to create this contact is known as a Schottky metal. Examples of Schottky metals are nickel (Ni) and nickel silicide (Ni_2Si). A typical Schottky junction betavoltaic battery is shown in Fig. 11.14.

The potential barrier developed in the Schottky contact is called a Schottky barrier. Schottky junction betavoltaics are low cost and easier to fabricate. They are typically useful for semiconductors that are difficult to grow a *p*-type layer on the substrate such as GaN. However, the power output of a Schottky junction betavoltaic battery is low

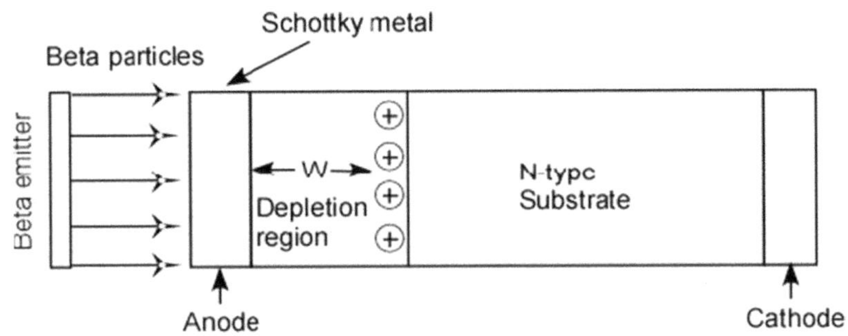

Figure 11.14 Schottky betavoltaic battery design.

because most of the beta particles are absorbed in the Schottky metal before reaching the depletion region.

Betavoltaics have a similar operating principle to that of photovoltaics where a photon is converted into electricity. One of the major differences between betavoltaics and photovoltaics are that each high-energy beta particle can create thousands of EHPs, whereas photons from sunlight are of much lower energy and generate only one EHP per photon absorption. The other major differences between betavoltaics and photovoltaics are that high-energy beta particles can penetrate deeply into a Si wafer [31], whereas the low-energy photon penetration depth depends on the absorption coefficient of the materials. On the other hand, the principle of operation for both betavoltaic batteries and radiation detectors is similar. However, a radiation detector as shown in Fig. 11.15 requires an application of an external high reverse bias voltage on the order of kilo-volts. This increases the internal electric field and creates a wide depletion region. Like betavoltaic batteries, a large depletion width is necessary to collect all the EHPs generated by the incident radiation beam. In radiation detectors, it is also useful to avoid any false positive signals. The collected charges in a radiation detector are registered in an external circuitry as a pulse signal for analysis, whereas a betavoltaic battery supplies a much higher number of charges to an external electrical load.

The design of a typical betavoltaic battery consists of an upper electrode, a p-type region (doped surface region), a depletion region, an n-type region (doped substrate), and a bottom electrode [9] as shown in Fig. 11.16A. The basic principle of this battery

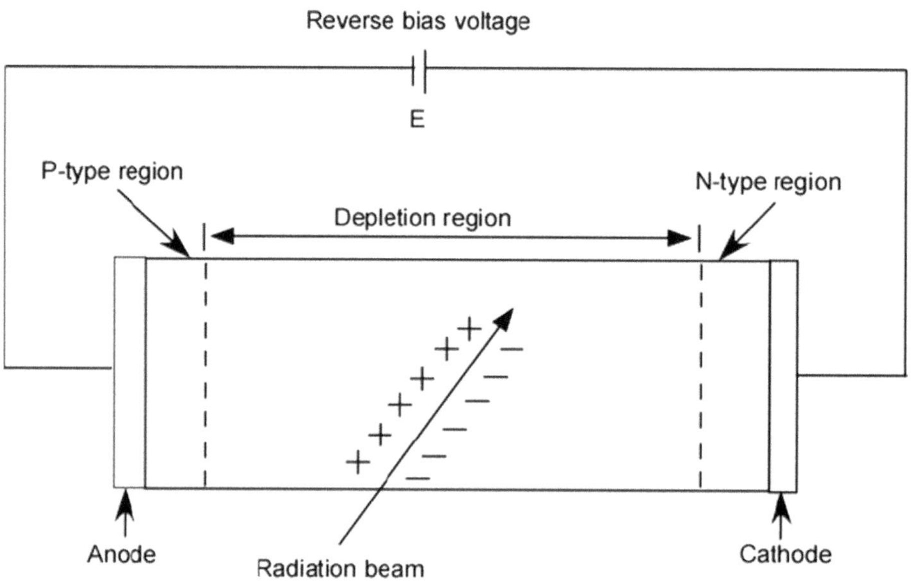

Figure 11.15 A typical radiation detector made of *p-n* junction semiconductor.

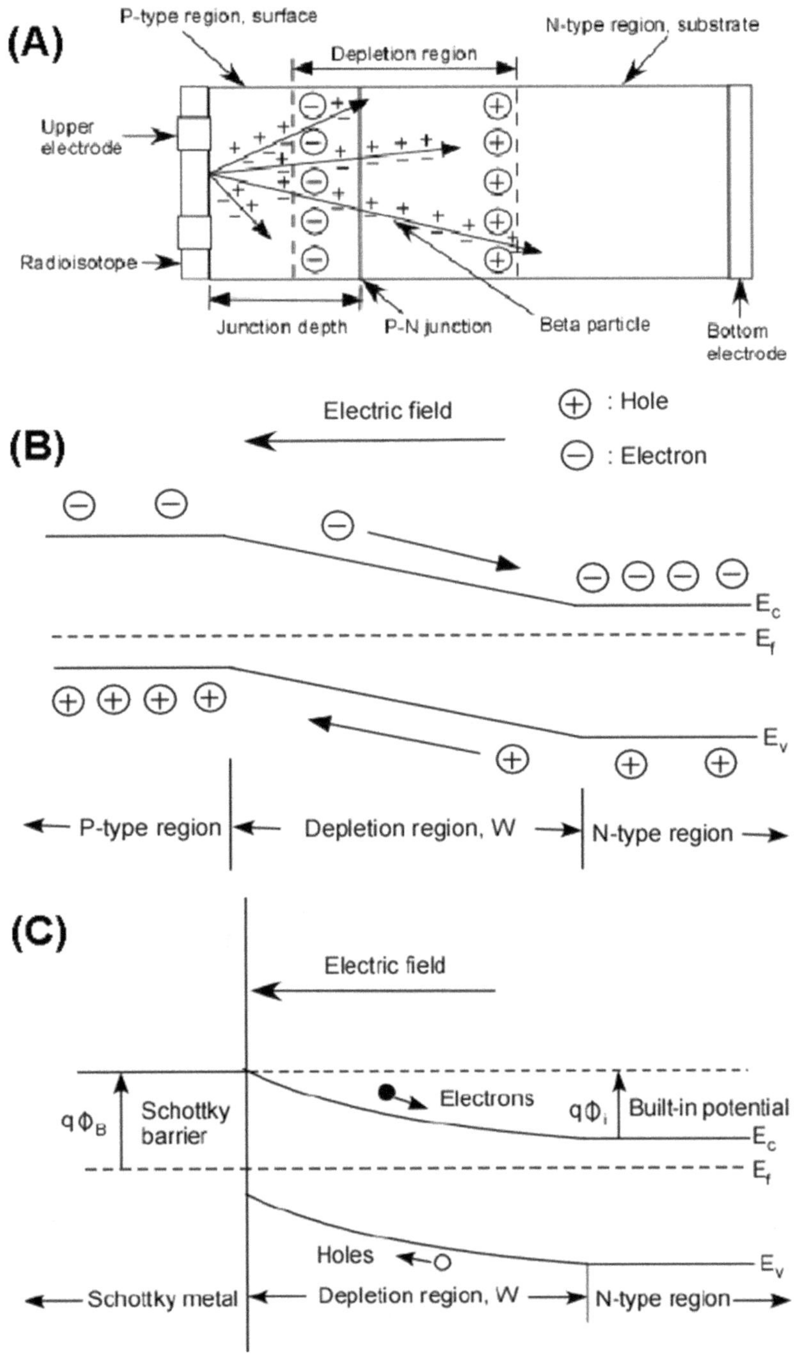

Figure 11.16 (A) EHP generation in a solid betavoltaic *p-n* junction battery design. (B) Electron and hole movement inside the depletion region of a *p-n* junction and (C) a Schottky junction. Adapted from [1].

is to generate EHPs in the semiconductor materials by the beta particles and collect them at the electrodes. The energetic beta particles emitted from the radioisotope enter the semiconductor, which in turn creates EHPs through collisions, excitations, and ionization. The number of EHPs generated depends on the incident energy of the individual beta particles. Each beta particle generates several tens of thousands of EHPs [13]. Betavoltaic batteries convert a low number of high-energy beta particles into a high number of low-energy EHPs. To achieve maximum power from a betavoltaic battery, it is important to maximize the separation of the EHPs being generated while minimizing recombination. Recombination of the EHPs contributes to energy loss in the betavoltaics. However, electron energy may also be lost due to collisions with other electrons or due to lattice vibrations resulting in a temperature increase. The separation mechanism of EHPs needs to be analyzed for maximum collection efficiency of the EHPs. If the radioisotope is placed on the p-type region of the semiconductor, electrons with higher energies from the *p*-type region will cross the depletion region to enter the n-type region. The EHPs generated on either the *p*-type or *n*-type region need to reach the depletion region through diffusion to be collected. Therefore, the goal is to form as many EHPs in the depletion region as possible, as opposed to outside the depletion region. Hence, a thin *p*-type layer is typically used in the design.

The EHPs generated in the depletion region are swept across to each side due to the built-in potential of the space charge as shown in Fig. 11.16B and C for a *p-n* junction and Schottky junction, respectively. The electric field in the depletion region sends the electrons and holes to the *n*-type and *p*-type regions, respectively. The electrode on the *p*-type region that collects holes is called the anode [28,30,36,37]. Similarly, the electrode on the *n*-type region that collects electrons is called the cathode. However, this contradicts the common notation used for batteries, where the negative and positive terminals of a battery are called the anode and the cathode, respectively [38]. On the other hand, if the radioisotope is placed on the *n*-type side of the semiconductor, it works using similar principles as placing the radioisotope on the *p*-type side. However, very few researchers [3,31,39] have placed the radioisotope on the *n*-type side of the semiconductor where p-type substrates were used.

One of the first steps is to calculate the EHP generation profile inside the semiconductor material. However, this will be discussed in more detail in Section 11.6 on radioisotopes. Once it is known where and how many EHPs are formed from the beta particles, the collection process of EHPs is then important to understand for optimization of betavoltaic battery design. Not many researchers have provided a good model for the charge transport within the semiconductor to the electrodes, but the few that did are discussed subsequently.

One of the design requirements of betavoltaic batteries is to increase the depletion width to increase the charge collection efficiency. Charge collection efficiency also depends on the particular charge carrier's drift length, which is related to carrier mobility, electric field intensity, and carrier lifetime [20]. The improvement of semiconductor films by minimizing the dislocation density will also increase the charge collection efficiency. However, it is difficult to create thick, high-quality crystal layers for wide-bandgap semiconductors due to fabrication limitations. Creating a wider depletion region in GaN, which is a wide-bandgap semiconductor, is limited by difficulties with *p*-type doping. GaN typically has an n-type conductivity due to residual impurities. It is hard to create intrinsic and *p*-type layers for GaN. For example, Cheng et al. [18] gradually developed *p-i-n* GaN from *p-n* GaN and *p-u-n* GaN. The unintentionally doped (u-type) layer is the layer grown with nominal doping that has an electron concentration over 5×10^{16} cm^{-3} due to residual impurities. To create an intrinsic layer, the electron concentration of the u-type material needs to be reduced. This was achieved by doping the u-region with Fe as a compensator. An intrinsic region thickness of 0.9 μm is achieved in this process with a Mg-doped p-type layer thickness of 0.05 μm and Si-doped n-type layer thickness of 1 μm to form *p-i-n* GaN. The collection efficiency of the EHPs generated within the depletion region is almost 100%.

It was assumed in a simplified model based on collection efficiency of EHPs [31] that all the EHPs generated in the depletion region and within one minority carrier diffusion length from the depletion region mostly contribute to create a current. The EHP collection probability for the carriers generated outside the depletion region is then approximated by Eq. (11.1) [9,11,12]:

$$CE = 1 - \tanh\frac{d}{L}, \tag{11.1}$$

where *CE* is the collection probability of an EHP, *d* is the distance from the depletion region boundary, and *L* is the minority carrier diffusion length for either the *n*-type or *p*-type material as appropriate. This indicates that the junction depth and the width of the depletion region need to be adjusted according to the penetration depth of the beta particles. The energy of the beta particles should be deposited in the depletion region as much as possible for maximum EHP collection [10]. The collection efficiency of the EHPs generated outside the depletion region depends on the distance from the depletion region. Therefore, the wider the depletion region is, the greater is the number of EHPs that would be collected. Besides the simplified model, the collection EHPs can also be calculated by the drift-diffusion model [3]. The collection of EHPs generated outside the depletion region is solved by diffusion to the depletion region.

The width of the depletion region is an important parameter for betavoltaic battery design. The width of the depletion region for a *p-n* junction can be calculated by Eqs. (11.2) and (11.3) [9]:

$$W = \sqrt{V_{bi}\left(\frac{2\varepsilon_r\varepsilon_0}{q}\right)\left(\frac{N_a + N_d}{N_a N_d}\right)}, \tag{11.2}$$

$$V_{bi} = \frac{kT}{q}\left(ln\frac{N_a N_d}{n_i^2}\right), \tag{11.3}$$

where V_{bi} is the built-in potential barrier, ε_r and ε_o are the dielectric constants in the region and in vacuum, respectively, N_a and N_d are the doping concentrations of the *p*-type (acceptor) and *n*-type (donor) regions, and n_i is the intrinsic carrier concentration for a pure, undoped semiconductor.

From basic semiconductor physics, the width of the depletion region for a Schottky junction can be expressed by Eq. (11.4) [13]:

$$W = \sqrt{\frac{2\varepsilon_s V_{bi}}{qN_D}} = \sqrt{\frac{2\varepsilon_s\left(\varphi_B - \frac{kT}{q}ln\left(\frac{N_C}{N_D}\right)\right)}{qN_D}}, \tag{11.4}$$

where ε_s is the relative dielectric constant of the semiconductor, V_{bi} is the built-in potential barrier, q is the electron charge, N_D is the dopant concentration of the semiconductor material, φ_B is the Schottky barrier height, k is the Boltzmann constant, T is the temperature in Kelvin, and N_C is the effective density of states in the conduction band of the semiconductor.

From these equations, it can be seen that the dopant concentration is one of the vital parameters for battery design. The dopant concentration affects depletion width, short circuit current, and open circuit voltage [9,12]. A lower dopant concentration increases the depletion width and minority carrier diffusion length, which in turn increases the charge collection efficiency. A lower concentration is favorable for increasing the short circuit current of a betavoltaic battery. However, it is not a favorable condition for increasing the open circuit voltage. A lower dopant concentration also increases the leakage current [3], which in turn decreases the open circuit voltage. Leakage current results from all types of recombinations including the random motion of the carriers that can overcome the built-in potential barrier and recombine. The built-in potential barrier decreases with lower dopant concertation. Therefore, the dopant concentration needs to be optimized to achieve the best battery performance.

In an analysis by Tang et al. [9], heavy doping on the order of magnitude of 10^{18} cm^{-3} to 10^{19} cm^{-3} in the *p*-type region and light doping on the order of magnitude of 10^{16} cm^{-3} to 10^{17} cm^{-3} in the n-type region were found optimal for Si with Ni-63 at a shallow junction depth of about 0.3 μm for maximum power output. A larger difference in the doping concentration on each side of the junction increases the width of the depletion region. However, a significant difference in doping concentration can lead to reduced power output due to increased leakage current [40], which results from increased recombination due to the lattice mismatch and defects. A small change in leakage current, which is usually several nanoamperes, has a large impact on the performance of betavoltaic batteries. The reason is the current generated due to irradiation in the battery is in the range of nanoamperes to microamperes, whereas the current range for a solar cell is 1−100 mA. Thus, an increase in leakage current can rapidly decrease the power output by a larger percentage in the case of a betavoltaic cell. This effect can be minimized by introducing interlayers in the semiconductor resulting in a higher conversion efficiency. The leakage current in a semiconductor such as GaAs consists of perimeter surface recombination and bulk junction recombination [41]. The leakage current can be reduced by forming perimeter depletion layers (PDLs) that create an isolation effect. It was found that a larger PDL is formed using a p^+-p-n$^+$ junction compared with three other investigated junctions such as n^+-p-p^+, p^+-n-n^+, or n^+-n-p^+, where p^+ and n^+ regions refer to more heavily doped regions. The minimum leakage current of approximately 10^{-11} A was found for the p^+-p-n^+ junction.

Aside from the leakage current, the charge collection can be hampered from a wide depletion region. If the minority carrier diffusion lengths are smaller than the width of the depletion region, charge collection will be reduced. For example, a comparison [41] of ideal and experimental short circuit current showed that only half of the EHPs were collected due to the minority carrier diffusion lengths being half of the depletion width. It was concluded that a multijunction structure would be a better design instead of a wider depletion region with a single junction. The suggested multijunctions are six to 10 for GaAs when the radioisotope is Ni-63. Another approach to resolve problems associated with a wider depletion region is to increase the concentration of energy deposition in a narrow region. An analysis compared both the ratio of energy deposition on the top layer and the range of beta particles in homojunction and heterojunction semiconductors. A heterojunction of Si/SiC showed the best energy concentration in the depletion region in a study of homojunctions of Si, GaN, GaAs, SiC, and InGaP and heterojunctions of these semiconductors with Si [19]. The results for a Si/SiC junction indicated 24.6% of total energy deposition in the top layer thickness of 0.3 μm with a penetration depth of 2.1 μm for beta particles from Ni-63. It reduced the depletion region thickness to 1.8 μm with a maximum energy deposition in the depletion region instead of in the top layer of the device. The radiation tolerance of the device was improved by introducing SiC, as it has better radiation tolerance than that of Si. Radiation tolerance will be discussed further in the next section.

The maximum power output and conversion efficiency of betavoltaics can be improved by two different methods: using a high-energy radioisotope source and improving or developing a new type of *p-n* junction [42]. The performance of a betavoltaic is also dependent on the operating temperature [27]. Diffusion length, intrinsic carrier concentration, and carrier mobility in semiconductors are all a function of temperature. Diffusion length and leakage current are directly related to the short circuit current and open circuit voltage of the device. The leakage current is further dependent on the intrinsic carrier concentration. An increase in temperature decreases device performance by reduced open circuit voltage with a low fill factor. The short circuit current is negligibly affected by change in temperature. The power output and efficiency of betavoltaics decrease almost linearly with an increase in temperature.

An equivalent circuit model of a betavoltaic cell as shown in Fig. 11.17A consists of a current source, diode, shunt resistance, series resistance, and a load resistance that represent, respectively, a radioisotope, a semiconductor, resistance due to leakage and carrier-recombination, internal resistance of the diode including electrode contact, and a load to receive power. The generated current can be expressed by Eqs. (11.5) and (11.6) [43]:

$$I = I_\beta - I_{bk} - I_{sh}, \tag{11.5}$$

$$I = I_\beta - I_0\left[exp\left(\frac{q(V + IR_s)}{nkT}\right) - 1\right] - \frac{V + IR_s}{R_{sh}}, \tag{11.6}$$

where I_β is the radioisotope generated current, I_0 is the leakage current, R_s is the series resistance, R_{sh} is the shunt resistance, q is the electron charge, n is the ideality factor that defines how closely it follows ideal diode current-voltage characteristics ($n = 1$ for an ideal diode), k is the Boltzmann constant, and T is the absolute temperature. Open circuit voltage and short circuit current are found from the equation by setting $I = 0$ and $V = 0$ respectively. From the equivalent circuit model, it can be seen that open circuit voltage depends on the shunt resistance and has a proportional relationship with it. Similarly, the short circuit current has a reciprocal relationship with the series resistance. In an ideal case, the series and shunt resistance are zero and infinite, respectively, for betavoltaic cells.

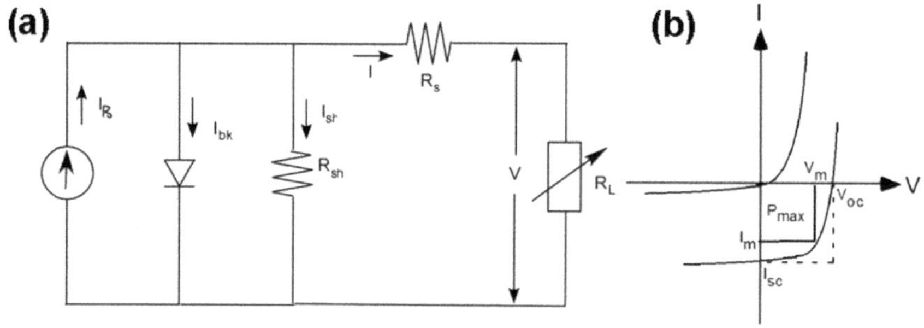

Figure 11.17 (A) Equivalent circuit model of a *p-n* junction betavoltaic battery and (B) current-voltage characteristics of a betavoltaic battery.

Fig. 11.17B shows the current-voltage relationship of a betavoltaic battery in comparison with an ideal diode. Current is negative for betavoltaic batteries as it supplies the power. The fill factor is an important parameter that determines the proximity of the battery output at full operational load to the theoretical maximum power output. Fill factor is defined by Eq. (11.7):

$$FF = \frac{P_{max}}{P_{th.\ max}} = \frac{I_m V_m}{I_{sc} V_{oc}}, \tag{11.7}$$

where I_m, V_m are the maximum current and voltage and I_{sc}, V_{oc} are the short circuit current and open circuit voltage.

The design of electrodes also affects the performance of a betavoltaic battery. The presence of an electrode in between the radioisotope and the *p*-type region works as a dead layer to the betavoltaic batteries. It causes some beta particles to be absorbed in it and most to be reflected back. As a result, the effective energy deposited in the semiconductor is much less than the incident energy [44]. To overcome this loss, the cross-sectional area of a high-Z metal such as Au [32] can be reduced or replaced by a low-Z metal such as Al. A well-designed interdigit device, such as a comb-shaped electrode [40] with optimum interspacing instead of continuous metal, increases the short circuit current by reducing the backscattering yield of the radioisotope [23].

In summary, the semiconductor material in a betavoltaic battery design needs optimization to maximize power output and efficiency. This includes doping concentrations, junction depth, width of the depletion region, minority carrier diffusion lengths, leakage current, internal defects, junction type, interlayer structure, temperature effect, and electrodes. Besides the design parameters of semiconductors, the radiation tolerance of semiconductors also needs to be considered to maintain consistent battery performance. This topic is taken up in the next section.

11.5 Radiation damage in betavoltaic batteries

Betavoltaic battery performance can suffer from degradation of the semiconductor materials due to irradiation over time. When this occurs, the power output deteriorates, and the battery may fail prematurely. Different measurements and calculations have been used to determine radiation damage. These include deterioration of the power output, normalized power output decay, capacitance voltage, deep level transient spectroscopy, minority carrier diffusion length, charge carrier concentration, open circuit voltage, and radiation damage factor. However, a decay in the power output may not be an indication of radiation degradation as the radioactive source also decays based on its half-life. Therefore, the source decay also needs to be considered when measuring radiation damage by means of power output.

Radiation damage by beta particles to the semiconductor material depends upon both the particle energy emitted by the radioisotope and the radiation hardness of the semiconductor material. The probability of radiation damage in materials can be minimized by reducing the energy of incident beta particle radiation and increasing the strength of the atomic bonds in the materials [11]. In the process of the creation of radiation-induced defects, atomic bonds in the material are broken after absorbing enough radiation energy and then the released atom diffuses inside the material. Wide-bandgap materials have a higher bond strength with minimal diffusion. The lower diffusivity then increases the self-annealing effect of the materials. However, due to the low diffusivity of the wide-bandgap semiconductors, it is also difficult to dope them sufficiently to form a *p-n* junction.

Gallium nitride is a wide-bandgap semiconductor, which is one of the popular radiation-tolerant materials used in betavoltaic batteries. It can also be used in space applications due to its radiation hardness. Ionascut-Nedelcescu et al. [45] determined the radiation hardness properties of GaN in an experiment where it was irradiated by electrons with an energy range of 300−1400 keV at room temperature to determine the threshold energy for damage. The electron threshold energy of GaN was found to be 440 keV with a measured atomic displacement energy of 19 ± 2 eV for Ga. The atomic displacement energy is a measure of the minimum kinetic energy required to displace an individual atom from its regular crystal lattice site to a defect position. There is no displacement energy found for the nitrogen atoms, which indicates self-annealing is occurring. Self-annealing is a recombination of vacancy-interstitial pairs. If the distance between a vacancy-interstitial pair is small, the recombination takes place by diffusion and it is temperature dependent.

Other radiation-hard semiconductor materials are SiC, GaAs, $Al_{0.7}Ga_{0.3}N$, a-Si:H, AlGaAs, indium phosphide (InP), InGaP, and diamond. However, GaN is chosen over SiC by Mohamadian et al. [12] in their battery design for its slightly better radiation hardness and its larger heat capacity. The radiation hardness of GaN is also much higher than that of GaAs. It requires two orders of magnitude higher radiation fluence at the same energy to degrade GaN compared to GaAs. The reason why GaN has better radiation-tolerant properties over many other semiconductor materials can be explained by its high bond strength over its minimal atomic displacement. Furthermore, it can be attributed to the presence of a high density of nitride materials, which have a low atomic number. Low atomic number materials reduce the interaction of core electrons in the lattice with the high energy radiation electrons. As a result, fewer defects occur due to radiation. For example, $Al_{0.7}Ga_{0.3}N$ with a wide bandgap of 5.8 eV has a radiation resistance six times higher than Si, since Al and N have lower atomic numbers [11]. In addition to higher radiation tolerance, the use of wide-bandgap semiconductors will increase the power conversion efficiency [11] since their leakage current is very low. However, an arbitrary increase of the bandgap will also reduce the conductivity of the semiconductors, which will hamper the charge collection in the semiconductor. Among all other radiation-tolerant semiconductors, SiC and GaN are the most popular wide-bandgap semiconductors used in betavoltaic batteries.

There are some experimental results reported to investigate the radiation damage of semiconductors. They showed that there was no radiation damage for semiconductors irradiated by beta particle energies lower than the radiation threshold energy of the semiconductors. For example, some researchers reported no evidence of radiation damage for a Si *p-n* junction with Ni-63 (66.9 keV; max beta energy) [46], a *p-i-n* junction of SiC irradiated with P-33 (248.5 keV) [21], and a 4H−SiC *p-n* junction with Ni-63 (66.9 keV) [32] observed for 6 months, 3 months, and 10 days, respectively. However, radiation damage was observed in SiC, InP, GaN, and Si by some researchers [11,47] when the semiconductors were irradiated by high-energy electrons. In an experiment by Rybicki [47], both SiC and InP were irradiated by 1-MeV electrons. The result was that the radiation resistance of SiC was not much better than that of InP.

In another experiment [11], a set of Si solar cells and n-type GaN were irradiated by Co-60 (317 keV) beta particles and gamma rays with a dose of 10 and 100 Mrad. The open circuit voltage and photoluminescence peak were then measured for the Si and GaN to analyze the radiation damage. The open circuit voltage is related to the minority carrier lifetime in terms of diffusion length. A maximum reduction factor of five for the minority carrier lifetime with a degradation voltage of 1.6% was observed for GaN, whereas a reduction factor of 52 for the minority carrier lifetime with a degradation voltage of 25% was measured for Si solar cells.

Therefore, GaN was found to be a better radiation-tolerant material compared to Si, and it showed very little degradation even with a very high radiation dose from Co-60. On the other hand, a-Si:H is claimed to have superior radiation hardness and is used in space applications for solar cells. Maturation of a-Si:H solar cells made it possible to have increased radiation hardness over crystalline semiconductors especially when operated at a low annealing temperature of 70°C. However, it was reported in an experiment by Deus [48] that a-Si:H had higher degradation than that of AlGaAs under a tritium (18.59 keV) gas atmosphere when the batteries were observed for 46 days. The efficiency decreased by 94% and 69% for a-Si:H and AlGaAs, respectively. The radiation damage was high due to the high diffusion of H-3, which was about one H-3 atom per three Si atoms for thin-film a-Si:H. The type and structural form (or phase state) of the radioisotope is believed to be the reason for the degradation. It was suggested that the radiation damage could be minimized by using tritiated titanium thin film instead of a free gas for the radioisotope form.

There are some techniques that can be employed to increase the radiation tolerance of semiconductors. For example, introduction of an intrinsic region in the junction [49] and a low doping concentration such as 2×10^{12} cm^{-3} in the intrinsic region [19] has been shown to improve the radiation degradation of the semiconductor materials in betavoltaic batteries. Noncrystalline structures like a liquid semiconductor may also limit the radiation damage [50].

In betavoltaic batteries, the radiation damage to the semiconductor depends on the beta particle energy, the atomic bond strength, and the migration barriers of vacancy and interstitial of the semiconductor, and the size of the atoms in the crystal lattice structure that are interacting with the impinging high-energy beta particles. Wide-bandgap materials such as GaN and SiC showed much better radiation tolerance due to their aforementioned qualities. However, there is a tradeoff for using wide-bandgap semiconductors, which must be considered. The charge (EHPs) collection is difficult in wide-bandgap semiconductors due to their low diffusivities.

11.6 Radioisotopes for betavoltaic batteries

Not all the beta particles from the radioisotope reach the semiconductor due to self-shielding (or self-absorption). The energy deposited in the semiconductor depends on the volume of the radioisotope. The radioisotope thickness needs to be optimized to minimize self-shielding effects. The radioactivity available for use by the battery can be represented by the actual activity and the apparent activity. Actual activity is the inherent activity of the radioisotope depending on its specific activity and mass. Apparent activity is the effective activity that is emitted by the source and would be measured by a radiation detector. Apparent activity differs from actual activity due to the self-absorption of the beta particles inside the source before they can exit [28], and it is always less than the actual activity. The apparent activity is given by Eqs. (11.8) and (11.9) [30,33]:

$$\Phi = \frac{C}{\mu_m}\left(1 - e^{-\mu_m t_m}\right), \tag{11.8}$$

$$\mu_m = 0.017/E_{max}^{1.43}, \tag{11.9}$$

where C is the specific activity (mCi/mg), μ_m is the mass attenuation coefficient (cm^2/mg) of the radioactive source material, t_m is the mass thickness (mg/cm^2) of the radioactive source, and E_{max} is the maximum beta particle energy of the radioisotope in MeV. The variation in actual and apparent activity with mass thickness of Ni-63 is shown in Fig. 11.18 for a surface area of 0.05×0.05 cm^2.

The radioisotope mass thickness (g/cm^2) is an important parameter for battery design because the surface activity (mCi/cm^2) of a radioisotope increases initially with an increase in mass thickness (g/cm^2) but then saturates due to the self-absorption effect, as shown in Fig. 11.18. For Ni-63, the difference in actual and apparent activity becomes 32% at a mass thickness of 1 mg/cm^2 in this case. However, a mass thickness of 1 mg/cm^2 results in an activity that is 56% of the saturation value with a mass reduction of 86%. This

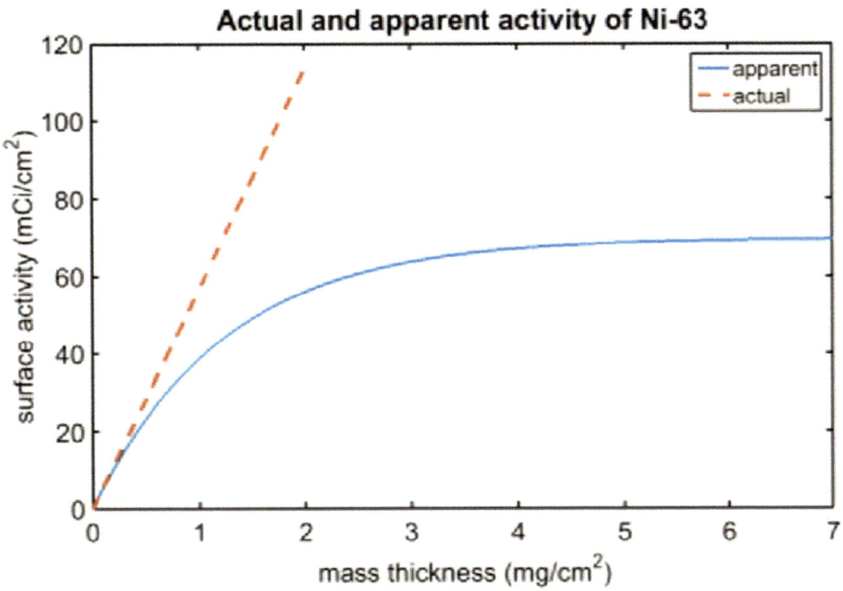

Figure 11.18 Plot of actual and apparent activity of Ni-63 versus its mass thickness.

analysis is useful for designing a radioisotope source with maximum utilization and reduced cost when the surface area of a betavoltaic battery is limited in size. Designing a betavoltaic battery with multiple betavoltaic cells stacked in series or parallel, depending on the application, can be an effective way to reduce the self-shielding effect. However, increasing the surface area of the radioisotope source and semiconductor can be another effective method of betavoltaic battery design to reduce the self-shielding effect. One way to increase the surface area is to modify the surface geometry of the semiconductor.

Aside from a planar surface, there are different surface geometries that can be used with semiconductors to increase the effective surface area, such as using inverted pyramids, a porous surface, or a hydrogenated amorphous material, as shown in Fig. 11.19. They offer a larger surface area for the beta particles to enter the material, resulting in greater energy deposition. A larger surface area will therefore generate higher power in betavoltaics [11].

To take full advantage of the modified semiconductors with an increased surface area, the recesses need to be filled with the radioisotope. However, the use of a solid planar radioisotope without filling the recesses in the semiconductor can still increase the energy conversion efficiency somewhat due to better trapping of the beta particles inside the recesses resulting in less backscattering from the semiconductor surface. Using a liquid state for the radioisotope makes it much easier to fill the recesses of the modified surface geometry semiconductor. An experiment [46] was conducted using a radioisotope source of ^{63}NiCl/HCl solution with a bulk micromachined inverted pyramid Si

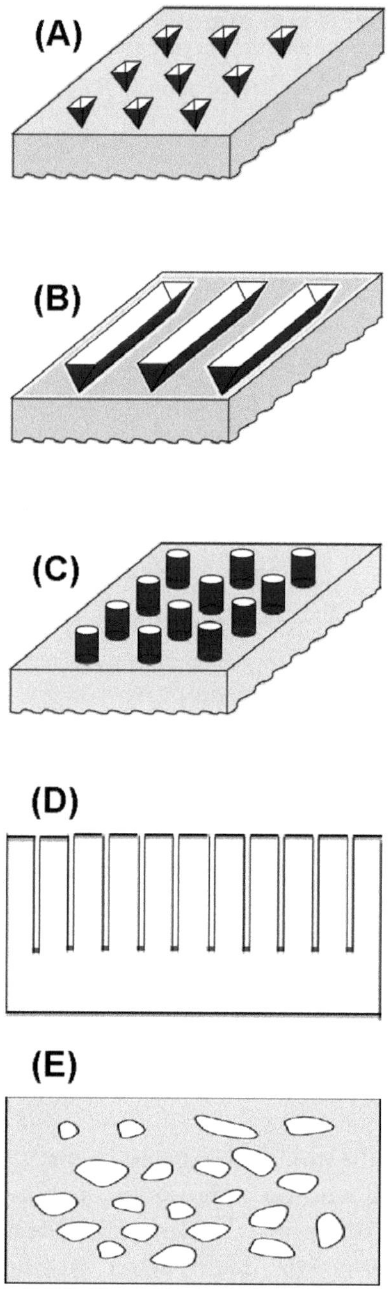

Figure 11.19 Different surface geometries: (A) inverted pyramid, (B) V-channel, (C) cylindrical holes, (D) porous, and (E) amorphous.

p-n diode, which is similar to the technique used in solar cells. The surface area with the inverted pyramid geometry is increased by a factor of 1.85 compared to a planar surface. In comparison with the planar Si *p-n* diode for the same amount of radioactivity, the short circuit current, open circuit voltage, and power output for the inverted pyramid array *p-n* Si diode increased by 18.67%, 11.3%, and 33.33% respectively.

Pores that are on the scale of a nanometer act like one-dimensional wells for electrons. This mechanism increases the existing bandgap of the semiconductor by $E_g + \Delta E$ ($\Delta E = \Delta E_c + \Delta E_v$). For example, the introduction of pores in Si increases its bandgap energy to 2 eV from 1.12 eV [51]. By trapping electrons in the well, it will reduce the backscattering effect resulting in a higher power output with a higher conversion efficiency. As an example, the maximum power output was increased by 42% using porous Si in comparison to using nonporous Si, where the nano-pores protruded into the depletion region [42]. A mathematical model developed by Zhao et al. [52] for the energy conversion of a solid radioisotope by a three-dimensional *p-n* Si diode showed that the short circuit current increased with an increase in the cylindrical hole-radius. However, the open circuit voltage decreased with an increase in cylindrical hole-radius, and the net power decreased compared with the two-dimensional model with no holes. The cylindrical holes on the surface of the *p-n* diode did not improve the power due to an increase in the dark current. However, a simulated semiconductor [19], which had 10^6 cylindrical holes with a depth-to-diameter ratio of 10:1 on the surface with Ni-63 embedded in the holes, showed an 18.15 times increase in energy deposition for only a five times magnification of the effective surface area. A decrease in the cylindrical hole diameter further increased the energy deposition. A cylindrical surface area was found to be more efficient than the inverted pyramid and planar surface area configurations. The power density for a cylindrical *p-n* junction also increases with the height-to-diameter ratio (HDR) of the holes. However, the height of a cylinder is limited by the thickness of any MEMS components, such as less than 1 mm. A maximum HDR of about 1150 was obtained [29] within this limitation for Ni-63 with a *p-n* junction of GaN using Monte Carlo N-Particle (MCNP) is a transport code analysis with a cylinder height and diameter of 970 and 0.8342 μm respectively. Besides different shapes of a crystalline semiconductor, a noncrystalline semiconductor like amorphous silicon (a-Si), which is a low-cost thin-film semiconductor, can increase the conversion efficiency by 1.2%, as shown in an experimental work by Deus [48]. Amorphous silicon is expected to have a higher conversion efficiency due to a low thickness and high intrinsic shunt resistance.

The penetration depth of the beta particles in a semiconductor determines the effective thickness and junction depth needed for the semiconductor material in a battery design. Different methods have been used to calculate the penetration depth of the beta particles in semiconductors, such as MCNP, the Katz-Penfold maximum range equation, the continuous slowing down approximation, the Kanaya-Okayama model, collision and radiative stopping power equations, and empirical equations.

Results, as shown in Table 11.1, vary widely for combinations of radioisotopes and semiconductor materials. The combination of a high-density semiconductor with a low-energy radioisotope will result in a shorter penetration distance for generating all the EHPs [11]. A thin thickness for a semiconductor can meet the design requirement if the density of the material is high [12].

The MCNP code for transport of beta particles can be used to calculate beta particle deposition energy and penetration depth in the semiconductor very accurately due to its high precision [9]. The deposition energy is required to estimate the number of EHPs generated in the semiconductor. The generation rate of EHPs for unit thickness of the semiconductor can be calculated using MCNP by Eq. (11.10) [9]:

$$G(x) = \frac{AE(x)}{E_{EHP}}, \tag{11.10}$$

where A is the activity of the radioisotope, $E(x)$ is the energy deposition estimated by MCNP as a function of the distance into the semiconductor, and E_{EHP} is the average energy required to generate one EHP, which is a property of the material. However, the generation rate of EHPs can be approximated without MCNP by Eq. (11.11) [11]:

$$G = \frac{A(1-f)E_{beta}}{E_{EHP}}, \tag{11.11}$$

where E_{beta} is the single beta particle energy, and f is the backscatter coefficient. Not all beta particles are transmitted into the semiconductor due to backscattering, which is taken into account by the backscatter coefficient f. Aside from backscattering, the generation of EHPs can be affected by the directionality of the radioactive beta particles. Beta particles from the radioisotope travel randomly in all directions. They may escape instead of reaching the semiconductor because of this random directionality. For example, if the radioisotope is placed on one side of the semiconductor, then almost 50% of the beta particles may not contribute in generating EHPs because they are heading in the opposite direction toward the backside of the radioisotope source. This would be another reason to use stacked betavoltaic cells where one can capture the beta particles going in both directions. If the cells are not stacked, some improvement in radioisotope source efficiency can be made by placing a scattering plate on the backside of the source to redirect some of the beta particles toward the semiconductor.

Beta particle energy deposition in the thickness direction of a semiconductor becomes almost insignificant beyond half of the beta particle stopping range in the material. This is because the trajectory of a beta particle is randomly scattered, as shown in Fig. 11.20 (for distance in micrometers). Therefore, very few beta particles penetrate along the thickness direction in a semiconductor material. As a result, the average penetration depth for energetic beta particles is less than the stopping range. In fact, the

Table 11.1 Stopping range of beta particles in semiconductors.

Radioisotope	Semiconductor	Energy deposition (keV)	Penetration depth (μm)	Method	References
Ni-63	Si	80% of total deposition	5	MCNP	[9]
		99.9% of total deposition	20		
Ni-63	Si	17.4	2.2	Katz-Penfold maximum range equation	[31]
Ni-63	Si	17.1	3	Continuous slowing down approximation (CSDA)	[46]
		66.7	43		
Ni-63	Si	17.1	2.68	MATLAB/SIMULINK for ionization and radiation loss theory	[52]
		66.7	27		
Ni-63	Si	17.4	3.76	Theoretical analysis	[3]
Ni-63	GaN	17.4	1.3	MCNP	[13]
		100% full spectrum	13		
		95% full spectrum	8		
Ni-63	GaN	17.1	1.52	MCNP	[53]
Ni-63	GaN	17	1.136	MCNP	[29]
		67	16.2		
Ni-63	GaN	17.4	1.5	SRIM	[44]
Ni-63	GaN (Schottky)	8.3	1.7	SRIM	[20]
Pm-147	GaN	62	11.8	MCNP	[10]
		224.7	100		
		62	9.5	Katz and Penfold	
		224.7	81.9		
Ni-63	SiC	17	3	Empirical equation	[23]
Ni-63	4H−SiC		3		[32]
Ni-63	4H−SiC (Schottky)	17.1	2	Katz and Penfold	[37]
Ni-63	GaAs	80% of energy deposition	5	Kanaya-Okayama (K−O) model	[40]
		100% of energy deposition	15		
Ni-63	Air	17.1	6000	Calculated by ionization and radiation loss theory	[30]
	Ni		0.73		
	Cr		2.52		
	Au		0.58		
Ni-63	Most solids	17.3	<10		[54]
H-3	Si	5.7	0.2		[17,55]
H-3	a-Si:H		0.5		[48]
S-35	Se (liquid)	49	50		[50]

Sources are given in table.

rule of thumb is that the penetration depth is about half of the stopping range [56]. In addition, the majority of the beta particle energy for low-energy radioisotopes such as Ni-63 is accumulated near the surface of a semiconductor material, and it becomes negligible beyond the penetration depth [10]. An MCNP calculation is a more precise method to calculate the penetration depth. Penetration depth changes with different beta particle energies. It is to be noted that the beta particles are emitted in a continuous energy spectrum during radioactive beta decay, as shown in Fig. 11.21. Using MCNP, a full beta particle energy spectrum can be considered instead of only monoenergetic beta particles. Therefore, a full spectrum analysis of beta particle energy using MCNP is important to get a better estimate of the penetration depth.

The radioisotope is the power source that drives the radioisotope battery. Betavoltaics use beta-emitting radioisotopes to create the battery mechanism. It is a generic practice to select a radioisotope for betavoltaics that has a maximum beta particle energy less than 250 keV. This energy value is lower than most of the electron radiation damage threshold values of the semiconductors. For example, the threshold beta particle energy for radiation damage to a Si crystalline structure is about 250 keV [9,57]. However, beta emitters with higher energy can be used for wide-bandgap semiconductors. For example, beta emitters with a maximum energy of 440 keV can be used with GaN due to its high threshold energy for radiation damage [45].

Safe handling of the radioisotope would be required when making the batteries to avoid skin penetration from high-energy beta particles. Usually, the threshold energy of beta particles for skin penetration is about 70−150 keV [58,59]. Nevertheless, eyes are very sensitive to beta particles even at much lower energies. Wearing safety glasses and protective shielding clothes should be considered a safety precaution for handling radioisotopes even at lower energies. Some of the beta emitters that were identified in the literature for betavoltaic applications [10,21,42] are listed in Table 11.2 in order of low to high average beta particle energy. When two isotopes are listed in the same row, the first

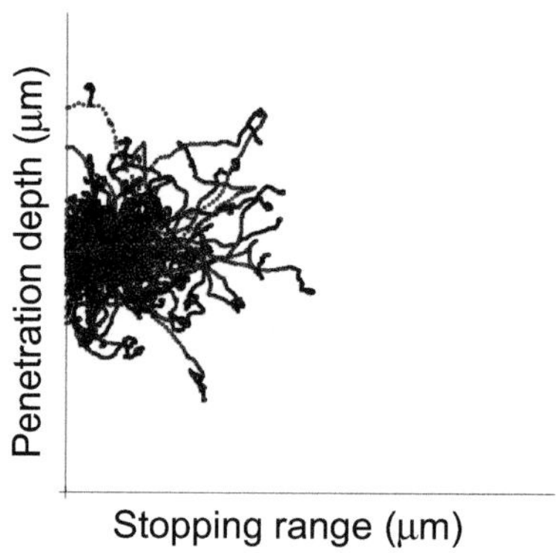

Figure 11.20 Stopping range versus penetration depth for beta particles (in micrometers).

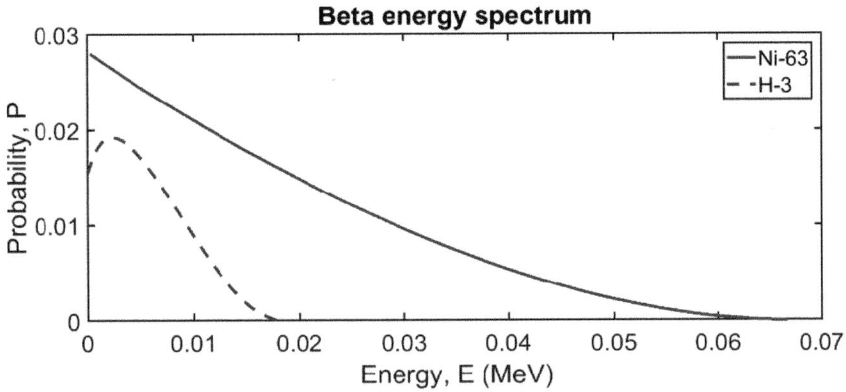

Figure 11.21 Full beta energy spectrum of Ni-63 and H-3.

isotope is the parent and the second one is the radioactive daughter isotope. Otherwise, the daughter decay products are nonradioactive.

To increase the betavoltaic power output, a radioisotope with high specific power should be selected. The average specific power is directly proportional to the average energy of the beta particles emitted but inversely proportional to the half-life of the radioisotope. Selection of half-life depends on the service life requirement of the battery application. Service life can be longer than the half-life. For example, the service life can be three times the half-life if the application load requires only 12.5% of the battery's initial power output. It is ideal in betavoltaic battery design to choose a radioisotope with medium or high half-life and high beta energy. However, very high beta energy can

Table 11.2 Radioisotopes listed in the literature.

β-emitter	Half-life	Average energy (keV)	Maximum energy (keV)	Average specific power (mW/g)
[a]H-3	12.32 y	5.69	18.59	324.914
[a]Ru-106	1.02 y	10.03	39.4	196.948
[a]Ni-63	100.2 y	17.42	66.94	5.796
[a]S-35	87.37 d	48.76	167.33	12339.148
Pm-147	2.62 y	61.93	224.6	340.367
[a]P-33	25.35 d	76.43	248.5	70701.623
Ca-45	162.61 d	76.86	255.8	8129.21
Cs-137	30.08 y	187.1	1175.63	96.241
[a]Sr-90	28.79 y	195.8	546	160.238
Y-90	64 h	933.6	2280.1	3.011E+06
Tl-204	3.783 y	244.05	763.76	669.977
Kr-85	10.76 y	250.7	687.4	581.465

[a]Pure beta emitters; no X-rays or gamma rays are emitted.
Source: *National Nuclear Data Center, Brookhaven National Laboratory (https://www.nndc.bnl.gov/), accessed October 1, 2021.*

cause radiation damage in semiconductors. The maximum beta energy needs to be below the threshold energy of radiation damage for the selected semiconductor. Not all the radioisotopes in Table 11.2 are an ideal choice for betavoltaic battery design. Only H-3, Ni-63, S-35, Pm-147, P-33, and Sr-90 have been used in the manufacture of betavoltaic battery devices. Ni-63 is the most widely used radioisotope for betavoltaic batteries and is a pure beta emitter. However, it has a very low specific power of 5.8 mW/g compared with the other beta emitters. The specific power of Ni-63 is very low because of its long half-life and low average beta energy. Both S-35 and P-33 have very short half-lives. S-35 was only used with liquid semiconductors because of its compatibility. H-3 has a good specific power of about 325 mW/g with a half-life of 12.3 years. However, it is a gas and hard to handle as well as difficult to obtain in high concentrations, which gives it a low effective specific activity. In general, it is used in the form of a tritiated metal to avoid low gas density and radiation hazards from a gas leak. Sr-90 is a pure beta emitter that decays to Y-90, which is also a beta emitter. However, Y-90 has a very short half-life of 64 h with a very high average beta energy of 933.6 keV. This high energy for its beta particles may cause significant radiation damage to the semiconductors. Pm-147 has a reasonable half-life of 2.6 years with a specific power of 347 mW/g. It seems like a very good radioisotope candidate for betavoltaic batteries, and its maximum beta energy (224.6 keV) is low enough to minimize radiation damage. Pm-147 does give off some low-energy gamma rays with low probability that do not impact the battery design. Thus, use of Pm-147 will primarily be based on if it meets the required service life of the battery application.

Some beta emitters like Sr-90 and Si-32 decay into a daughter, which also emits beta particles. In most cases, the daughter radioisotope has a much shorter half-life and a much higher average beta particle energy than that of its parent. In this particular case, the activity of the daughter will already be in secular equilibrium with the parent. In secular equilibrium, the average specific power of a mixture becomes higher than that of the parent radioisotope alone. The average specific power also increases due to the high average energy and short half-life of the daughter radioisotope with a similar mass to the parent radioisotope. This could be an advantage for increasing the power output of a betavoltaic battery. However, the resulting high energy of the daughter radioisotope beta particles can cause significant damage to the semiconductor materials. Along with a high average energy beta particle, some daughter radioisotopes also emit high-energy X-rays and gamma rays. This type of radioisotope either needs to be avoided or requires special shielding.

The following list of radioisotopes from the National Nuclear Data Center (NNDC) was generated to investigate other possible beta emitters not considered in the literature with a maximum energy of less than 2000 keV. Table 11.3 lists these other possible radioisotopes that might potentially be used in betavoltaic batteries. This table does not take into account the feasibility of producing these isotopes or their availability. The radioisotopes are listed in order of increasing average beta particle energy of the parent isotope.

Many of the beta emitters in Table 11.3 have half-lives less than a year. Ar-42 has a half-life 12 times higher than that of Pm-147 and is a pure beta emitter. It has also a high specific power of 357 mW/g, which is more than that of Pm-147. However, it is a noble gas and cannot be used in a solid state. In addition, it is in secular equilibrium with its daughter K-42. However, K-42 emits very high energy beta particles. Thus, this will rule out using Ar-42. Ce-144 also looks like a good choice with a high specific power of

Table 11.3 Other potential radioisotopes for betavoltaic batteries.

β-emitter	Half-life	Average energy (keV)	Maximum energy (keV)	Average specific power (mW/g)
Os-194	6 y	16	96.6	29.12
Ir-194	19.28 h	800	2233.8	3.97E+06
Sm-151	90 y	19.63	76.6	3.06
Tm-171	1.92 y	24.77	96.4	159.88
Eu-155	4.753 y	47	252.7	135.21
Th-234	24.1 d	47.9	273	6566.09
[a]C-14	5700 y	49.47	156.48	1.31
Hg-203	46.594 d	57.87	492.1	4731.19
Ru-103	39.247 d	63.8	763.4	12213.97
[a]Si-32	153 y	69.55	227.2	30.12
[a]P-32	14.27 d	695.03	1710.66	11.78E+05
Ce-144	284.91 d	82.1	318.7	1548.17
Pr-144	17.28 m	1208	2997.5	5.41E+08
Sb-125	2.76 y	86.7	766.7	533.025
W-188	69.78 d	99	349	5836.159
Re-188	17.00 h	763	2120.4	4.43E+06
Zr-95	64.03 d	117	1123.6	14885.75
Nb-95	34.99 d	43.43	925.6	10111.62
Fe-59	44.50 d	118	1565.2	34792.31
Hf-181	42.39 d	121	1029.8	12196.92
W-185	75.1 d	126.9	432.5	7063.89
Ce-141	32.51 d	145.3	580.7	24525.83
Tb-160	72.3 d	210	1835.1	14042.47
[a]Ar-39	269 y	218.8	565	44.22
[a]Ar-42	32.9 y	233	599	357.507
K-42	12.36 h	1430.5	3525.45	5.118E+07
Tm-170	128.6 d	317	968	11215.49
Sn-123	129.2 d	523.1	1403.6	25470.35
Sr-89	50.56 d	587.1	1500.9	100977.80
Y-91	58.51 d	603	1544.3	87654.31

[a]Pure beta emitters; no X-rays or gamma rays are emitted.
Source: *National Nuclear Data Center, Brookhaven National Laboratory (https://www.nndc.bnl.gov/), accessed October 1, 2021.*

1565 mW/g if the application only needs a service life of 9—30 months. But it also has a daughter in secular equilibrium that emits a very high energy beta particle. Other radioisotopes that have stable daughters that may be possible candidates for betavoltaic batteries are Tm-171, Eu-155, and Sb-125.

In conclusion, the self-shielding effect of the radioisotope needs to be considered in betavoltaic battery design. The self-shielding effect can be reduced by increasing the surface area of the radioisotope, which will increase the power output of the battery for the same amount of radioactivity. A modified surface geometry of the semiconductor such as inverted pyramid and cylindrical holes is useful to increase the power output. In addition, the analysis of penetration depth of beta particles in the semiconductor is very important for betavoltaic battery design. An MCNP analysis using the full beta energy spectrum of the radioisotope will provide a better estimate of the penetration depth. Furthermore, selection of a radioisotope for betavoltaic batteries is an important choice to make [11]. The most important factor for selection of a radioisotope is the power output requirement [12]. For example, an average energy deposition of 3.3 eV will generate one EHP in Si. When Ni-63 is used as a radioisotope source with Si, it generates almost 5000 EHPs per average beta particle energy of 16.7 keV. Similarly, Pm-147 generates 19,000 EHPs per average beta particle energy of 62 keV. From this, it is predicted that the maximum power output using Pm-147 is more than three times higher than that of using Ni-63 with the same amount of radioactivity [42]. Other factors for the selection of the radioisotope involving more tradeoffs are the application service life, radiation damage, cost, availability, and shielding. Factors like the beta particle energy spectrum, the penetration depth of the particles in the semiconductor, the specific activity, and the radioisotope half-life determine the performance of the betavoltaics. Instead of using a real radioisotope in experiments, some researchers have used monoenergetic electrons produced by a scanning electron microscope (SEM) [32,54,60]. This technique is useful for testing battery design, such as optimizing the junction depth and the depletion region width of a battery by checking the output [31]. However, an actual radioisotope is needed to characterize actual conversion efficiencies and power outputs.

11.7 Betavoltaic batteries: Results and analysis

A summary of betavoltaic battery results will be useful to understand the state-of-the art technology. The performance of a betavoltaic battery is measured by its open circuit voltage, short circuit current, maximum power output, efficiency, and fill factor. The performance metrics for both simulated and experimental results of betavoltaic batteries are summarized in Table 11.4 by type of semiconductor material. A range of minimum to maximum values is then provided for each performance metric. Although the short circuit current and the maximum power output in Table 11.4 are shown in terms of nA/cm^2 and nW/cm^2, respectively, most of the batteries have sizes on the scale of few mm^2 or few μm^2. Some researchers reported short circuit current and maximum power output on the scale of nA/cm^2 and nW/cm^2 respectively, whereas others reported the net short circuit current and maximum power output in terms of nA or μA and nW or μW respectively. The net short circuit current and maximum power output were then converted to units of nA/cm^2 and nW/cm^2, respectively, using the reported sizes of the batteries for the sake of comparison. Some researchers provided the net short circuit current and maximum power output without reporting the size of the battery.

For these results, it was difficult to do a comparison, and they were not included in Table 11.4. However, they will be included in the results discussion. The betavoltaic battery results for net short circuit current (nA) and maximum power output (nW) are much lower than that of their short circuit current density (nA/cm^2) and maximum power output (nW/cm^2) due to their size often being much less than a square centimeter. In addition, the net results are low as they are reported for single cells only to do the comparison. However, the open circuit voltage, short circuit current, and maximum power output will increase when multiple cells are used in series or parallel for specific applications. The efficiency and fill factor will not be changed for multiple cells.

Table 11.4 Summary of results from the literature, discussed in Section 11.7.

Semiconductor material	Open circuit voltage (V)	Short circuit current (nA/cm^2)	Maximum power output (nW/cm^2)	Overall efficiency (%)	Fill factor
Si	0.25 to 0.54	3.92 to 3200	1.04 to 500	4.94 to 13.7	0.79 to 0.81
[a]Si	0.0008 to 0.46; mostly 0.12	4.44 to 1490; mostly below 55	0.12 to 4.6	0.01 to 1.75 mostly below 1	0.27 to 0.83
GaN	0.46 to 3.16; mostly above 2	7.5 to 5000; mostly below 1600	2.3 to 4927	4.1 to 25.4	0.57 to 0.94
[a]GaN	0.025 to 1.65	1.2 to 89.2; mostly below 20	1.75 to 14.26	0.21 to 1.13	0.14 to 0.55
4H–SiC	0.72 to 2.1	16.8 to 24.31	6.17	14.5	0.51
[a]4H–SiC	0.26 to 0.49	3.46 to 29.44	0.513 to 4.85	<0.5 to 1.2	0.34 to 0.59
[a]SiC	0.95 to 2.06	5.25 to 2300; mostly below 31	3.5 to 2400; mostly below 10	0.52 to 1.99	0.47 to 0.75
[a]a-Si:H	0.02 to 0.46	5.3 to 637	0.26 to 136	0.1	0.59
[a]GaAs	0.28 to 0.32	0.3 to 28	0.05 to 0.07		0.58 to 0.61
[a]Se (liquid)	0.34	102.4	15.58	1.19	
[a]AlGaAs	0.47	937	259		

[a]Experimental results.

Only overall conversion efficiencies are considered for comparison among different types of efficiencies. The overall conversion efficiency of a betavoltaic battery is calculated by Eq. (11.12):

$$\eta = \frac{P_{max}}{P_{theory}} = \frac{FFI_{sc}V_{oc}}{1.6 \times 10^{-19} \times 3.7 \times 10^{7} AE_{avg}}, \tag{11.12}$$

where P_{max} is the maximum power output of a betavoltaic battery, and P_{theory} is the theoretical maximum thermal power of the radioisotope used in the battery, the activity A is in mCi, and the average beta particle energy E_{avg} is in eV. The overall conversion efficiency can also be defined by Eq. (11.13):

$$\eta = \eta_s \times \eta_c \quad , \tag{11.13}$$

where η_s and η_c are the source efficiency and semiconductor conversion efficiency, respectively. Source efficiency is defined by the ratio of the energy deposited in the semiconductor to the actual energy of the radioisotope. It takes into account the self-shielding effect of the radioisotope, all the geometry factors and losses including the fraction of beta particles that are not directed toward the semiconductor. The semiconductor efficiency is determined by the maximum power output of the battery divided by the rate of the energy deposited in the semiconductor. It takes into account recombination and heat losses within the semiconductor. Table 11.4 shows the overall conversion efficiency values for betavoltaic batteries. It is noted that the semiconductors were placed only on one side of the radioisotopes for making betavoltaic batteries, so almost 50% of the beta particles did not reach the semiconductors. It is noted that the reported efficiencies could be doubled if the radioisotopes were sandwiched by the semiconductors. Thus, a proper overall conversion efficiency calculation for a single cell should only use half the radioisotope activity.

Results for different semiconductor materials are listed in Table 11.4. Note that 4H−SiC and SiC are shown as different categories since that is how they were reported in the literature. However, 4H−SiC is a polytype of SiC, and n-type 4H−SiC substrates were used to fabricate different junctions of SiC. To summarize all the experimental results from Table 11.4, it can be seen that the ranges of open circuit voltage, short circuit current, maximum power output, efficiency, and fill factor from experimental results are 0.01−2.06 V, 0.3−2300 nA/cm^2, 0.05−2400 nW/cm^2, 0.01%−1.99%, and 0.14 to 0.83, respectively.

The betavoltaic battery performance results listed in Table 11.4 cannot be directly compared to each other. It shows the range of results obtained for different semiconductors, different radioisotopes, and different activities used in betavoltaic battery design. The battery performance does not depend only on the semiconductor material but also on the type of junction, the radioisotope, and the amount of radioactivity, which were not incorporated in Table 11.4 to keep it readable. However, some key results can still be highlighted from Table 11.4. It shows that the theoretical models (those rows without the ‡ symbol) tend to overpredict the experimental results. For example, the higher ranges of open circuit voltage and short circuit current for Si are 0.54 V and 3200 nA/cm^2, respectively, whereas the open circuit voltage and short circuit current from experiments are mostly below 0.12 V and 55 nA/cm^2, respectively. It also shows that wide-bandgap semiconductors such as GaN and SiC provide higher open circuit voltage and efficiency compared with narrow-bandgap semiconductors such as Si. However, the short circuit currents from narrow-bandgap semiconductors are higher than that of wide-bandgap semiconductors. Despite the lower short circuit current, the power output is still generally higher for the wide-bandgap semiconductors.

A comparison of bandgaps for different semiconductor materials used in betavoltaic batteries is shown in Table 11.5. Table 11.5 shows that Si is a narrow-bandgap semiconductor with a bandgap of 1.12 eV, whereas GaN is a wide-bandgap semiconductor with a bandgap energy of 3.4 eV. A brief summary and comparison of results are given in the following subsections based on the semiconductor material in the betavoltaic batteries.

11.7.1 Si betavoltaic battery

The open circuit voltage, short circuit current density, power output density, and efficiency from one MCNP simulation for a Si semiconductor were about 0.25 V, 600 nA/cm^2, 100 nW/cm^2, and 5%, respectively [9]. However, the experimental results for a Si semiconductor were much less than the simulation results. For experimental results, the reported power output range was from 0.0025 nW [31] to 44 nW [42]. Some power output results were reported in the form of power density, which is good for a comparative analysis. The power density ranged from 0.25 nW/cm^2 [46] to 4.6 nW/cm^2 [17]. Similarly, short circuit current ranged from 1.31 nA [19] to 550 nA [42], and some reported current densities ranged from 4.44 nA/cm^2 [46] to 1.49 μA/cm^2 [17]. A wide range of open circuit voltages was reported from 0.0008 V [31] to 0.46 V [27]. The conversion efficiency ranged from 0.01% [27] to 1.75% [42]. The fill factor was reported only for a temperature analysis. In that case, the fill factor varied from 0.27 to 0.83 [27]

Table 11.5 Bandgap of different semiconductor materials; literature source included with value.

Semiconductor materials	Si	GaN	4H−SiC	a-Si:H	GaAs	Se	AlGaAs
Bandgap (eV)[a]	1.12 [10]	3.4 [10]	3.3 [32]	1.75 [48]	1.42 [61]	1.77 [50]	1.55−1.97 [61]

[a]At room temperature.

with changes in temperature. Overall, the open circuit voltage, short circuit current density, and efficiency of Si betavoltaic batteries from the experimental results were about 0.1 V, less than 60 nA/cm^2, which was mostly around 4 nA/cm^2 to 30 nA/cm^2, and mostly below 1%, respectively. The maximum power density for a Si semiconductor betavoltaic obtained from experimental results was 4.6 nW/cm^2.

Most of the betavoltaic battery research was carried out on Si semiconductors, as Si is readily available and well-studied. It is widely used in the Integrated Circuit (IC) industry and in MEMS with developed micromachining technology [31]. Si is also easy to manufacture with relatively lower cost than other semiconductors. Different types of radioisotopes such as Ni-63, Pm-147, and H-3 were used with the Si semiconductors. Most of the reported Si betavoltaic batteries were planar *p-n* junctions. Different surface geometries such as porous, cylindrical holes, and inverted pyramid were investigated for Si semiconductors to increase the power output. Porous and inverted pyramid surfaces were analyzed experimentally [17,42,46], whereas a surface with cylindrical holes was analyzed analytically [52]. Using a porous surface instead of a planar surface magnified the surface area, which increased the entrapment of beta particles due to a decrease in the backscattering effect. As a result, the power output of the betavoltaic battery with a porous surface with a *p-n* junction improved significantly [42] due to an improvement in open circuit voltage and short circuit current for the same type of solid radioisotope. An improvement in the power output was also observed for an inverted pyramid array surface with a liquid radioisotope [46] that had the potential to take full advantage of the surface magnification. An inverted pyramid array Si *p-n* junction semiconductor with liquid Ni-63 was reported [19] to increase the efficiency compared to a planar Si p-i-n junction semiconductor from 0.042% to 1%. Using a porous surface or inverted pyramids with an increased surface area makes it possible to put the radioisotope into the pores. This tends to improve the efficiency and power output of the betavoltaic battery due to the increased surface area of the source as discussed in Section 11.6. In addition, higher energy and source activity for the beta particles tend to give more power. However, this does not take into account possible radiation damage effects. Porous Si *p-n* junctions were used with Pm-147 [42] and H-3 [17]. Compared with a nonporous Si p-n junction in an experiment using Pm-147, the porous Si *p-n* junction increased the power output by 13 nW (42%) with an increase in open circuit voltage and short circuit current of 0.01 V (9%) and 0.14 μA (34%), respectively [42]. The battery performance for porous Si *p-n* junction improved due to the increase in bandgap energy with the introduction of nano-pores and better absorption of beta particles with reduced back-scattering. The beta particle energy of Pm-147 is higher than that of H-3. A 1.53% greater efficiency was reported [42,62] for a porous Si *p-n* junction with Pm-147 than that of an H-3 betavoltaic battery as the pores do not have better retention of the H-3 gas. However, in another experiment [17], tritium was used in three different forms: gas, surface-occluded, and solid scandium tritide (ScT). Among the three different forms of tritium sources, gaseous tritium at 0.8 atm pressure with a porous Si *p-n* junction provided the highest power output density of 4.6 nW/cm^2. The efficiency for the gaseous tritium source was 0.09%, which is 0.04% greater than the next best ScT source.

Betavoltaic battery performance is also affected by temperature. A temperature analysis showed [27] that the open circuit voltage, the short circuit current, the maximum power output, the efficiency, and the fill factor for betavoltaic batteries made of Ni-63 with a *p-n* junction of Si improved by a significant margin at a very low temperature due to the reduction in leakage current. This analysis shows that the battery performance degrades at higher temperatures, which may result from the decay heat and conversion losses. Hence, multiple cell batteries, passive cooling mechanisms may need to be incorporated in the design to minimize degradation in performance due to temperature effects.

A very wide range of betavoltaic battery performance for crystalline Si semiconductors resulted from different types and quality of semiconductor junctions, various radioisotope sources with different amounts of radioactivity, and different design parameters for batteries. However, an increased surface area such as nano-pores or inverted pyramid array of Si would be a better design due to their better performance in betavoltaic batteries.

11.7.2 GaN betavoltaic battery

Gallium nitride is a wide-bandgap semiconductor with a bandgap of 3.4 eV that attracted the second-most attention after that of Si semiconductors by the researchers. Overall, GaN betavoltaic batteries showed a much higher open circuit voltage and a better efficiency than that of the planar Si betavoltaic batteries. The open circuit voltage, short circuit current density, and efficiency for non-Schottky junction GaN betavoltaic batteries using MCNP simulation were about 2 V [11] to 3 V [10], 1 $\mu A/cm^2$ [12] to 1.6 $\mu A/cm^2$ [10], and 13% [10], respectively, where a non-Schottky junction refers to other types of junctions such as *p-n*, *p-u-n*, and *p-i-n*. One analytical result reported [11] very high efficiency of about 25% for GaN. From MCNP simulations, the reported short circuit currents were on the scale of a microampere except for a Schottky junction, which was on the scale of a nanoampere. Besides Ni-63, some other radioisotopes such as Pm-147 and H-3 were used in the simulations. However, Ni-63 was the only radioisotope used in the experiments for GaN with a radioactivity density of 3–20 mCi/cm^2. Unlike in the simulation, the actual experimental results using GaN suffered from poor output and efficiency. In fact, most of the experimental results were not optimized. Several experimental results for GaN betavoltaic batteries are listed in Table 11.6.

Some researchers [13,20] focused on using a Schottky junction for GaN betavoltaic batteries due to the fabrication limitation of a thick, high-quality junction in GaN semiconductors. A Schottky junction for a GaN betavoltaic battery is easy to fabricate with an n-type GaN material and a metal electrode. However, it generates much lower output than that of a non-Schottky junction GaN semiconductor. The open circuit voltage, short circuit current density, and efficiency for a Schottky junction GaN betavoltaic battery using MCNP and semiconductor simulation were about 0.5 V, 7.5 nA/cm^2, and 2.25%, respectively [13]. However, the open circuit voltage, short circuit current density, and efficiency obtained from experimental results for Schottky junction GaN betavoltaic batteries were about 0.1 V, 1.2 nA/cm^2, and less than 0.32%, respectively as shown in Table 11.6. The actual device results were much lower than that predicted from simulation. Most of the beta particles were absorbed and reflected in the dead layer, which is the Schottky electrode. The experimental results also suffered from a very small depletion layer width of 0.07 μm [20]. Nevertheless the Schottky junction GaN betavoltaic battery results for open circuit voltage and efficiency were comparable to that of planar Si betavoltaic batteries. However, the short circuit current density from the Schottky junction GaN betavoltaic battery was lower than that for a Si betavoltaic battery.

Some researchers were able to manufacture *p-n* and *p-i-n* junctions of GaN semiconductors for betavoltaic batteries as shown in Table 11.6. The open circuit voltage, short circuit current density, and efficiency for non-Schottky junction GaN betavoltaic batteries from experimental results were about 1.6 V, less than 20 nA/cm^2, and below 1.2%, respectively. Cheng et al. [63] reported a low open circuit voltage of 0.025 V and a short circuit current of 2 nA/cm^2 for a betavoltaic battery made of a GaN *p-n* junction with Ni-63 and compared experimental results with simulation results. They showed discrepancies between the two results of about 95% for both open circuit voltage and short circuit current. The experimental results were much worse than the optimized simulation results for several reasons. First, an actual junction depth of only 0.4 μm was achieved during GaN fabrication, but it was 1.3 μm in the simulation. Second, the fabricated GaN semiconductor had a cross-sectional area of 1 cm^2, which was large

Table 11.6 Some GaN experimental results

Semiconductor (junction type)	Radioactivity (mCi/cm^2)	Open circuit voltage (V)	Short circuit current (nA/cm^2)	Maximum power output (nW/cm^2)	Overall efficiency (%)	Fill factor
GaN (Schottky) [20]	Ni-63 (3)	0.1	1.2		<0.32	
GaN (*p-n*) [63]	Ni-63 (11)	0.025	2		<1	
GaN (*p-i-n*) [53]	Ni-63 (12.5)	0.47	17	2.72	0.21	0.34
GaN (*p-i-n*) [64]	Ni-63 (12.5)	1.65	12	10.69	0.82	0.54

Source: References given in the table.

enough to generate crystalline defects in the fabrication process and contributed to a large recombination current. Finally, the poor crystalline quality of the p-type region of the GaN semiconductor increased the leakage current. The open circuit voltage was then improved to 0.47 V [53] for a GaN p-i-n junction as shown in Table 11.6. Although it was called a p-i-n junction, it was actually a p-u-n junction. A u-type region in GaN is an unintentionally doped region with a high background electron carrier concentration (like an n-type region) due to impurities. However, the introduction of a u-type region of 0.3 μm increased the depletion width, which in turn improved the battery output. It was further improved to 1.65 V [64] by introducing an intrinsic region of 0.6 μm and creating a p-i-n junction. The p-type, i-type, and n-type regions of a GaN were created by doping with Mg, Fe, and Si respectively. It is still challenging to grow a thick p-type region using GaN. A non-Schottky junction GaN betavoltaic battery [64] showed much higher open circuit voltage than a Si betavoltaic battery [19], but the short circuit current density was lower, for Ni-63 with equivalent activity. However, the power density of a non-Schottky junction GaN battery was almost two times higher than the maximum power density for a Si betavoltaic battery. The p-i-n junction for GaN betavoltaic batteries showed better battery performance when compared with other types of junctions for GaN. The maximum efficiency of a GaN betavoltaic battery was 1.13% [43], which is much lower than its theoretical efficiency due to fabrication limitations.

11.7.3 SiC betavoltaic battery

Hexagonal (4H)−SiC is a hexagonal polytype of SiC that has a bandgap of 3.3 eV. The availability of high-quality substrates and the advancement in epitaxial growth technology has made 4H−SiC a better candidate for betavoltaics. 4H−SiC is a semiconductor material that has one of the best qualities for electronic device fabrication [23]. It has some good properties such as high thermal conductivity, temperature stability, radiation hardness, and good electronic mobility that are useful in betavoltaics [32]. A Schottky barrier of 4H−SiC can be a promising design option for betavoltaics [37]. It is less expensive with easier fabrication. It has already been used for detecting solar-blind ultraviolet (UV) with a response only to the UV and not to visible light [23].

Simulation, analytical, and experimental work has been done for both Schottky and *p-n* junctions of 4H−SiC betavoltaic batteries. All of them used Ni-63 as the radioisotope source. A Schottky junction using 4H−SiC [37] showed better performance than that of a Schottky junction GaN betavoltaic battery [20] for the same radioisotope of Ni-63 with almost similar activity. Maximum open circuit voltages of about 1 V [23] and 0.5 V [30] were achieved for Schottky junction 4H−SiC betavoltaic batteries from an SEM beam of electrons and from a Ni-63 radioisotope experimental analysis, respectively. In experimental analysis, it generated a maximum short circuit current density of about 30 nA/cm^2 [30] with an efficiency of about 1% [30,37]. The Schottky junction for 4H−SiC showed an almost similar power density [30] compared to the maximum power density of a Si semiconductor [17]. A *p-n* junction using 4H–SiC [32] showed a slightly higher power density than the maximum power density obtained from a Si semiconductor [17]. The experimental results of 4H–SiC *p-n* junction [32] showed almost 160% greater open circuit voltage than that of a 4H–SiC Schottky junction [37] with about 30% reduction in short circuit current for Ni-63 with almost similar activity. However, a net increase in power density of 50% was found for a 4H–SiC *p-n* junction over that of a Schottky junction. In addition to 4H–SiC betavoltaic batteries, some SiC betavoltaic batteries were reported, as shown in Table 11.7. However, p-type epilayer growth on a 4H–SiC substrate is expensive.

A much larger open circuit voltage of about 2 V was achieved for both *p-i-n* and *p-n* junctions in SiC compared with other semiconductor materials as shown in Table 11.7. Short circuit current densities of 2000 nA/cm^2 [21] and 30 nA/cm^2 [54] were reported corresponding to SiC *p-i-n* and p-n junctions. The reported short circuit current density and power density for a SiC *p-i-n* junction [21] on the scale of $\sim 1\ \mu A/cm^2$ and $\sim 1\ \mu W/cm^2$ respectively are very attractive for betavoltaic battery applications. However, this high short circuit current density was achieved using the P-33 radioisotope that has a very short half-life of about 25 days, whereas the other one was achieved using Ni-63 that has a long half-life of about 100 years. On the other hand, a SiC *p-i-n* junction [21] showed about a 20% increase in open circuit voltage compared to a GaN *p-i-n* junction semiconductor [64]. However, it showed an efficiency of 0.56% [21], which was lower than that of a GaN *p-i-n* junction semiconductor [64]. The battery results for SiC with Ni-63 were not as high as it was when used with P-33, as shown in Table 11.7. In SiC betavoltaic batteries, a wafer substrate with thickness of usually 300–500 μm is used [66] that has too many defects. The battery results improved significantly for a 4H–SiC *p-n* junction semiconductor with Ni-63 [54] compared with a 4H–SiC *p-i-n* junction semiconductor with Ni-63 [65], as shown in Table 11.7, due to substrate thickness down to 50 μm using a wafer-thinning technology.

Comparison of the SiC betavoltaic battery results in Table 11.7 shows that the use of substrate-thinning technology improved the results significantly. The SiC betavoltaic battery results as shown in Table 11.7 are found to be better than the GaN betavoltaic

Table 11.7 Some SiC experimental results.

Semiconductor (junction type)	Radioactivity (mCi/cm^2)	Open circuit voltage (V)	Short circuit current (nA/cm^2)	Maximum power output (nW/cm^2)	Efficiency (%)	Fill factor
SiC (*p-i-n*) [21]	P-33 (820)	2.04	2000	2100	0.56	0.51
SiC (*p-n*) [54]	Ni-63 (1.5)	1.9	30			
SiC (*p-i-n*) [65]	Ni-63 (4)	0.98	12.75	8	1.99	0.64
	(2.4)	0.95	5.25	3.5	1.5	0.75

battery results shown in Table 11.6. It is difficult to compare the results when radioisotopes, activity, and semiconductor junctions are not the same. Further investigation is required to do a better comparison.

11.7.4 GaAs betavoltaic battery

GaAs has a bandgap of 1.42 eV, which is slightly higher than Si. Different types of junctions such as *p-i-n*, p^+-*p*-n^+, and p^+-*p-i-n*-n^+ were investigated for GaAs betavoltaic batteries [40,41]. They generated an open circuit voltage of about 0.3 V [40,41], which is slightly higher than that of a Si betavoltaic battery [42,46]. However, it showed a very wide range of short circuit current density ranging from 0.3 to 28 nA/cm^2, as shown in Table 11.4, that makes it difficult to compare with other betavoltaics. A betavoltaic battery made of a *p-n* junction of AlGaAs with H-3 in the gaseous state [48] showed much better initial results than that of a GaAs betavoltaic battery [40,41]. The AlGaAs battery generated almost one and a half times higher open circuit voltage (0.5 V) than that of the GaAs betavoltaic batteries. The short circuit current density obtained from the AlGaAs battery was almost on the scale of 1 μA/cm^2, and the power density was about 60 times higher than the maximum power density from a Si betavoltaic battery using H-3 in gaseous state [17]. However, the battery performance degraded rapidly due to high diffusion of tritium into the semiconductor that created defects.

11.7.5 Amorphous silicon and other betavoltaic batteries

Besides crystalline betavoltaic batteries, some noncrystalline betavoltaic batteries such as liquid selenium and hydrogenated amorphous silicon (a-Si:H) were used by the researchers. Tritium or a tritiated metal was used as a radioisotope for the noncrystalline semiconductors except for the liquid semiconductor, where a radioactive sulfur (S-35) was used as the radioisotope.

Like crystalline semiconductors, a-Si:H is a semiconductor material with low cost and good optoelectronic properties [55]. The presence of hydrogen in it reduces the dangling bond defects, which in turn reduces the localized states. The experimental work on a-Si:H by Kosteski et al. [55] showed very low open circuit voltage of about 0.02 V and short circuit current less than 1 nA using a tritiated layer of intrinsic semiconductor material. The results were not consistent with the experimental work on a-Si:H by Deus [48]. Although both works were done on a *p-i-n* junction of a-Si:H with the radioisotope H-3, Deus [48] achieved a higher open circuit voltage of 0.46 V and a higher current density of 637 nA/cm^2 with a power density of 136 nW/cm^2. However, the use of the radioisotope H-3 in gaseous form degrades the battery output rapidly. A wide range of open circuit voltage from 0.02 to 0.46 V and a short circuit current density from 5.3 to 637 nA/cm^2 make it difficult to compare hydrogenated amorphous Si semiconductors with other semiconductor materials.

In an experimental analysis [50], a selenium (Se) semiconductor was used with S-35 as a radioisotope source that is compatible with liquid semiconductors. Selenium-sulfur is a liquid semiconductor created by a eutectic mixture of $Se_{65}S_{35}$, which has a reduced melting point of 105 from 221°C. The selenium betavoltaic battery then was operated at 140°C. Liquid betavoltaics produced the open circuit voltage of 0.34 V similar to GaAs [40] but almost 79% and 83% less than that of the experimental p-i-n junction of GaN (1.65 V) with Ni-63 and the p-i-n junction of SiC (2.04 V) with P-33 betavoltaics respectively. The open circuit voltage of the liquid betavoltaic using a Schottky junction was much lower than that of a p-i-n junction of GaN and SiC. However, it generated a short circuit current and power densities of 104.4 nA/cm^2 and 15.58 nW/cm^2, which were much higher than that of GaAs [40,41] and about nine times higher in current density and slightly higher in power density than that of the p-i-n junction GaN with Ni-63 betavoltaic battery [64]. Although it produced better results, S-35 has a short half-life of about 87 days, which is one reason for its high short circuit current and power. Ni-63 and H-3 with a better half-life were used for GaAs and a-Si:H, respectively. On the other hand, liquid betavoltaics have the advantage of embedding the radioisotope with the semiconductor material, which will increase the beta particle absorption. However, the device may require external energy to maintain its high operating temperature of 140°C. A leak-tight package of the device is also required to avoid a hazardous condition.

Aluminum gallium nitride ($Al_{0.7}Ga_{0.3}N$) is a wide-bandgap semiconductor with a bandgap of 5.8 eV [11], which is even much higher than the wide-bandgap semiconductor GaN. In a theoretical analysis, an $Al_{0.7}Ga_{0.3}N$ semiconductor showed very high output with the highest efficiency of 27.4%. However, no $Al_{0.7}Ga_{0.3}N$ betavoltaic battery has been built experimentally.

11.7.6 Experimental results and analysis: conclusions and lessons learned

Analytical and simulation results usually predicted a better output than that obtained in experimental results. Part of that may be that the experimental batteries were not optimized, and part of that may be from fabrication limitations or poor crystal structure due to lattice defects. A betavoltaic battery output depends on the semiconductor material, the type of junction, the radioisotope, the amount of radioactivity, the type of electrodes, and the overall design parameters. A Schottky junction 4H–SiC betavoltaic battery with Ni-63 showed better output than a Schottky junction GaN betavoltaic battery with Ni-63. It also showed higher open circuit voltage than a Si betavoltaic battery. Similarly, a p-i-n junction SiC betavoltaic battery with P-33 showed much higher output compared to other betavoltaic batteries. However, that result is partly due to the short half-life of P-33. From the actual built device, the best open circuit voltage of 2.06 V, the best short circuit current density of 2300 nA/cm^2, and the best maximum power density of 2400 nW/cm^2 were reported for the betavoltaic battery made of a p-i-n junction of SiC using P-33 with an activity of about 230 mCi [21]. On the other hand, the best efficiency of 1.99% was reported for the betavoltaic battery made of a p-i-n junction of SiC using Ni-63 with an activity of 0.16 mCi [65]. However, more analysis is required to validate the aforementioned comparisons and to draw any conclusions.

In conclusion, use of a wide-bandgap semiconductor in a betavoltaic battery increases the open circuit voltage, power output, and efficiency. Higher energy beta particles and larger activity increase the short circuit current. The summary of results in Table 11.4 is not a very direct comparison, but it does give an indication of the wide range in performance of actually built betavoltaic batteries and a comparison with the theoretical models. It suggests that a better theoretical model and better betavoltaic battery performance optimization are needed.

11.8 Recent advances in betavoltaic batteries

The conventional design of betavoltaic batteries consists of a layer of radioisotope coupled with a layer of *p-n* junction semiconductor diode. This type of design is known as planar betavoltaics. Two-layer planar betavoltaic battery design is not efficient. As beta particles emit isotropically from the source, half the percentage of the radioisotope energy cannot be harvested by this design. The experimental results based on two-layer planar betavoltaic design reported power on the order of nanowatts with a very low efficiency. The power generated in the range of a few nanowatts makes it difficult to have wide varieties of applications. To increase the power, a design of multiple alternating layers of radioisotopes and semiconductors was investigated. This type of design can be referred to as three-dimensional betavoltaic battery design. The thicknesses of radioisotope and semiconductor layers and the total number of layers are designed and optimized based on applications. The other three-dimensional betavoltaic battery design is based on increased coupling surfaces of radioisotopes and semiconductors. The coupling surface between a radioisotope and semiconductor can be increased by adding three-dimensional features to the surface such as holes and pillars. The dimensions and the number of holes or pillars are determined by the design and optimization based on battery applications. Furthermore, the increased surface coupling design can be stacked up to increase the power that has the potential to meet the power requirements for many remote applications. The research focus in betavoltaic batteries have been shifted from two-dimensional to three-dimensional design. Recent advancements in three-dimensional betavoltaic battery design for high power output and efficiency are described next.

Bormashov et al. [67] reported the manufacturing of a prototype for 200 single conversion cells based on Schottky diamond diode coupled with approximately 24% pure Ni-63 radioisotope. The maximum power output of 10.63 $\mu W/cm^3$ was obtained from the prototype. The open circuit voltage, short circuit current, and overall efficiency of the prototype battery were 1.02 V, 1.27 μA, and 1.25%, respectively. The open circuit voltage and the efficiency are usually much lower for Schottky diode than the *p-n* junction diode. The low potential barrier and narrow junction region make it difficult to collect all the generated EHPs efficiently in Schottky diode. Furthermore, beta particles energy deposition in the Schottky metal that is needed for the Schottky junction cannot be utilized in this design. Schottky junction diamond was used for the design as the n-type diamond was difficult to fabricate to form a *p-n* junction diamond. Although the theoretical design parameter for the Schottky junction diamond thickness is 5 μm for Ni-63, the thickness of 15 μm was used in the prototype due to the manufacturing limitations. A higher power output of approximately 40 $\mu W/cm^3$ can be achieved by this design with a highly enriched Ni-63 source. A *p-n* junction diamond-based radioisotope battery will further increase the power output and efficiency. Shimaoka et al. [68] reported *p-n* junction diamond for betavoltaic batteries with a very high semiconductor conversion efficiency of 28% based on the electron beam irradiation of 5–20 keV.

A radioisotope source optimization model was developed by Alam et al. [69] for stacking multilayers of radioisotopes and semiconductors to design a three-dimensional betavoltaic battery. The design was demonstrated for various tritiated compounds and semiconductors. The model was validated by experimental results [70] from a planar betavoltaic battery made of titanium tritide with 4H–SiC. A form factor approach was used to optimize the radioisotope source thickness in multilayer volumetric three-dimensional design. The maximum theoretical power outputs of 3.06 and 3.95 mW/cm^3 were estimated from the model for titanium tritide and beryllium tritide with 4H–SiC respectively.

Hogan et al. [71] reported a modeling and optimization for a three-dimensional, pillar-shaped GaN-based betavoltaic battery coupled with $^{63}NiCl_2$ radioisotope solution. GaN is grown by a bottom-up approach that has the advantage of low dislocation density. The depletion region extends into the vertical pillars from the core shell in this three-dimensional structure. The higher surface area along with the larger depletion region volume increases the charge collection and power output. The liquid radioisotope solution provides a better conformal coupling in this design. The semiconductor design parameters such as pillar gap, width, and height were varied, and results were compared to the GaN planar structure. The design parameters were optimized based on beta particles energy deposition using MCNP and charge collection in semiconductor using Technology Computer Aided Design (TCAD). The optimum gap between the pillars was 2 μm and the pillar width was also 2 μm in the design. Power outputs of 7.5 and 11.3 $\mu W/cm^2$ were achieved for 4- and 8-μm pillar heights respectively. The power output increased for 8-μm pillar height as the activity increased by 1.96 times compared with the 4-μm pillar height. Both three-dimensional designs based on 4- and 8-μm pillar heights showed a significant increase in power outputs compared with two-dimensional planar design. Power outputs for three-dimensional pillar designs were 3.75 and 5.65 times higher for 4- and 8-μm pillar heights, respectively, than the two-dimensional planar design. Since the increase in power output for 4-μm pillar height was achieved by approximately half the activity than that of planar design, it was recommended as an economically viable design.

Russo et al. [72] investigated different geometry shapes of three-dimensional pillars and holes such as cylindrical, rectangular, hexagonal, and ridges based on SiC with tritiated nitroxide compounds. The optimum power output of 4.54 mW/cm^3 was obtained for rectangular pillar geometry. Wagner et al. [73] developed an optimization model utilizing Monte Carlo particle transport code and semiconductor charge transport code. They demonstrated a significant increase in power output and efficiency for hexagonal nanowire semiconductors coupled with Ni-63, H-3, and tritiated compounds compared with the traditional planar design. Their analysis showed that wide-bandgap semiconductors may not provide the expected higher power output if the surface recombination velocity of the semiconductor is high. They obtained higher power output of 5.6 $\mu W/cm^2$ for 10-μm nanowire heights of GaAs coupled with tritium.

Table 11.8 shows that the estimated power output for three-dimensional betavoltaic batteries made of tritiated compounds with SiC to be approximately 4 mW/cm^3. However, this has not been demonstrated experimentally yet. The experimental result based on Schottky junction diamond was on the order of microwatts per cubic centimeter. Schottky junctions are expected to provide lower power outputs as beta particles absorbed in the metal electrode. A larger three-dimensional volumetric betavoltaic battery based on *p-n* junction needs to be built and compared with the simulation results to understand the fabrication limitations.

11.9 Principles of betavoltaic battery design

Based on the discussions in previous sections, we propose that the following guidelines be considered when designing a betavoltaic battery.

1. The selection of a radioisotope source for a betavoltaic battery depends on the specific application. It is driven by the service life and specific power requirements of the application.
2. The choice of radioisotope is limited by its availability, its half-life, the average and maximum energy of its beta particles, whether significant gamma rays are emitted during decay, and by its physical form or phase.
3. The maximum beta particle energy of a radioisotope needs to be below the threshold energy for the radiation damage of the chosen semiconductor material.

Table 11.8 Results comparison for three-dimensional volumetric design.

Radioisotopes	Semiconductors	Power output (mW/cm^3)	Method(s)
Ni-63 (24% pure)	Diamond (Schottky)	0.011	Experimental
Titanium tritide	4H–SiC	3.06	Modeling and simulation
Beryllium tritide	4H–SiC	3.95	Modeling and simulation
Tritiated nitroxide	4H–SiC	4.54	Modeling and simulation

4. The self-shielding effect of the radioisotope and the geometry of its interface with the semiconductor need to be considered to maximize the energy deposition in the semiconductor materials. Electrode design needs to be optimized with an interdigit structure and low-Z metal so that backscattering and energy absorption of the beta particles from the radioisotope source in the dead layer (electrode) can be minimized, especially for Schottky junctions.
5. The efficiency of a betavoltaic battery can be increased (almost doubled) by placing the semiconductors on each side of the radioisotope to form a double-cell sandwich.
6. A wide-bandgap semiconductor material has higher radiation tolerance. Therefore, a radioisotope with higher beta particle energy can be chosen for a wide-bandgap semiconductor in betavoltaic battery design. However, the minority carrier diffusion lengths of EHPs in wide-bandgap semiconductor materials are lower, which makes it difficult to collect them.
7. The penetration depth of the beta particles is important to design the junction depth and the depletion region.
8. Use of full beta energy spectrum is important to determine the penetration depth. An MCNP analysis provides a better estimate of the penetration depth and energy deposition due to its high accuracy and precision.
9. The width of the depletion region for a betavoltaic battery design needs to be optimized so that the majority of the beta particles deposit their energy in the depletion region. A wide depletion region is a favorable condition for a betavoltaic battery design. The width of the depletion region depends on the doping concentrations of the semiconductor.
10. A depletion region cannot be made too wide to introduce crystal mismatch, which in turn will increase the leakage current. Furthermore, if the width of the depletion region is much larger than the minority carrier diffusion length, most of the EHPs will not be collected.
11. A lower doping concentration increases the minority carrier diffusion length, which in turn increases charge collection efficiency.

12. Leakage current is dependent on the doping concentration. The leakage current increases with a decrease in doping concentration.
13. The open circuit voltage depends on the leakage current, which in turn depends on the intrinsic carrier concentration. Open circuit voltage decreases with an increase in leakage current and intrinsic carrier concentration.
14. Intrinsic carrier concentration depends on temperature. It increases with an increase in temperature. Open circuit voltage, fill factor, efficiency, and power output decrease with the increase in temperature. However, the temperature effect on short circuit current is negligible. Some type of cooling or low-temperature operation will increase the battery performance. Thus, cell packaging becomes important.
15. A wide-bandgap semiconductor increases the efficiency, the open circuit voltage, and the power output but reduces the short circuit current [74]. However, a narrow-bandgap semiconductor such as silicon is still used due to its wide availability, lower costs, and more developed fabrication and processing techniques.
16. A higher activity for the radioisotope increases the short circuit current.
17. A high-quality crystalline structure for the semiconductor material and surface passivation are important to reduce recombination losses.
18. It is important that the electrodes have a good ohmic contact with low contact resistance.
19. A better theoretical model on EHP transport in the semiconductor will improve the design of actual betavoltaic batteries.
20. Radioisotope source thickness can be optimized by using a Monte Carlo source model [75] to minimize self-shielding effect and maximize power output. Higher surface area for the same source thickness will increase the power output. Higher source thickness can be used for a low-density radioisotope source. However, low-density materials have lower specific activity per unit volume.
21. Surface recombination velocity of a semiconductor is an important design parameter. As most EHPs are generated near the coupling surface of the semiconductor, higher surface recombination velocity reduces the EHP collection significantly [76].

Please note that a number of guidelines conflict with each other. Therefore, specific optimization of parameters is required for each application.

11.10 Betavoltaic batteries for space applications

The advantages of nuclear batteries are higher energy density, long lifetime, and reliability. However, nuclear batteries have low power density and efficiency. The tradeoffs can be balanced by finding applications where benefits outweigh drawbacks. Furthermore, the betavoltaic battery design can be improved and optimized for specific applications. The choice of radioisotopes, semiconductors, coupling, and semiconductor parameters optimization will improve the battery design to fulfill the service requirements of the applications. Space applications are one of the areas where betavoltaic batteries can be an alternative depending on the power requirements. Radioisotope thermoelectric generators (RTGs) are employed in space for applications where solar power is not available and high power is needed. RTGs are not a viable option where payload is important due to their higher weights. Betavoltaic batteries are lightweight and better suited for small power applications. CubeSats are small satellites that require several watts of power [77], and betavoltaic batteries with three-dimensional design have the potential to meet these power requirements. The disadvantages of solar-powered satellites are power interruption during solar eclipse and parts of the orbit where no sunlight is available. The deeper space mission is also limited by the availability of the sunlight. Furthermore, solar power is dependent on the intensity that varies with the distance from the Sun and the angle of photon incidence as the direction cosine normal to the solar panel is the effective sunlight to generate solar power. All these issues can be mitigated by using betavoltaic batteries on satellites without adding any significant payload. As the betavoltaic batteries are on-chip power devices, they can be a reliable source of power for deep space missions.

A comparison to solar power is shown in Table 11.9 to assess whether the power output from betavoltaic batteries is enough for small satellites. The theoretical power outputs from titanium tritide and beryllium tritide using SiC for alternate layers of three-dimensional design were considered for this comparison. It shows that the theoretical efficiency for beryllium tritide with SiC can be more than 10%. Since all the semiconductor physics were not considered and all the EHPs were assumed to be collected in SiC, this overall efficiency was the maximum theoretical estimate for alternate layers design. The power from beryllium tritide with SiC using half the maximum theoretical efficiency was estimated to be 366 W for the solar panel dimensions of 546 × 548 × 620 mm [78]. Although half of the maximum theoretical efficiency was considered, the power output from beryllium tritide with SiC was three times higher than the solar power for the same solar panel dimensions. Therefore, smaller panels can be utilized for betavoltaics to generate equivalent solar power. This comparison shows that three-dimensional betavoltaic battery design can be a viable option for small satellites

Table 11.9 A theoretical comparison of solar power to betavoltaic power.

Radioisotopes	Theoretical maximum power (mW/cm^3)	Theoretical efficiency (%)	Solar panel dimensions (mm)	Solar power (W)	Betavoltaic power for half the theoretical efficiency (W)
Titanium tritide	3.06	9.33	546 × 548 × 620	120	284
Beryllium tritide	3.95	12.4			366

for deep space missions where solar power is not available. However, three-dimensional design of this size of solar panel is conceptual and has yet to be demonstrated by overcoming manufacturing challenges [79].

11.11 Conclusions

High energy density, long service life, and miniaturization to fit on a chip give betavoltaic batteries some advantages over chemical batteries. The working principle of a betavoltaic battery has both similarities and differences with photovoltaic cells and radiation detectors. It converts the kinetic energy of beta particles from a beta-emitting radioisotope into electrical energy using semiconductors. The design of a betavoltaic battery involves optimization of both the radioisotope and the semiconductor material. The radioisotope for a betavoltaic battery is selected based on its maximum beta particle energy and the type of application. The energy deposition in semiconductors is maximized by minimizing self-absorption in the radioisotope source and by increasing the surface geometry of the semiconductor interface.

The primary focus of betavoltaic battery design is to increase the beta particle energy deposition in the depletion region of the semiconductor. An MCNP analysis is useful to determine the penetration depth and energy deposition in semiconductors. The junction depth and the width of the depletion region are then based on the MCNP analysis. EHPs generated in the depletion region have a higher collection probability at the electrodes. A wide depletion region can accommodate most of the energy deposition in it and increase the charge collection. A low doping concentration increases the depletion region, which in turn increases the short circuit current by increasing the charge collection. However, a low doping concentration also increases the leakage current, which in turn reduces the open circuit voltage. Therefore, an optimization of the doping concentration is necessary to obtain maximum power output. In addition, increasing the effective shunt resistance and lowering the series resistance of the semiconductor improves the power output. Good crystal quality is required to increase the shunt resistance. The thickness of a semiconductor needs to be based on the minority carrier diffusion lengths and the width of the depletion region to decrease the series resistance. Increasing the width of the depletion region may introduce some crystal lattice mismatch that decreases the shunt resistance, which in turn increases the leakage current. It is difficult to obtain a wider depletion region for wide-bandgap semiconductors due to the limitation of semiconductor fabrication techniques. However, wide-bandgap semiconductors provide higher open circuit voltage, power, and efficiency. In addition, they have high radiation resistance, which can be further increased by modifying the junction such as introducing an intrinsic region.

Different types of junctions such as *p-n*, *p-i-n*, and Schottky junctions were investigated for betavoltaic batteries. Schottky junctions were mainly used for wide-bandgap semiconductors GaN and SiC, as they are easier to manufacture compared with a *p-n* junction. The power output obtained from a Schottky junction of 4H–SiC was similar to the *p-n* junction of Si, and better than that of a Schottky junction of GaN. However, non-Schottky junction betavoltaic batteries showed better results. The wide range of performance of betavoltaic batteries suggests that most of the experimental research devices were not optimized. In addition, the choice of radioisotope and the semiconductor design parameters need to be similar for better comparison.

The recent trends of research in betavoltaics are to explore more wide-bandgap semiconductors such as diamond for higher efficiency, and three-dimensional semiconductor design and fabrication to increase surface area for higher power output. The stacking and multilayer volumetric betavoltaic battery designs are essential to increase the power output on the order of milliwatts and several watts. This will make betavoltaic batteries a suitable and attractive choice for many widespread applications. In summary, while much research has been conducted in the area of betavoltaic battery design, there is still more significant progress that can be made. A betavoltaic battery prototype to demonstrate high power density on the order of milliwatts to several watts is needed to make a revolutionary progress and real-life applications.

References

[1] M. Prelas, M. Boraas, F. De La Torre Aguilar, J.-D. Seelig, M.T. Tchouaso, D. Wisniewski, Nuclear Batteries and Radioisotopes, Springer International Publishing, 2016, pp. 524.

[2] K.E. Bower, Y.A. Barbanel, Y.G. Shreter, G.W. Bohnert, Polymers, Phosphors, and Voltaics for Radioisotope Microbatteries, CRC Press, Boca Raton, FL, 2002, p. 504.

[3] K. Wu, C. Dai, H. Guo, A theoretical study on silicon betavoltaics using Ni-63, in: 2011 6th IEEE International Conference on Nano/Micro Engineered and Molecular Systems, 2011, pp. 724–727, https://doi.org/10.1109/NEMS.2011.6017456.

[4] G.N. Yakubova, Nuclear Batteries with Tritium and Promethium-147 Radioactive Sources, University of Illinois, 2010 available at: https://www.ideals.illinois.edu/items/16917.

[5] L. Popa-Simil, Advanced space nuclear reactors from fiction to reality, Phys. Proc. 20 (2011) 270–292, https://doi.org/10.1016/j.phpro.2011.08.025.

[6] J. Yang, T. Caillat, Thermoelectric materials for space and automotive power generation, MRS Bull. 31 (3) (2006) 224–229.

[7] M. Sychov, A. Kavetsky, G. Yakubova, G. Walter, S. Yousaf, Q. Lin, et al., Alpha indirect conversion radioisotope power source, Appl. Radiat. Isot. 66 (2) (2008) 173–177.

[8] H. Li, A. Lai, Self-reciprocating radioisotope-powered cantilever, J. Appl. Phys. 92 (2) (2002) 1122–1127.

[9] X. Tang, D. Ding, Y. Liu, D. Chen, Optimization design and analysis of Si-63Ni betavoltaic battery, Sci. China Technol. Sci. 55 (4) (2012) 990–996.

[10] X. Tang, Y. Liu, D. Ding, D. Chen, Optimization design of GaN betavoltaic microbattery, Sci. China Technol. Sci. 55 (3) (2012) 659–664.

[11] C. Honsberg, W.A. Doolittle, M. Allen, C. Wang, GaN betavoltaic energy converters, in: Proc. 31st IEEE Photovoltaic Specialists Conference, 2005, pp. 102–105, https://doi.org/10.1109/PVSC.2005.1488079.

[12] M. Mohamadian, S. Feghhi, H. Afarideh, Conceptual design of GaN betavoltaic battery using in cardiac pacemaker, in: Proceedings of 13th International Conf on Emerging of Nuclear Energy Systems (ICENES), Istanbul, Turkey, 2007.

[13] H. San, S. Yao, X. Wang, Z. Cheng, X. Chen, Design and simulation of GaN based Schottky betavoltaic nuclear micro-battery, Appl. Radiat. Isot. 80 (2013) 17–22.

[14] W. Pfann, W. Van Roosbroeck, Radioactive and photoelectric *p-n* junction power sources, J. Appl. Phys. 25 (11) (1954) 1422–1434.

[15] P. Rappaport, The electron-voltaic effect in *p-n* junctions induced by beta-particle bombardment, Phys. Rev. 93 (1) (1954) 246.

[16] P. Rappaport, Radioactive Battery Employing Intrinsic Semiconductor, U.S. Patents, 1956. U.S. Patent 2745973A, May 15, 1956.

[17] B. Liu, K.P. Chen, N.P. Kherani, S. Zukotynski, Power-scaling performance of a three-dimensional tritium betavoltaic diode, Appl. Phys. Lett. 95 (23) (2009) 233112.

[18] Z. Cheng, X. Chen, H. San, Z. Feng, B. Liu, A high open-circuit voltage gallium nitride betavoltaic microbattery, J. Micromech. Microeng. 22 (7) (2012) 074011.

[19] H. Gao, S. Luo, H. Zhang, H. Wang, Z. Fu, Demonstration, radiation tolerance and design on a betavoltaic micropower, Energy 51 (2013) 113–122.

[20] M. Lu, G. Zhang, K. Fu, G. Yu, D. Su, J. Hu, Gallium nitride Schottky betavoltaic nuclear batteries, Energy Convers. Manag. 52 (4) (2011) 1955–1958.

[21] C.J. Eiting, V. Krishnamoorthy, S. Rodgers, T. George, et al., Demonstration of a radiation resistant, high efficiency SiC betavoltaic, Appl. Phys. Lett. 88 (6) (2006) 064101.

[22] M. Lewis, S. Seeman, Performance Experience with prototype betacel nuclear batteries, Nucl. Technol. 17 (2) (1973) 160–167.

[23] A. Sciuto, G. D'Arrigo, F. Roccaforte, M. Mazzillo, R.C. Spinella, V. Raineri, Interdigit 4H-SiC vertical Schottky diode for betavoltaic applications, IEEE Trans. Electron. Dev. 58 (3) (2011) 593–599.

[24] L. Olsen, Betavoltaic energy conversion, Energy Convers. 13 (4) (1973) 117–127.

[25] J. Kwon, J.D. Robertson, World's smallest nuclear battery, Am. Ceram. Soc. Bull. 88 (2009) 13–14.

[26] A. Lal, J. Blanchard, Daintiest dynamos [nuclear microbatteries], IEEE Spectrum 41 (9) (2004) 36–41.

[27] G. Wang, R. Hu, H. Wei, H. Zhang, et al., The effect of temperature changes on electrical performance of the betavoltaic cell, Appl. Radiat. Isot. 68 (12) (2010) 2214–2217.

[28] D.-Y. Qiao, X.-J. Chen, Y. Ren, W.-Z. Yuan, A micro nuclear battery based on SiC Schottky barrier diode, J. Microelectromech. Syst. 20 (3) (2011) 685–690.

[29] M. Mohamadian, S. Feghhi, H. Afariadeh, Geometrical optimization of GaN betavoltaic microbattery, in: Proceedings of the 7th WSEAS International Conference on Power Systems, Beijing, China, 2009.

[30] D.-Y. Qiao, W.-Z. Yuan, P. Gao, W.-Z. Yao, et al., Demonstration of a 4H SiC betavoltaic nuclear battery based on Schottky barrier diode, Chin. Phys. Lett. 25 (10) (2008) 3798.

[31] B. Ulmen, P.D. Desai, S. Moghaddam, G.H. Miley, S.I. Masel, Development of diode junction nuclear battery using ^{63}Ni, J. Radioanal. Nucl. Chem. 282 (2) (2009) 601–604.

[32] M.V.S. Chandrashekhar, C.I. Thomas, H. Li, M.G. Spencer, A. Lal, Demonstration of a 4H SiC betavoltaic cell, Appl. Phys. Lett. 88 (3) (2006) 033506.

[33] N. Tsoulfanidis, S. Landsberger, Measurement and Detection of Radiation, 5th Edition, CRC Press, Boca Raton, FL, USA, 2021, p. 610.

[34] C.A. Klein, Bandgap dependence and related features of radiation ionization energies in semiconductors, J. Appl. Phys. 39 (4) (1968) 2029–2038.

[35] M.V.S. Chandrashekhar, C.I. Thomas, M.G. Spencer, Measurement of the mean electron-hole pair ionization energy in 4H SiC, Appl. Phys. Lett. 89 (4) (2006).

[36] C.D. Cress, B.J. Landi, R.P. Raffaelle, Modeling laterally-contacted nipi-diode radioisotope batteries, IEEE Trans. Nucl. Sci. 55 (3) (2008) 1736–1743.

[37] X.-.Y. Li, Y. Ren, X.-J. Chen, D.-Y. Qiao, W.-Z. Yuan, ^{63}Ni Schottky barrier nuclear battery of 4H-SiC, J. Radioanal. Nucl. Chem. 287 (1) (2011) 173–176.

[38] G. Pistoia, Batteries for Portable Devices, Elsevier, 2005, p. 309.

[39] J. Chu, X. Piao, L. Jian, H. Lin, Research of radioisotope microbattery based on β-radio-voltaic effect, J. Nanolithogr. MEMS, MOEMS 8 (2) (2009) 021180.

[40] D.-R. Li, L. Jiang, J.-H. Yin, Y.Y. Tan, N. Lin, Betavoltaic battery conversion efficiency improvement based on interlayer structures, Chin. Phys. Lett. 29 (7) (2012) 078102.
[41] H. Chen, L. Jiang, X. Chen, Design optimization of GaAs betavoltaic batteries, J. Phys. Appl. Phys. 44 (21) (2011) 215303.
[42] H. Guo, H. Yang, Y. Zhang, Betavoltaic microbatteries using porous silicon, 20th International Conference on Micro Electro Mechanical Systems (MEMS), in: IEEE, 2007, pp. 867–870, https://doi.org/10.1109/MEMSYS.2007.4433006.
[43] Z.-J. Cheng, H.-S. San, X.-Y. Chen, B. Liu, Z.H. Feng, Demonstration of a high open-circuit voltage GaN betavoltaic microbattery, Chin. Phys. Lett. 28 (7) (2011) 078401.
[44] M. Lu, G. Wang, C.S. Yao, Gallium nitride for nuclear batteries, Adv. Mater. Res. 343 (2012) 56–61.
[45] A. Ionascut-Nedelcescu, C. Carlone, A. Houdayer, H.J. von Bardeleben, J.-L. Cantin, S. Raymond, Radiation hardness of gallium nitride, IEEE Trans. Nucl. Sci. 49 (6) (2002) 2733–2738.
[46] H. Guo, A. Lal, Nanopower betavoltaic microbatteries, 12th International Conference on Solid-State Sensors, Actuators and Microsystems 1 (2003) 36–39, https://doi.org/10.1109/SENSOR.2003.1215247.
[47] G. Rybicki, C. Vargas-Aburto, R. Uribe, Silicon carbide alphavoltaic battery, in: Twenty Fifth IEEE Photovoltaic Specialists Conference, 1996, pp. 93–96, https://doi.org/10.1109/PVSC.1996.563955.
[48] S. Deus, Tritium-powered betavoltaic cells based on amorphous silicon, in: Twenty-Eighth IEEE Photovoltaic Specialists Conference, 2000, pp. 1246–1249, https://doi.org/10.1109/PVSC.2000.916115.
[49] C.D. Cress, InGaP alpha voltaic batteries: Synthesis, modeling, and radiation tolerance, J. Appl. Phys. 100 (11) (2006) 114519.
[50] T. Wacharasindhu, J.W. Kwon, D.E. Meier, J.D. Robertson, Radioisotope microbattery based on liquid semiconductor, Appl. Phys. Lett. 95 (1) (2009) 014103.
[51] L.T. Canham, Silicon quantum wire array fabrication by electrochemical and chemical dissolution of wafers, Appl. Phys. Lett. 57 (10) (1990) 1046–1048.
[52] Z. Zhao, Z. Cheng, S. Yao, Z. Deng, Y. Li, H. San, Effect of three-dimensional cylindrical hole array on energy conversion efficiency of radioisotope battery, in: 6th IEEE International Conference on Nano/Micro Engineered and Molecular Systems, 2011, pp. 992–995, https://doi.org/10.1109/NEMS.2011.6017522.
[53] Z. Cheng, Z. Zhao, H. San, X. Chen, Demonstration of a GaN betavoltaic microbattery, in: 6th IEEE International Conference on Nano/Micro Engineered and Molecular Systems, 2011, pp. 1036–1039, https://doi.org/10.1109/NEMS.2011.6017533.
[54] S. Tin, A. Lal, Ultra-high efficiency high-power density thinned-down silicon carbide betavoltaics, in: Proceedings of the PowerMEMS, 2009, pp. 189–192. Washington D.C.
[55] T. Kosteski, N.P. Kherani, P. Stradins, F. Gaspari, et al., Tritiated amorphous silicon betavoltaic devices, IEE Proceedings: Circuits, Devices and Systems 150 (4) (2003) 274–281.
[56] J.E. Turner, Atoms, Radiation, and Radiation Protection, John Wiley & Sons, 2008.
[57] L. Romano, V. Privitera, C. Jagadish, Defects in Semiconductors, Elsevier Science, 2015.
[58] I. Gusev, A. Guskova, F.A. Mettler, Medical Management of Radiation Accidents, 2nd Edition, CRC Press, Boca Raton, FL, USA, 2001, pp. 640.
[59] J. Shapiro, Radiation Protection: A Guide for Scientists, Regulators, and Physicians, La Editorial, UPR, 2002, 4th Edition, Harvard University Press, Cambridge, MA, USA, 2002, 688 pp.
[60] S.-K. Lee, S.-H.Son K. Kim, J.-W. Park, et al., Development of nuclear micro-battery with solid tritium source, Appl. Radiat. Isot. 67 (7) (2009) 1234–1238.
[61] S. Adachi, GaAs, AlAs, and $Al_xGa_{1-x}As$: Material parameters for use in research and device applications, J. Appl. Phys. 58 (3) (1985) R1–R29.
[62] W. Sun, N.P. Kherani, K.D. Hirschman, L.L. Gadeken, P.M. Fauchet, A three-dimensional porous silicon *p-n* diode for betavoltaics and photovoltaics, Adv. Mater. 17 (2005) 1230–1233.
[63] Z. Cheng, H. San, Y. Li, X. Chen, The design optimization for GaN-based betavoltaic microbattery, in: 5th International Conference on Nano/Micro Engineered and Molecular Systems, IEEE, 2010, pp. 582–586, https://doi.org/10.1109/NEMS.2010.5592469.

[64] Z. Cheng, H. San, Z.-H. Feng, B. Liu, X. Chen, High open-circuit voltage betavoltaic cell based on GaN pin homojunction, Electron. Lett. 47 (12) (2011) 720—722.
[65] H. Guo, Y. Shi, Y. Zhang, Y. Zhang, J. Han, Fabrication of SiC p-i-n betavoltaic cell with 63Ni irradiation source, in: 2011 IEEE International Conference of Electron Devices and Solid-State Circuits, 2011, pp. 1—2, https://doi.org/10.1109/EDSSC.2011.6117636.
[66] S. Tin, A. Lal, A radioisotope-powered surface acoustic wave transponder, J. Micromech. Microeng. 19 (9) (2009) 094014.
[67] V.S. Bormashov, S.Yu. Troschiev, S.A. Tarelkin, A.P. Volkov, et al., High power density nuclear battery prototype based on diamond Schottky diodes, Diam. Relat. Mater. 84 (2018) 41—47.
[68] T. Shimaoka, H. Umezawa, K. Ichikawa, J. Pernot, S. Koizumi, Ultrahigh conversion efficiency of betavoltaic cell using diamond *p-n* junction, Appl. Phys. Lett. 117 (10) (2020) 103902.
[69] T.R. Alam, M.G. Spencer, M.A. Prelas, M.A. Pierson, Design and optimization of radioisotope sources for betavoltaic batteries, Int. J. Energy Res. 42 (7) (2018) 2564—2573.
[70] C. Thomas, S. Portnoff, M. Spencer, High efficiency 4H-SiC betavoltaic power sources using tritium radioisotopes, Appl. Phys. Lett. 108 (1) (2016) 013505.
[71] K. Hogan, M. Litz, F. Shahedipour-Sandvik, 3D GaN-based betavoltaic device design with high energy transfer efficiency, Appl. Radiat. Isot. 145 (2019) 154—160.
[72] J. Russo, M.S. Litz, I.I.W. Ray, et al., Planar and textured surface optimization for a tritium-based betavoltaic nuclear battery, Int. J. Energy Res. 43 (9) (2019) 4370—4389.
[73] D. Wagner, D. Novog, R. LaPierre, Design and optimization of nanowire betavoltaic generators, J. Appl. Phys. 127 (24) (2020) 244303.
[74] T.R. Alam, M.A. Pierson, M.A. Prelas, Design principles of narrow and wide bandgap-based betavoltaic batteries, IEEE Trans. Electron. Dev. 65 (12) (2018) 5518—5524.
[75] T.R. Alam, M.A. Pierson, M.A. Prelas, A Monte Carlo radioisotope source model for betavoltaic batteries, Transactions of the American Nuclear Society 117 (1) (2017) 471—472.
[76] T.R. Alam, M.A. Pierson, M.A. Prelas, Beta particle transport and its impact on betavoltaic battery modeling, Appl. Radiat. Isot. 130 (2017) 80—89.
[77] S.S. Arnold, R. Nuzzaci, A. Gordon-Ross, Energy budgeting for CubeSats with an integrated FPGA, in: 2012 IEEE Aerospace Conference, 2012, pp. 1—14, https://doi.org/10.1109/AERO.2012.6187240.
[78] E. Agasid, R. Burton, R. Carlino, G. Defouw, A.D. Perez, A.G. Karacalıoğlu, B. Klamm, A. Rademacher, J. Schalkwyk, R. Shimmin, J. Tilles, S. Weston, State of the Art Small Spacecraft Technology, NASA/TP—2018—220027, updated November 2020, December 2018, p. 202. Available at: https://www.nasa.gov/sites/default/files/atoms/files/soa2018_final_doc-6.pdf https://www.nasa.gov/smallsat-institute/sst-soa-2020. accessed 09.21.21.
[79] M. Spencer, T. Alam, High power direct energy conversion by nuclear batteries, Appl. Phys. Rev. 6 (3) (2019) 031305.

PART THREE

Near-Earth and deep-space missions

CHAPTER TWELVE

Solar array designs for deep space science missions

Carolyn R. Mercer
NASA Glenn Research Center, Cleveland, OH, United States

12.1 Introduction

Photovoltaics have been used to generate electrical power since the beginning of the space age, when the benefits of harnessing the Sun's energy to power spacecraft were recognized almost immediately after the invention of silicon photovoltaic cells at Bell Laboratories in 1954. In 1958 the US Naval Research Laboratory's Vanguard I was the first spacecraft to make use of solar power. Six solar cells mounted to the body of the satellite as shown in Fig. 12.1A produced only a single watt of power but demonstrated that mission life could be extended to years instead of the then-standard few days provided by batteries.

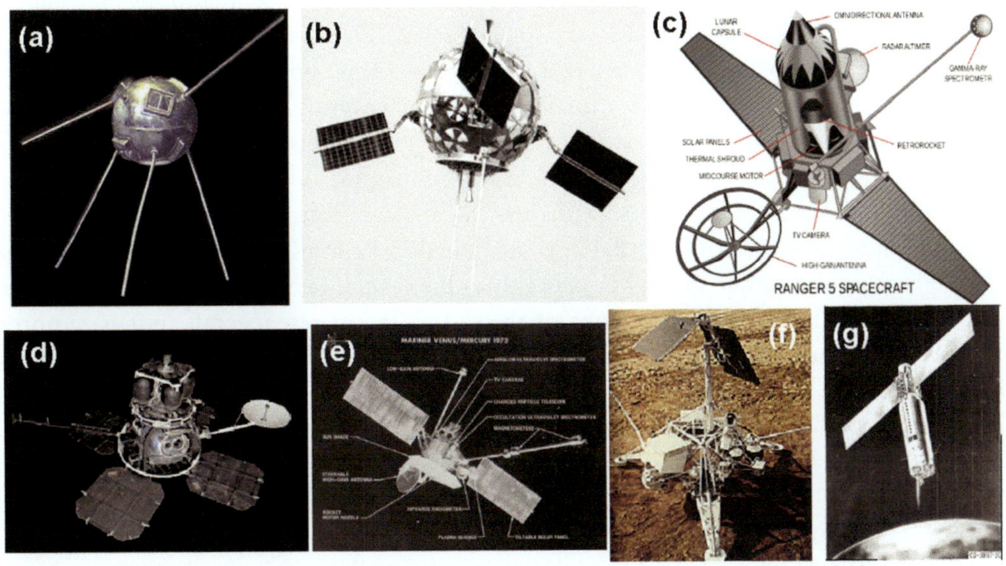

Figure 12.1 Representative progression of solar arrays used in deep space science missions (launch years): (A) Vanguard I (1958), (B) Pioneer P-3 (1959), (C) Ranger I (1961), (D) Lunar Orbiter I (1966), (E) Mariner 10 (1973), (F) Surveyor 3 (1966), (G) SERT II (1970). *Courtesy: NASA.*

Photovoltaics for Space
ISBN 978-0-12-823300-9, https://doi.org/10.1016/B978-0-12-823300-9.00006-6

By the very next year, the Pioneer 5 (and subsequent Pioneer P-Series) spacecraft produced enough power to operate six scientific instruments by deploying four 60 cm × 60 cm solar "paddles" in flight as shown in Fig. 12.1B. These paddles were mounted on struts distributed equidistantly around the spherical spacecraft, with the solar cells on each paddle facing a different plane so that at least two would capture sunlight for any orientation of the spin-stabilized spacecraft. Do we have an explanation of spin-stabilized elsewhere? Pioneer 5 mapped the magnetic field between Earth and Venus, and the P-Series were intended to enter lunar orbit, but none were successful.

The Ranger series, beginning in 1961, was a series of missions to image the Moon in advance of the Apollo landing. Each spacecraft carried six television cameras that required a stable platform to provide clear images. Spin stabilization was therefore no longer sufficient, and a gas jet attitude control system was developed to provide three-axis stabilization. This enabled a planar configuration for the solar arrays since the spacecraft could be continuously oriented toward the Sun for full solar illumination upon the solar cells. The Block 1 and 2 Ranger spacecraft (Ranger 1 through 5) each had a total of 2200 solar cells mounted on trapezoidal panels that deployed after launch, as shown in Fig. 12.1C, producing 135 W to power the cameras and spacecraft. Each Block 3 spacecraft (Ranger 9 and 10) had a total of 9792 cells on two rectangular panels with a total of 2.3 m^2 surface area, producing 200 W.

In 1966 AIMP-D (Explorer 33) became the first US mission to approach the Moon, flying in a highly elliptical Earth orbit, coming within 40,000 km at nearest approach (perilune). Its spacecraft design was similar to that of Pioneer 5, including spin stabilization and four paddles that produced a total of 38 W. Five weeks later, Lunar Orbiter 1 became the first US mission to enter a truly lunar orbit, with a perilune of 189 km. Within the span of 2 years, five Lunar Orbiters successfully mapped out 99% of the Moon with a resolution of at least 60 m. Like Ranger, gas jets were used to provide the three-axis stabilization needed for high-resolution imaging. This in turn enabled the use of deployable, fixed-orientation solar panels, in this case four square panels that folded down in flight, as shown in Fig. 12.1D, producing 375 W total spacecraft power [1]. The choice to use four, rather than two, panels may have been because at 1.65 m tall, Lunar Orbiter 1 was almost half the height of the 3.1 m Ranger spacecraft, but it required almost double the power. Had the Lunar Orbiter used only two panels, they would likely have needed to be hinged or else extend above the spacecraft bus when stowed for launch.

In 1973 the Mariner 10 mission to fly by Venus and Mercury employed the two-wing, gimbaled solar array design shown in Fig. 12.1E and found in common usage today. The driving design challenge was protecting the spacecraft from the heat loads generated at 0.3 AU during the Mercury flybys, where solar irradiation is 11 times that at Earth. A large sunshade, also shown in Fig. 12.1E, protected the spacecraft bus and instruments, but the solar arrays necessarily needed to remain exposed to the Sun. The first design considered was to rigidly swing the dual wings back toward the aft end of the spacecraft (known as a "V-tilt") to reduce the surface area normal to the Sun and therefore minimize radiative heat loads while still generating power. However, this created an unacceptable thermal path back to the bus, so instead the decision was made to rotate the arrays about their long axis, a configuration then referred to as the "sail" configuration and now used almost universally for gimbaled arrays. For Mariner 10, this configuration also allowed the gas jets used to control roll and yaw to be mounted on the solar array tips, providing a long moment arm that reduced the amount of gas needed to effect orientation changes [2]. Even with this conservative arrangement, an oscillation problem occurred during flight that depleted much of the compressed gas needed for control. To conserve the remaining gas, flight controllers oriented the solar panels and high-gain antenna so that solar radiation pressure could spin the spacecraft to stabilize it during cruise [2]. While Mariners 1 through 4 experimented with solar sailing, including a Dacron "sail" to balance solar pressure on the unequal solar panels on Mariner 1 [3] and solar vanes on tips of Mariner 4's solar panels [4], Mariner 10 was the first spacecraft to use solar wind as a major means of spacecraft orientation. This approach enabled a second and third flyby of Mercury. Of note is that the large solar arrays used on Mariner 10 (5.1 m^2 total area) [5] extended well above the spacecraft bus when stowed [2]. The power was used to drive six instruments, with the largest power draw of these by a television (or video) camera that consumed 67 W [6]. The substantially taller Centaur upper stage, compared to the Agena used to launch the Ranger and Lunar Orbiter series, allowed for this single panel per wing configuration.

Landed spacecraft are now often powered by solar arrays, and the first was Surveyor 1 launched in 1966. Within 2 years, seven Surveyor missions were launched based on the same single-panel solar array design shown in Fig. 12.1F that provided power not only on the surface but also during the cruise phase to the Moon. The solar panel was mounted on a mast that provided four-axis rotational control, as shown in Fig. 12.1F. For launch, the solar panel was stowed vertically, and then after launch was positioned horizontal to the spacecraft deck so the solar cells were oriented toward the Sun and along the thrust vector for the cruise period to the Moon [7]. Upon landing near the equator at the beginning of a lunar day, the solar panel was rotated up to near vertical to point to the Sun low in the horizon, using a secondary sun sensor embedded within the solar panel to help locate the Sun. The panel tracked the Sun's motion, lowering to horizontal at lunar noon, and back to near-vertical facing the other horizon for sunset. At sundown the solar panel was positioned near-horizontally to minimize the solar panel output at the next dawn, allowing time for the batteries and electronics to warm up gently. In addition to providing primary power, the mast-mounted solar panel was used to shadow the lander, significantly reducing thermal loads on the television camera and other subsystems. The Surveyor solar panel had a surface area of 0.855 m^2 and generated between 55 and 90 W of power depending on the lunar surface temperature over the course of the day [7].

Finally, while not a deep-space mission, the 1970 SERT 2 satellite used solar arrays to demonstrate electric propulsion for later deep-space use. Twin wings with six panels each to allow for compact stowage generated a total of 1500 W to power two ion thrusters (Fig. 12.1G). While the thrusters were gimbaled the solar arrays were not, and the spacecraft flew in a Sun-synchronous Earth orbit using gravity gradient control augmented by gyroscopes to maintain Solar pointing at all times [8].

The spacecraft described thus far illustrate a progression toward increasing power requirements and toward use in new environments that place new demands and constraints on solar arrays. However, spacecraft are designed to accomplish specific mission objectives, and simpler designs are used wherever possible to reduce cost and complexity. In 1965, after the Ranger series had launched, the Pioneer 6 through 9 spacecraft were developed to create the world's first space-based solar weather network. Each spacecraft carried field and particle instruments needed to map out the solar wind, solar magnetic field, and cosmic rays. These instruments, like those on Pioneer 5, did not need to point at a fixed target, so spin stabilization was all that was needed. To remove the need for a power-hungry omnidirectional antenna, the Pioneer 6 to 9 spacecraft used an orientable spin-stabilized control system whose spin axis was oriented with a simple gas jet [9]. This permitted the use of a cylindrical spacecraft with body-mounted solar cells on the curved surface, with an antenna mounted at one end of the cylinder.

This general progression to ever larger solar arrays to power ever more sophisticated spacecraft in ever harsher environments, punctuated by missions that emphasize lower cost and complexity, is a trend that continues to this day. While Mariner 10 ventured close to the Sun, recent missions such as Juno and Europa Clipper are expanding the use of solar power deep into the solar system. Tables 12.1—12.3 list NASA's deep space robotic science missions (conducting primarily planetary science) grouped by fundamental solar array type: body mounted following in the path of Vanguard I and Pioneer 6; multiwing, both fixed following in the path of the Pioneer, Ranger, AIMP, and Lunar Orbiters, and gimbaled following in the path of Mariner 10; and single wing following in the path of Surveyor I. Table 12.4 lists rideshare missions conducting planetary science. Within each table the missions are listed in chronological order by destination. Sections 12.2—12.4 summarize solar array design considerations for these missions, and Section 12.5 rideshare missions encompasses these design types. Finally, future trends in solar array configurations are discussed in Section 12.6.

The focus of this chapter is the configuration of the solar arrays, not the photovoltaic cells themselves. In almost every case, the arrays were configured using commercially available cells with the highest efficiency at the AM0 standardized sun light spectrum outside Earth's atmosphere, even though that cost was higher than lower efficiency cells. This is because, in general, solar array fabrication costs are dominated by the labor needed to install the arrays rather than the cost of the cells The use of high-efficiency cells reduces the mass and stowed volume of the arrays that has a ripple effect on the

Table 12.1 Missions with body-mounted solar arrays.

Destination	Mission (body mounted)	Year(s)
Sun to Earth-Moon System		
Lunar orbit	Lunar Prospector	1998
Lunar orbit	LADEE	2013
LEO	NRL's Vanguard 1	1958
Solar orbit	Pioneer 6–9	1965–68
Venus orbiter	Pioneer 12	1978
Venus multiprobe bus	Pioneer 13	1978
Comet tour	ISEE-3/ICE	1978
NEO	Contour	2002
Mars		
Mars	Mars Pathfinder: Cruiser	1996
Mars surface	Mars Pathfinder: Sojourner	1996
Mars	MER: Cruiser	2003
Mars surface	MER: Spirit/Opportunity	2003
Mars	MSL: Cruiser	2011
Mars	Mars 2020: Cruiser	2020
Mars surface	Mars 2020: Ingenuity	2020

entire spacecraft mass, which almost always directly scales with spacecraft cost. Finally, all of the missions discussed are NASA's, and all dates refer to launch dates unless otherwise noted. All mission destinations refer to the design mission since several missions did not successfully reach their destination.

12.2 Modern missions using multiple wings or paddles

While the gimbaled wings configuration may be considered to be "classic," the specific details of solar array designs were set by the unique mission requirements. To reduce costs, several later missions rigidly affixed their solar arrays to the spacecraft bus instead of providing gimbaled sun-tracking.

12.2.1 Classic gimbaled wings

As shown in Fig. 12.2, the Magellan, Deep Space 1, Clementine, and Dawn missions all use dual solar arrays gimbaled about their long axis to control their orientation with respect to the Sun, tracing this design heritage straight to Mariner 10. In 1989 Magellan returned to Venus, using aluminized Kapton reflectors on the solar arrays, as well as solar off-pointing, to keep the temperature below 115°C. In addition to covering the entire back surface of the arrays, these solar reflectors also covered about 35% of the front surface [10]. Magellan also carried a synthetic aperture radar that required 200 W to

Table 12.2 Missions with multiwing solar arrays.

Design destination	Mission (symmetric deployed)	Configuration	Year(s)
Sun to Earth-Moon System, including Near-Earth Asteroids (NEAs)			
Lunar orbit	Pioneer P-Series	Four fixed paddles	1959[b]
HE Earth orbit	Ranger 1–9	Two fixed wings	1961–65
HE Earth orbit	AIMP-D (Explorer 33)	Four fixed wings	1966
Lunar orbit	Lunar Orbiter 1–5	Four fixed wings	1966–67
Lunar orbit	AIMP E (Explorer 35)	Four fixed paddles	1967
Lunar orbit	Clementine	Two wings	1994
Lunar orbit	GRAIL A&B	Two fixed wings	2011
Venus flyby	Pioneer 5	Four fixed paddles	1960
Venus flyby	Mariner 1 and 2	Two fixed wings	1962
Venus flyby	Mariner 5	Four fixed wings	1967
Venus/Mercury flyby	Mariner 10	Two wings	1973
Venus orbit	Magellan	Two wings	1989
433 Eros rendezvous	NEAR Shoemaker	Four fixed wings	1996
9969 Braille	Deep Space 1	Two wings	1998
Wild 2	Stardust	Two fixed wings	1999
Sun-Earth L1	Genesis	Two fixed wings	2001
Mercury orbit	MESSENGER	Two wings	2004
NEA (Bennu)	OSIRIS-REx	Two wings[a]	2016
NEA (Didymos)	DART	Two wings[a]	2021
Mars to Jupiter System			
Mars flyby	Mariner 3 and 4, 6 and 7	Four wings	1964, 1969
Mars orbit	Mariner 8 and 9	Four wings	1971
Mars orbit	Viking 1 and 2	Four wings	1975
Mars orbit	Mars Global Surveyor	Two wings[a]	1996
Mars surface	Mars Pathfinder: Lander	Three fixed wings	1996
Mars	Mars Polar Lander: Cruiser	Two fixed paddles	1999[b]
Mars surface	Mars Polar Lander	Two fixed wings[a]	1999[b]
Mars orbit	MRO	Two wings[a]	2005
Mars	Phoenix: Cruiser	Two fixed paddles	2007
Mars surface	Phoenix: Lander	Two fixed wings[a]	2007
Mars orbit	Maven	Two wings[a]	2013
Mars	Insight: Cruiser	Two fixed wings	2018
Mars surface	Insight: Lander[a]	Two fixed wings[a]	2018
Main asteroid belt	Dawn	Two wings	2007
Jupiter	Juno	Three wings[a]	2011
Jupiter system	Lucy	Two wings[a]	2021
Main asteroid belt	Psyche	Two wings[a]	2023–4[c]
Europa	Europa Clipper[a]	Two wings	2025[c]

[a]Denotes unique configuration described in Section 12.5.
[b]Failed
[c]Planned.

Table 12.3 Missions with single-panel solar arrays.

Destination	Mission (single panel)	Year(s)
The Moon		
Lunar surface	Surveyor lander 1–7	1966–1968
Lunar orbit	LRO	2009
Mars		
Mars orbit	Mars Observer	1992[a]
Mars orbit	Mars Climate Orbiter	1998[a]
Mars orbit	Mars Odyssey	2001
Earth Orbit		
Earth orbit	Deep Impact/EPOXI	2005
Earth orbit	Kepler	2009
Earth orbit	NEOWISE	2009

[a]Failed.

Table 12.4 Rideshare missions.

Destination	Mission (rideshare)	Config	Launch year
LEO			
LEO	GENE-SAT (CSLI)	Body-Mount	2006
LEO	Pharma-SAT (CSLI)	Body-Mount	2009
LEO	O/OREOS (CSLI)	Body-Mount	2010
LEO	Q-PACE (CSLI)	Body-Mount	2021
The Moon			
Lunar orbit	PFS 1 and 2 (Apollo 15 and 16)	Body-Mount	1971, 1972
Lunar impact	LCROSS (LADEE)	Body-Mount	2009
Lunar orbit	LunaH-Map (EM-1)	Two wings	2022[a]
Lunar orbit	Lunar Trailblazer (IMAP)	Two wings	2025[a]
Near-Earth			
1996 FG3 1991 VH	Janus A (Psyche) Janus B (Psyche)	Two wings	2022[a]
Mars			
Mars flyby	MarCO A and B (Insight)	Two wings	2018

[a]Planned.

operate [11]. These requirements resulted in two panels each with an area of 6.25 m^2, almost two and a half times as large as those on Mariner 10 [11].

From the 1970s to the early 1990s, the size of planetary science missions grew to be consistently about 1000 kg or more. The 1994 Clementine mission was designed to

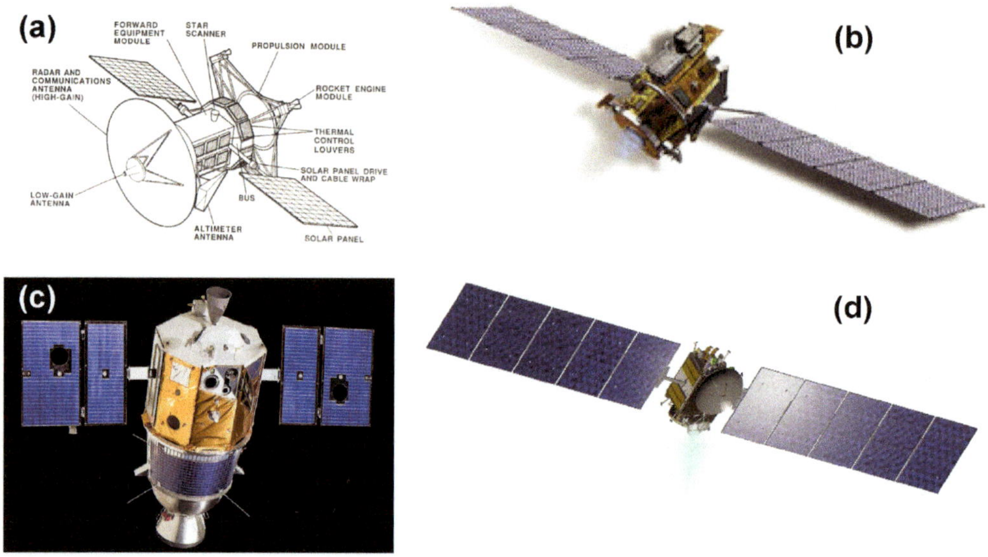

Figure 12.2 Deep space spacecraft using classic gimbaled wings. (A) Magellan (1989), (B) Deep Space 1 (1998), (C) Clementine (1994), (D) Dawn (2007). *Courtesy: NASA.*

demonstrate that missions could be accomplished with small, low-cost spacecraft. The 424-kg spacecraft relied on a launch vehicle upper stage for transport to the Moon. Its twin solar arrays were the second science mission to use gallium arsenide (GaAs) solar cells, following NASA's 1992 Solar Anomaly and Magnetospheric Explorer (SAMPEX) mission [12]. The success of the mission led to the initiation of NASA's Discovery program 2 years later to continue with competed, lower cost missions.

The 1998 Deep Space 1 solar electric propulsion technology demonstration mission built upon the legacy of SERT 2, and it was the first to use a solar concentrator array to provide primary power in space, and the first to use two types of solar cells on a single wing (dual and triple junction). The refractive, thin-film concentrators passively deployed as the wings were unfurled to provide a 7X linear solar concentration, which reduced the area of the cells and cover glass material by nominally the same amount. The two concentrator arrays produced a total 2.5 kW of power [13] needed to drive the maximum throttle level of the ion engines (2.3 kW) [14]. In 2007 the Dawn mission made operational use of the solar electric propulsion system demonstrated by Deep Space 1. Its five-panel twin wings with triple-junction cells were sized to generate an end-of-life power of 10.3 kW at 1 AU, enough to power the thrusters in the main asteroid belt (3 AU) to explore the asteroids Vesta and Ceres. Although it did not employ the concentrator array earlier demonstrated, the InGaP/InGaAs/Ge cells were screened for best performance at low intensity conditions [15].

12.2.2 Fixed wings or paddles

In the early 1990s, NASA established the Discovery program to reduce the time and cost needed to develop planetary science missions, using competitively selected missions led by a principal investigator. Near-Earth Asteroid Rendezvous (NEAR) Shoemaker was the first Discovery mission selected, followed shortly thereafter by Stardust and Genesis. As mentioned previously, each of these three missions rigidly affixed their solar arrays to the spacecraft bus instead of providing gimbaled sun-tracking to reduce costs. Gravity Recovery and Interior Laboratory (GRAIL) launched nearly a decade later, but it adopted a similar solar array design to enable a dual-spacecraft mission within the constrained cost cap. These spacecraft are shown in Fig. 12.3.

The 1996 NEAR Shoemaker was able to use this lower cost, simpler fixed-panel design because the trajectory used to reach its mission target of asteroid 433 Eros permitted the Sun-spacecraft-Earth angle to never reach more than about 30 degrees except for brief periods that could be covered by battery power. The solar arrays were sized to accommodate the roughly 10% loss caused by off-pointing, resulting in a total needed area of roughly 9 m^2 to generate 2000 W with GaAs cells. To further simplify the spacecraft design, this area was distributed among four rectangular panels, each of which had only to swing up and latch for deployment after launch. This robust design performed well during the rendezvous with 433 Eros; the spacecraft survived the touchdown and provided data for 2 weeks after landing. Although 433 Eros is considered a

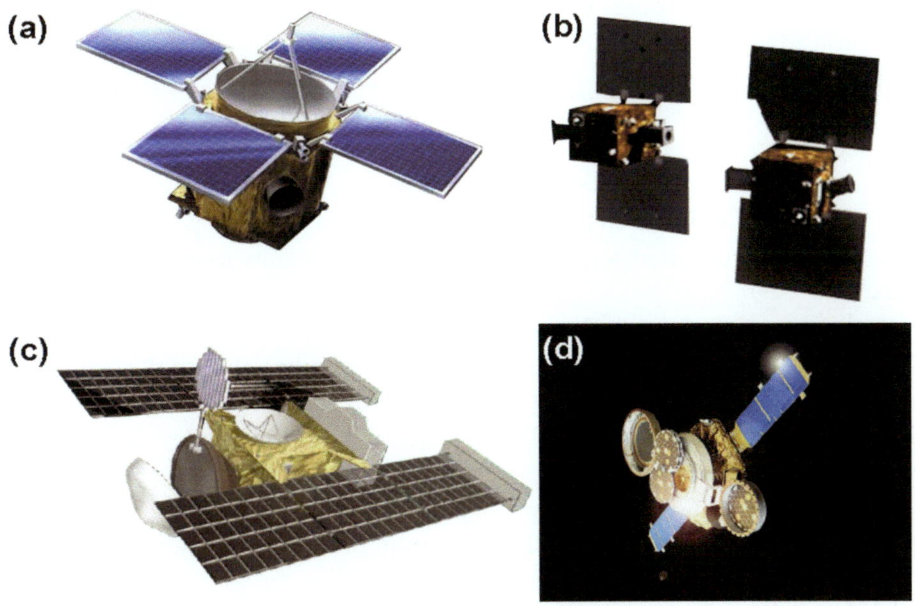

Figure 12.3 Deep space spacecraft using fixed wings or paddles. (A) NEAR Shoemaker (1996), (B) Grail (2001), (C) Stardust NeXT (1999), (D) Genesis (2001). *Courtesy: NASA.*

near-Earth object, NEAR Shoemaker rendezvoused with it while it was about 2.3 AU from Earth, making it the first spacecraft to use solar cells for primary power beyond Mars orbit as well as the first to orbit and land on an asteroid.

Launched in 2001, the GRAIL mission studied the interior of the Moon by measuring the relative changes in the position of its two identical spacecraft as they flew over regions of differing lunar gravity. The mission was inspired by the success of NASA's GRACE mission that mapped Earth's gravity using the same twin spacecraft technique. While the GRACE spacecraft employed body-mounted solar arrays on a trapezoidal bus, each GRAIL spacecraft used twin single-panel wings deployed after launch and rigidly latched open. In this case the choice of solar array design lies not in the unique demands of the mission, but rather in the reuse of existing hardware by the spacecraft bus providers to keep costs low. For GRACE, Astrium GmbH reused their CHAMP spacecraft bus, and for GRAIL, Lockheed Martin reused much of their XSS-11 spacecraft developed for the US Air Force, including the pair of fixed-wing, single-panel arrays. GRAIL's solar arrays were sized to allow for data acquisition at solar beta angles >49 degrees [16].

The 1999 Stardust mission was the first to return samples from a known comet. Its capture of material from comet Wild 2's tail took place 2.72 AU from the Sun, setting a distance record at the time for a solar-powered spacecraft. Over the course of the mission the solar irradiance varied from 13% of Earth's at its furthest distance, to 29% at the comet, and back to 100% for the return to Earth, with temperatures ranging from −61°C to +79°C. Rather than building a custom peak power tracking system, to save costs a hybrid direct energy transfer system coupled with a solar array switching unit was used instead, with solar cell strings sized optimally for the comet encounter, and then reconfigured in flight to increase the voltage generated near Earth [17]. Another unique feature of the Stardust solar arrays was the Whipple shields mounted on the ends of each array, as shown in Fig. 12.3C, with a central shield to protect the spacecraft bus. (Can we find a reference here for Whipple shields?) The spacecraft was oriented so the cometary material impacted the shields to protect the spacecraft from high-speed particle impacts, with the solar arrays oriented perpendicular to the particle flow to minimize their frontal surface area.

Although launched 2 years after Stardust, the 2003 Genesis mission to capture solar wind material lasted just under 3 years and so holds the distinction of the first mission to return to Earth a robotically captured sample collected beyond the Moon. Genesis flew in a halo orbit around the Sun-Earth Lagrange point 1. The trajectory during cruise and science operations oriented the spacecraft to be sun-pointing except when executing trajectory maneuvers, allowing for a spin-stabilized, fixed-wing design. The single-panel solar arrays stowed for launch in a triangular configuration similarly to Mariner 10.

12.2.2.1 Landers

The 1996 Mars Pathfinder mission delivered Sojourner to Mars, the first robotic rover to another planet. The Pathfinder cruise stage is described in Section 12.4.1 and the rover in Section 12.4.2; they and the lander all used solar arrays for primary power. Mars

Pathfinder used an airbag landing system pioneered in 1966 by the Soviet Luna 9 lunar lander, complete with triangular "petals" that unfurled close to the surface once the craft was settled on the surface. However, unlike the Luna 9 lander, Mars Pathfinder lined the petals with solar cells to power a communications relay between Sojourner and Earth (Fig. 12.4A).

The 1999 Mars Polar Lander, as its name describes, landed very close to the southern Mars pole. At that latitude the Sun stays about 30 degrees above the horizon for continuous illumination during the summer, and the Mars Polar Lander expected to operate there for 337 days, or roughly half of a Mars year (although it actually failed during the descent stage). Two primary solar arrays and two auxiliary solar panels were configured to capture this polar light, as shown in Fig. 12.4B. The two primary arrays consisted of two panels each. The top panels were rigidly mounted to the lander structure at an angle of 30 degrees to directly face the summer Sun. The bottom panels, folded down after landing, were positioned horizontally to prevent potential terrain clearance problems like interference with large rocks. The horizontal orientation still allowed the capture of nearly 90% of the incident radiation. The auxiliary panels, added to provide needed power while keeping the primary arrays small enough to stow within the conical backshell portion of the entry aeroshell, were also canted at 30 degrees. The two auxiliary panels were located on the north side of the lander to capture the Sun, leaving room on the spacecraft's southern (shadowed side) for a robotic arm. In total, the primary and auxiliary panels provided nominally 200 W. In addition to orienting the

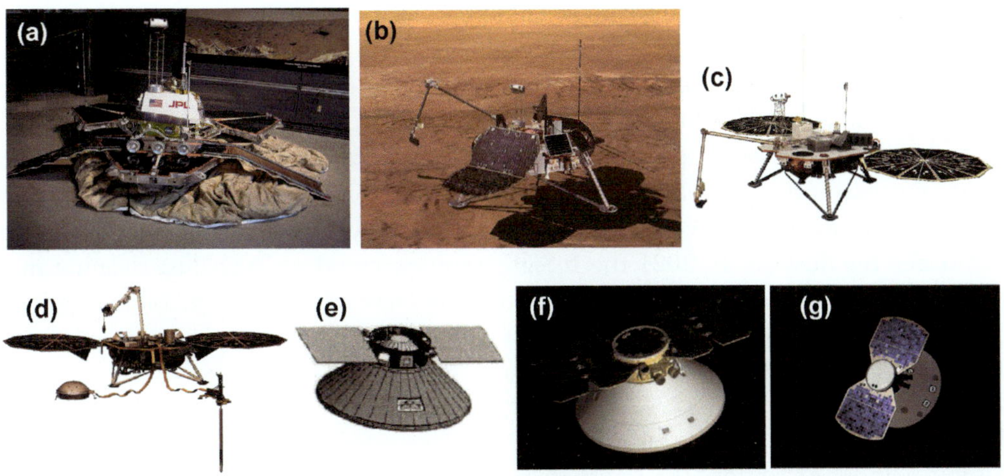

Figure 12.4 Fixed wings or paddles on deep space cruisers and landers. (A) Mars Pathfinder Lander (1996), (B) Mars Polar Lander (1999), (C) Phoenix (2007), (D) InSight (2018), (E) Mars Polar Lander (1999), (F) Phoenix (2007), (G) InSight (2018). *Courtesy: NASA.*

panels directly at the low-horizon Sun, the 30-degree tilt also reduced dust accumulation.

After the failure of the Mars Polar Lander, NASA initiated the Scout program to develop lower cost, PI-led missions to Mars, similar to the Discovery program. The first Scout mission was Phoenix, launched in 2007. Like the Mars Polar Lander, Phoenix was designed to conduct environmental science in the polar regions during the summer months. However, the Phoenix lander used a new Ultraflex solar array design that affixed solar cells onto a flexible, rather than rigid, substrate, and with a unique circular design that provided stiffness through radial spokes. This design allowed for large arrays (each 2.1 m diameter, 7 m^2 total area) with reduced mass, that provided about 338 W at Mars noon [18] using advanced triple-junction photovoltaic cells [19], substantially more than provided by the geometrically constrained Mars Polar Lander. These fixed arrays were oversized to compensate for cosine losses stemming from the mismatch between the horizontal arrays and varying Sun angle, allowing for a straightforward horizontal deployment without risk of ground clearance interference with large rocks and boulders (Fig. 12.4C).

Launched in 2018, the InSight mission adopted the core UltraFlex design demonstrated on Phoenix, which by then was also in use as the primary power system for the Cygnus commercial resupply spacecraft for the International Space Station (Fig. 12.4D). However, the UltraFlex arrays on InSight are larger, and furthermore the equatorial landing site allows for the capture of more solar radiation. This makes the InSight lander the largest solar-generated electrical power source ever landed beyond Earth. Furthermore, it has surpassed even radioisotope-powered systems in the total amount of generated electricity, reaching 4588 W-hr in 2018 and still going strong. The design driver for the large arrays is the requirement to provide power for science operations, not just keep-alive power, during the Mars nights to provide a continuous record of seismic activity and interior structure during the course of 1 year. Therefore, in addition to providing primary power for the spacecraft, this solar-generated electricity must also charge a bank of batteries needed to operate the full instrument suite during the Mars night. Furthermore, the solar arrays are oversized to compensate for expected dust accumulation, with the expectation that periodic cleaning occurs via "dust devil" events that sweep the dust off. In 2021 the InSight team executed a clever dust cleaning maneuver by using spacecraft's robotic arm to deposit a few relatively large soil particles on the array surface to enhance the natural dust devil effect. It would be interesting here to include a picture of the dust devils on Mars.

12.2.2.2 Cruisers

Almost without exception, the cruise stages needed to deliver landed payloads to their destination have been solar powered using body-mounted panels or strut-supported fixed panels. The exceptions are Surveyor 1, which used the lander itself to power the transit to the Moon, and Dragonfly, a nuclear-powered mission to Titan. The former did not need a cruise stage because of the simplicity of the mission, and the latter will be traveling out beyond 9 AU from the Sun where the solar irradiance is too low to be practical for use with solar arrays.

For the United States, each solar-powered cruise stage carried landed assets to Mars. In each case the cruise stage was not required to operate sophisticated, power-hungry instruments or communicate high data rates during flight, and it either burned up in the Mars atmosphere or suffered an uncontrolled landing, ending the need for the cruise vehicle once the payload was delivered to the surface. The design of these spacecraft is correspondingly simple. The cruise stages for Mars Polar Lander, Phoenix, and InSight employed twin paddles, extending just slightly beyond the diameter of the spacecraft, as shown in Figs. 12.4E–G, respectively, somewhat reminiscent of Lunar Orbiter 1.

12.2.3 Missions with unique solar arrays: fixed and gimbaled

Beyond generating electrical power using the Sun's energy, solar arrays on deep-space missions have also been innovatively used for aerobraking, tilted for thermal management and collision avoidance, and other unique features.

12.2.3.1 Aerobraking configurations

The solar arrays on the 1996 Mars Global Surveyor (MGS) were configured similarly to the two-wing spacecraft described in Section 2.1, but the mission placed three major requirements on the solar arrays that had not been encountered before, that led to the unique tilt-wing design shown in Fig. 12.5A, named the "V-tilt" when considered by Mariner 10. First, there were severe cost constraints under NASA Administrators Dan Goldin's "better, cheaper, faster" paradigm; this manifested in the first use of two types of solar cells on a single array to provide the required power within the cost constraint: then state-of-the-art GaAs/Ge cells populated the inboard solar panels providing roughly 55% of the spacecraft power, while lower-cost and lower-mass Si cells were used on the outer panels providing the balance of the roughly 2100 W at 1 AU (1.5×10^9 km) [20]. Second, to reduce electromagnetic noise from the spacecraft bus, MGS was the first US

Figure 12.5 Solar arrays used for aerobraking. (A) Mars Global Surveyor (1996), (B) Mars Reconnaissance Orbiter (2007), (C) Maven (2013). *Courtesy: NASA.*

planetary mission to mount its magnetometers at the tips of the solar arrays. This required a circuit layout designed to minimize induced magnetic moments [20]. Finally, and most significantly, the MGS mission required a low-altitude, near-circular Mars orbit to produce high-resolution maps of the planet's entire surface. Such an orbit requires a very large amount of delta-V, which would require a significant amount of propellant to achieve propulsiveness, which would drive up the spacecraft's mass and cost.

In a bold technologic step forward, MGS used its solar arrays for aerobraking to achieve the desired orbit. MGS methodically lowered its orbit by making multiple passes around the planet, each time lowering its periapsis and generating aerodynamic drag forces on the extended wings. To maintain aerodynamic stability while simultaneously reducing the thermal load on the solar arrays during each pass, the solar arrays were canted back by 30 degrees. In addition, Kapton panels added to the tips of each solar array increased the surface area of each wing, serving as drag flaps to increase frictional forces to further decelerate the spacecraft [21]. MGS was the first mission to employ aerobraking as a critical maneuver. While Magellan had demonstrated the feasibility of the technique at mission end, its solar arrays had not been designed for that technology demonstration. After MGS and Mars Odyssey (to be covered in Section 12.3), aerobraking was similarly used by the Mars Reconnaissance Orbiter in 2007 to reduce its orbit from over 45,000 km to around 450 km for close observations of the red planet, returning to a two-wing design but without any tilt and instead using a 120-degree in-plane separation with the high-gain antenna occupying the symmetric third position, as shown in Fig. 12.5B [22].

In 2013 the Mars Atmosphere and Volatile Evolution (MAVEN) mission employed a unique "gull wing" solar array design, as shown in Fig. 12.5C, to provide atmospheric control as it traversed the upper atmosphere of Mars to study the history of Mars' atmosphere. The outer panel of each wing is bent up at 20 degrees to use aerodynamic forces to stabilize the spacecraft [23]. In addition, similar to the MGS, Maven's magnetometers were mounted at the solar array tips to reduce electromagnetic noise from the spacecraft bus.

12.2.3.2 Tilt-back configurations

The MESSENGER spacecraft used a solar array design similar to Magellan for its 2004 mission to orbit Mercury, using reflectors and solar off-pointing, although GaAs cells and thermally conductive composite face sheets were used together with off-pointing by folding the arrays back along the spacecraft body [24]. This allowed for operations at temperatures up to 150°C and survivability up to 275°C [25]. In addition, a large ceramic heat shield protected the spacecraft bus (Fig. 12.6A).

The Parker Solar Probe, launched in 2018, was designed to operate even closer to the sun, with a closest solar approach of about six million kilometers of the surface of the sun, where the incident solar flux is 650 kW/m^2, 475 times the solar intensity at Earth orbit. Although the main body of the spacecraft was kept behind a shadow shield to avoid

Figure 12.6 Tilt-back solar arrays. (A) MESSENGER (2004), (B) Parker Solar Probe (2021), (C) OSIRIS-ReX (2016). *Courtesy: NASA.*

heating, the solar arrays were designed to operate even at closest approach. To operate in this high-flux environment, the Parker Solar Probe uses several techniques [26]. First, like MESSENGER and earlier spacecraft, it implemented off-pointing of the main solar arrays by folding the arrays back along the spacecraft body as it approached the sun (Fig. 12.6B). During the period of close approach to the sun, a secondary solar array using concentrator solar cells was used. Both the main and the high-concentration solar arrays used water cooling to transfer heat to thermal radiators, that were shielded from direct solar illumination so they could radiate absorbed heat to space.

The OSIRIS-ReX mission launched in 2016 to collect a sample from asteroid Bennu and return it to Earth. To increase the ground clearance and potential dust accumulation during the touch-and-go sample collection maneuver, the arrays were swung backward to create a "Y-wing" configuration, as shown in Fig. 12.6C (what was called a "V-tilt" on Magellan 10, and as used on MESSENGER and Parker Solar Probe).

12.2.3.3 Other configurations

In 2001, Juno launched to become the first solar-powered spacecraft to explore Jupiter, reaching a full 5.5 AU from the Sun. Reminiscent of the Pioneer spacecraft, Juno used spin stabilization to reduce propellant during the long mission. This led to Juno's unique solar array configuration: the total area needed to be 60 m^2 to provide the 420 W needed at Jupiter where the solar irradiance is 25−30 times less than at Earth and temperatures reach −130°C, but the arrays needed to be very stiff to maintain a highly accurate spin. The symmetric three-wing design aligned with the hexagonal-shaped bus, as shown in Fig. 12.7A, represented the choice that met both requirements and allowed for an unconstrained view for the science instruments [27].

The arrays were not gimbaled to track the sun, but each individual array can be articulated by a small amount for vehicle control and stability [28]. Two magnetometers were mounted at the tip of one of the solar arrays, placing them 10 m and 12 m, respectively, from the body of the spacecraft for magnetic cleanliness, and also taking

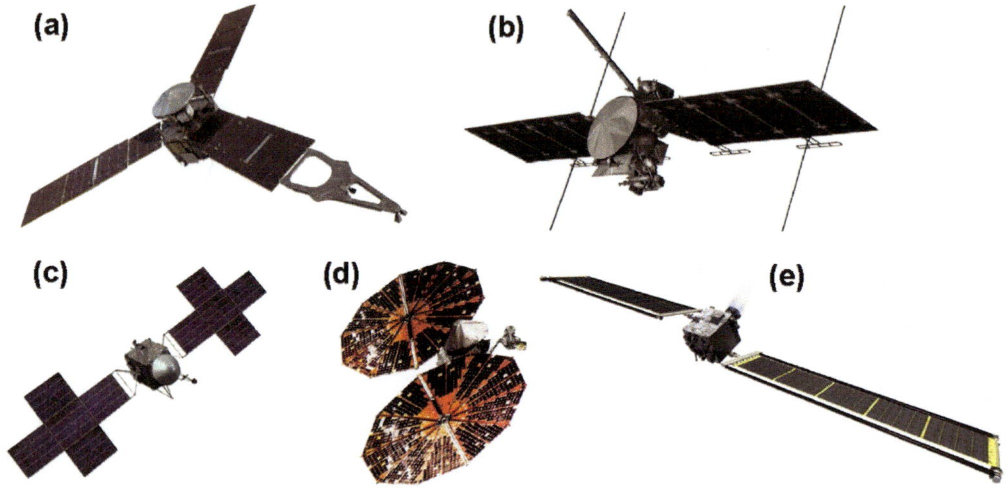

Figure 12.7 Deep-space science missions with unique solar arrays. (A) Juno (2001), (B) Europa Clipper (2025), (C) Psyche (2022), (D) Lucy (2021), (E) DART (2022). *Courtesy: NASA.*

advantage of the spacecraft's spin to separate environmental fields from those that rotate along with the spacecraft [29]. To maintain spin balance, that solar array was shortened, so the overall length of all three wings remained the same. Triple junction cells screened for low intensity performance were used, and the spacecraft stayed predominantly in a high-altitude orbit to minimize radiation exposure.

Europa Clipper, scheduled to launch in 2024, is notable because it has the largest solar arrays that will fly beyond Earth orbit, with a total area of 102 m^2 to provide 728 W of power at the end of its mission to survey Jupiter's moon Europa. The arrays have to survive the high-temperature environment of a Venus gravity assist, but the triple-junction cells (all 28,120 of them) are optimized for low-intensity, low-temperature operation at 5.46 AU (3.4% of AM0 irradiance) and $-150°C$, and the extreme radiation environment at Jupiter requires oversizing the array to compensate for significant degradation over the life of the mission. Clipper will carry an ice-penetrating radar, whose antennae are mounted to the solar arrays, as shown in Fig. 12.7B. The solar panels are strengthened to carry the load, the wiring is designed to provide ground and conductivity, and an eddy current damper is included within the solar array to keep induced magnetic fields low [30].

The Psyche mission to be launched in 2022 uses the cruciform solar array configuration shown in Fig. 12.7C that is uncommon for planetary science but very common for commercial telecommunications satellites orbiting the Earth. As the commercial space industry has exploded, a new trend is for the incorporation of more commercial, instead of custom, subsystems to reduce costs. Although maybe not so new, the 1992 Mars Observer mission used a similar strategy, as described in Section 12.3.1. There is also an interdependence of technologies developed for NASA missions, becoming commercialized and returning as commercial subsystems.

The circular, flexible-substrate solar arrays used on the 2007 Phoenix Mars lander, described in Section 12.2.2.1 were developed under Small Business Innovative Research (SBIR) contracts for NASA and in turn have been flying since 2013 on the commercial Cygnus cargo spacecraft to resupply the International Space Station under NASA's commercial crew and cargo program. They were used with small modifications on the InSight Mars lander in 2018 (see Section 12.2.2.1), and now the Lucy mission, launched in 2021, is also using these UltraFlex solar arrays, albeit much larger, as shown in Fig. 12.7D. Lucy will tour several Trojan asteroids in Jupiter's orbit, and like the two missions to the gas giant just mentioned, it also requires a large solar array area to produce 500 W at over 5 AU. Lucy's solar arrays have a diameter of 7.3 m, substantially larger than those used on Insight or Cygnus, and the scale-up required a substantial amount of engineering, particularly to route the large number of electrical wires through the central hub [31].

The Double Asteroid Redirect Mission (DART) to be launched in 2021 (was it launched?) will impact one body of a binary asteroid to study a deflection technique that may be used should a future asteroid arrive on a trajectory and with a mass that could seriously harm life on Earth. This technology demonstration was designed to minimize mass and stowed volume for potential flight as a rideshare. DART uses a roll-out solar array (Fig. 12.7E) that employs a flexible substrate for the solar cells, substantially reducing the solar array mass and stowed volume. In addition to being the first mission to fly these novel arrays as a primary power source, several reflector-based solar concentrators will be incorporated into the array to demonstrate 2X concentration on inverted metamorphic multijunction (IMM) solar cells for significantly improved performance on future low-intensity low-temperature (LILT) missions to the outer planets [32].

12.3 Missions using single solar panels

12.3.1 Deployable single panels

While most spacecraft use symmetrically located solar arrays to reduce the need for navigation corrections caused by asymmetric solar pressure, several planetary science missions found that even if power loads require the use of deployable solar arrays, using a single array can substantially reduce mission costs. The cost savings can be realized simply because of the reduced number of deployment mechanisms, yokes, and gimbals, but also because the reuse of prior Earth-orbiting designs can reduce development costs (such as for Mars Observer and the Lunar Reconnaissance Orbiter) and because a single wing can serve as an effective drag shield for aerobraking, which significantly reduces mission costs by eliminating requirement propellant mass that would otherwise be used to achieve final orbit (such as for Mars Climate Orbiter and Mars Odyssey). Missions with deployable single wings are listed in Table 12.2 and shown in Fig. 12.8.

The Mars Observer mission, launched in 1992, was the first of NASA's missions to Mars after the Viking program concluded in the 1970s. Cost containment was a priority, and a fixed-price contract was awarded to a telecommunications satellite provider with the expectation of maximum reuse of existing subsystems. Although most of the subsystems were off-the-shelf, the mission itself was unique. The solar array, while derived from a commercial telecom satellite, was atypically located on one side of the spacecraft to provide sufficient fields of view for radiative coolers required for the scientific instruments. The six-panel solar array shown in Fig. 12.8A was intended to be fully deployed once in Mars orbit to map geologic features, with just four panels deployed during the cruise phase to Mars [33]. The spacecraft never entered Mars orbit, most likely because of a propulsion system failure [34].

The Lunar Reconnaissance Orbiter mission (LRO) was designed as a low-cost mission to the Moon. A single solar array, as shown in Fig. 12.8B, was used to minimize the number of mechanisms and deployables. Launched in 2009, solar cell technology had advanced sufficiently so that the triple-junction cells on the single array still generated almost 2 kW of power, enough to power the seven scientific instruments needed to map the Moon including a radar, spectrometer, and neutron detector.

After the failure of the solar array structure on MGS, a different approach was taken for the Mars Climate Orbiter. As shown in Fig. 12.8C, a single solar array was designed to do double duty as the generator of primary power and also as a drag shield during

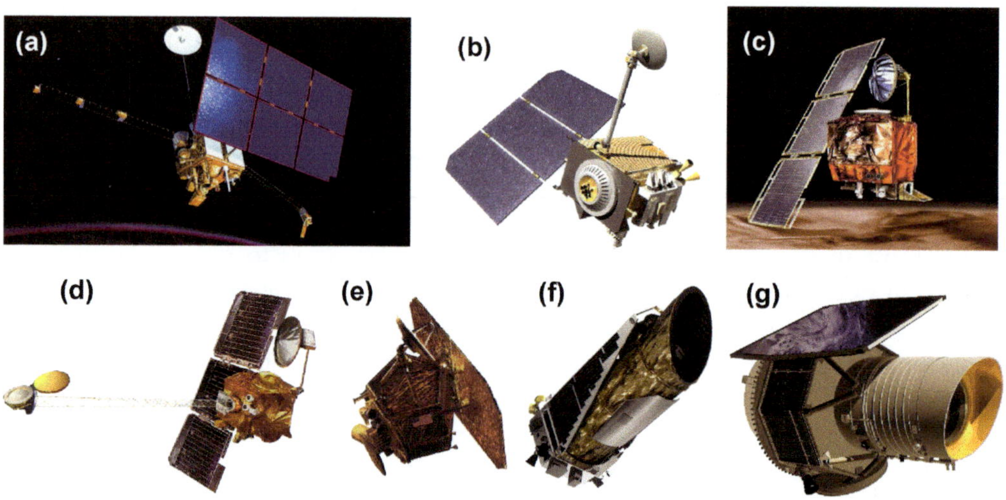

Figure 12.8 Deep space science missions using single solar arrays. (A) Mars Observer (1992), (B) Lunar Reconnaissance Orbiter (2009), (C) Mars Climate Orbiter (1998), (D) Mars Odyssey (2001), (E) EPOXI Deep Impact (2005), (F) Kepler (2009), (G) NEOWISE (2009). *Courtesy: NASA.*

aerobraking. During cruise, the array was deployed perpendicularly from the spacecraft bus to capture solar energy. During aerobraking, the array was oriented perpendicular to the direction of travel, so its surface area was used to slow the spacecraft. During this phase, primary spacecraft power was provided by batteries. Although Mars Climate Orbiter failed to achieve its mission, the failure was not caused by the solar arrays, and the follow-on mission, Mars Odyssey, adopted the same design for the solar power system, as shown in Fig. 12.8D. While the solar array design was the same for Mars Climate Orbiter [35], a more conservative aerobraking profile was adopted to allow for dust storms and wider atmospheric variations [36].

12.3.2 Missions utilizing single, fixed-panel solar arrays

The Deep Impact mission rendezvoused with comet Tempel 1 and released an impactor to expose materials on the comet's surface. To avoid deposition of ice and debris on the solar array, the array was mounted on one side of the spacecraft, as shown in Fig. 12.8E, and the spacecraft was oriented to keep the array directly opposite the comet upon impact. After observing the ejecta, the spacecraft continued as an extended mission to fly by a second comet, Hartley 2, and was renamed EPOXI.

While Earth-orbiting telescopes are not representative of typical planetary science missions, the 2009 Kepler mission is exploring planets that orbit stars outside of our galaxy (Fig. 12.8F), and NEOWISE, the extended mission for the Wide-Field Infrared Survey Explorer astrophysics mission, is studying the distribution of asteroids near Earth. Typical of other space-based telescopes, they orbit the Earth and operate by pointing at one point in space for an extended duration, needing to neither point toward the Earth nor toward the Sun. This configuration results in one side of the spacecraft continuously pointing to the Sun, so one solar panel on that side is the most efficient design, as shown in Fig. 12.8G. These solar arrays are mounted with a gap between the array and the spacecraft to provide a thermal break, so they are not technically body-mounted but are nonetheless similar in the simplicity of the solar panel design.

12.4 Missions using body-mounted solar arrays

The initial phase of US solar-powered solar system exploration concluded with the 1978 launches of the Pioneer 12 and 13 to Venus. Both of these spacecraft went "back to the future" with body-mounted solar cells, as shown in Fig. 12.9A. Pioneer 13 dropped four probes that descended through the dense Venusian atmosphere to the surface of Venus, while Pioneer 12 served as an orbiting data relay station. Both were based on the same design to reduce cost, and the absence of protruding solar arrays enabled Pioneer 13 to descend to about 110 km altitude while acquiring atmospheric data and releasing its large and three small probes. The first mission to fly past a comet was International

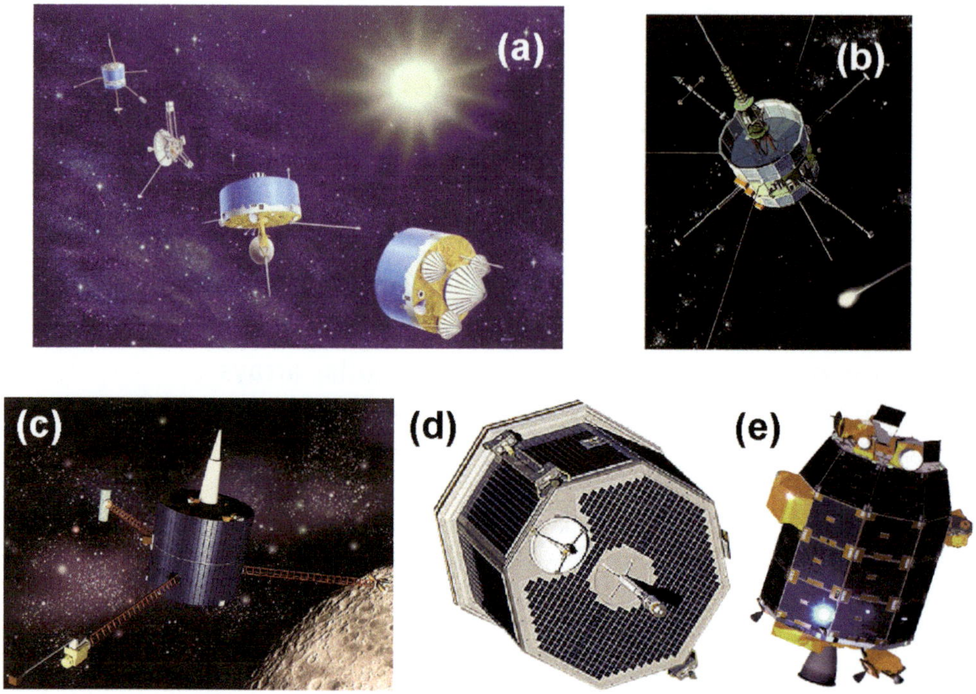

Figure 12.9 Deep space science missions using body-mounted solar arrays. (A) Pioneer 6–9, 10–11, 12, and 13 (left to right), (B) ICEE-3/ICE (1978), (C) Lunar Prospector (1998), Contour (2002), LADEE (2009). *Courtesy: NASA.*

Comet Explorer (ICE), the extended mission of the heliophysics International Sun-Earth Explorer (ISEE-3), employing a very similar design as Pioneer 12/13, as shown in Fig. 12.9B.

In 1998, the Lunar Prospector mission carried five instruments to measure gamma rays, electrons, neutrons, alpha particles, and the magnetic field at the Moon. None of these instruments required imaging, and each needed to be mounted far from the electrically noisy spacecraft bus. The design solution to carry these instruments was to mount them on radial booms while keeping the solar panels mounted on the spacecraft surface. This also enabled the use of relatively low-cost spin stabilization attitude control since the arrays did not need to be kept pointing at the Sun. After 19 months of lunar orbit, Prospector intentionally impacted the Moon to create a dust cloud that was studied from Earth. The penetrating cone shown in Fig. 12.9C was used to increase the penetration depth.

Selected as a Discovery mission just after Stardust, the 2002 Comet Nucleus Tour (CONTOUR) mission was designed to analyze material flowing from comets Encke, Schwassmann-Wachmann-3, and d'Arrest, taking it on a path ranging 0.75 to 1.5 AU from the Sun. Like Stardust, CONTOUR used a Whipple shield to protect the spacecraft from high-velocity particle impacts, but unlike Stardust the solar panels on CONTOUR were mounted directly on the body of the spacecraft to offer additional protection (Fig. 12.9D). As was the case for Lunar Prospector, this allowed for spin stabilization during the mission's long cruise periods, reducing both spacecraft operations and Deep Space Network tracking costs. Of note was the mission's first use of an indirect launch mode to reduce launch costs, maintaining an elliptical Earth orbit for several weeks before injecting into the heliocentric orbit needed for its first comet encounter.

The 2009 Lunar Atmosphere and Dust Environment Explorer (LADEE) employed a body-mounted solar panel configuration, as shown in Fig. 12.9E, for its mission to impact the Moon. Like DART, LADEE was originally considered a rideshare mission, and a key objective was to keep costs down. The spacecraft design was based on the Modular Common Spacecraft Bus developed by the NASA Ames Research Center to provide a nominally off-the-shelf spacecraft bus for multiple lunar missions. Thirty panels of silicon solar cells generated 295 W, and the body-mounted design simplified the spacecraft by eliminating deployment and articulation mechanisms, and spin stabilization simplified thermal design by eliminating hot and cold sides.

12.4.1 Cruisers

As described in Section 12.2.2.2, the designs of spacecraft used to deliver landed assets are relatively simple. The cruise stages for Mars Pathfinder, Mars Exploration Rover, Mars Science Laboratory, and Mars 2020 used body-mounted solar panels mounted directly to the aft end of the transit vehicle, as shown in Fig. 12.10, providing power for avionics and communications during transit in an economical design.

12.4.2 Rovers

The tiny Sojourner rover on Mars Pathfinder, the first to operate on another planet, was designed to be low cost, and its design was accordingly simple (Fig. 12.11A). A single solar panel affixed to the entire upper surface of the vehicle accommodated over 200 GaAs/Ge cells on the 0.22 m^2 panel to generate 35 W of electrical power at noon on Mars, roughly equivalent to the power generated by the AIMP lunar orbiter. While not traversing, excess electrical energy from the solar array was used to resistively heat a warm electronics box, although three radioisotope heater units provided the main source of nighttime warmth.

The solar panels on the Spirit and Opportunity rovers flown on the 2003 Mars Exploration Rover mission folded out to extend beyond the surface of the rover chassis as shown in Fig. 12.11B and stayed in a fixed horizontal position. They provided roughly 140 W at Mars noon, supplying power to acquire imagery and spectroscopic

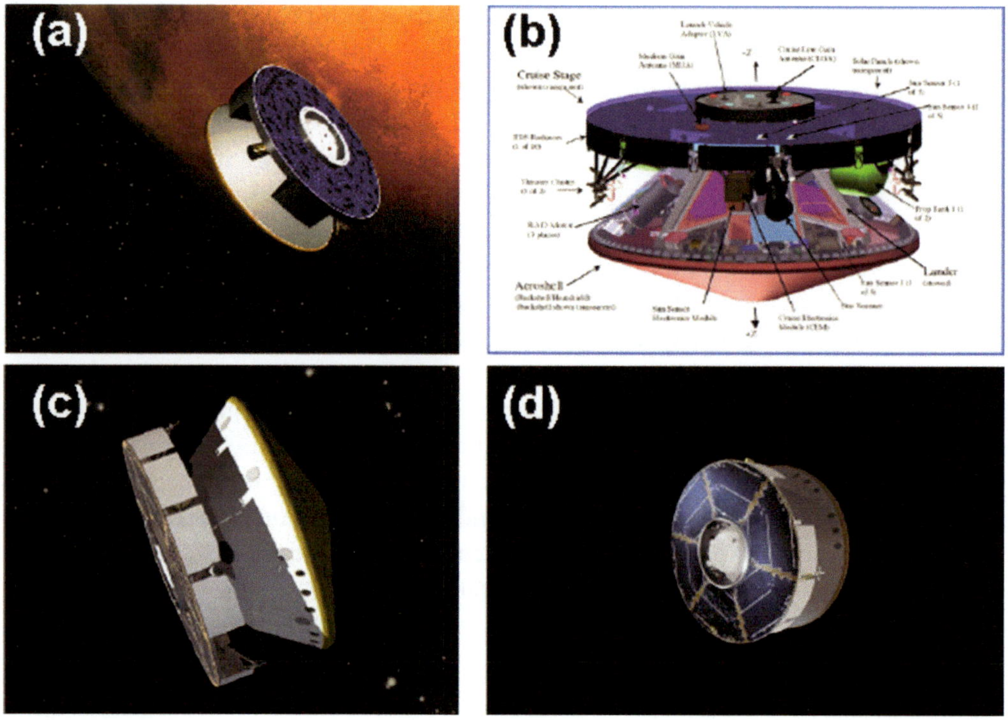

Figure 12.10 Deep space cruisers using body-mounted solar arrays. (A) Mars Pathfinder (1996), (B) Mars Exploration Rover (2003), (C) Mars Science Laboratory (2011), (D) Mars 2020 (2020). *Courtesy: NASA.*

measurements of the Mars surface. Opportunity holds the current planetary roving record of 45 km over its 14-year life.

The Mars 2020 mission delivered the tiny Ingenuity helicopter in addition to the nuclear-powered Perseverance rover. Ingenuity is the first aircraft to sustain controlled flight on another planet. It is powered by a single solar panel mounted on the top of its rotor, as shown in Fig. 12.11C. Because low mass is so critical to flight in the thin atmosphere at Mars, Ingenuity is the first spacecraft to be powered solely by inverted metamorphic solar cells that provide high-efficiency energy conversion (32%) on a flexible, low-mass substrate. In addition, the solar panel served a dual use as the reflector for a monopole omnidirectional antenna used for the helicopter to communicate with Perseverance [37].

The Volatiles Investigating Polar Exploration Rover (VIPER) is scheduled for a 2023 launch to map out water ice deposits directly from the ground in the lunar South Pole. Its solar arrays are mounted on the vertical sides of the rover to capture sunlight that rises at most 10 degrees above the horizon (Fig. 12.11D) [38].

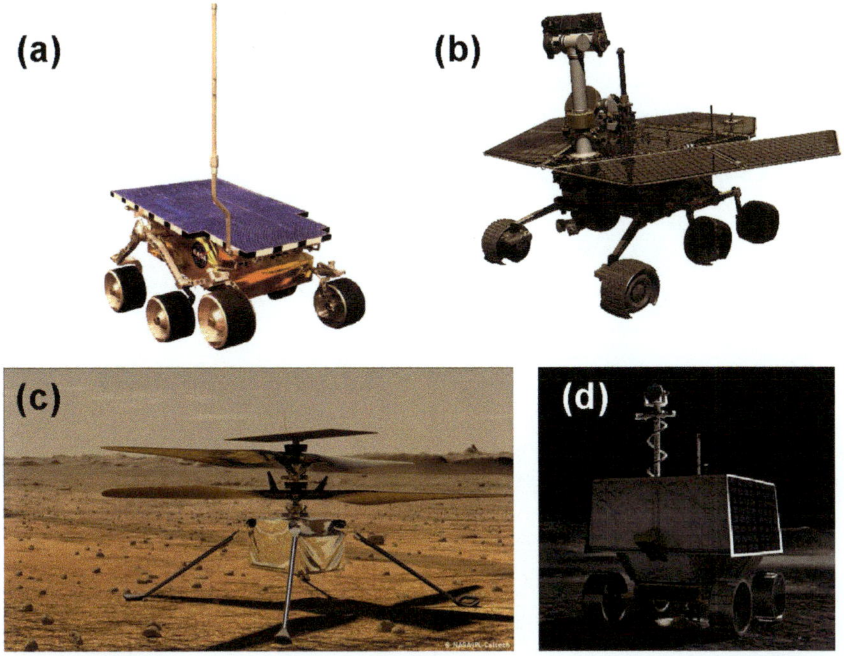

Figure 12.11 Deep space rovers using body-mounted solar arrays. (A) Mars Pathfinder "Sojourner" (1996), (B) Mars Exploration Rovers "Spirit" and "Opportunity" (2003), (C) Ingenuity (2020), (D) VIPER (2023). *Courtesy: NASA.*

12.5 Rideshare missions

The final class of planetary science missions are those that ride as secondary missions, sharing a launch vehicle with another, primary, mission. These are listed in Table 12.4 and shown in Fig. 12.12. Although these missions flew in low-Earth orbit and not deep space, the four CubeSat missions studied topics of interest to planetary sciences. GENE-SAT, Pharma-SAT, and O/OREOS studied the effects of the space environment on organic matter to understand how life may survive elsewhere in the Solar System. Q-PACE was designed to study how particles adhere and fragment in zero gravity to better understand the formation of planets and planetary ring systems. All four of these spacecraft were low-cost CubeSats employing body-mounted solar panels, harkening back to the earliest spacecraft.

However, rideshares are not new. Both Apollo 15 and 16 carried 35-kg lunar orbiters known as the Particles and Fields Subsatellites (PFS-1 and PFS-2) to measure magnetic fields, lunar gravity variations, and proton and electron flux densities. They were manually deployed by the Apollo astronauts and used body-mounted solar panels, as shown in Fig. 12.13A, just as CubeSats often do today.

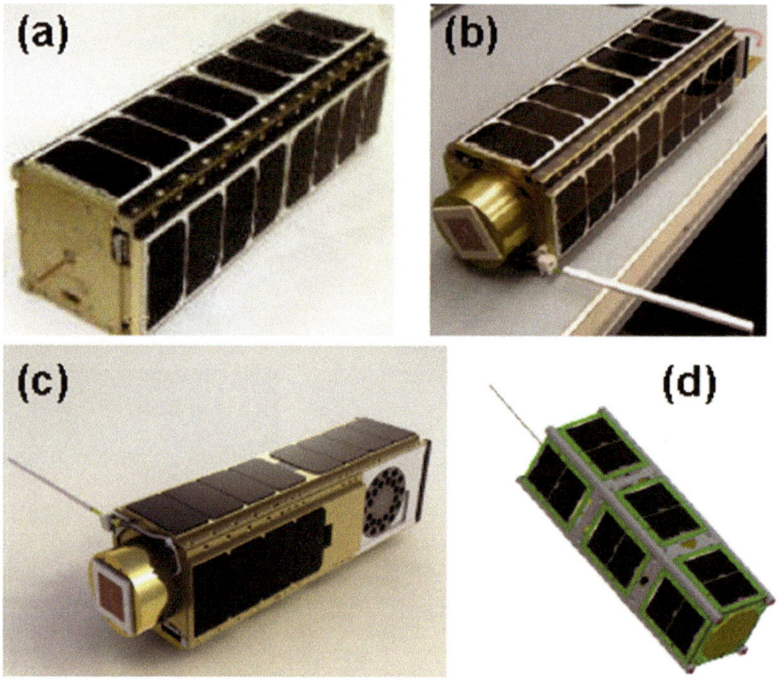

Figure 12.12 Solar panels on LEO planetary science missions. (A) GENE-SAT (2006), (B) Pharma-SAT (2009), (C) O/OREOS (2010), (D) Q-PACE (2021). *Courtesy: NASA.*

The Lunar Crater Observation and Sensing Satellite (LCROSS) mission flew as a rideshare to the Moon with the LADEE mission. LCROSS's innovative design incorporated the launch vehicle's secondary payload adapter ring as its primary structure, with its fuel tank mounted within the ring, and all of its subsystems mounted to the ring's six ports. One of these ports was allocated for the solar panel, as shown in Fig. 12.13B. This single wing was oriented to the Sun during the cruise phase. The primary objective of LCROSS was to observe ejecta created by the impact of the launch vehicle's upper stage into the Moon to determine the presence or absence of water ice in permanently shadowed regions [39].

Similarly, the twin MarCO spacecraft were rideshares with InSight to Mars. Each 6U CubeSat followed the same ballistic trajectory as InSight and was used to image the Insight landing. Also like the other rideshares described in this section, MarCo was designed to maximize the use of commercially available subsystems and components. Commercially available solar panels deployed in the two-wing, single-panel per wing orientation shown in Fig. 12.13C were used.

LunaH-Map will orbit the Moon after delivery there by Artemis-1 in 2022. It will use electric thrusters to enter a highly elliptical orbit with a periapsis at the South Pole to

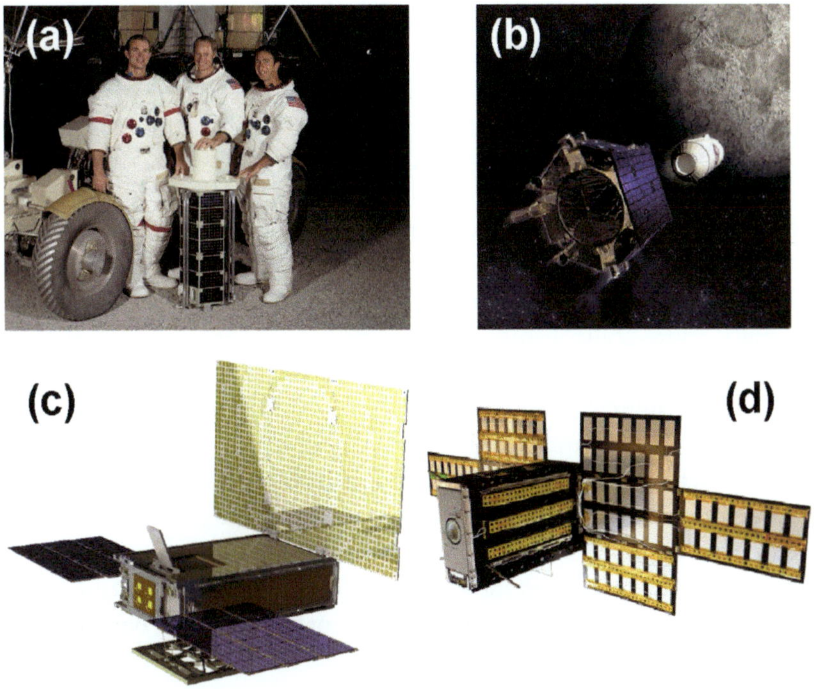

Figure 12.13 Solar panels on deep space planetary rideshare missions. (A) Apollo 15 PFS-1 (1971), (B) LCROSS (2009), (C) MarCO (2018), (D) LunaH-Map (2022). *Courtesy: NASA.*

map out concentrations of hydrogen. To power the thrusters and on-board neutron spectrometer the commercially available multipanel solar arrays were used as shown in Fig. 12.13D.

In contrast, Janus and Lunar Trailblazer are under development to independently arrive at their destination via a rideshare with a primary mission headed to a different location. The twin Janus spacecraft will visit asteroids 1996 FG3 and 1991 VH, respectively, after sharing a launch with the Psyche mission that will use electric propulsion to visit its eponymous asteroid. Similarly, Lunar Trailblazer will get itself to lunar orbit after sharing a launch with the IMAP mission headed to the Sun-Earth L1 point to study the Sun. Both Janus and Lunar Trailblazer will employ twin, gimbaled solar arrays.

12.6 Future trends

The missions described above show the increasing use of solar power in regions that are ever more inhospitable to photovoltaics: in high or low temperature, high radiation, low intensity, and dusty environments with long periods of eclipse. Research into photovoltaics capable of operations in the low-intensity, high-temperature, caustic environment of the Venus atmosphere [40,41] could enable very long duration missions to understand atmospheric and geologic processes on Earth's nearest planetary neighbor. Progress in concentrator technology such as the reflectors to be demonstrated on DART coupled with IMM cells hold the promise for efficient solar array usage out to Saturn [42], especially when coupled with the flexible blanket arrays used on Lucy and DART. These arrays can be scaled to provide up to 150 kW at 1 AU to enable the delivery of large payloads to Mars for human exploration [43], which should in turn should be ample for missions to Saturn and even beyond. Just as the solar arrays on OSIRIS-REx tilted out of the way of asteroid Bennu and MGS's arrays tilted back to provide aerodynamic stability, more controllability of solar arrays may be used to, for instance, fully retract an array for close approach to a surface [44] or for increased structural dynamic control for precise interactions with planetary atmospheres, further blurring the distinction between aerodynamics and astrodynamics begun with MGS and Maven.

The development of photovoltaics optimized for the solar spectrum at individual planetary bodies such as Mars or Titan could increase capabilities at those locations [45]. Novel configurations such as vertical solar arrays or a network of tent-like panels could provide power for polar rovers and infrastructure for *in situ* resource utilization [46]. And while it seems that losses would preclude the efficient use of power beaming, it is possible that solar arrays could be mounted on surface assets stationed in permanently dark regions to receive power from orbiting lasers [47].

Rideshare missions also may dramatically change the way planetary science missions are conducted. As the small spacecraft industry evolves, the use of commercial-off-the-shelf components, including solar arrays, could reduce mission costs and allow for a more frequent cadence of small planetary science missions. While it seems that the most efficient solar cells will still be used to reduce overall costs, it may be that low-efficiency cells with ultra-low manufacturing costs could contribute to reduced mission costs to enable very low-cost deep-space rideshare missions.

12.7 Conclusion

Conventional wisdom once held that solar arrays are not for use far from the Sun, in high radiation environments, near the poles on planetary surfaces, for missions that span long nights or often are exposed to dust, or in a subsurface application. Except for the latter issue, these barriers have been broken down. Juno, Europa Clipper, and Lucy are solar-powered missions to Jupiter, surmounting its low-intensity, low-temperature, and (for Juno and Clipper) high-radiation environment. Mars Polar Lander and Phoenix each took a different approach to solar power at the Mars south pole, and VIPER plans to use side-mounted solar panels at the southern lunar pole. InSight is taking seismic measurements during the days and nights on Mars and continues to operate after heavy dust storms like Opportunity did before, relying on natural dust dispersion from winds.

The achievements of these spectacular missions are grounded in the solar array designs of the past. The earliest spacecraft progressed from simple body-mounted cells, to fixed fold-out panels or paddles, ending up in little over a decade with the deployable and gimbaled multipanel wings that are often used today. While this gimbaled design is now nearly ubiquitous in commercial telecommunications satellites, unique mission requirements drive unique solutions for deep space planetary science missions. Even as we extend the use of solar arrays deeper into the Solar System, echoes of these early designs persist and all robotic US planetary science missions have employed some variant on these themes, employing unique variations as mission requirements demanded. Solar arrays became larger as their use at distances from the Sun has increased, and in a modern twist, the recent emergence of low-cost rideshare missions has roughly mirrored the progression of solar array designs on those earliest spacecraft, starting again with body-mounted cells and now emerging with gimbaled wings.

As shown by the missions described in this chapter, solar arrays have enabled the robotic exploration of the solar system and the space environment, and the reach of solar-powered missions continues to increase. Advances in solar cells and array structures promise to enable missions low into the atmosphere of Venus and perhaps out to Saturn and even beyond. By harnessing the power of the Sun, missions can last a very long time as evidenced by the 14-year operation of the Opportunity rover on Mars and NASA's oldest surviving spacecraft, Pioneer 6, launched in 1965.

In addition to providing primary power, solar arrays have been tasked with additional spacecraft functions. Surveyor 1's mast-mounted solar panel provided shadowing for thermal control of its lunar lander spacecraft; the Mars Exploration Rover landers supported off-ramps for the Spirit and Opportunity rovers; and Maven's gull-wings provided aerodynamic control in the Mars atmosphere. Beginning with Mariner 3, spacecraft have been designed so the solar arrays help to balance solar radiation pressure, and Mariner 10 relied on this feature to control the spacecraft to minimize propellant usage. Solar arrays have been used as critical drag shields for aerobraking on MGS and Mars Reconnaissance Orbiter, and several, beginning with MGS, used their solar array structures to position magnetometers far from their electromagnetically noisy spacecraft bus. The future of solar-powered exploration of our solar system has never been brighter.

Acknowledgments

The author thanks Rob Manning and Andreea Boca at JPL, Geoffrey Landis at NASA Glenn Research Center, and Mike Amato and Florence Tan at NASA Goddard Space Flight Center for fruitful and interesting discussions.

References

[1] The Boeing Company, Lunar Orbiter I: Extended-Mission Spacecraft Subsystem Performance, NASA CR-870, NASA HQ, Washington, D.C, September, 1967. Available at: https://ntrs.nasa.gov/api/citations/19670026411/downloads/19670026411.pdf.

[2] J.A. Dunne, E. Burgess, The Voyage of Mariner 10, NASA SP-424, 1978. Available at: https://history.nasa.gov/SP-424/contents.htm.
[3] NASA, Mariner 1, NASA Space Science Data Coordinated Archive, Version 5.1.12, November 17, 2021. https://nssdc.gsfc.nasa.gov/nmc/spacecraft/display.action?id=MARIN1.
[4] J.R. Scull, Guidance and control of the Mariner planetary spacecraft, in: 1st IFAC Symposium on Automatic Control in the Peaceful Uses of Space (June 21-24, 1965, Stavanger, Norway) 2 (1) 97-107.
[5] NASA, Mariner 10, NASA Space Science Data Coordinated Archive, Version 5.1.12, https://nssdc.gsfc.nasa.gov/nmc/spacecraft/display.action?id=1973-085A. (Accessed 17 November, 2021).
[6] NASA Ames Research Center, Science Instrument Survey, NASA TM-108628, May 1973. Available at: https://archive.org/details/nasa_techdoc_19930074143.
[7] NASA Space Science Data Coordinated Archive, Surveyor 1 Website, Version 5.1.12, Oct. 28, 2021. https://nssdc.gsfc.nasa.gov/nmc/spacecraft/display.action?id=1966-045A (Accessed 18 November 2021).
[8] R.G. Goldman, G.S. Gurski, W.H. Hawersaat, Description of the SERT II Spacecraft and Mission, NASA Technical Memorandum TM X-52862, January 1, 1970. Available at: https://ntrs.nasa.gov/citations/19700024438.
[9] W.R. Corliss, The Interplanetary Pioneers, Volume 2: System Design and Development, NASA SP-279, January 1, 1972. Available at: https://ntrs.nasa.gov/citations/19730009155.
[10] NASA JPL, Magellan Mission to Venus, Chapter 4: The Magellan Spacecraft, https://www2.jpl.nasa.gov/magellan/guide4.html. (Accessed 19 September 2021).
[11] C. Young (Ed.), The Magellan Venus Explorer's Guide, JPL Publication 90-24, August 1990. Available at: https://www2.jpl.nasa.gov/magellan/guide.html.
[12] R. Dobson, J. Kukulka, G. Dalermanji, L. Roufberg, A. Ahmad, J. Lyons, GaAs/Ge Solar Panels for the SAMPEX Program, AIAA-92-1063 AIAA Aerospace Design Conference, Irvine CA, USA, February 3-6, 1992, https://doi.org/10.2514/6.1992-1063.
[13] D.M. Murphy, The Scarlet Solar Array: Technology Validation and Flight Results. https://www.yumpu.com/en/document/view/35962023/the-scarlet-solar-array-pds-small-bodies-node. (Accessed 2 May 2021).
[14] NASA Glenn Research Center, Deep Space 1 Ion Propulsion System Operation Sequence and Status, May 21, 2008. https://www.nasa.gov/centers/glenn/about/history/ds1opseq.html (Accessed 2 May 2021).
[15] N.S. Fatemi, S. Sharma, O. Buitrago, P.R. Sharps, R. Blok, M. Kroon, C. Jalink, et al., Performance of High-Efficiency Advanced Triple-Junction Solar Panels for the LILT Mission Dawn, 31st IEEE Photovoltaic Specialists Conference, Lake Buena Vista, FL USA, January 3–7, 2005. Available at: https://trs.jpl.nasa.gov/handle/2014/40632.
[16] M.T. Zuber, C.T. Russell, D.E. Smith, D.H. Lehman, T.L. Hoffman, S.W. Asmar, M.M. Watkins, GRAIL: Mapping the Moon's interior, Space Sci. Rev. 178 (1) (2013) 3–24.
[17] S. Gasner, K. Sharmit, P. Stella, C. Craig, S. Mumaw, The Stardust solar array, in: Proc. of the 3rd World Conference on Photovoltaic Energy Conversion, 1, May 11–18, 2003, pp. 646–649. Osaka, Japan, https://ieeexplore.ieee.org/document/1305364.
[18] A. Techsystems, ATK Ultraflex Solar Arrays Deploy and Provide Critical Power to the Mars Phoenix Lander. http://www.spaceref.com/news/viewpr.html?pid=25506. (Accessed 2 May 2021).
[19] P.M. Stella, N. Mardesich, R.C. Ewell, R.L. Mueller, S. Endicter, D. Aiken, K. Edmondson, C. Fetzer, Multijunction Solar Cell Technology for Mars Surface Applications, 4th World Conference on Photovoltaic Energy Conference, May 7–12, 2006, pp. 1907–1910, https://doi.org/10.1109/WCPEC.2006.279869. Waikoloa, HI, USA.
[20] P.M. Stella, R.G. Ross, B.S. Smith, G.S. Glenn, K.S. Sharmit, Mars Global Surveyor (MGS) high temperature survival solar array, in: Conference Record of the 25th IEEE Photovoltaic Specialists Conference, May 13-17, 1996, pp. 283–288, https://doi.org/10.1109/PVSC.1996.564001. Washington, D.C. USA.
[21] NASA, Mars Global Surveyor: An Explanation of How Aerobraking Works, https://mars.nasa.gov/mgs/confrm/aerobexp.html. (Accessed 2 May 2021).

[22] NASA, Mars Reconnaissance Orbiter: Electrical Power. For further details see: https://mars.nasa.gov/mro/mission/spacecraft/parts/electricalpower. (Accessed 2 May 2021).
[23] NASA, MAVEN website, (Updated June 1, 2022). Available at: https://mars.nasa.gov/maven/ (Accessed July 23, 2022).
[24] G. Dakermanji, J. Jenkins, C.J. Ercol, The MESSENGER spacecraft solar array design and early mission performance, in: IEEE 4th World Conference on Photovoltaic Energy, Waikoloa, HI, May 7–12, 2006 as U.S. Government funded work, it is publicly Available at: https://messenger.jhuapl.edu/Resources/Publications/Dakermanji.et.al.2006.pdf.
[25] C.J. Ercol, G. Dakermanji, B. Le, The MESSENGER spacecraft power subsystem thermal design and early mission performance, AIAA 2006-4144, in: 4th International Energy Conversion Engineering Conference and Exhibit (IECEC), June 26-29, 2006, p. 6. San Diego, CA, as U.S. Government funded work, it is publicly Available at: https://messenger.jhuapl.edu/Resources/Publications/Ercol.et.al.2006.pdf.
[26] G.A. Landis, P.C. Schmitz, J. Kinnison, M. Fraeman, et al., in: Solar power system design for the Solar Probe+ mission, AIAA2008-5712, 6th International Energy Conversion Engineering Conference, July 28–30, 2008, Cleveland, OH USA, Available at: https://doi.org/10.2514/6.2008-5712.
[27] NASA, Juno's Solar Cells Ready to Light up Jupiter Mission, July 15, 2011 updated August 7, 2017, https://www.nasa.gov/mission_pages/juno/launch/Juno_solarpower.html (Accessed 2 May 2021).
[28] NASA, Juno Spacecraft Information, https://spaceflight101.com/juno/spacecraft-information. (Accessed 2 May 2021).
[29] J.E.P. Connerney, M. Benn, J.B. Bjarno, T. Denver, J. Espley, J.L. Jorgensen, P.S. Jorgensen, et al., The Juno magnetic field investigation, Space Sci. Rev. 213 (February 2017) 39–138.
[30] G.A. Landis, J. Fincannon, Study of power options for Jupiter and outer planet missions, in: 2015 IEEE 42nd Photovoltaic Specialist Conference (PVSC), 14-19 June 2015, pp. 1–5, https://doi.org/10.1109/PVSC.2015.7356136. New Orleans, LA USA.
[31] S. Clark, Scientists Optimistic Planetary Probes Won't Face Coronavirus Launch Delays, Spaceflight Now, April 29, 2020. https://spaceflightnow.com/2020/04/29/scientists-optimistic-planetary-probes-wont-face-coronavirus-related-launch-delays (Accessed 2 May 2021).
[32] E. Gaddy, Transformational Solar Array Final Report, NASA Contract NNC16CA19C, https://ntrs.nasa.gov/api/citations/20170010684/downloads/20170010684.pdf. (Accessed 2 May 2021).
[33] RCA Astro-Electronics, Mars Observer: Phase 0 Safety Review Data Package TG-27886, NASA Contractor Report CR-180526, November 17, 1986. Available at: https://ntrs.nasa.gov/citations/19870011586.
[34] Mars Observer Mission Failure Board Report: Volume 1, December 31, 1993. Available at: https://spacese.spacegrant.org/Failure%20Reports/Mars_Observer_12_93_MIB.pdf
[35] M.A. Johnson, W.H. Willcockson, Mars Odyssey aerobraking: the first step towards autonomous aerobraking operations, in: 2003 IEEE Aerospace Conference Proceedings, Big Sky, MT, USA, March 8–15 2003, pp. 8_3503–3508_3510. IEEE-AC paper #1169, https://ieeexplore.ieee.org/document/1235536.
[36] J.C. Smith, J.L. Bell, Mars Odyssey Aerobraking, AIAA/AAS Astrodynamics Specialist Conference, August 5-8, 2002, Monterey, CA, 2001, https://doi.org/10.2514/6.2002-4532 (AIAA 2002-4532).
[37] J.L. Green, N. Chahat, Talking to Ingenuity and Other Space Robots, Gravity Assist Podcast, Season 5, Episode 2, Available at: https://podtail.com/podcast/gravity-assist/gravity-assist-talking-to-ingenuity-and-other-spac/.
[38] NASA, VIPER's Mission Operations. https://www.nasa.gov/viper/lunar-operations. Updated October 12, 2021. (Accessed 17 November 2021).
[39] K. Ennico, M. Shirley, A. Colaprete, L. Osetinsky, The Lunar Crater observation and sensing satellite (LCROSS) payload development and performance in flight, Space Sci. Rev. 167 (2012) 23–69, https://doi.org/10.1007/s11214-011-9753-4.
[40] G.A. Landis, Power systems for Venus surface missions: a review, Acta Astronaut. 187 (October 2021) 492–497.

[41] G.A. Landis, E. Haag, Analysis of solar cell efficiency for Venus atmosphere and surface missions, AIAA-2013-4028, in: AIAA 11th International Energy Conversion Engineering Conference, San Jose CA, July 15-17, 2013, https://doi.org/10.2514/6.2013-4028.
[42] A. Boca, R. Warwick, B. White, R. Ewell, A data-driven evaluation of the viability of solar arrays at Saturn, IEEE J. Photovoltaics 7 (4) (July 2017) 1159—1164.
[43] C.R. Mercer, M.L. McGuire, S.R. Oleson, M.J. Barrett, Solar Electric Propulsion Concepts for Human Space Exploration, NASA TM-2016-218921, AIAA-2015-4521, AIAA Space 2015 Conference, Pasadena CA, August 31-September 2, 2015. Available at: https://ntrs.nasa.gov/citations/20160003683.
[44] J. Moore, Simple Reliable Retractable Lunar Lander Solar Array, NASA SBIR Contract Phase II, 2020-2022. https://techport.nasa.gov/view/103057. (Accessed 17 November 2021).
[45] P.M. Beauchamp, J.A. Cutts, Solar Power Technologies for Future Planetary Science Missions, Report JPL D-101316, JPL-Caltech, Pasadena, CA USA, December 2017. Available at: https://solarsystem.nasa.gov/resources/548/solar-power-technologies-for-future-planetary-science-missions/.
[46] G.A. Landis, T.W. Kerslake, P.P. Jenkins, D. Scheiman, Mars solar power, in: NASA/TM—2004-213367, November 2004, Also: AIAA 2nd International Energy Conversion Engineering Conference, August 16—19, 2004. Providence, RI, Available at: https://ntrs.nasa.gov/api/citations/20040191326/downloads/20040191326.pdf.
[47] J. Grandidier, P. Jaffe, W.T. Roberts, M.W. Wright, A.A. Fraeman, C.A. Raymond, A. Austin, et al., Laser Power Beaming for Lunar Night and Permanently Shadowed Regions, Whitepaper #302 submitted to the Planetary Science and Astrobiology Decadal Survey 2023-2032. Published by the Bulletin of the AAS, March 18, 2021. Available at: https://baas.aas.org/pub/2021n4i302/release/1?readingCollection=7272e5bb.

CHAPTER THIRTEEN

Lunar science based on Apollo solar cell measurements

S. Aranya[1,2] G. Gopkumar[1] and T.E. Girish[1]
[1]Department of Physics, University College, Trivandrum, Kerala, India
[2]Department of Physics, Sree Narayana College, Trivandrum, Kerala, India

13.1 Introduction

Apollo manned missions to the Moon have made significant contributions to lunar science and exploration of its material resources [1]. Lunar dust detection experiments on the Moon between the years 1969 and 1976 using silicon (Si) solar cells and different temperature probes is relevant in this context [2,3]. Recently we have attempted to investigate weather changes in Trivandrum (Kerala) India using solar panel output variations. The output of solar cells often reflects their environment and serves as an environmental probe; a recent study from Indonesia documents the value of this approach [4]. In this chapter we present our studies on the analysis of the results of Apollo 14 dust detector experiments on the Moon conducted between the years 1971 and 1976 from a lunar science perspective. This work is a continuation of our earlier studies on the performance of solar cells in the lunar environment [5,6].

13.2 Solar cells used for the Apollo 14 dust detector experiment

This experiment was conducted with the Apollo 14 Lunar Module by the principal investigator, J.R. Bates. The purpose of the Lunar Dust Detector Experiment (also referred to as the Dust, Thermal, and Radiation Engineering Measurements package or DTREM) on the Apollo Lunar Surface Experiments Package (ALSEP) was to assess the long-term effects of the lunar environment on silicon solar cells by measuring the power output reduction caused by high-energy solar/cosmic ray particle and ultraviolet radiation damage and dust accumulation. The experiment also included measurement of lunar surface temperatures through measurements of reflected infrared brightness and solar cell temperatures [7].

The DTREM had two components: a sensor package, which comprised a G-10 fiberglass box and solar cells with a mass of 0.159 kg, mounted on the top northwest corner of the central station sun shield; a printed circuit board, 0.118 kg, located within the central station that interfaced with a preamplifier; and the power distribution unit of the ALSEP data subsystem. Three 10-ohm-cm *n*-on-*p* structure single-crystal silicon

Photovoltaics for Space
ISBN 978-0-12-823300-9, https://doi.org/10.1016/B978-0-12-823300-9.00009-1

solar cells, each 1 × 2 cm, were mounted on the top horizontal Kovar sheet facing upward in this experiment.

Voltage from the solar cells ranged from 0–70 mV during sunlight conditions, depending on sun angle and individual cell characteristics. Other details of these solar cells were as follows:

1. AX04 solar cell: no cover glass and not preirradiated;
2. AX05 solar cell: provided with same glass cover as AX06 and preirradiated on the ground with 10E15 electrons of 1 MeV energy;
3. AX06 solar cell: provided with 0.15-mm-thick blue filtered glass cover for protection against particles and ultraviolet radiation.

The Apollo 14 ALSEP central station was located at 3.6440 S latitude, 17.4775 W longitude (IAU Mean Earth Polar Axis coordinate system). For the Apollo 14 central station there was a loss of signal and reacquisition six times between March 1, 1975 and August 7, 1977. The DTREM was turned off when ALSEP operations were terminated on September 30, 1977 [8].

13.3 Details of solar cell data used in this study

The original data were stored on 16-mm microfilm reels in the form of data tables, which are now available as digital image files in the National Aeronautics and Space Administration (NASA) Space Science Data Coordinated Archive (NSSDCA) [9,10] located at NASA's Goddard Space Flight Center in Greenbelt, MD, USA. For each day, the data tables consist of voltages from each cell and temperatures from the three thermistors during daytime conditions. DTREM data consist of the following:

1. internal temperature (AX01);
2. external infrared temperature (AX03);
3. bare solar cell voltage raw output (AX04);
4. 0.15-mm covered irradiate solar cell raw voltage output (AX05);
5. 0.15-mm covered solar cell raw voltage output (AX06);
6. the solar cell voltages were corrected for temperature as a function of time.

The data tables were made available to us through request to NSSDCA in 2011 through a staged NSSDCA site and transferred by file transfer protocol. Data was supplied to us every 54 s for each day, with coverage from May 1971 to February 24, 1976.

13.4 Variations of output of Apollo 14 solar cells within a lunation period

We have studied output variations (daily maximum voltages) of AX04, AX05, and AX06 solar cells within a period of 59 lunations between May 1971 to February 1976. Practically, the length of a lunation period is equal to 15 Earth days. Typical examples of these variations are shown for three different lunations in Figs. 13.1–13.3. One interesting result we want to emphasize is that Apollo 14 solar cell output variations were not observed for *all days* within a lunation period. During the fourth lunation of 1971 the solar cell output was observed for 13 out of 15 days and during lunation of 1972 for only 12 out of 15 days.

However, we noted that solar cell output variations were observed for all 15 days during the initial lunation of 1974. The day of maximum solar cell output in a lunation period can be identified as the lunar noon. Since one lunation is equivalent to 14.5 days near the lunar equator, we chose seven days before and seven days after each lunar noon day in a completed lunation period. Based on these criteria the AX04 solar cell output for the fourth lunation period in 1971 is plotted (Fig. 13.4), which consists of 15 days starting from the lunar astronomical sunrise day. From this figure, we can find that solar cell output was not observed for the first 2 days during this lunation. During lunations in 1972, solar cell output was absent for first 2 days in the dawn side and last day in the dusk side during a lunation period. We will discuss this in detail in Section 13.7.

During the fourth lunation period in 1971, the Apollo 14 solar cells are in the beginning of life (BOL) mode of operation and not subjected to any significant degradation. During this lunation the third day output of the three solar cells is found to be about 60% of the noon maxima. If we assume that the Apollo solar cell output variations mimic solar radiation changes at BOL conditions, then we can infer diurnal

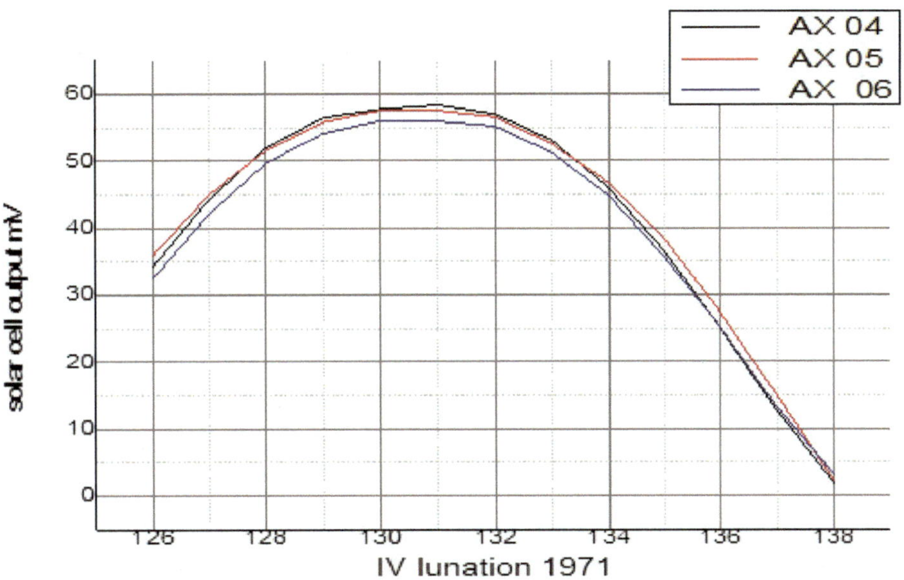

Figure 13.1 Variations of the output (daily maximum voltages) of the three Apollo 14 solar cells AX04, AX05, and AX06 during the fourth lunation of 1971 on the Moon (Earth days 126–138). Solar cell output is only observed for 13 days in this lunation.

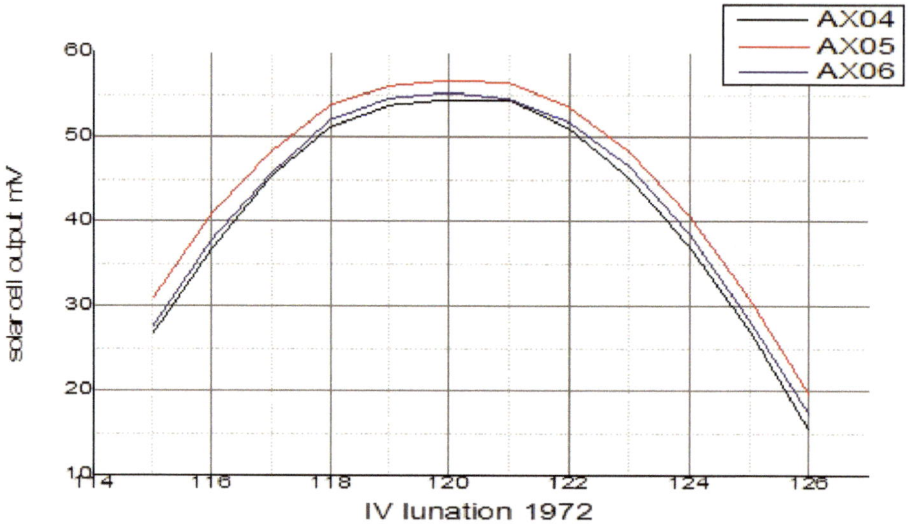

Figure 13.2 Variations of the output (daily maximum voltages) of the three Apollo 14 solar cells AX04, AX05, and AX06 during the fourth lunation of 1972 on the Moon (Earth days 115–126). Solar cell output is observed only for 12 days in this lunation.

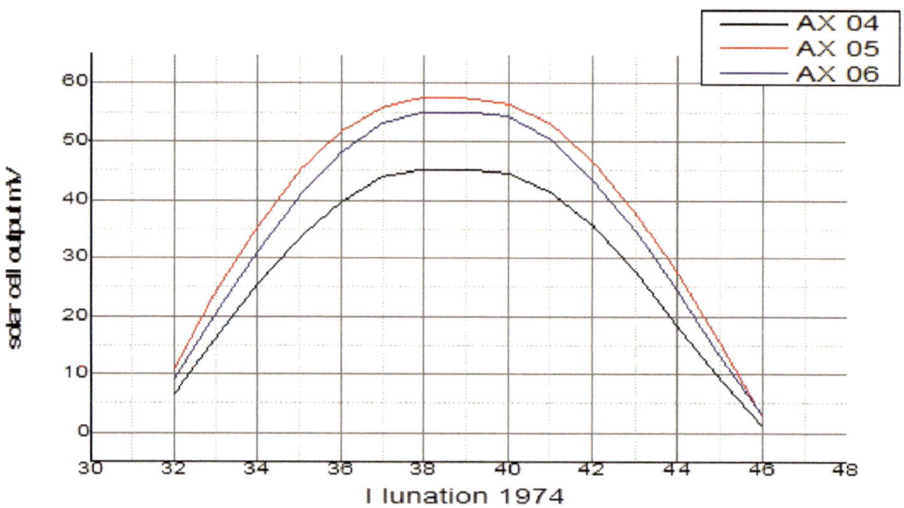

Figure 13.3 Variations of the output (daily maximum voltages) of the three Apollo 14 solar cells AX04, AX05, and AX06 during the first lunation of 1974 on the Moon (Earth days 32–46). Solar cell output is observed for all the 15 days in this lunation.

solar radiation power variations near the Moon's equator from the former. The result of this inference is shown in Fig. 13.5. Here, we have assumed that the average noon maxima of solar radiation power near the lunar equator is 1368 W/m^2 [11].

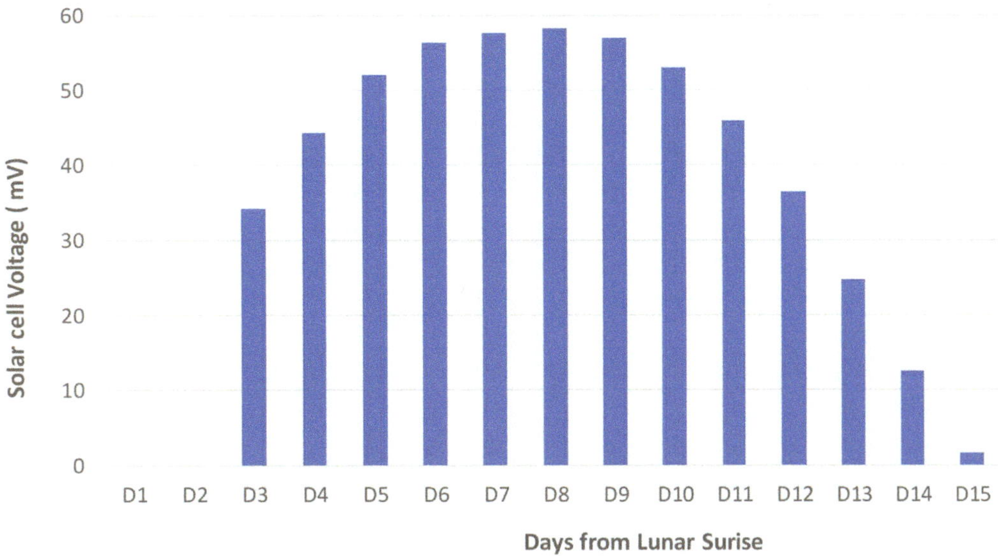

Figure 13.4 Variations of the output of the Apollo 14 solar cell AXO4 during the fourth lunation on the Moon in 1971 as a function of days from astronomical lunar sunrise. Here eighth day is the lunar noon where the solar cell output is at maxima. Solar cell output on the Moon is found to be absent during the first 2 days of this lunation.

13.5 Annual variations of the solar cell output on the Moon

We have studied lunation-to-lunation changes in the noontime maximum output of different solar cells on the Moon for each year between 1971 and 1975. The results for

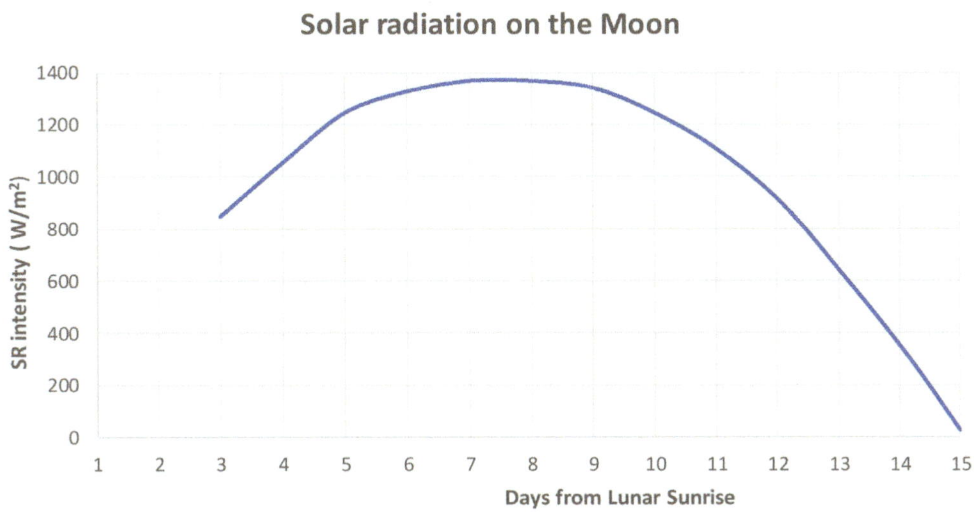

Figure 13.5 Inferred solar radiation intensity near the lunar equator inferred from the variations of output of the Apollo 14 solar cells during the fourth lunation period in 1971.

the years 1973 and 1974 are shown in Figs. 13.6 and 13.7 respectively. Cyclic changes in Sun—Moon distance cause an annual variation in the solar cell output observed on the Moon. During the years 1973—74, the lunar noontime solar cell output maxima shows a minima during lunar aphelion period (June—July months as observed from Earth), and the same shows a maxima during lunar perihelion period (December month as observed from Earth). The annual variations in noontime maximum output voltages for covered Apollo solar cells (AX05 and AX06) during 1973 is found to be 7.4%, and during the year 1974 the same is found to be 8.55%. This is close to the value of 8% given in the Apollo mission summary report of NASA [1]. If the maximum solar radiation power near the lunar equator is assumed to be 1368 W/m^2 [11], then the annual variation in solar radiation intensity inferred from Apollo solar cell output observations during 1973 is 101.23 W/m^2 and during 1974 is 116.96 W/m^2.

13.6 Solar proton events: Long-term degradation of Apollo 14 silicon solar cells

In Fig. 13.8 we show lunation-to-lunation change in the noontime voltages of three Apollo solar cells from May 1971 to February 1976. We find that long-term degradation in solar cell output is significant only for the uncovered AX04 solar cell

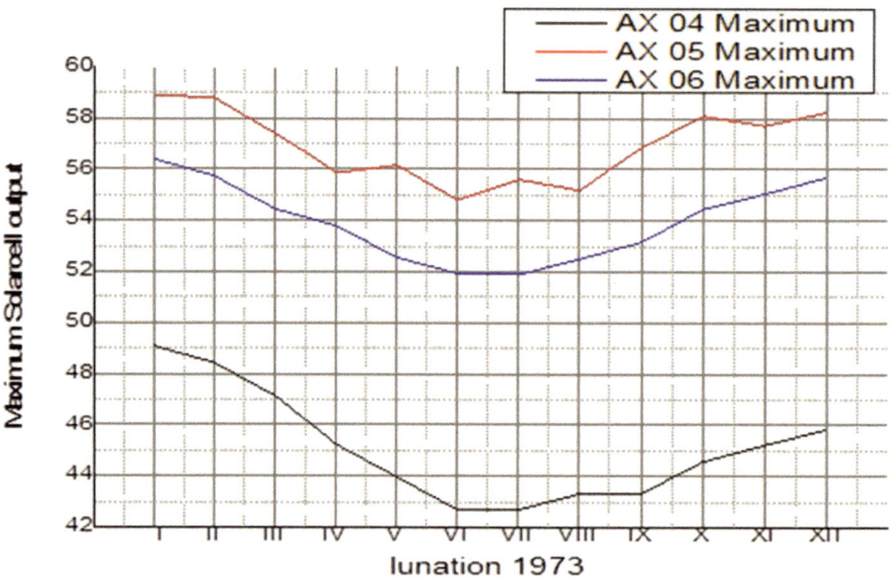

Figure 13.6 Annual variations of the noon maxima of the three Apollo 14 solar cells during different lunations during 1973. The maxima and minima in solar cell output are observed during the lunar perihelion and aphelion periods.

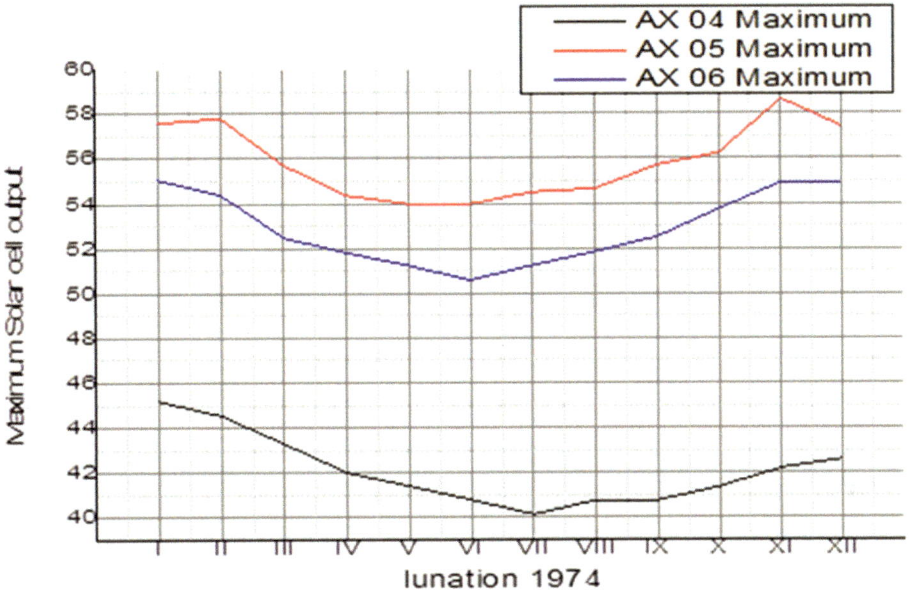

Figure 13.7 Annual variations of the noon maxima of the three Apollo 14 solar cells during different lunations during 1974. The maxima and minima in solar cell output are observed during the lunar perihelion and aphelion periods.

during this period. Solar energetic particle (mainly protons) (SEP) events can affect performance on the Moon. The NOAA classification of the SEP on the "S" scale [12] is shown in Table 13.1 along with possible effects on solar cells expected for each SEP category. The SEP flux in the heliosphere is measured in proton flux units or pfu (1 pfu = 1 particle per unit area per unit time per unit solid angle: number of particles cm^{-2} s^{-1} sq. $radian^{-1}$) In Fig. 13.9 we have plotted the occurrences of SEP's during 1971–76 belonging to different categories such as weak, minor, moderate, strong, and severe. Fluxes for the weak category are <10 pfu. From this figure, we observe that SEP potentially affected Apollo solar cell performance on the Moon only for the years 1971 and 1972.

13.7 Discussion and analysis

We suggest that Apollo 14 solar cell output variations may be a probe to infer solar radiation changes near the lunar equator. The amplitude of annual (seasonal) variations in inferred solar radiation changes from Apollo 14 solar cell observations is found to be in agreement with recent theoretical models and summary of published Apollo mission science results [1,9]. It is necessary to consider in detail the Apollo 14 solar cell output changes within a lunation period or lunar day. During May 1971 to Feb 1976, we found

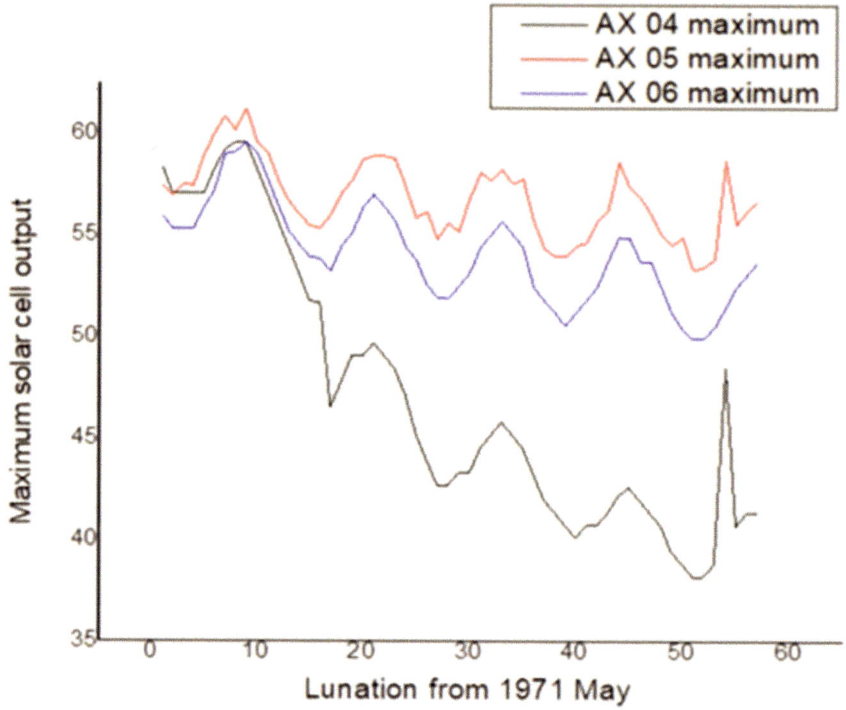

Figure 13.8 Long-term variations of the lunation-to-lunation changes in the noontime maxima of the output of three different Apollo solar cells during the years 1971–76. The first lunation corresponds to the fourth lunation period in the year 1971.

Table 13.1 NOAA intensity classification of solar proton events in proton flux units (pfu) (particles $cm^{-2}s^{-1}sq.\ radian^{-1}$) and associated effects on solar cells.

Scale	Description	pfu	Possible effects on solar cells in space environments
S1	Minor	10	No significant effects on solar cells
S2	Moderate	10^2	No significant effects on solar cells
S3	Strong	10^3	Possible small reduction in efficiency of solar cells
S4	Severe	10^4	Possible degradation of solar cell efficiency
S5	Extreme	10^5	Permanent damage to solar cells

Source: NOAA.

that the solar cell output was observed on the Moon only for 12–14 days for a majority of the lunation periods. The length of a lunar day varies between 14.5 and 15 days near the lunar equator [13]. Thus we can expect solar cell output observations for 15 days in an ideal situation with half of the day illuminated for the 15th day even if the lunar day length is 14.5 days. The number of days of observation of Apollo 14 solar cell output per lunation period is 15 days for one lunation period, 13 days for 54% of the lunations,

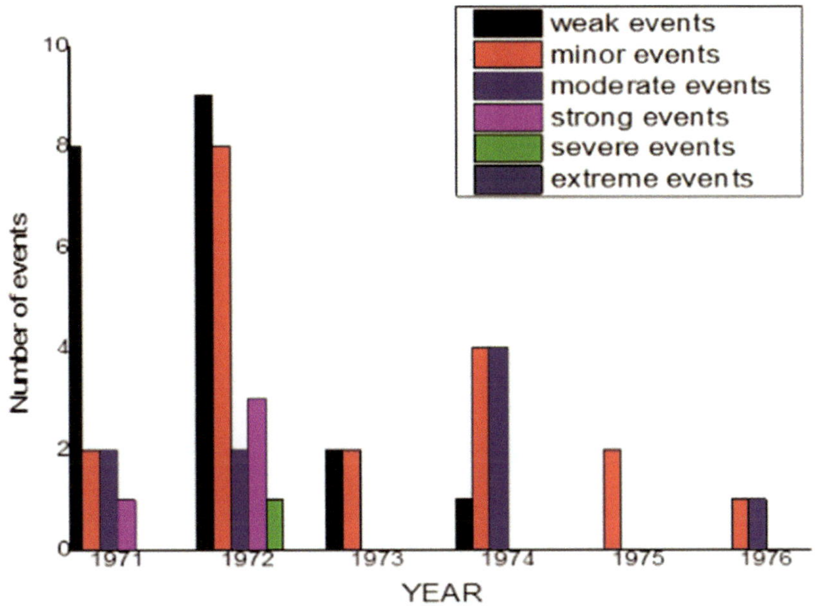

Figure 13.9 Frequency of occurrences of the solar particle events (solar proton events) with different ranges of fluences from weak to severe during the years 1971–76.

12 days for 10% of the lunations, 14 days for 25% of the lunations, and 5–10 days for 8.5% of the lunations. Missing days at the beginning of a lunation are most likely caused by the time necessary to fully charge the batteries, which may take up all the solar cell output for the first few days. The last category was found to be associated with missing solar cell observations due to technical issues such as telemetry problems [8].

Apollo 12 and 14 solar cell experiment sites are known to be located close to the lunar equator (3.6°S latitude). For modeling diurnal variations of Apollo 14 AX05 solar cell output (current) on the Moon, Bates and Feng [2] have assumed that the length of a lunar day is 14 Earth days. The same length is also adopted by O'Brien for his investigations of Apollo 12 solar cell output variations during the first lunation period of that experiment in 1969 [14]. We now consider two examples: (a) the first lunation period of 1974 where 15 days of solar cell output is observed and (b) the third lunation period of the same year (Fig. 13.10) where we could find 14 days of observation of output for the three solar cells. In Table 13.2, we have given the Sun elevation angles and Apollo 14: AX05 solar cell output during the dawn, noon, and dusk periods on the Moon along with specific remarks for these lunation periods. The lunation-to-lunation change in the first day maximum voltage (dawn values) of AX05 solar cell between the years 1971 and 1976 is plotted in Fig. 13.11. The observed variations in the dawn voltages may indicate fluctuations in sun angles associated with these observations and/or possible time variations of lunar orbital parameters from one lunation to another.

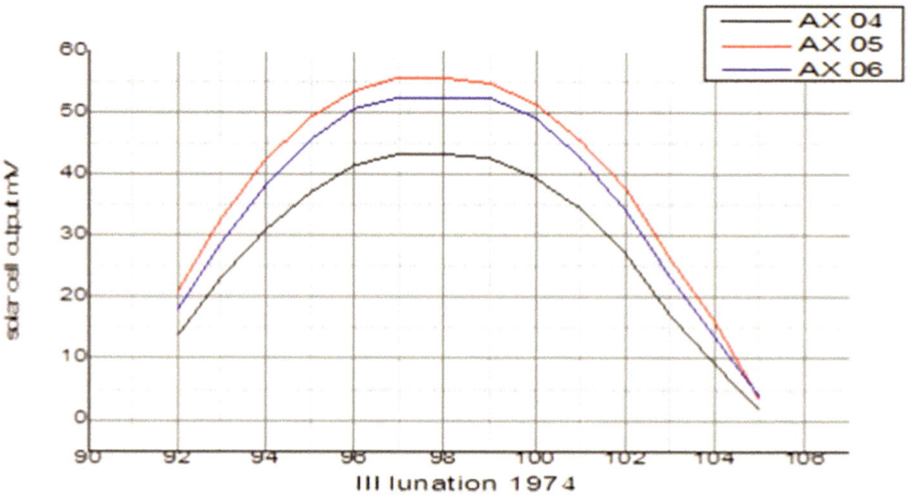

Figure 13.10 Variations of the output (daily maximum voltages) of three Apollo 14 solar cells during the third lunation period of 1974.

After the termination of solar cell experiments on the Moon as a part of Apollo missions during the year 1977, the Chang'e 3 mission was the first to continue dust detection experiments on the Moon using solar cells and photo diodes. The solar radiation measurements on Mare Imbrium region in the midlatitude of the Moon (44.12°N, 340.49°E) were carried out by this robotic lunar mission for one year from December 2013 to November 2014 [15–17]. Two results that we could infer from the published data from this Chinese lunar mission are as follows: (a) solar cell output is observed only for 12 days per lunation period, and (b) the amplitude of annual variation of solar cell current on the Moon measured by Chang'e 3 triple-junction solar cells is about 15% (nearly double of the value observed near lunar equator). Both these results suggest that both diurnal and annual variations of solar radiation intensity on the Moon depend on the lunar latitude of observation [18].

13.8 Conclusions

Theoretical models of solar radiation variations on the Moon have not included any consideration of the variability of the length of the lunar day on the Moon near the lunar equator. The dawn (sunrise) temperature is found to vary between 87.8 and 92.9 K and dusk (sunset) temperature is found to vary between 146 and 152 K on the Moon the near the lunar equator [19]. Thus, we need to investigate also the possibility of the poor LILT performance of Apollo 14 solar cells during the dawn and dusk periods on the Moon to account for the missing solar cell output for 1–3 days during a lunation period. The Apollo 14 solar cell output variations during 1971–76 and Chang'e 3 observations during

Table 13.2 Sun angles and AX04 solar cell output for the dawn, noon, and dusk periods on the Moon during the first lunation and third lunation periods for 1974. Specific remarks are also given.

Period of Apollo 14 solar cell observation on the Moon	Sun elevation angle (degrees)	AX05 solar cell voltage (mV)	Remarks
First lunation, 1974 Day 32 Dawn −9 h, 49 m	3.16	2	First observation on the sunrise day on the moon
First lunation, 1974 Day 38 Noon maximum	80.57	57.58	Solar cell output maxima during 1 day before the astronomical noon day on the moon
First lunation, 1974 Day 46 Dusk to early hours	11.85	2.73	Solar cell output on the sunset day on the moon
Third lunation, 1974 Day 92 Dawn	20.53	21	This observation is during second day from the sunrise. No output during sunrise day on the moon
Third lunation 1974 Day 97 Noon maximum	80.28	55.71	Solar cell output maxima during 1 day before the astronomical noon day on the moon
Third lunation 1974 Day 105 Dusk 4 h, 12 m	11	2	Last observation of the AX05 solar cell output on the sunset day on the Moon

Source: NASA.

2013−14 strongly indicate the possibility of both time variations and latitudinal dependence of the length of the lunar day (number of sunlit days in a lunation period; see Table 13.2).

The degradation in winter maximum of the solar cell voltage between 1971 and 1976 for different Apollo solar cells are AX04-uncovered (30.4%), AX05-covered and preirradiated (5%), and AX06-covered (9.2%). The values reported by Hollick and O'Brien who have used normalized digital counts of the output of Apollo 14 solar cells are slightly different from our estimates [3]. The low degradation rate of AX05 solar cells suggests that covered and preirradiated silicon solar cells may be used for photovoltaic power generation on Mars if dust protection is provided due to the similarities in the radiation environment of these two planetary bodies.

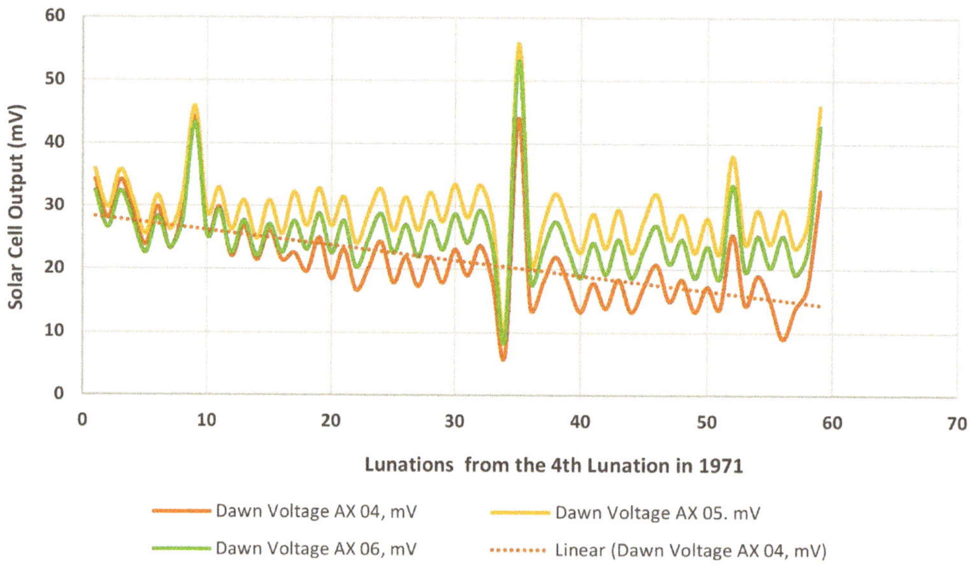

Figure 13.11 Lunation-to-lunation changes in the dawn (first sunlit day during lunar daytime) daily maximum voltages of three different Apollo 14 solar cells during the years 1971–76. The linear trend line for the uncovered AX04 solar cell dawn voltage variations is also shown here. The first lunation corresponds to the fourth lunation period during 1971.

It is interesting to note that lunar aphelion happens near summer solstice on Earth, and perihelion happens near winter solstice on Earth [1]. The annual variations in the solar cell output observed on the Moon during the years 1973–74 by the Apollo 14 mission are found to be correlated with the reported annual variations in noontime maximum surface temperature on the Moon during the same years. The results of amplitude of annual variations in lunar solar cell temperature from the Apollo 15 mission [20] are as follows: 1972 (5.8 K), 1973 (6 K), and 1974 (6 K). The model by Huang [21] predicts a lunar surface temperature change of 1 K for a change of solar radiation intensity by 14.3 W/m^2.

There are several reports on the variations of lunar orbit and allied dynamical parameters based on theoretical studies or experimental observations such as lunar laser ranging [22–25]. In this chapter, we have discussed how solar cell output variations on the Moon can be effectively used to probe solar radiation conditions and the vagaries of conditions on the lunar surface. To illustrate this idea we have used both Apollo and Chang'e 3 lunar mission observations. Future lunar lander missions may be planned to include solar cell experiments with this lunar science perspective.

Acknowledgments

The authors wish to thank NASA Goddard Space Flight Center's NASA Space Science Data Coordinated Archive (https://nssdc.gsfc.nasa.gov) for supplying Apollo 14 solar cell data tables for our investigations.

References

[1] Apollo Program Summary Report (JSC-09423), National Aeronautics and Space Administration — Lyndon B. Johnson Space Center, Houston, TX, USA: NASA-TM-X-68725. April 1975, 526 pp.

[2] J.R. Bates, P.H. Fang, Results of solar cell performance on lunar base derived from Apollo missions, Sol. Energy Mater. Sol. Cell. 26 (1992) 79—84.

[3] M. Hollick, B.J. O'Brien, Lunar weather measurements at three Apollo sites 1969—1976, Space Weather 11 (11) (November 2013) 651—660, https://doi.org/10.1002/2013SW000978.

[4] M.A.M. Ramli, E. Prasetyono, R.W. Wicaksana, N.A. Windarko, K. Sedraoui, Y.A. Al-Turki, On the investigation of photovoltaic output power reduction due to dust accumulation and weather conditions, Renew. Energy 99 (December 2016) 836—844.

[5] T. E Girish, S. Aranya, Photovoltaic power generation on the Moon: Problems and prospects, in: V. Badescu (Ed.), Moon (Prospective Energy and Material Resources), Springer, Berlin, Germany, 2012, pp. 367—376, https://doi.org/10.1007/978-3-642-27969-0_16.

[6] V.M. Kumar, T.N. Sathyan, N.G. Nisha, S. Aranya, T.E. Girish, Inferring stellar activity variations near habitable extrasolar planets using dynamical effects, in: D. Banerjee, J. Jiang, K. Kusano, S. Solanki (Eds.), Proceedings of the International Astronomical Union, Volume 13, Symposium S340: Long-Term Datasets for the Understanding of Solar and Stellar Magnetic Cycles, February 2018, Cambridge University Press, November 27, 2018, pp. 297—298, https://doi.org/10.1017/S1743921318001655.

[7] https://nssdc.gsfc.nasa.gov/nmc/experiment/display.action?id=1971-008C-12, Apollo 14 Lunar dust detector experiment, accessed April 3, 2021.

[8] J.R. Bates, W.W. Lauderdale, H. Kernaghan, ALSEP Termination Report, NASA Reference Publication 1036, NASA, 1979, 165 pp.

[9] https://nssdc.gsfc.nasa.gov/nmc/dataset/display.action?id=PSPG-00356, Apollo 14 DRTEM solar cell data, accessed April 3, 2021.

[10] D.R. Williams, H.K. Hills, P.T. Taylor, E.J. Grayzeck, E.A. Guinness, Restoration of Apollo Data by the Lunar Data Project/PDS Lunar Data Node: An Update, 2018. https://core.ac.uk/download/pdf/42698571.pdf.

[11] Y. Song, X. Wang, S. Bi, J. Wu, S. Huang, Effects of solar radiation, terrestrial radiation and lunar interior heat flow on surface temperature at the nearside of the Moon: Based on numerical calculation and data analysis, Adv. Space Res. 60 (5) (September 2017) 938—947, https://doi.org/10.1016/j.asr.2017.05.013.

[12] https://www.swpc.noaa.gov/noaa-scales-explanation, NOAA Space Weather Scales for SEP's, accessed on 3rd April 2021.

[13] B.E. Schaefer, The length of the lunar month, J. Hist. Astron. 23 (17) (February 1992) S32—S42.

[14] B. O'Brien, Direct active measurements of movements of lunar dust: rocket exhausts and natural effects contaminating and cleansing Apollo hardware on the Moon in 1969, Geophys. Res. Lett. 36 (9) (2009) L09201, https://doi.org/10.1029/2008GL037116.

[15] H.Y. Zhang, Y. Wang, L.P. Chen, et al., In-situ lunar dust deposition amount induced by lander landing in Chang'E-3 mission, Sci. China Technol. Sci. 63 (January 2020) 520—527, https://doi.org/10.1007/s11431-019-1434-y.

[16] D. Li, Y. Wang, H. Zhang, et al., In situ measurements of lunar dust at the Chang'E-3 landing site in the northern Mare Imbrium, JGR Planets 124 (8) (August 2019) 2168—2177, https://doi.org/10.1029/2019JE006054.

[17] D. Li, Y. Wang, H. Zhang, X. Wang, Y. Wang, et al., In situ investigations of dust above the lunar terminator at the Chang'E-3 landing site in the Mare Imbrium, Geophys. Res. Lett. 47 (17) (September 2020), https://doi.org/10.1029/2020GL089433.

[18] M. Kaczmarzyki, M. Gawronski, G. Piatkowski, Global database of direct solar radiation at the Moon's surface for lunar engineering purposes, E3S Web Conf. 49 (2018) 00053, https://doi.org/10.1051/e3sconf/20184900053, 12 pp.

[19] C.J. Cremers, R.C. Birkebak, J.E. White, Lunar surface temperatures from Apollo 12, Moon 3 (3) (1971) 346—351.

[20] K. Peters, M.G. Langseth, Long Term Temperature Observations on the Lunar Surface at Apollo Sites 15 and 17, Technical Report No. 3-CU-3-75 (Contract NAS 9-6037), National Aeronautics and Space Administration, USA, October 1975, 36 pp.

[21] S. Huang, Surface temperatures at the nearside of the Moon as a record of the radiation budget of Earth's climate system, Adv. Space Res. 41 (2008) 1853–1810.

[22] L. Iorio, On the anomalous secular increase of the eccentricity of the orbit of the Moon, Mon. Not. Roy. Astron. Soc. 415 (2011) 1266–1275.

[23] B.G. Bills, R.D. Ray, Lunar orbital evolution: a synthesis of recent results, Geophys. Res. Lett. 26 (1999) 3045–3048.

[24] D.A. Pavlov, J.G. Williams, V.V. Suvorkin, Determining parameters of Moon's orbital and rotational motion from LLR observations using GRAIL and IERS-recommended models, Celestial Mech. Dyn. Astron. 126 (2016) 61–88, https://doi.org/10.1007/s10569-016-9712-1.

[25] J. Chapront, M. Chapront-Touzé, G. Francou, A new determination of lunar orbital parameters, precession constant and tidal acceleration from LLR measurements, Astron. Astrophys. 387 (May 2002) 700–709, https://doi.org/10.1051/0004-6361:20020420.

CHAPTER FOURTEEN

Space photovoltaics for extreme high-temperature missions

Geoffrey A. Landis
NASA John Glenn Research Center, Photovoltaic & Electrochemical Systems Branch, Cleveland, OH, United States

14.1 Introduction

Solar arrays for use on the surface of the Earth must be designed to withstand an extremely degrading environment: surrounded by a highly oxidizing atmosphere, intermittently exposed to corrosive liquid water, subject to wind loading, abrasion by sand and dust, and occasionally impacted by hail. Solar arrays for space are not subject to these effects, but instead have a different set of environmental hazards, including more extreme temperature cycles, particulate and ultraviolet radiation in space, micrometeoroid damage, and exposure to a flux of atomic oxygen in low-Earth orbit. Over the years since the first solar cells were sent into space on Vanguard 1 in 1958, space solar array technology has advanced to make photovoltaic cells resistant to these degradation mechanisms.

As each of the degradation mechanisms has been conquered, the range of environments in which we use photovoltaics has expanded outward [1], from initial use of arrays in low-Earth orbit, to satellites flying in the high-radiation environment of the Van Allen radiation belts, and outward to the dusty environment of Mars, the low-intensity (or illumination)/low-temperature (LILT) conditions in the asteroid belt and beyond, and with the success of the solar-powered Juno mission [2], even in the cold and high-radiation environment in Jupiter's orbit, with proposals made for photovoltaic power even farther beyond [3,4]. It is also, however, of interest to expand the range of solar arrays in the opposite direction: inward toward the Sun. The environment close to the Sun has its own unique challenges, with one of the most obvious being high temperatures.

Extending the temperature range of operation for solar arrays is highly desirable for extending the range of operation of space missions to the near-Sun environment [5–7]; interestingly, high temperatures help prevent arcing of solar arrays [8]. Achieving high-efficiency and reliable operation in these temperature regimes is a difficult technologic challenge. Existing solar cells lose performance at the high temperatures encountered in Mercury orbit and inward toward the Sun. For future missions designed

Photovoltaics for Space
ISBN 978-0-12-823300-9
https://doi.org/10.1016/B978-0-12-823300-9.00012-1

to probe environments close to the Sun, it is desirable to develop array technologies for high (light) intensity and high temperature (HIHT). Approaches to solar array design for near-Sun missions include thermal management at the systems level to optimize efficiency at elevated temperature or the use of techniques to reduce the incident solar energy to limit operating temperature. An additional problem is found in missions that involve a range of intensities, such as the Parker Solar Probe mission [9,10], which ranges from a starting distance of 1 astronomical unit (1 AU, roughly the distance from Earth to the Sun) from the Sun to a minimum distance of 9.5 solar radii, or 0.044 AU. During the mission, the solar intensity ranges from one to about 500 times the intensity at air mass zero (AM0). This requires a power system to operate over nearly three orders of magnitude of incident intensity. Another, more difficult problem is the environment of the surface of Venus [11]. In this case, the photovoltaics must operate at high temperature and low intensity, with the additional challenge of a modified spectrum shifted toward the red end of the spectrum [12] and a high-pressure, corrosive atmosphere.

14.2 Solar cell operating temperature and efficiency

If future missions designed to probe environments close to the Sun will be able to use photovoltaic power generation, solar cells that can function at high temperatures under high light intensity and high radiation conditions must be developed. The significant problem is that solar cells lose performance at high temperatures. In radiative equilibrium, the operating temperature of a solar cell depends on the fourth root of the incident intensity, as well as the ratio of solar absorptivity alpha (α) to thermal emissivity epsilon (ε). According to radiative balance, if I is the incident intensity, the thermal radiation from the array must equal the absorbed solar radiation:

$$\alpha I = (\varepsilon_f + \varepsilon_r)\sigma T^4 \tag{14.1}$$

where T is the operating temperature in absolute (Kelvin) units, and the subscripts f and r indicate the emissivity from the front and rear sides of the cell to account for the fact that the array can radiate waste heat away from both front and back sides (ε_r should be left out for an array with no backside radiation). For convenience, here we define the absorptivity α as the net energy absorption, incorporating a factor of $(1-\eta)$ to account for the fact that the fraction of incident energy that is converted to electricity is not radiated by the cell. Thus, the equilibrium operating temperature T is

$$T = \left[\left(\alpha/\left((\varepsilon_f + \varepsilon_r)\sigma\right)\right)I\right]^{1/4} = cI^{1/4} \tag{14.2}$$

where the constant c is defined as

$$c = \left[\alpha/\left((\varepsilon_f + \varepsilon_r)\sigma\right)\right]^{1/4} \tag{14.3}$$

Note that since front and back sides of the cell need not have identical radiative properties in the infrared, the value of ε is the average of front and back side properties. The intensity depends on distance from the Sun:

$$I = I_o/r^2 \tag{14.4}$$

where I_o is the solar intensity at Earth's distance from the Sun, 1 AU, and distance r is measured in AU. This equilibrium temperature is shown as a function of distance from the Sun in Fig. 14.1 for various values of the ratio of the α to ε (here ε is the average of the front and back emissivity).

For the case of a planetary orbiter, the temperature is somewhat greater since the solar array is heated not only by the incident solar flux, but also by solar flux reflected by the planet (known as "albedo"), as well as infrared emitted by the planet. This additional heating becomes more significant as the orbital altitude decreases. In the lowest orbit limit, this results in about twice the incident intensity, so an absolute temperature increases by a factor of the fourth root of 2 ($\sim$1.19).

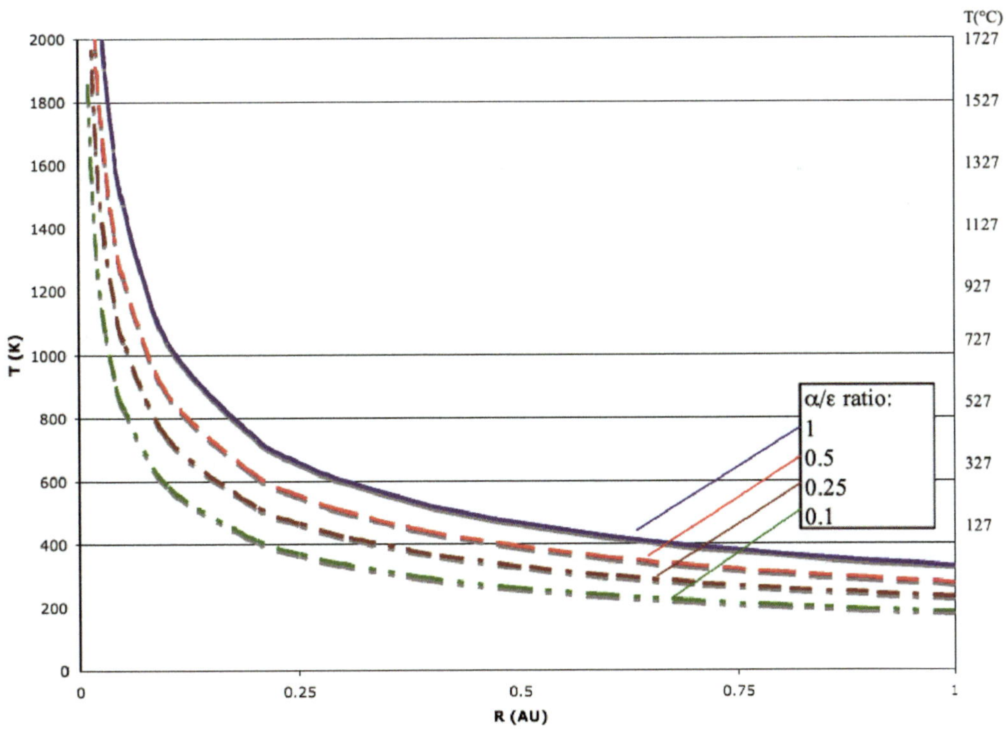

Figure 14.1 Temperature as a function of distance from the Sun, for various values of α/ε ratio. *Courtesy NASA.*

A solar cell's (unnormalized) temperature coefficient of efficiency κ is defined (Eq. 14.5) as the change of conversion efficiency η per unit temperature,

$$\kappa = d\eta/dT \tag{14.5}$$

and the power P at a temperature T can be compared to the power output from a reference temperature (Eq. 14.6), typically 27°C (300K) by a linear extrapolation:

$$P = (\eta_{300K} + \kappa\Delta T)I \tag{14.6}$$

In general, the temperature coefficient κ is negative, corresponding to decrease in performance with increasing temperature [13]. The temperature coefficient is not a constant but varies slightly with both intensity and temperature [14], discussed subsequently, but these variations for the moment can be ignored in a first-order analysis.

From this, the power output from a cell in radiative equilibrium is a nonlinear function of the intensity. Defining η_o as the efficiency linearly extrapolated to 0 K, which is calculated from the efficiency at 27°C as

$$\eta_o = \eta_{300K} - 300\kappa \tag{14.7}$$

Eq. (14.8) results:

$$P = I(\eta_o + \kappa T) = I\left(\eta_o + \kappa c I^{1/4}\right) = I\eta_o + \kappa c I^{5/4} \tag{14.8}$$

Fig. 14.2 shows the calculated curve of power output as a function of temperature, comparing a high-efficiency silicon solar cell with a wide-bandgap solar cell, in the case of the linear assumption. For any given solar cell technology, there exists an incident intensity above which the solar cell output decreases with increased intensity. In the linear model of Eq. (14.7), this intensity is as follows:

$$I_{peak-output} = (-\eta_o/\kappa c)^4 \tag{14.9}$$

14.3 Temperature coefficient(s)

Temperature coefficients of solar cells have been analyzed by Fan [13] and others [14,15]. The power output of a solar cell can be composed as the product of three factors, short-circuit current density (J_{sc}), the open-circuit voltage (V_{oc}), and the curve fill-factor (FF):

$$P = (J_{sc})(V_{oc})(FF) \tag{14.10}$$

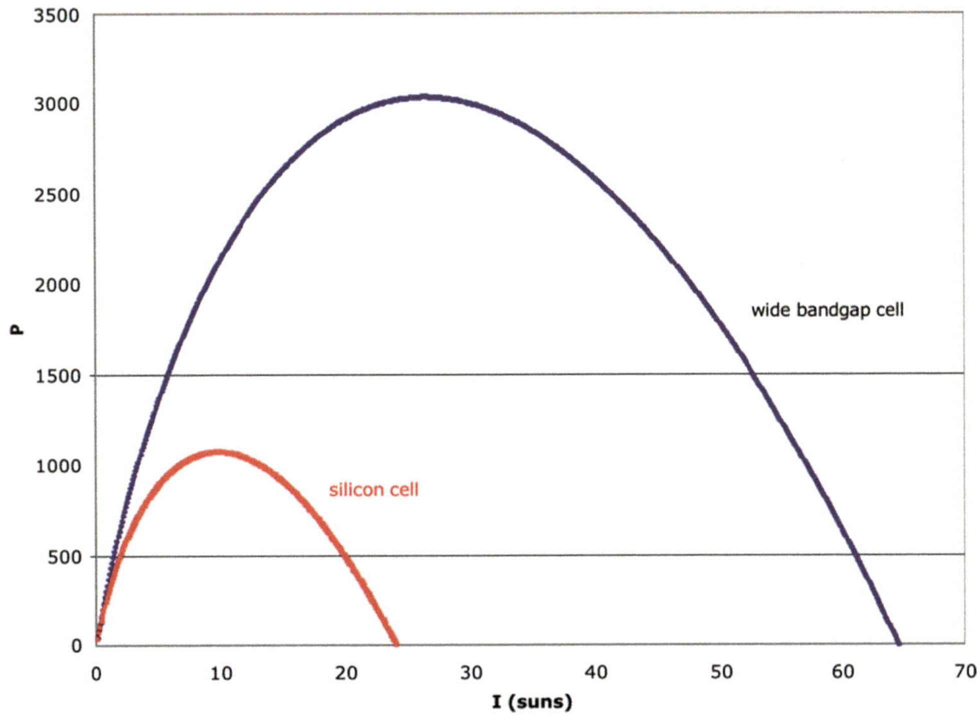

Figure 14.2 Curve of power output (normalized) as a function of intensity, assuming linear temperature coefficient. *Courtesy NASA.*

The variation of solar cell performance as a function of temperature can be approximated in terms of linear temperature coefficients:

$$J_{sc} = \frac{dJ_{sc}}{dT}(T - T_o) \tag{14.11}$$

$$V_{oc} = \frac{dV_{oc}}{dT}(T - T_o) \tag{14.12}$$

$$FF = \frac{dFF}{dT}(T - T_o) \tag{14.13}$$

where T is the temperature of interest, and T_o is the nominal test temperature (often set to 28°C).

Note that since these three factors all have different temperature dependence, the maximum power point voltage of a solar cell will also vary with temperature. For an optimized system, the power management will need to account for this and adjust the operating point to maximize power if the spacecraft operates at a range of distances from the Sun. Alternately, the operating point voltage can be selected for the most power-critical phase of the mission, and losses associated with operation away from the maximum power point accepted for other phases of the mission, where power is less critical. Of the three components of temperature coefficient, voltage will decrease with increasing temperature, while short-circuit current will increase (due to bandgap narrowing). For well-optimized cells, the V_{oc} temperature coefficient contributes the largest amount to the total variation of efficiency with temperature.

Temperature coefficients are also often expressed in the form of normalized temperature coefficients, where the normalization is done by dividing by the value at T_0, so the variation is expressed as a fraction of the initial value. Using normalized temperature coefficients, the overall temperature coefficient will simply be the sum of the J_{sc}, V_{oc}, and FF components. The temperature coefficient is, in general, nonlinear. For this analysis we have been assuming a linear dependence of efficiency on temperature, but for a more detailed analysis, the nonlinearity must be accounted for in more detailed modeling.

Outside of a narrow temperature range, temperature coefficient for multijunction cells can be quite nonlinear. In particular, for multijunction cell technologies, the voltage component of the temperature coefficient is the sum of the voltage dependence of the individual subcells. At elevated temperature, the lowest bandgap subcells will drop to zero output, and beyond this point, the temperature coefficient will depend only on the remaining cells; this results in a discontinuous decrease in the temperature coefficient at higher temperature, as the lowest bandgap subcells in the series drop out. Likewise, since the bottom cells see a spectrum that has been filtered by the top cells in the cascade, and the bandgap of these top cells varies with temperature, the current component of the temperature coefficient can be nonlinear.

A temperature coefficient is also dependent on the illumination intensity [14—16]. The largest factor in temperature dependence is the dV_{oc}/dT term, which depends on the difference between the open circuit voltage of the cell and the bandgap voltage [6]. Since this decreases as the V_{oc} increases, κ decreases in magnitude proportionally to the V_{oc}, which increases as the logarithm of the intensity. This increases the output at high intensities slightly over the constant κ model.

The theoretical temperature coefficient of V_{oc} can found from the ideal solar cell equation (Eq. 14.14). As derived by Fan [13], for an ideal cell the temperature coefficient is as follows:

$$\frac{\partial V_{oc}}{\partial T} = \frac{1}{T}\left[V_{oc} - \frac{E_q}{q} - \frac{3kT}{q}\right] + \frac{1}{q}\frac{\partial E_q}{\partial T} + \frac{kT}{qJ_{sc}}\frac{\partial J_{sc}}{\partial T} \tag{14.14}$$

where T is temperature, E_g is the band gap energy in eV, k is Boltzmann's constant, and J_{sc} is the short-circuit current density of the solar cell. The factor q, electron charge, converts the bandgap from units of electron-volts into units of volts. The first factor, in general, is large compared to the others, so we can usually ignore the second two factors. Thus, the temperature coefficient is proportional to the difference between the open circuit voltage V_{oc} and the bandgap E_g/q. V_{oc} is always less than the bandgap, so the temperature coefficient of V_{oc} is always negative.

The question of how to improve the temperature coefficient of solar cells has been addressed in earlier studies [5–7]. Since the fractional loss of V_{oc} with temperature decreases in magnitude as bandgap increases [13], photovoltaic cells from wide-bandgap materials can operate at higher intensity (so higher temperatures) than cells from narrow-bandgap materials [5–7]. Any advance that improves the open circuit voltage (to bring the voltage closer to the bandgap voltage) will also decrease the temperature coefficient.

For an array designed to operate at high temperature, it is also a requirement that the solar cell not physically degrade at high temperatures, for example, from degradation of the semiconductor or from ohmic contacts diffusing through the junction. Ohmic contacts that are stable at high operation temperatures have been demonstrated for GaAs [17–19] and $GaInP_2$ [20,21], but work remains to be done for other semiconductor

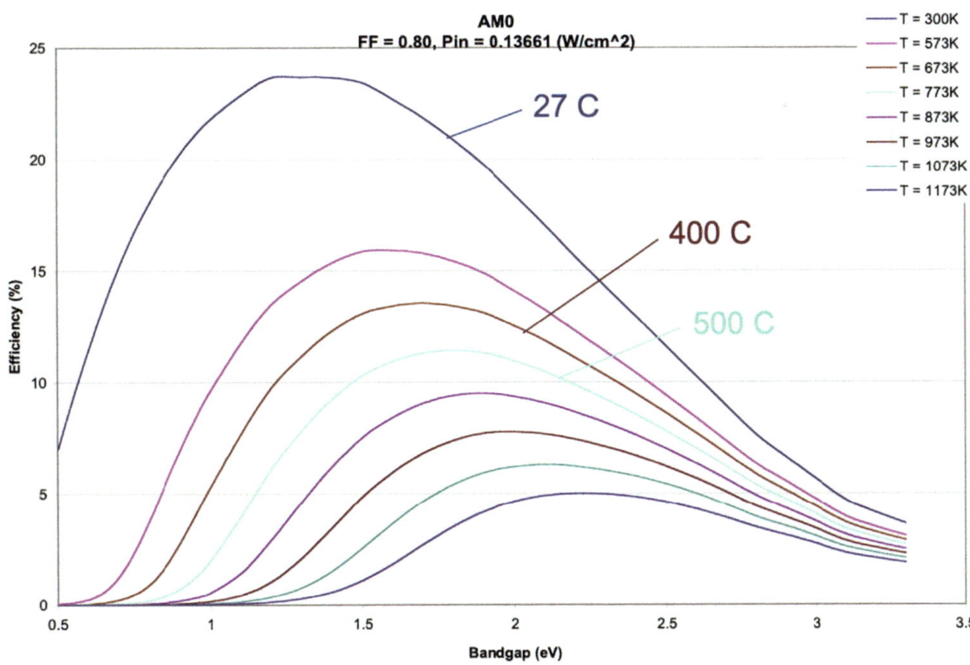

Figure 14.3 Modeled efficiency of a single-junction solar cell as a function of the semiconductor bandgap, for temperatures ranging from 27 to 900°C. *Courtesy NASA.*

technologies. Likewise, other components of the system, such as the cover glass adhesive and the interconnect, must be stable at high-temperature operation.

From the basic semiconductor diode calculation and the calculation of temperature coefficient as a function of bandgap, it is possible to calculate the effect of the solar cell bandgap on the efficiency as a function of temperature. Fig. 14.3 shows this calculation for a conventional single-junction cell. As is clear, the efficiency drops as the temperature increases. It is also notable that the optimum bandgap rises with increased operating temperature since higher bandgap solar cells have higher open circuit voltage, and thus lose a lower fraction of their voltage as the temperature increases.

14.4 Approaches to solar arrays for near-Sun missions

Approaches to solar arrays for near-Sun missions include modifying any of the terms governing temperature or efficiency of the cell: I, the incident intensity, α, solar absorption, or ε, the emissivity of the cell, or κ, the temperature coefficient. Possible approaches include the following:

1. high epsilon/low alpha coatings
2. array off-pointing (i.e., array normal points at angle to Sun)
3. partially populated array (with missing cells replaced with mirrors)
4. Spectrally selective reflective coatings
5. solar cells designed to operate at high temperature, with low temperature coefficient κ
6. solar arrays incorporating added thermal radiators

Since light that is not absorbed by the solar array does not contribute to the waste heat rejected by the radiator, it is advantageous to reflect light in the wavelengths not used by the cell. Most importantly, infrared light of energy less than the bandgap energy of the bottom subcell contains significant energy but makes no contribution to the output power. Blue-red rejection filters could enhance the performance of the radiators by reflecting this unused band. The three subcells of a multijunction cell contribute roughly equally to the temperature coefficient, although the bottom cell is only a small contributor to the overall efficiency. Thus, it is likely that even more heat rejection could be accomplished if the bottom (Ge) cell were omitted entirely from the stack, essentially returning the technologic approach toward the earlier dual-junction cell technology. This would allow all the energy of wavelength longer than about 850 nm to be reflected away by an infrared-rejection (dichroic) filter. Since the controlling parameter for the radiator sizing is the efficiency of conversion of the light absorbed by the cell, not the overall solar conversion efficiency, some reflection of the above-bandgap spectral range is acceptable, so the filter does not need a sharp cutoff at the band edge. This means that a thin metal film (for example, gold) or a transparent conductor could be a good choice for infrared rejection.

14.5 Solar arrays with constant power at variable heliocentric distance

An additional problem is presented in that most missions involve a range of intensities, requiring a power system capable of operating all the way from Earth departure through a close solar pass. There are several approaches to such a power system. One straightforward approach would be to utilize several different arrays, each one optimized for a different heliocentric distance. For example, a large solar array could be used for the highest solar distances, and then discarded, or else folded away in the shadow behind the spacecraft, so that a smaller solar array optimized for higher intensity could be used. Another approach would be to progressively use increasing amounts of off-pointing, so that the area exposed to the Sun decreases with distance. The problem of adapting to multiple intensity regimes is particularly difficult if the mission is required to operate both at a near-solar approach and also at distances farther from the Sun than the Earth, for example, missions using a gravity-assist at Jupiter, solar sail missions, and missions that follow a comet from distant reaches of the solar system through close approach to the Sun.

One approach is to design the array to produce a constant amount of power regardless of distance from the Sun. Fig. 14.4 shows a design that makes use of the fact that the angular diameter of the Sun increases as the spacecraft approaches the Sun. A Fresnel lens array focuses light onto a light pipe to transmits the light through a shield of multiple

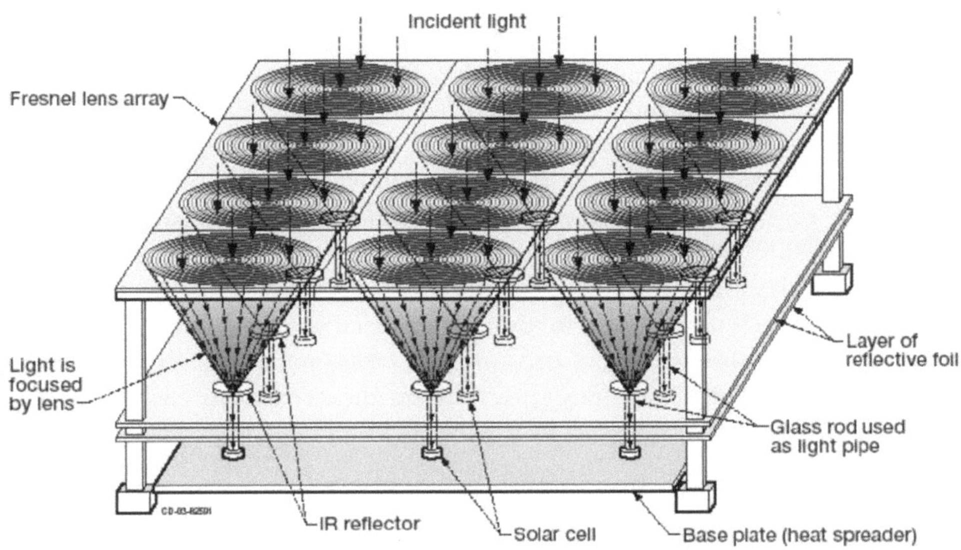

Figure 14.4 Concept for a concentrator solar array designed to have constant power output for variable heliocentric distances. *Courtesy NASA.*

layers of foil. At high distances from the Sun, the fact that the array is a concentrator array minimizes the effect of cell operation under LILT conditions. As the array approaches the Sun, the light pipe acts as an aperture stop. As the image of the solar disk at the focal plane gets larger, the added incident light is reflected by the mirrored plate. The solar cells are mounted on a heat spreader that allows the full backside surface to serve as a thermal radiator. The portions of the collector exposed to the sunlight are refractory material (silica or high-alumina glass for the optical elements; titanium or carbon for the structural components). The solar cells, located behind the multiple layers of metal-foil insulation, are not exposed to high temperature. This array can thus be designed to put an amount of solar intensity on the cells exactly equal to that of the peak defined above.

14.6 Thermal conversion for near-Sun missions

As a spacecraft approaches the Sun, the temperature of a flat plate exposed to the Sun increases. This brings out a natural question of whether it is possible to convert this heat, using a heat engine, rather than using photovoltaic arrays. Thermal power conversion is a demonstrated technology, and there exists a choice of conversion technologies, including thermoelectric conversion, a technology with comparatively low efficiency, but high spaceflight heritage, or Stirling conversion, which has the advantage of higher efficiency but no spaceflight heritage.

Stirling converters have demonstrated 38% conversion efficiency [22] operating at 850 hot-end, 90°C cold-end temperatures, which is slightly over 50% of the theoretical (Carnot) efficiency. System efficiency, accounting for other losses, is on the order of 26%. These efficiency numbers are roughly comparable to the best photovoltaic technologies. The efficiency scales with temperature directly with the Carnot efficiency, proportional to the difference between the hot- and cold-end temperature. Thermal conversion approaches would require a heat rejection radiator at the cold end of the system, which must be not exposed to the Sun, either in the shadow of an absorber or the thermal shield, or else edge-on to the incident flux. The radiator must be sized to radiate the waste energy at the cold-side temperature.

Fig. 14.5 shows the block diagram of the energy flow in such a thermal system. Since the power produced is proportional to the conversion efficiency η, and the radiator area proportional to the waste power radiated, which is proportional to $(1-\eta)$, the radiator area A_r (Eq. 14.15), which depends sensitively on the conversion efficiency and the radiator temperature, is

$$A_r = [(1-\eta)/(\eta)]/\left(\varepsilon\sigma T^4\right) P_{electric} \tag{14.15}$$

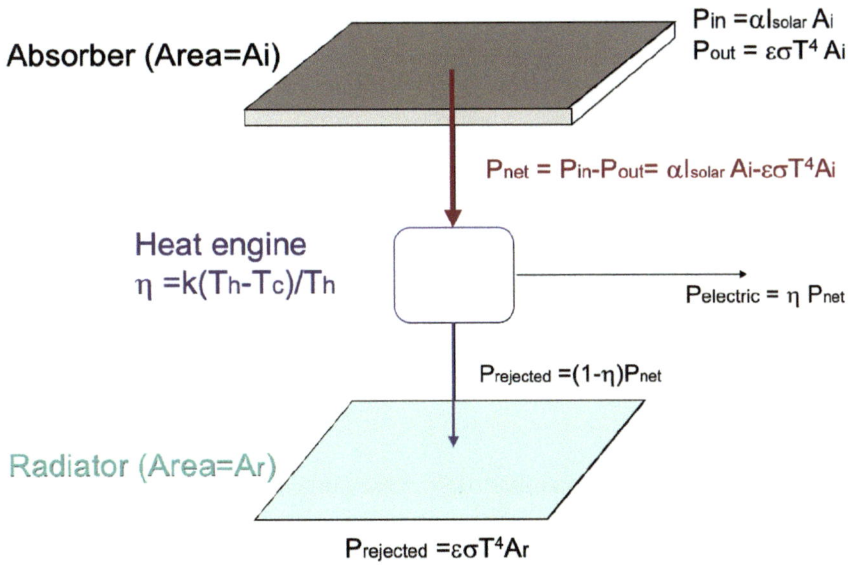

Figure 14.5 Thermal conversion energy flow block diagram. *Courtesy NASA.*

Emissivity ε for thermal radiators is typically very close to 1. If the radiator is two sided, the area A includes both front and rear sides. The same equation for radiator size can be used if the thermal radiator is used to cool a solar array instead of a heat engine.

The thermal system chosen must operate over a wide range of intensities, thus temperatures. This can represent an engineering challenge for many missions; for example, the Solar Probe is only at close approach to the Sun for a duration on the order of 10 h. The proposal to operate a thermal conversion system, incorporating a radiator with pumped cooling to achieve the cold-side temperature, brings up the possibility of using a similar cooling loop to keep a solar array within operating temperature limits, thus reducing the requirement for high-temperature operation. This approach was adopted for the Parker Solar Probe (Section 14.8).

14.7 Earlier near-Sun missions

Early missions to operate in a near-Sun environment include the Soviet Venera missions and the US Mariner 10 mission to Venus and Mercury. More recently, the solar-powered Mercury Surface, Space Environment, Geochemistry and Ranging (MESSENGER) mission successfully reached and orbited Mercury, and the ESA BepiColumbo mission is on its way to Mercury. Due to the high temperatures to be experienced, starting with spacecraft Venera 2 and Venera 3, launched in November 1965, the Soviet Venus missions utilized gallium arsenide solar cells [23], despite the fact that GaAs-based cells were at a lower state of development in the 1960s and would not supplant silicon cells for conventional space missions until the very end of the 1980s.

Figure 14.6 Solar panel for the MESSENGER mission, showing that two-thirds of the array area is covered with mirrors, and only one-third of the area (dark stripes in the image) has solar cells. *Courtesy NASA.*

The reason for their choice was that the wider bandgap of GaAs (1.4 eV, compared to silicon at 1.1) gave it a significantly lower temperature coefficient, so better performance at Venus. The Mariner series took a different approach. The Mariner 10 spacecraft kept the (at the time) industry-standard silicon cell technology, but it employed the technique of gimbaling the solar arrays away from the Sun-pointing direction to minimize the temperature [1]. This reduces the incident solar flux by a factor of cos(ø), where ø is the off-pointing angle, so it reduces the temperature by a factor of $\cos^{(1/4)}(ø)$.

As an example of a mission with solar panels designed to operate at high solar flux using both limiting the incident flux to the array and modifying solar absorption (α), the MESSENGER mission to orbit Mercury [24,25] had designed an array with mirrored panels (Fig. 14.6) to reflect two-thirds of the incident solar energy to limit operating temperature. The mirrors incorporate a high-emissivity silica coating, allowing efficient thermal radiation [26]. In addition to the reflectors, the MESSENGER arrays also used off-pointing to fold the solar arrays back along the spacecraft body to reduce the incident flux. These panels were successfully operated during the full 10.5 year mission, including the cruise to Mercury and for over 4 years in Mercury orbit. The mission reached a minimum distance of 0.3 AU from the Sun, and the arrays had been qualified in thermal vacuum testing to be capable of operating as close as 0.25 AU from the Sun, at an intensity of 16 times the intensity at Earth orbit. Likewise, the BepiColumbo spacecraft utilizes a strategy to tilt its solar arrays off the Sun-pointing axis to reduce the incident solar flux to keep the temperature within limits, up to an angle of over 70 degrees at a distance of 0.3 AU [27].

14.8 Parker Solar Probe

The highest solar intensity encountered by a mission flown to date is that seen by the Parker Solar Probe. The mission concept for the Solar Probe mission incorporated multiple Venus flybys to drop the probe into a low-perihelion orbit. The primary mission uses seven Venus flybys to bring its orbit incrementally closer to the Sun, making 24 near-Sun passes over a period of 6.9 years, with a final objective of reaching a perihelion distance of roughly 9.6 solar radii, or 0.044 AU, significantly inside the orbit of Mercury. This will be about eight times closer than any spacecraft has previously approached the Sun.

An enabling technology for the mission was to design a power system that could not only operate at the highest intensity portion of the mission, at and near the perihelion pass close to the Sun, but also provide power during the cruise and multiple Venus flybys. The array design therefore had to provide power at all distances from 1 AU down to 0.044 AU [9]. A solar array strategy was designed to meet these criteria [10]. The strategy adopted was to use two solar arrays:

1. *Conventional solar array*: The primary array is used from Earth to slightly inside the Mercury orbit. Since the MESSENGER array is already demonstrated in flight, the engineering assessment concluded that only a small amount of cell and materials development is needed. The array can be off-pointed from the Sun by progressively folding them back along the spacecraft body, to allow operation in to approximately 0.25 AU, an intensity of 16 suns. At this distance the main solar array is folded in behind the shadow shield, where it is not exposed to high temperatures, and a second power supply is used.
2. *High-intensity solar array:* A secondary solar array (Fig. 14.7) was then incorporated to power the mission at the high-intensity portion of the mission, operating inside 0.25 AU. Since at this distance the intensity was high, the secondary solar array could be much smaller. This power supply used high-efficiency triple-junction solar cells

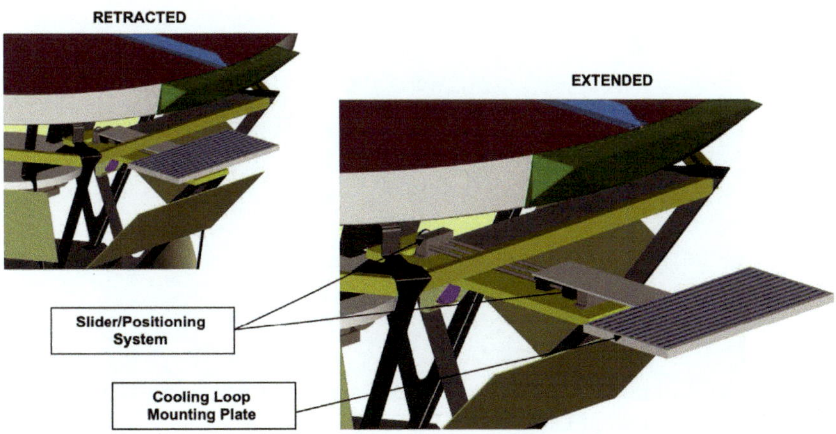

Figure 14.7 The early concept for the secondary (high-intensity) solar array for the Solar Probe+ mission, designed to be retracted behind a knife-edge shadow shield as the spacecraft approaches the closest distance from the Sun. *Courtesy NASA.*

designed for high solar concentrations. To keep the solar cells within the nominal temperature limits as the solar intensity increased, a liquid cooling loop was incorporated to pump coolants through thermal radiators mounted behind the solar shade on the outside of the spacecraft.

At the closest approach to the Sun, the secondary array was progressively slid behind a knife-edge shadow shield, moving the array into a partial shadow (penumbra). This was done both to keep the array power constant as the solar intensity increased and also as a way to reduce the thermal load that needed to be sent to the radiators. Details of the solar array design process are provided by Landis et al. [10]. This power system design solved the spacecraft's problems; it was presented to the review board, who approved the mission to go forward.

The final spacecraft design, now renamed Parker Solar Probe, differed from the initial design in several ways, although keeping the concept of a main solar array for operation near aphelion and a smaller high-intensity array for operation near perihelion. The secondary high-intensity solar arrays were relocated from their original position to put them at the end of the primary solar arrays, although set at an angle to the primary arrays. In this new position, the sliding mechanism was now unnecessary, and the rotation that angled the primary arrays back now also served as the mechanism to move the secondary arrays. At close distances to the Sun, the primary arrays are folded back behind the shadow shield, leaving only the secondary arrays, at the tip, exposed. As the spacecraft approaches yet closer to the Sun, additional angling slides (and rotates) the secondary arrays behind the knife-edge that progressively puts them in partial shadow from the full

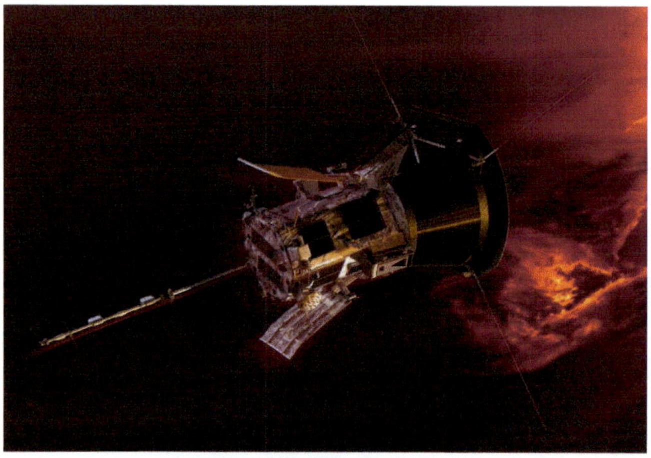

Figure 14.8 Artist's conception of Parker Solar Probe approaching the Sun, with main solar arrays feathered back. This spacecraft is designed to be operated on photovoltaic power as close as 0.044 AU from the Sun. The flat solar shield protects the instrument package from the thermal environment near the Sun. *Courtesy: NASA/Johns Hopkins APL/Steve Gribben.*

intensity of the solar flux. The net result incorporates the functionality but is mechanically much simpler than the initial concept.

An artist's conception of the revised design can be seen in Fig. 14.8, where the conventional arrays are shown feathered backward behind the shadow shield, protecting them from the peak solar intensity, and the secondary arrays, located at the tips of the primary arrays, are angled slightly to view the Sun. A second improvement to the design was in the coolant loop. In the revised design, the coolant loop is not just circulated behind the high-intensity panel, but it is also circulates behind the main solar array. This serves three functions: during the perihelion approach to the Sun, the area of the main array serves as additional radiator area; during the solar approach portion of the mission, before (and after) the nearest passage, the cooling reduces the load on the solar array; and during the portion far from the Sun, the solar input to the main array actually serves to keep the cooling liquid from freezing.

Gaddy et al. provide details of the Solar Probe mission arrays [28]; solar power output calculations are described in a subsequent publication [29]. The mission, now renamed Parker Solar Probe in honor of physicist Eugene Parker (1927–2022), who predicted the existence of the solar wind, launched in 2018. As of early 2022, it has successfully made five Venus flybys and 11 perihelion passes close to the Sun, with the power system operating perfectly.

14.9 Photovoltaic power at Venus

The surface of Venus is a target of great interest to science, but it is the most hostile operating environment of any of the solid-surface planets in the solar system [30]. The surface of Venus has been explored by a number of missions from Earth, including the Russian Venera missions that landed probes on the surface and the American Pioneer missions that flew both orbiters and atmospheric probes to Venus, but the longest survival time of any mission to the surface of Venus, to date, has been only 2 hours on the surface of Venus [11]. Four effects make the surface of Venus a challenging environment when considering solar power [12]:

(1) temperature
(2) solar intensity
(3) solar spectrum
(4) corrosive environment

The greatest difficulty is the surface temperature of Venus, averaging 452°C, with little difference between daytime and nighttime. As on Earth, temperature decreases with elevation, and the tops of Venus' mountains are slightly cooler: at the top of Maxwell Montes (10.4 km above mean elevation), the temperature is "only" 390°C. Secondarily, Venus is continually covered with a thick layer of clouds. The surface does not ever get a direct view of the Sun, and the solar intensity at the surface is about 2% of the intensity above the atmosphere. The light level is equivalent to the light level during a rainy day on Earth. The thick atmosphere also filters the sunlight, with the surface solar spectrum depleted in blue wavelengths due to Rayleigh scattering. The surface pressure is 92 bars, equivalent to pressure a kilometer under the ocean, and the atmosphere is primarily of carbon dioxide. The atmosphere at the surface also contains significant amounts of anhydrous sulfur compounds, such as SO_3, which are corrosive. At the temperatures and pressures of the surface, stability against chemical attack is a significant concern.

These factors combine to multiply the challenges of power on the surface. The low light intensity alone reduces power availability, and the reduction of performance of solar cells due to temperature exacerbates this difficulty. This puts the solar cell operation into the low-intensity, high-temperature (LIHT) regime [31], a somewhat different regime from the HIHT region discussed previously.

For high-temperature operation, as discussed before, a high-bandgap solar cell material would be preferred, but the blue-deficient spectrum puts a limit on the availability of short-wavelength photons. Nevertheless, an analysis by Landis and Haag [12] showed that photovoltaic power systems, although limited in power, could produce power on the surface. Further work by Grandidier et al. [20,31] extended this work and tested cells to verify performance at Venus temperatures. Their lifetime testing showed degradation beginning at about 7 weeks of exposure at 465°C [31].

The problem of encapsulation of cells to protect against the corrosive effects of the Venus atmosphere, however, has yet to be addressed. While the glass used for solar cell covers is unaffected by the atmosphere or the temperature, preliminary testing under Venus conditions has shown that conventional silicone adhesives used to adhere cover glass to solar cells will blacken under Venus conditions. A possible solution would be direct bonding of a thermal expansion coefficient-matched glass to the cell, taking care that the bonding must be completely sealed, since even small amounts of leakage would be likely to corrode the cell, the metallization, and the electrical interconnects.

14.10 Conclusions

Operation of photovoltaic arrays at the high solar intensity conditions of near-Sun missions presents unique challenges to solar cell technology. If the temperature of an array is allowed to vary with the thermal radiation limit, the power produced by an array will increase as the incident flux increases up to some limit determined by the temperature coefficient and thermal properties of the solar cell, and above that output, power will decrease as incident intensity increases. A number of approaches have been developed to mitigate the effects of high-temperature operation. Several of these have been successfully demonstrated to enable solar-powered spacecraft to explore the near-Sun planets such as Mercury and Venus as well as the Sun itself.

References

[1] C.R. Mercer, Solar array designs for deep-space science missions, in: S.G. Bailey, A.F. Hepp, D.C. Ferguson, R.P. Raffaelle, S.M. Durbin (Eds.), Photovoltaics for Space: Key Issues, Missions and Alternative Technologies, Elsevier, 2022, pp. 349–378.

[2] S. Dawson, P. Stella, W. McAlpine, G. Smith, in: JUNO Photovoltaic Power at Jupiter, in: 10th International Energy Conversion Engineering Conference, 30 July–1 August 2012, Atlanta, GA, USA, https://doi.org/10.2514/6.2012-3833.

[3] A. Boca, R. Warwick, B. White, R. Ewell, A data-driven evaluation of the viability of solar arrays at Saturn, IEEE J. Photovoltaics 7 (4) (2017) 1159–1164.

[4] G.A. Landis, J. Fincannon, Study of power options for Jupiter and outer planet missions, in: Proc. 42nd IEEE Photovoltaic Specialists Conference, New Orleans LA, June 14–19, 2015, https://doi.org/10.1109/PVSC.2015.7356136.

[5] D. Merritt, S. Houlihan, R. Raffaelle, G. Landis, Wide bandgap space solar cell development, in: Proc. 21st. IEEE Photovoltaic Specialist's Conference, 2005, pp. 552–555.

[6] G.A. Landis, P. Jenkins, D. Scheiman, R. Raffaelle, Paper AIAA-2004-5578, Extended temperature solar cell technology development, in: Proc. 2nd International Energy Conversion Engineering Conference, August 16–19, 2004, Providence, RI, USA, https://doi.org/10.2514/6.2004-5578.

[7] G.A. Landis, R. Raffaelle, D. Merritt, High temperature solar cell development, in: Proc. 19th European Photovoltaic Science & Engineering Conference, 2004, pp. 3659–3661.

[8] D.C. Ferguson, New frontiers in spacecraft charging, IEEE Trans. Plasma Sci. 40 (2) (February 2012) 139–143, https://doi.org/10.1109/TPS.2011.2172635.

[9] J.D. Kinnison, A.A. Dantzler, Y. Guo, S. Vemon, C. DeBoy, M.M. Donegan, D. Drewry, et al., The Solar Probe+ Mission: a new concept for close solar encounters, in: Proc. 59th International Astronautical Congress, Glasgow, Scotland, Sept. 29–Oct. 3, 2008, vol 3, IAC, 2008, pp. 1596–1604.

[10] G.A. Landis, P.C. Schmitz, J. Kinnison, M. Fraeman, L. Roufberg, S. Vernon, M. Wirzburger, Power system design for the solar Probe+ mission, in: 6th International Energy Conversion Engineering Conference, Cleveland, OH, USA, July 28–30, 2008, https://doi.org/10.2514/6.2008-5712.

[11] G.A. Landis, Power systems for Venus missions: a review, Acta Astronaut. 187 (Oct. 2021) 492–497.

[12] G.A. Landis, E. Haag, Analysis of solar cell efficiency for Venus atmosphere and surface missions, in: AIAA 11th International Energy Conversion Engineering Conference, San Jose, CA, USA, July 15–17, 2013. Available at: https://doi.org/10.2514/6.2013-4028.

[13] J.C.C. Fan, Theoretical temperature dependence of solar cell parameters, Sol. Cells 17 (1986) 309–315.

[14] M.Y. Feteha, G.M. Eldallal, The effects of temperature and light concentration on the GaInP/GaAs multijunction solar cell's performance, Renew. Energy 28 (2003) 1097–1104.

[15] M.A. Green, General temperature dependence of solar cell performance and implications for device modelling, Prog. Photovoltaics 11 (5) (August 2003) 333–340.

[16] G.A. Landis, D.J. Belgiovane, D.A. Scheiman, Temperature coefficient of multijunction space solar cells as a function of concentration, in: Proc. 37th IEEE Photovoltaic Specialists Conference, Seattle WA, June 19–24, PVSC, 2011, pp. 1583–1588.

[17] S.P. Tobin, M.B. Spitzer, C. Bajgar, L. Geoffroy, C.J. Keavney, Advanced metallization for highly efficient solar cells, in: Proc. 19th IEEE Photovoltaic Specialists Conference, 1987, pp. 70–75.

[18] M. B Spitzer, J.E. Dingle, R.P. Gale, P. Zavracky, M. Boden, D.H. Doyle, Gallium arsenide concentrator solar cells with highly stable metallization, in: Proc. 20th IEEE Photovoltaic Specialists Conference, 1988, pp. 930–933, https://doi.org/10.1109/PVSC.1988.105840.

[19] M.B. Spitzer, J.E. Dingle, High Temperature Metallization System for Contacting Semiconductor Materials, U.S. Patent 5,075,763 (Dec. 24, 1991).

[20] E.E. Perl, J.S.D.J. Friedman, N. Jain, P. Sharps, C. McPheeters, Y.S.M.L. Lee, M.A. Steiner, (Al) GaInP/GaAs tandem solar cells for power conversion at elevated temperature and high concentration, IEEE J. Photovoltaics 8 (2) (2018) 640–645.

[21] J. Grandidier, A.P. Kirk, P. Jahelka, M.A. Stevens, P.K. Gogna, D. Crisp, M.L. Osowski, T.E. Vandervelde, H.A. Atwater, J.A. Cutts, Photovoltaic operation in the lower atmosphere and at the surface of Venus, Prog. Photovoltaics Res. Appl. 28 (6) (June 2020) 545–553, https://doi.org/10.1002/pip.3214.

[22] J. Chan, J. Wood, J. Schreiber, Development of advanced Stirling radioisotope generator for space exploration, in: Space Technology and Applications International Forum Conference, 2007. NASA/TM-2007-214806.

[23] V.M. Andreev, GaAs and high-efficiency space cells, in: Chapter ID-1, Practical Handbook of Photovoltaics, second ed., Academic Press, 2012, pp. 399–416.

[24] A.G. Santo, R.E. Gold, R.L. McNutt Jr., S.C. Solomon, C.J. Ercol, R.W. Farquhar, T.J. Hartka, et al., The MESSENGER mission to Mercury: Spacecraft and mission design, Planet. Space Sci. 49 (14–15) (December 2001) 1481–1500.

[25] J. McAdams, R.W. Farquhar, A.H. Taylor, B.G. Williams, MESSENGER mission design and navigation, Space Sci. Rev. 131 (1–4) (August 2007) 219–246, https://doi.org/10.1007/s11214-007-9162-x.

[26] G. Dakermanji, J. Jenkins, C.J. Ercol, The MESSENGER spacecraft solar array design and early mission performance, in: Proc. 2006 IEEE 4th World Conference on Photovoltaic Energy Conversion, vol 2, 2006, pp. 1919–1922.

[27] C.G. Zimmermann, C. Nömayr, M. Kolb, A. Caon, A solar cell design for the bepi Colombo high intensity—high temperature mission, in: 37th IEEE Photovoltaic Specialists Conference, 2011, pp. 3713–3718.

[28] E. Gaddy, R. Decker, M.K. Lockwood, L. Roufberg, G. Knutzen, D. Marsh, The Solar Probe Plus solar array development and design, in: Proc. 35th IEEE Photovoltaic Specialist's Conference, Honolulu, HI June 20–25, 2010, 2010, pp. 717–722, https://doi.org/10.1109/PVSC.2010.5617077.

[29] E. Gaddy, M. Butler, M.K. Lockwood, G. Martin, L. Roufberg, C. Vigil, A. Boca, B. Richards, R. Stall, M. Schurman, Predicting the solar probe plus solar array output, in: Proc. 40th IEEE Photovoltaic Specialist Conference, Denver CO, June 8–13, 2014, 2014, pp. 2155–2160, https://doi.org/10.1109/PVSC.2014.6925352.

[30] A. Bermudez-Garcia, P. Voarino, Olivier Raccurt, Environments, needs and opportunities for future space photovoltaic power generation: A review, Appl. Energy 290 (2021) 116757.

[31] J. Grandidier, A.P. Kirk, M.L. Osowski, P.K. Gogna, S. Fan, M.L. Lee, M.A. Stevens, et al., Low-intensity high-temperature (LIHT) solar cells for Venus atmosphere, IEEE J. Photovoltaics 8 (6) (2018) 1621–1626.

CHAPTER FIFTEEN

Space photovoltaic concentrators for outer planet and near-Sun missions using ultralight Fresnel lenses

Mark J. O'Neill
Mark J. O'Neill, LLC, Fort Worth, TX, United States

15.1 Introduction and summary

For the past several years, NASA has supported the development of advanced space photovoltaic concentrator technology using three key elements [1–5]:

1. Ultralight, robust, color-mixing, flat Fresnel lens optical elements: the latest lenses are strengthened with either a ceria-doped glass superstrate to support the silicone prisms forming the lens or an embedded mesh in the silicone lens itself.
2. Advanced multijunction solar cells of two types: triple-junction germanium-based solar cells or inverted metamorphic multijunction (IMM) solar cells with three or more junctions to enhance conversion efficiency.
3. Waste heat radiators made from graphene, a material with unprecedented in-plane thermal conductivity. The latest radiators also have new features: the graphene can be deposited in a tapered thickness pattern onto the back side of a reflective aluminum foil using innovative methods, and the bi-material radiator can mitigate both low-intensity, low-temperature (LILT) effects and high-intensity, high-temperature (HIHT) effects for deep-space and near-sun missions, respectively.

Both 4X line-focus concentrators requiring only single-axis sun tracking (with radiator/solar cell articulation for large longitudinal incidence angles) and 25X point-focus concentrators requiring two-axis sun tracking have been developed and tested at the fully functional prototype level. This chapter will first provide a brief history of space photovoltaic (PV) concentrator technology and discuss the basic optical and thermal constraints on the technology. Next, the latest versions of these line-focus and point-focus concentrators will be described. Finally, the performance, mass, and cost advantages that they offer for various space missions, including outer planet missions and near-Sun missions, will be examined and quantified.

Photovoltaics for Space
ISBN 978-0-12-823300-9, https://doi.org/10.1016/B978-0-12-823300-9.00007-8

15.2 A brief history of space PV concentrator technology

In the late 1970s, Sandia National Laboratories in Albuquerque, NM, began to explore terrestrial PV concentrator technology, and by the mid-1980s, NASA and the US Department of Defense (DOD) began to explore this technology for space applications [6,7]. DOD was especially interested in the radiation hardness of space PV concentrator technology for planned deployments of military spacecraft related to the Strategic Defense Initiative, which was referred to in the popular media as the "Star Wars" program. The US DOD funded several significant development programs related to space PV concentrators including the Survivable Solar Power Subsystem Demonstrator and Survivable Concentrating Photovoltaic Array programs with major aerospace firms offering different technologies, from the TRW "mini-Cassegrainian" two-mirror concentrator to the General Dynamics Solar Linear Array Thermal System offset parabolic trough reflective concentrator to the Boeing "shot-glass" total internal reflection concentrator.

Concurrently, NASA funded the development of the mini-dome Fresnel lens refractive concentrator that was quickly adopted by Boeing as a better replacement for their previous "shot-glass" approach. Boeing also developed a very high efficiency concentrator cell that used two mechanically stacked cells, gallium arsenide (GaAs) on top to convert short and midrange wavelengths and gallium antimonide (GaSb) on the bottom to convert infrared wavelengths of the solar spectrum.

A flight experiment, Photovoltaic Array Space Power Plus Diagnostics (PASP-Plus) of a dozen different leading contenders for next-generation space photovoltaic technologies was jointly developed by NASA and DOD which flew in 1994—95 [8]. The 50X silicone mini-dome lens array using the Boeing dual-junction cell (shown in Fig. 15.1) demonstrated the highest efficiency and lowest degradation rate of the dozen arrays flown, leading to renewed interest in refractive concentrator technology for space power applications.

In the mid-1990s, NASA and DOD provided funding for the development of the 8X solar concentrator array using a refractive linear element technology (SCARLET) linear

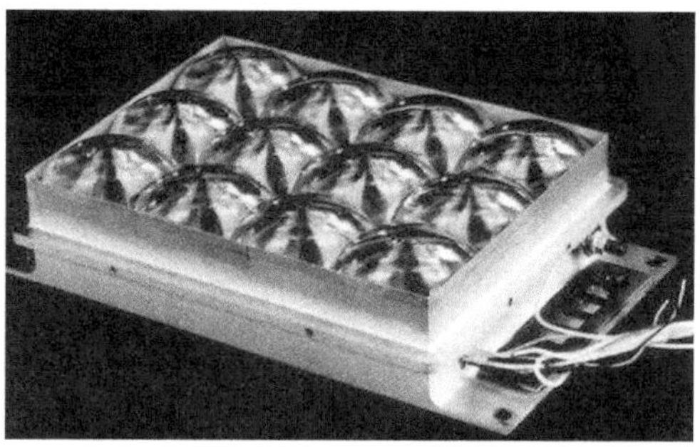

Figure 15.1 Mini-dome lens array on PASP-Plus. *Source: author.*

Fresnel lens array, which used the newly emerging monolithic multijunction cells (gallium indium phosphide (GaInP)/GaAs/germanium (Ge)) in the focal lines of arched line-focus lenses. SCARLET was selected to power NASA's Deep Space 1 mission (shown in Fig. 15.2) to demonstrate solar electric propulsion using a NASA Solar Electric-Power Technology Application Readiness (NSTAR) ion engine to propel the spacecraft to encounters with both a comet and an asteroid during the 38-month mission. The mission was extremely successful from 1998 to 2001 [9], and the SCARLET array and the NSTAR ion propulsion system validated the technologies of both Fresnel lens—multijunction cell space photovoltaic concentrators and solar electric propulsion. Deep Space 1 won a number of awards, including the 2001 NASA Turning Goals into Reality award.

In the 1999—2001 timeframe, Hughes Space and Communications, renamed Boeing Space Systems, launched several (six commercial spacecraft and a classified number of DOD spacecraft) of their new, large 702-class communication satellites, which unfortunately had a flawed "concentrator" design that relied on reflective wings to add irradiance to dual-junction cell panels at about 1.5X concentration (shown in Fig. 15.3). The flawed design of these arrays caused cell overheating with excessive outgassing followed by condensation of the outgassing products onto the reflective wings. The array power output degraded rapidly on orbit leading to billions of dollars of spacecraft losses [10]. Litigation with customers and insurers went on for years after this set of failures, and the generic term "concentrators" suffered a critical blow. For a few years, no distinction was made by many potential customers between successful space concentrators like SCARLET and the failed Boeing 702 "concentrators."

Fortunately, NASA and DOD made the distinction between successful space concentrators and the flawed Boeing "concentrators," and in the 2000s it funded the development of advanced Fresnel lens photovoltaic concentrators including the

Figure 15.2 SCARLET array on Deep space 1. *Courtesy: NASA.*

Figure 15.3 Solar array on Hughes/Boeing 702 with reflective wings on both sides of PV panels. *Courtesy: Boeing.*

Stretched Lens Array, an improved follow-on to the SCARLET array. In the 2010s, NASA and DOD funded the development of even further improved Fresnel lens concentrator technology including the Deployable Space Systems' SOLAROSA (Solar Optical Lens Architecture Using Roll-Out Solar Array) and Orbital ATK's (now Northrop Grumman's) PFC-CTA (Point-Focus Concentrator on Compact Telescoping Array) [1–5].

The latest space Fresnel lens photovoltaic concentrators use robust ultralight lenses that are about 90% lighter than the SCARLET lenses to focus sunlight onto multi-junction cells that are 50%–60% more efficient than the SCARLET cells. Waste heat radiators now employ graphene, which is more than 10X more thermally conductive than the carbon fiber–reinforced composite sheet radiators used on SCARLET. Both line-focus and point-focus versions have been developed, as discussed in the following section.

Before describing these advanced concentrators, a brief discussion of the limits on concentration based on the second law of thermodynamics is needed. For a simple reflective or refractive optical concentrator, the second law imposes limits on achievable concentration to preclude heat transfer from a cooler body to a warmer body. These limits are well known in optics [11], following Eqs. (15.1 and 15.2):

$$\text{Line Focus Concentration} \leq \frac{\sin(\text{max incidence angle})}{\sin(\text{maxacceptance angle})} \tag{15.1}$$

$$\text{Point Focus Concentration} \leq \left[\frac{\sin(\text{maxincidence angle})}{\sin(\text{maxacceptance angle})}\right]^2 \tag{15.2}$$

Where max acceptance angle = solar disk angular radius + sun-pointing tolerance angle, and the max incidence angle = maximum incidence angle for rays intercepting receiver.

At 1 astronomical unit (AU) distance from the sun to the concentrator, the solar disk angular radius averages about 0.267 degrees over the year. For simple Fresnel lens photovoltaic concentrators without any secondary optics, the max incidence angle onto the cell is generally in the 15- to 30-degree range, the former value for flat Fresnel lenses and the latter number for arched or dome Fresnel lenses. Fig. 15.4 shows the curves for the above equations for various sun-pointing tolerance angles. For a typical mission, a sun-pointing tolerance angle of about ±2 degrees is often selected, for example, the SCARLET array on Deep Space 1 and the SOLAROSA and PFC-CTA platforms, discussed above. For a flat Fresnel lens concentrator with a 15-degree max incidence angle onto the cell, the second law limits for ±2 degrees are about 7X and 43X for

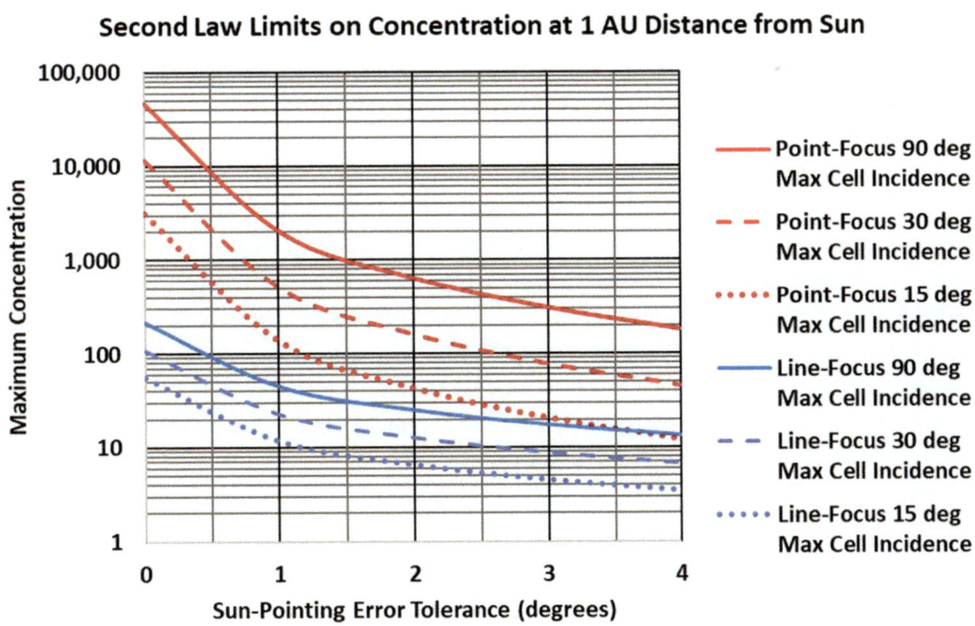

Figure 15.4 Second law limits on concentration for concentrators at 1 AU from the Sun. *Source: author.*

line-focus and point-focus concentrators, respectively. Real-world advanced Fresnel lens concentrators do not reach these second law limits because of errors and aberrations, but they come relatively close, as discussed in the following section.

Before describing these advanced concentrators, a brief discussion of waste heat radiator requirements is also needed, to explain why the basic building blocks of space photovoltaic concentrators need to be small in dimension. Since even the most efficient photovoltaic cells are presently well under 50% conversion efficiency, such cells produce more waste heat than electricity under sunlight irradiance, and even more waste heat when the cells are not operating at their peak power point. This waste heat must be dissipated efficiently to deep space by thermal radiation. For a 1-sun cell mounted to a rigid or flexible structure, both sides of the structure can radiate effectively to space. In contrast, for a Fresnel lens concentrator, the waste heat from the radiator to which the cell is mounted can radiate more effectively from the back side to deep space than from the front side, since the lens is between the radiator and deep space on the front side. This causes the average radiator temperature to be warmer for a concentrator cell than for a 1-sun cell. The effective radiation heat transfer is readily calculated for either type of solar array.

For a 1-sun array, the total radiation area is about the same as the total cell area. For a concentrator array, the total radiation area (on both the front and the back) is about the same as the total lens aperture area. For example, a point-focus Fresnel lens array of 10×10 cm lenses next to one another can have no more waste heat radiator area than 10×10 cm for each cell without interfering with neighboring radiators. Furthermore, the concentrator cell is much smaller than the available radiator area, so the waste heat must be conducted efficiently over the available radiator area to effectively use the full radiator area. This introduces a challenge for concentrators compared to 1-sun arrays, in that the radiator must be an efficient conductor of heat over the available radiator area.

Simplified example calculations of the radiator temperature distributions for line-focus and point-focus concentrators are shown in Fig. 15.5A and Fig. 15.5B, respectively. These example calculations are for a lens and radiator that are both 10 cm in width, with the abscissa the distance from the center of the cell outward to the edge. These calculations are also for a uniform 40-μm-thick graphene radiator for the line-focus concentrator and a uniform 100-μm-thick graphene radiator for the point-focus concentrator. A high Earth orbit with negligible albedo from the Earth is assumed. A waste heat load of 60% of the incident sunlight is also assumed, and this heat is placed at the center of the cell for the line-focus concentrator and in a small region near the center of the cell for the point-focus concentrator. The average radiator temperature is the same for both geometries since the same amount of waste heat must be dissipated by both radiators. If the radiation heat transfer coefficient is assumed constant at the average radiator temperature, the top curve has the form of a combination of hyperbolic cosine functions for an assumed linear heat conduction geometry, and the bottom curve has the

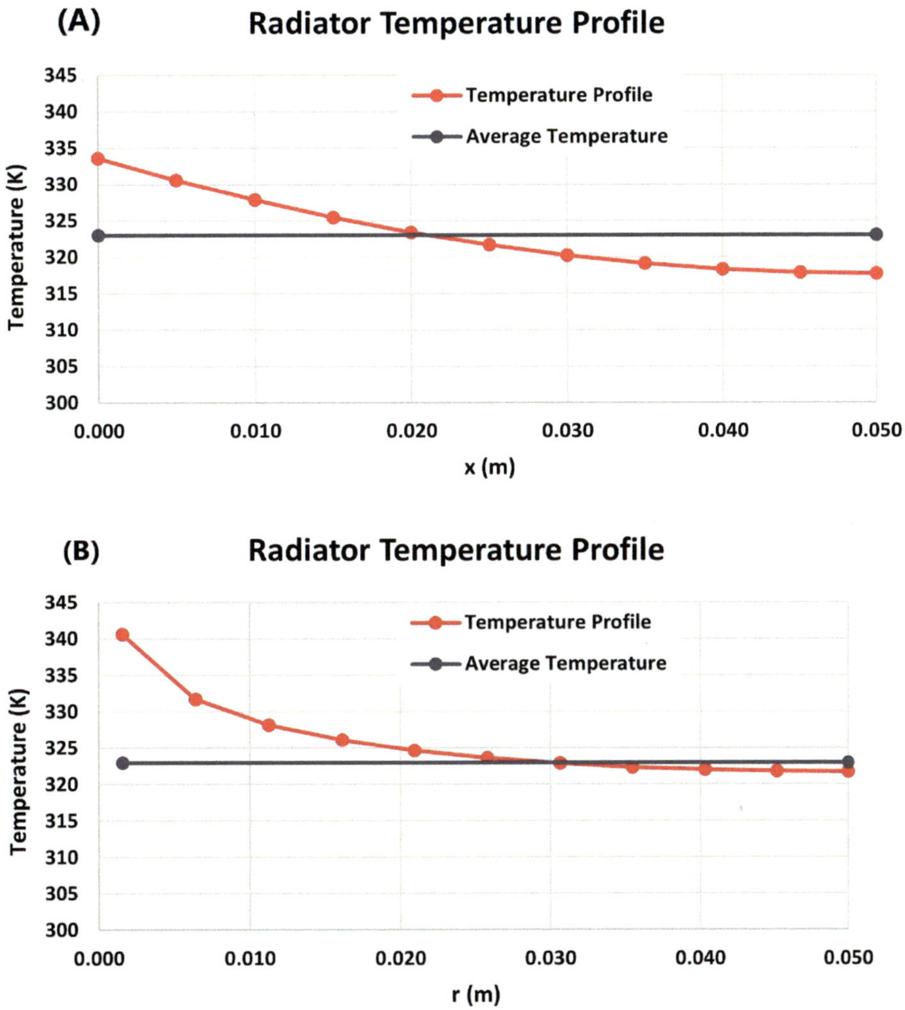

Figure 15.5 (A) Temperature profile for line-focus concentrator; (B) temperature profile for point-focus concentrator. *Source: author.*

form of a combination of Bessel functions for an assumed radial heat conduction geometry. Numeric analysis with correct treatment of the local temperature-based radiation confirms similar curves for both systems. The profile is steeper for point-focus (bottom) than for line-focus (top) due to the physical difference in radial heat conduction compared to linear heat conduction. Parametric analyses for both concentrators show that if the lens and radiator widths are doubled, for example, the radiator thickness must be increased by 4X, to maintain the same temperature profile. More generally, the required radiator thickness increases with the width of the lens and radiator squared. Small concentrator building blocks are therefore essential to maintain a lightweight radiator, as further explained below.

Since the radiator area is approximately equal to the lens aperture area, the larger the lens is, the greater is the amount of intercepted solar irradiance and the larger the amount of waste heat to be radiated to space. Compounding this problem, the larger the lens aperture area is, the larger is the distance over which waste heat must be efficiently conducted to effectively use the total available radiator area. These two effects lead to the fact that to maintain the same cell temperature for two different size concentrators, either point focus or line focus, the radiator thickness must increase with the square of the lens aperture width. For large lenses, this leads to impractical radiator thicknesses and associated masses.

Improvements in radiator materials over the past 25 years have allowed the lens dimensions to increase significantly. The mini-dome lens array discussed above used aluminum radiators, which limited the aperture width to less than 4 cm. The SCARLET array discussed above used carbon fiber—reinforced composite radiators, which led to an 8.5-cm aperture width. The latest advanced concentrators incorporate graphene into the radiators, allowing an increase in aperture width to 10 cm, as discussed in the following section.

However, before describing these advanced concentrators, a brief discussion is needed of another major advantage of space photovoltaic concentrators over conventional 1-sun photovoltaic cell arrays in the mitigation of electrostatic discharges and their corresponding damage. Dr. Dale Ferguson, author of Chapter 2 of this book [12], which addresses this topic in more detail, is a leading authority on this problem that has caused damage to numerous space photovoltaic arrays, especially over the past 25 years, as array power output and corresponding operating voltage levels have increased. Photovoltaic concentrators have an inherent advantage over 1-sun arrays in mitigating electrostatic discharge from solar cells to the surrounding environment in space because the concentrator cells are much smaller and more widely spaced from neighboring cells. The small cell size allows concentrator cells to be more thoroughly insulated, encapsulated, and isolated from surrounding objects and the space plasma, with very low performance penalty or mass penalty. The wide spacing impedes any sustained electrostatic discharge from one cell to a neighboring cell.

Ground tests have been performed of fully encapsulated and insulated concentrator cells in a simulated space plasma while biased to 1000 V and while subjected to simulated micrometeoroid impacts [13]. These tests proved the robustness of such concentrator cells in short-term tests. Longer term (months) ground tests have also been performed using underwater high-pot tests at very high voltage bias (2250 to 4500 V) to verify corona failure resistance [14]. Even earlier high-voltage (500 V) tests were performed in space on the mini-dome lens photovoltaic concentrator on PASP-Plus (Fig. 15.1) with no discharges [8].

High-voltage operation is important for future space photovoltaic arrays to minimize power harness conductor size and mass for very high-power arrays (e.g., 100 kW and larger) and to enable direct-driving of solar electric propulsion thrusters to minimize performance losses and mass additions of voltage step-up power management and distribution equipment. Ground tests have confirmed that high-voltage photovoltaic concentrators can reliably direct-drive electric thrusters [15].

15.3 Description of the latest space Fresnel lens photovoltaic concentrators

Fig. 15.6 shows the latest line-focus and point-focus concentrator modules, including the lens, photovoltaic cell assembly, and graphene-based radiator. Note that the two concentrator modules share many important features. The nominal 4X and 25X geometric concentration ratios (lens aperture area/cell active area) correspond to sun-pointing error tolerances of ± 2 degrees about the critical lateral axis (alpha) for line-focus and about both axes (alpha and beta) for point-focus.

To enable wide variation in sun-pointing error tolerance about the noncritical beta axis for the line-focus concentrator, a novel articulation scheme has been developed for the radiator/receiver relative to the lens, as shown in Fig. 15.7 [16]. This scheme allows excellent focusing for wide variations in the beta angle; Fig. 15.7 shows ray traces for both 0-degree and 50-degree beta angles. Actual lens/cell demonstrations have validated excellent optical performance over the full range from 0-degree to ±50-degree beta angles. This variation in spacing between lens and radiator/receiver is relatively simple to implement on both SOLAROSA and PFC-CTA platforms.

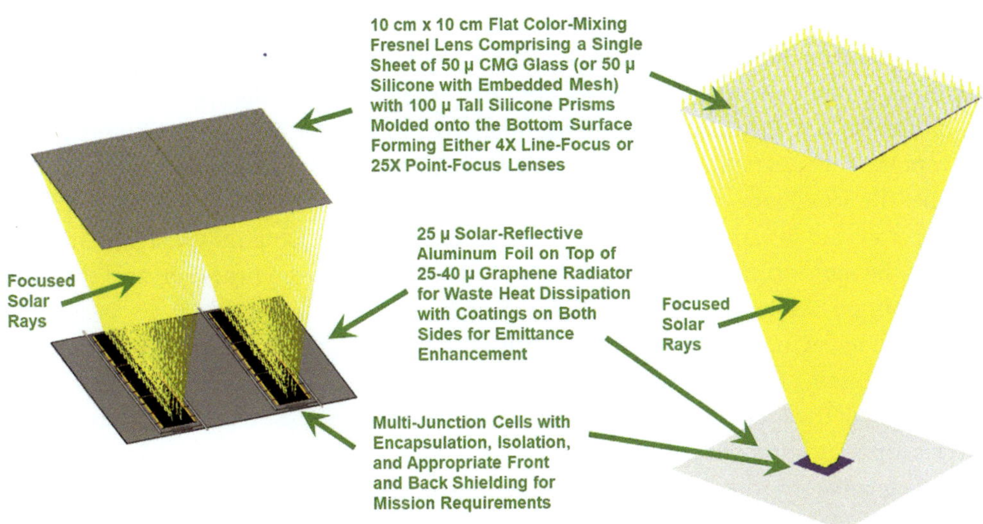

Figure 15.6 4X line-focus and 25X point-focus PV concentrators share many common features. *Source: author.*

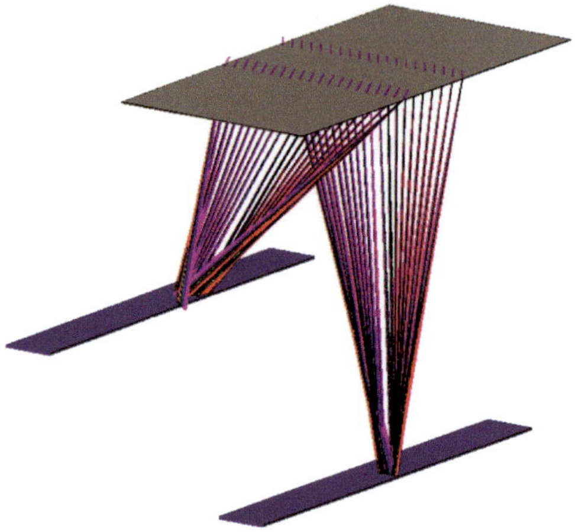

Figure 15.7 Linear receiver articulation. *Source: author.*

Low-intensity, low-temperature mitigation is accomplished by the optical concentration itself, as discussed by Landis et al. [17,18]. For the 4X line-focus concentrator, the cell irradiance at the focal plane remains about 1 AM0 sun out to 2 AU distance from the sun, where the incident solar irradiance is 1/4th of 1 AM0 sun. For the 25X point-focus concentrator, the cell irradiance remains about 1 AM0 sun out to 5 AU distance from the sun, where the incident solar irradiance is 1/25th of 1 AM0 sun.

Even deep space missions often use a near-Sun inner planet swing-by as a gravity assist maneuver to reduce the propellant needed to reach an outer planet, exposing the cells to both HIHT and LILT effects over the course of the mission. Mitigation of high temperatures in the high solar intensity (near Sun) environments can be accomplished by intentionally defocusing the lens by shortening (or lengthening) the spacing between the lens and solar cell, allowing part of the focused sunlight to miss the cell and instead intercept the reflective aluminum foil surrounding the cell. By reflecting away part of the sunlight, the irradiance on the cell can be maintained near its 1 AU values at <1 AU distance from the sun, thereby keeping the cell at a mild operating temperature, and thus allowing the same concentrator assembly to be used for both the near-Sun and the outer solar system power.

For a Venus swing-by, since Venus is about 0.7 AU from the sun compared to Earth at 1.0 AU, the solar irradiance is about twice as high near Venus compared to near Earth. Reflecting half the focused sunlight away will therefore keep the cell temperature and the power output at approximately the same value as at 1 AU with the cell in its normal position relative to the lens. This concentrator technology will be deployed and supported as a dual (lens blanket and radiator blanket) flexible-blanket array, and the HIHT mitigation will be accomplished by moving the two blankets slightly closer together at array to sun distances less than 1 AU.

The low-temperature portion of the LILT environment is of lesser importance than the low-intensity portion. Low temperatures are somewhat mitigated by the presence of the lens in front of the radiator, providing a thermal radiation barrier, which slightly increases the radiator and cell operating temperature. If desired, the temperature can be further mitigated by allowing the bi-material radiator to curl up due to differential thermal contraction, reducing the radiator view factor to deep space and keeping the cell temperature a little bit warmer still. The combined effect of the lens and *nyctinastic* (biomimetic, i.e., plant, movement in response to light or temperature) radiator can be to keep the cell about 20°C warmer near Jupiter (about 5 AU from the sun) than a 1-sun cell under the same conditions, if desired.

15.4 Recent lens developments

As discussed in previous publications [1–5], our (Mark O'Neill, LLC) lenses use a unique color-mixing approach wherein neighboring prisms in the Fresnel lens have their angles slightly tweaked to intentionally overlap various portions of the solar spectrum to produce a similar current concentration profile over the cell for each of the various junctions in the cell, thereby avoiding "chromatic aberration" power losses in monolithic multijunction solar cells.

Early versions of the concentrator lens were tested using only the weak silicone Fresnel lens material, with no mechanical substrate or reinforcement. While these lenses had sufficient strength in ground testing, space testing showed that the lenses embrittled in the UV/radiation environment, leading to tearing of the lens [19]. Thus, the current version of the Fresnel lenses incorporates additional mechanical strength [1–5]. For improved mechanical robustness of our lenses, we have recently developed two new styles of robust lenses that use different strengthening elements for the relatively weak silicone lens material:

1. A ceria-doped glass superstrate lens using 50- to 100-μm CMG glass from Qioptiq to support the 100-μm-tall silicone prisms on the inside surface of the lens
2. New high-transmissivity embedded meshes to support the 100-μm silicone lens, including 50-μm photo-etched titanium mesh.

For scalable mass-production of these two new styles of robust lenses, we have developed a molding process using disposable polymer lens molding tools, as described in Fig. 15.8. The only difficult step in the manufacturing process is the removal of the molding tool from the cured silicone lens assembly at the end of the process. We have recently developed a novel approach that eliminates this tool removal step. Instead of removing the tool, we dissolve the tool in a "green" solvent, exposing the final lens assembly. This

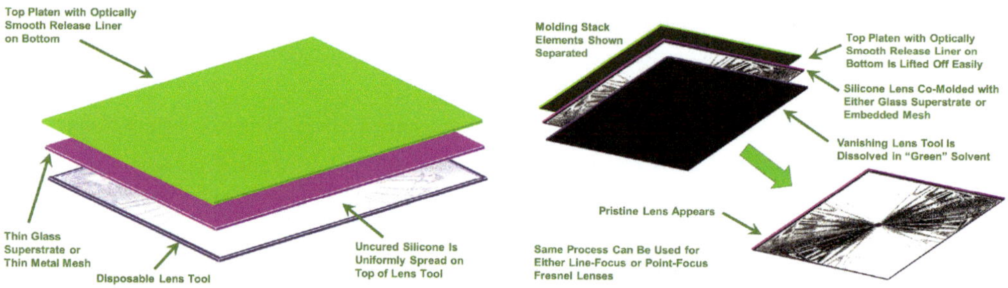

Figure 15.8 "Vanishing" lens molding tool manufacturing approach. *Source: author.*

eliminates damage to the lens assembly due to mechanical removal of the lens tool. Fig. 15.8 shows this novel process; recently, we have proven its technical feasibility by making lenses of both styles, as shown in Fig. 15.9.

We have had success with several different pairs of "vanishing" lens molding tool materials and appropriate green solvents, including the following:

1. polystyrene (PS) molding tools and limonene solvent
2. polymethyl methacrylate (PMMA) molding tools and anisole solvent

Molding tools of both polymer materials were made by 10X Technology using nickel electroform replicas of the master diamond-turned tool. We have developed a simple, scalable process to dissolve the polymer molding tool quickly and efficiently using a green solvent for both styles of robust lenses. Fig. 15.10A shows one of the glass superstrate lenses after the polymer molding tool was dissolved. The process is simple and

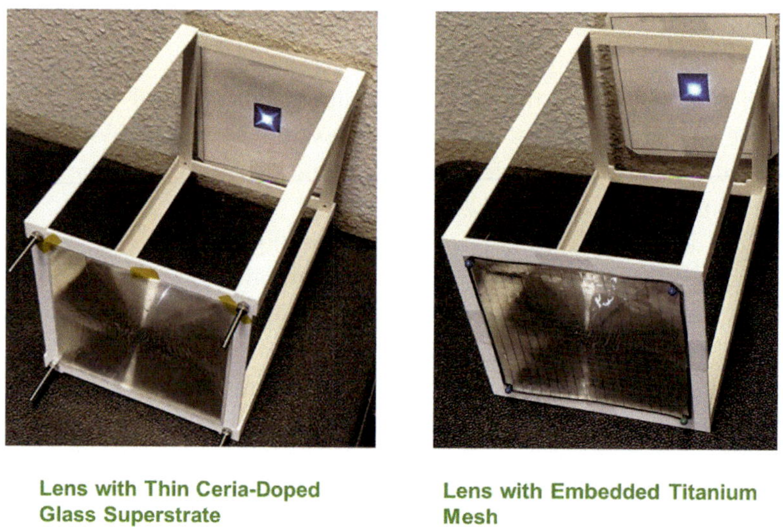

Figure 15.9 Prototype lenses (glass superstrate on left, embedded titanium mesh on right) made with "vanishing" lens molding tools, showing the focal spots. *Source: author.*

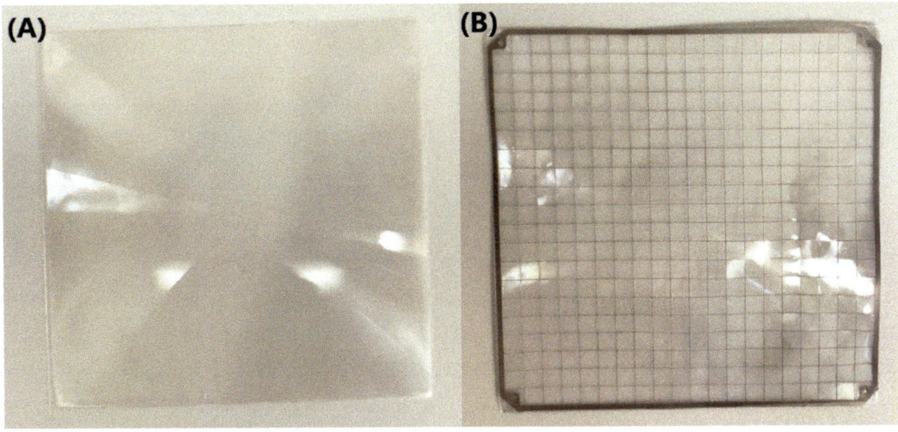

Figure 15.10 (A) Glass superstrate lens after dissolving; (B) embedded mesh lens after dissolving tool. *Source: author.*

rapid, taking only a few minutes to dissolve the tool. Fig. 15.10B shows one of the embedded mesh lenses after the polymer molding tool was dissolved. For the mesh lens, the process is equally simple and rapid, taking only a few minutes.

We made several of each type of lens (glass superstrate and embedded mesh), with 100% yield and no significant problems, to validate the "vanishing lens tool" process. The latest mesh-supported lenses are the lightest lenses we have ever made, with a typical mass of about 1.1 g for a lens aperture area of 10×10 cm $= 100$ cm^2. The robust, ultralight, and novel embedded mesh lenses are our current favorite approach. These mesh lenses require an ultraviolet rejection coating, which is well developed and proven in space, to prevent the silicone from yellowing under long-term exposure to very short wavelength solar ultraviolet radiation in space [20].

As described in previous publications [1–5], the embedded mesh lenses typically have about 85% optical efficiency compared to about 90% optical efficiency for the glass superstrate lenses, due to approximately 5% light blockage by the mesh. If such a 100 cm^2 mesh lens is used to focus sunlight onto a 30% efficient multijunction solar cell, the power output of the lens/cell combination would be about 3.44 W for an Earth-orbiting spacecraft (1 AM0 sun irradiance on lens). Thus, the mass/power parameter for the lens is only 0.32 g/W = 0.32 kg/kW. This ratio is often called the "alpha" parameter (not to be confused with the alpha angle described previously) and is extremely important for most space missions. The alpha parameter of the lens of course needs to be added to the alpha parameters for the solar cell package, the waste heat radiator, the array deployment, and support structure, etc., to fully quantify the mass/power ratio of the solar array system. Since the lens also increases the irradiance onto the solar cell by a net factor of about 21X (85% lens transmittance times 25X geometric concentration ratio), it also dramatically reduces the alpha parameter of the solar cell package compared to conventional 1-sun solar cell arrays.

15.5 Recent multijunction cell developments

SolAero has developed four-junction IMM cells (each 1.2 × 3.2 cm active area) on glass carriers for the 4X line-focus concentrator and assembled prototypes into three-cell parallel-connected receivers, as shown in Fig. 15.11. They also developed three-junction cells (each 2.0 × 2.0 cm active area) on Ge wafers for the 25X point-focus concentrator, as shown in Fig. 15.12. Both AM0 simulator measurements and outdoor AM1 measurements of these cells with the appropriate lenses verified expected gains in both current and power provided by the lenses compared to 1-sun measurements on the same cells. For future flight systems, IMM cells with four or more junctions will be used in both line-focus and point-focus concentrators. For the latter, the glass carrier will need to be replaced with a thinner or more thermally conductive carrier to provide

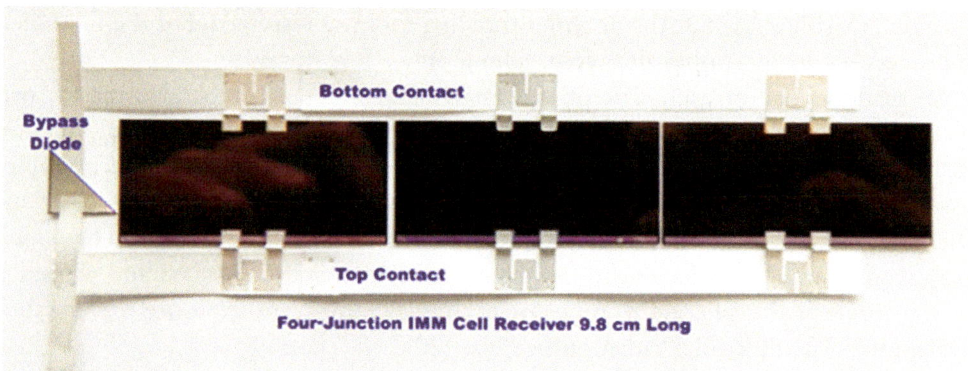

Figure 15.11 Three-cell receiver for 4X line-focus concentrator. *Source: author.*

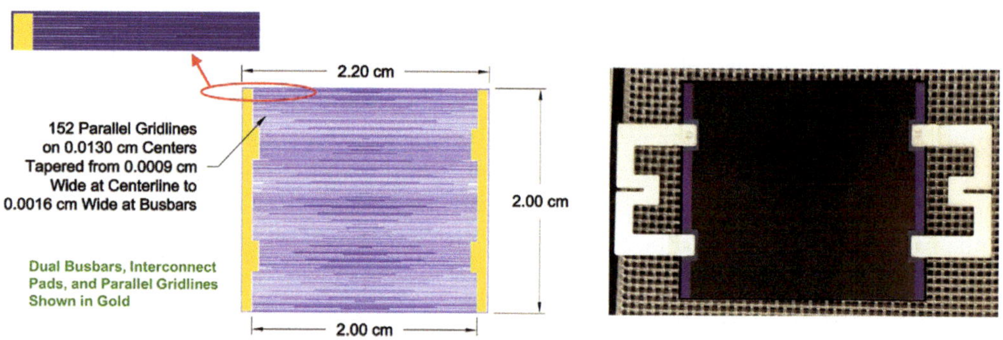

Figure 15.12 Cell for 25X point-focus concentrator. *Source: author.*

an acceptably small temperature difference between cell and radiator for the more concentrated point-focus irradiance distribution.

15.6 Recent graphene radiator developments

For space PV concentrators, cell efficiency is increased under operational conditions by conducting the waste heat away from the active cell area and radiating it to space from both the front and back sides of the concentrator assembly (Fig.15.6). In our system, this is achieved by the use of graphene, which has an exceptionally high in-plane thermal conductivity [1–5]. Under a recent NASA contract, our team worked with the University of Connecticut to develop novel methods of making and depositing bio-graphene onto aluminum foil, to function as a bi-material thermal conductor, to spread heat across the radiator surface.

This novel method of making and depositing graphene allows tapering of the graphene thickness to optimize the heat spreading over the radiator area. Fig.15.13 is a schematic of the basic concept, with thicknesses exaggerated and false colors. The tapering effect can also be achieved in an alternate way by using thin commercially available graphene sheet and bonding multiple graphene sheet layers together near the cell with thermally conductive adhesive, as Fig. 15.13 also shows.

We have recently conducted thermal analyses that show a major reduction in cell operating temperature can be achieved by tapering the graphene thickness rather than keeping the graphene thickness constant over the radiator. For the same total mass of

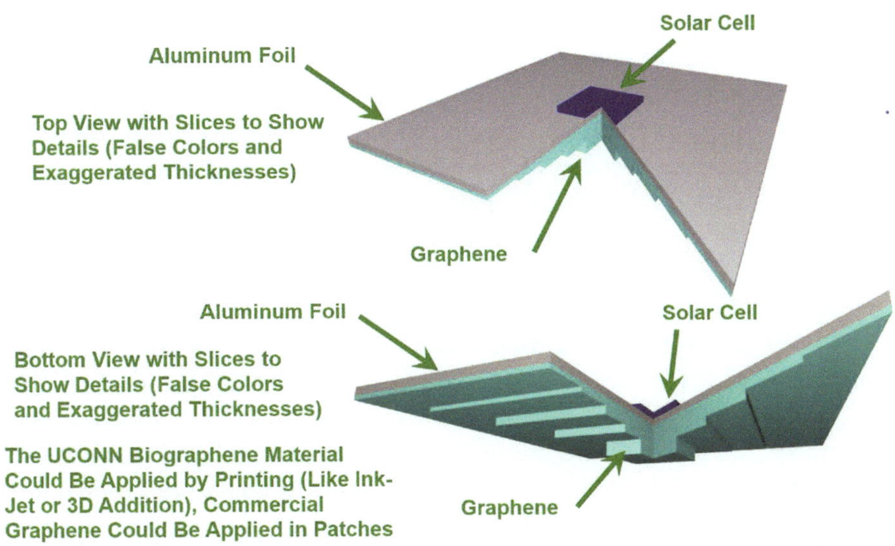

Figure 15.13 Crude schematic of graphene tapering concept. *Source: author.*

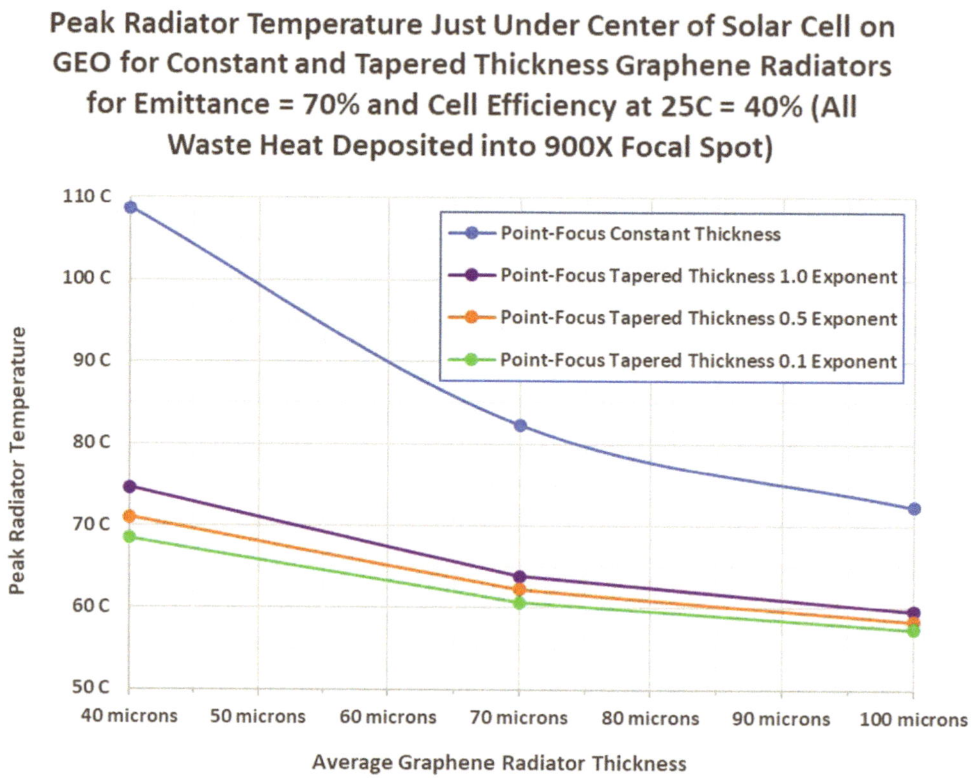

Figure 15.14 Thermal performance of tapered versus constant thickness graphene radiators on GEO. *Source: author.*

graphene, the peak radiator temperature on geosynchronous equatorial orbit (GEO) for the point-focus concentrator is significantly reduced by tapering the graphene thickness, as shown in Fig. 15.14. The various "exponent" curves correspond to different tapering profiles, all with the same average thickness identified on the abscissa.

For this analysis, the thickness of the graphene is varied from a maximum value at the center of the radiator to a zero value at the four adiabatic edges of the radiator. We varied the tapering curve from linear (1.0 exponent in Fig. 15.14) to nonlinear (other exponents in Fig. 15.14). Note that for a 40-μm average thickness, the peak temperature is about 35°C–40°C lower with tapering than without tapering, a substantial reduction with no mass penalty. For larger average graphene thicknesses, the peak temperature reduction is smaller, but still worthwhile.

Furthermore, we assumed the use of an emittance-enhancing coating on the graphene and aluminum foil, both of which have relatively low emittances in their uncoated form. Our team has recently tested a simple 25-μm coating of silicone on the graphene, which increased the emittance from 33% to 70% in testing by the Air Force Research

Lab in Albuquerque. The total mass of a 10 × 10 cm silicone-coated radiator using 25-μm aluminum foil and 40-μm average thickness graphene is only 2.1 g. Therefore, the alpha parameter for the bi-material radiator based on the 3.44 W of cell power output discussed in the previous section is only 0.61 g/W = 0.61 kg/kW.

Another advantage of the tapered graphene thickness is the greater radiation shielding of the solar cell from charged particles (electrons and protons) incident on the back side of the array. The greater thickness of graphene under the cell, desired for improved lateral thermal conduction, also provides reduced radiation degradation of the solar cell and improved end of life power. For all the tapered graphene thicknesses in Fig. 15.11, regardless of the taper exponent, the graphene is over 100 μm thick behind the entire solar cell. Together with the 25-μm aluminum foil and the 25-μm silicone coating, these materials provide over 125 μm of equivalent fused silica shielding for the back of the cell. The cell superstrate or carrier and the adhesive between the cell package and the radiator add to this equivalent fused silica shielding thickness for the solar cell.

If the multijunction solar cell is an inverted metamorphic (IMM) type of cell, the cell package mass will be dominated by the front and back shielding. As discussed above, the tapered graphene radiator provides over 125 μm of equivalent fused silica shielding for the back of the cell. If an equivalent 125 μm of fused silica shielding is also used on the front of the cell, this will add approximately 0.13 g of front shielding mass to the 2 × 2 cm solar cell. The alpha parameter, based on the previously discussed 3.44 W of cell power, for this front shielding is only 0.04 g/W (= 0.04 kg/kW).

15.7 Performance metrics and cost savings

Based on the mass discussions in the previous two sections, the combined alpha parameter for the mesh lens, the tapered graphene radiator, and the front cell shielding is 0.97 kg/kW. While this parameter is calculated from just the mass of these three elements (not including the array deployment and support structure, the harnessing, etc.), it still allows comparison to the shielded 1-sun solar cell used in a conventional solar array.

The inverse of this parameter is the specific power. For just the combined lens, radiator, and cell shielding, the specific power is about 1035 W/kg. If the same 2 × 2 cm cell with the same front and back shielding was used at 1-sun irradiance, it would produce about 0.16 W (assuming 30% efficiency in Earth orbit at 1 AM0 sun irradiance). The front and back shielding would have a mass of 0.13 g (front) + 0.13 g (back) = 0.26 g (total), as discussed above. Thus, the alpha parameter for 1-sun shielding would about 1.53 g/W = 1.53 kg/kW. The specific power for the 1-sun cell shielding alone would be about 656 W/kg, less than two-thirds that of the 25X concentrator lens, radiator, and cell shielding.

For many missions, however, long lifetime in a high-radiation environment is a design objective, and thicker radiation shielding is required. As shown in Fig. 15.15, as the thickness of the cell radiation shielding increases, the advantage of the concentrator grows rapidly, due to the much higher power output of the cell under concentration. Radiation shielding at low mass penalty is one of the hallmark attributes of concentrator technology.

This attribute is of particular importance for future NASA missions. Missions are being proposed to orbit or even land on the inner moons of Jupiter, for example, but this requires operation in an environment with low solar intensity, low operating temperature, and a high radiation intensity. For this application the ability to incorporate thick radiation shield on the cells with only modest mass increase can be an enabling technology [17,18].

Our team has worked closely with larger firms to incorporate the Fresnel lens photovoltaic concentrators into state-of-the-art deployment and support platforms. One recent NASA-funded program performed by Orbital ATK (now Northrop Grumman) looked in detail at the performance metrics and cost savings of the 25X point-focus concentrator on the compact telescoping array (CTA) [21]. Fig. 15.16 shows the basic solar array approach; the results were very positive, as summarized in Figs. 15.17 and 15.18.

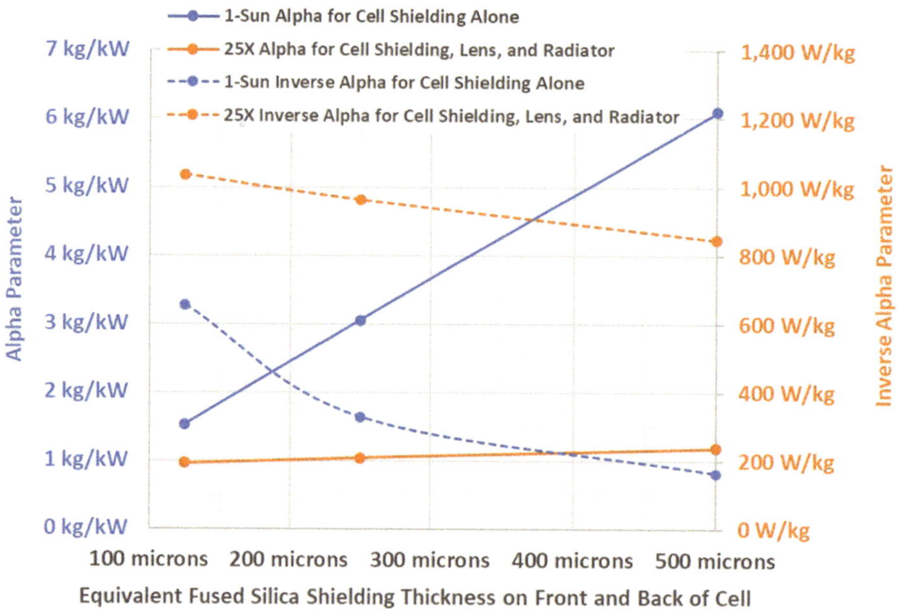

Figure 15.15 Alpha, in kg/kW, and specific power (inverse alpha) parameters for key elements of 25X concentrator versus 1-sun cells. *Source: author.*

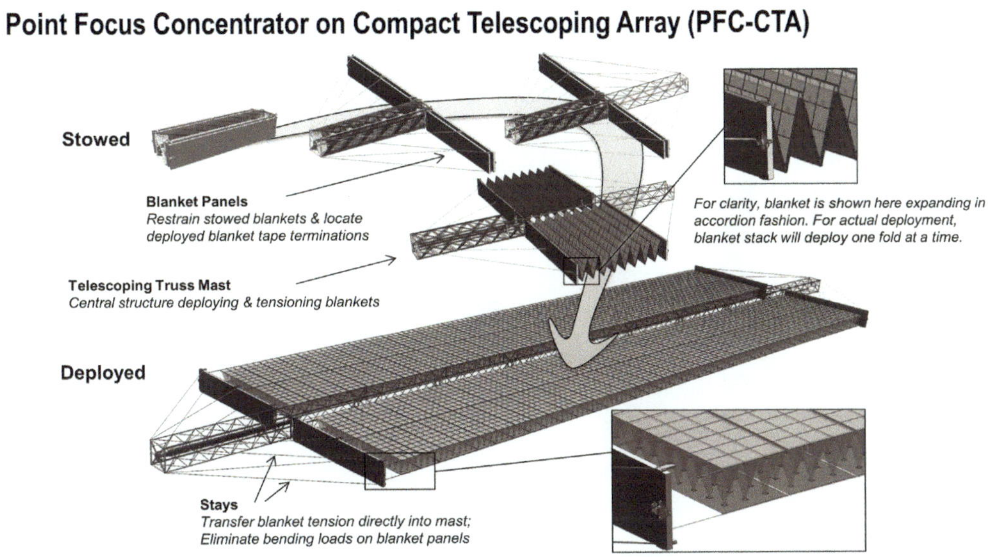

Figure 15.16 25X point-focus concentrator on compact telescoping array for NASA's Extreme Environment Solar Power project. *Source: author.*

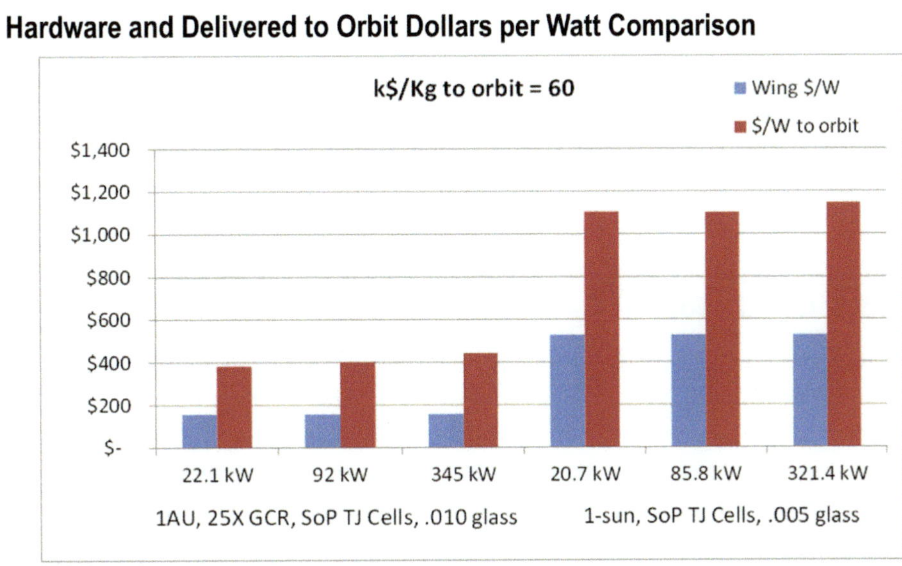

Figure 15.17 Comparison of 25X point-focus concentrator versus 1-sun cell versions of compact telescoping array (CTA): hardware and delivered-to-orbit dollars per watt. *Source: author.*

Stowed Power per Volume (PPV) and Specific Power Comparison

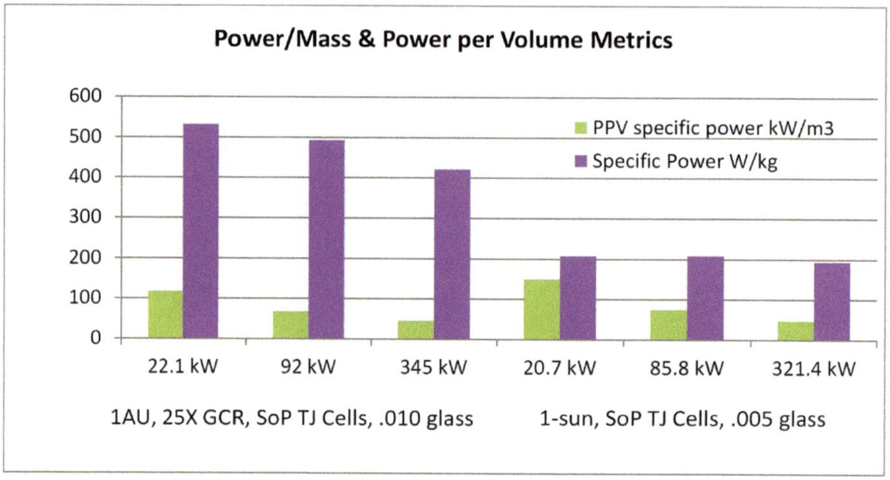

Figure 15.18 Comparison of 25X point-focus concentrator versus 1-sun cell versions of compact telescoping array (CTA): stowed power per volume and mass specific power. *Source: author.*

Note in Figs. 15.17 and 15.18 that the cost savings and mass savings are more than 50% for the concentrator array compared to 1-sun arrays for three different solar array power outputs from about 20 to 300 kW. Cost savings are shown in the left column chart for the solar array wing prior to launch (blue columns) and placed in orbit (red columns), including launch cost. The two primary reasons for the concentrator cost savings are the much smaller cell area and cost, and the much lower mass, reducing launch costs. The primary reason for the mass savings is the inherently lower combined mass of the lens, radiator, and cell shielding for the concentrator compared to just the cell shielding for the 1-sun cells, as discussed above [13–15].

15.8 Conclusions

Over the past several decades, space photovoltaic concentrator technology has undergone continuous development, refinement, and technology maturation. The latest concentrator technology offers many important advantages over more traditional 1-sun photovoltaic cell technology. These advantages include the following:

1. Major cost savings in terms of $/Watt at the array level, due to the use of much lower cost lenses to collect and focus the light onto much smaller cells with dramatic cost reduction for the expensive multijunction cells.
2. Major mass savings in terms of W/kg at the array level, due to the dramatic reduction in mass of the cell and shielding due to the much smaller cell size.

3. Elimination of the low-intensity problems (LILT) of 1-sun cells for deep space missions.
4. Major improvements in cell production capacity in terms of kW/year, since each wafer processed into small concentrator cells provides much higher on-orbit power than for the same wafer processed into 1-sun cells.
5. Major improvement in radiation hardness of cells for extreme environment missions, since the small concentrator cells can be heavily shielded with minor mass penalty.

As we have demonstrated throughout this chapter, space photovoltaic concentrator technology stands ready to be utilized for future space missions of all types. Indeed, for several specific destinations and/or applications, this technology could be mission enabling.

References

[1] M. O'Neill, A.J. McDanal, H. Brandhorst, K. Schmid, P. LaCorte, M. Piszczor, M. Myers, et al., Recent space PV concentrator advances: More robust, lighter, and easier to track, in: 42nd IEEE Photovoltaic Specialists Conference, 2015, https://doi.org/10.1109/PVSC.2015.7356142 (New Orleans, LA, USA).

[2] M. O'Neill, A.J. McDanal, H. Brandhorst, B. Spence, S. Iqbal, P. Sharps, et al., Space photovoltaic concentrator using robust fresnel lenses, 4-junction cells, graphene radiators, and articulating receivers, in: 43rd IEEE Photovoltaic Specialists Conference, June 5-10, 2016, https://doi.org/10.1109/PVSC.2016.7749833 (Portland, OR, USA).

[3] H.W. Brandhorst Jr., M.J. O'Neill, A.J. McDanal, B. Spence, et al., Space photovoltaic concentrator using robust Fresnel lenses, 4-junction cells, graphene radiators, and articulating receivers, in: Proc. 53rd AIAA/SAE/ASEE Joint Propulsion Conference, July 10-12, 2017 (Atlanta, GA, USA), 512−516 (AIAA 2017−4711).

[4] M. O'Neill, A.J. McDanal, M. Piszczor, T. Peshek, M. Myers, C. McPheeters, J. Steinfeldt, et al., Advanced development of space photovoltaic concentrators using robust lenses, multi-junction cells, & graphene radiators, in: 7th World Conference on Photovoltaic Energy Conversion (WCPEC-7), June 2018 (Waikoloa, HI, USA) available at: https://ntrs.nasa.gov/citations/20180005321.

[5] M. O'Neill, A.J. McDanal, G. Landis, R. Pricone, C. Kumar, M. Puglia, Space PV concentrators for outer planet and near-sun missions, using ultra-light fresnel lenses made with vanishing tools, in: 46th IEEE Photovoltaic Specialists Conference, June 16-21, 2019, https://doi.org/10.1109/PVSC40753.2019.8980908 (Chicago, IL, USA).

[6] V.E. Dudley, R.M. Workhoven, Concentrating Solar Collector Test Results, Collector Module Test Facility, DOE Sandia Laboratory, Albuquerque, NM, USA, 1978, p. 53. Sand 78-0815.

[7] R. Surampudi, et al., Solar Cell and Array Technology for Future Space Missions, Jet Propulsion Laboratory, Pasadena, CA, USA, 2002, p. 100. JPL Report D-24454.

[8] H.B. Curtis, D. Marvin, One year of flight data from the PASP Plus experiment, Prog. Photovolt. 4 (2) (March/April 1996) 139−146.

[9] M.J. O'Neill, M.F. Piszczor, M.I. Eskenazi, A.J. McDanal, P.J. George, M.M. Botke, H.W. Brandhorst, et al., Ultra-light stretched Fresnel lens solar concentrator for space power applications, in: Proceedings of SPIE - the International Society for Optical Engineering, vol 5179, 2003, pp. 116−126. Optical Materials and Structures Technologies, San Diego, CA, August 4-7, 2003, https://doi.org/10.1117/12.505801.

[10] D.C. Miller, S.R. Kurtz, Durability of Fresnel lenses: A review specific to the concentrating photovoltaic application, Sol. Energy Mater. Sol. Cell. 95 (8) (August 2011) 2037−2068.

[11] N. Shatz, J. Bortz, R. Winston, Thermodynamic efficiency of solar concentrators, Opt Exp. 18 (101) (April 2010) A5−A16.

[12] D.C. Ferguson, D.P. Engelhart, R.C. Hoffmann, V.J. Murray, E.A. Plis, Space solar arrays and spacecraft charging, in: S.G. Bailey, A.F. Hepp, D.C. Ferguson, R.P. Raffaelle, S.M. Durbin (Eds.), Photovoltacis for Space, Elsevier, Cambridge, MA, 2022, pp. 29−50.
[13] H.W. Brandhorst Jr., D.C. Ferguson, M.F. Piszczor, B.V. Vayner, M.J. O'Neill, Impact of solar array designs on high voltage operation in space, Valencia, Spain, in: Proc. 57th International Astronautical Congress, 9, October 2-6, 2006, pp. 5786−5792, https://doi.org/10.2514/6.IAC-06-C3.1.05. available at:.
[14] H. Brandhorst Jr, J. Rodiek, D. Ferguson, M. O'Neill, Stretched lens array (SLA): A proven and affordable solution to spacecraft charging in GEO, in: 22nd European Photovoltaic Solar Energy Conference, 2007 (Milan, Italy) available at: https://www.markoneill.com/SLA-Spacecraft-Charging-2007.pdf.
[15] H.W. Brandhorst, S.R. Best, J.A. Rodiek, M.J. O'Neill, M.F. Piszczor, Test results for a high-voltage multi-junction-cell concentrator array direct-driving an electric thruster, in: 34th IEEE Photovoltaic Specialists Conference, June 7-12, 2009, https://doi.org/10.1109/PVSC.2009.5411144 (Philadelphia, PA, USA).
[16] M.J. O'Neill, Fresnel lens solar concentrator configured to focus sunlight at large longitudinal incidence angles onto an articulating energy receiver, U.S. Patent US9660123B2 (May 23, 2017).
[17] G.A. Landis, J. Fincannon, Study of power options for Jupiter and outer planet missions, in: 42nd IEEE Photovoltaic Specialists Conference, June 14-19, 2015 (New Orleans, LA, USA), available at: https://ntrs.nasa.gov/citations/20160004679.
[18] G.A. Landis, et al., Photovoltaic power for Jupiter and beyond, 24th Space Photovoltaic Research & Technology Conference, September 20-22, 2016, Cleveland, OH, USA.
[19] P.P. Jenkins, D.C. Bentz, J. Barnds, C.R. Binz, S.R. Messenger, J.H. Warner, M.J. Krasowski, et al., TACSAT-4 Solar Cell Experiment: Two Years in Orbit, 10th European Space Power Conference (ESPC), Noordwijkerhout, the Netherlands, 2014 available at: https://www.markoneill.com/ESPC-TacSat-2014.pdf.
[20] M.J. O'Neill, Stretched fresnel lens solar concentrator for space power, with cords, fibers, or wires strengthening the stretched lens, U.S. Patent 10353187B2 (July 16, 2019).
[21] M.E. McEachen, D. Murphy, S. Meinhold, J. Spink, M. Eskenazi, M. O'Neill, Point-focus concentration compact telescoping array-Extreme Environments Solar Power Base Phase Final Report, Orbital ATK, NASA/CR-2017-219712, 2017. https://ntrs.nasa.gov/citations/20180000351.

CHAPTER SIXTEEN

Technological relevance and photovoltaic production potential of high-quality silica deposits on Mars

T.E. Girish, Thara N. Sathyan, and Varnana M. Kumar
Department of Physics, University College, Trivandrum, Kerala, India

16.1 Introduction

Exploration and utilization of in situ resources are crucial for future human habitation of Mars. NASA is planning a manned mission to this planet by 2030 [1]. Both solar and nuclear power options are actively being considered for this purpose. Ever since the Mariner 4 mission of the United States in 1964, there have been several successful flyby, orbiter, lander, and rover missions to Mars by different countries including India. No samples from Mars have been returned to Earth by these missions, but they have brought valuable information about mineral resources on this planet. White silica sands (primarily quartz) are known to be an important raw material on Earth to produce silicon semiconductor chips and solar cells. The knowledge of high-silica deposits in the rocks and sand dunes of Mars derived from Mars missions since 2001 needs particular mention here because of its technologic importance. It may enable our vision of building "silicon valleys" on Mars.

In this chapter we will first investigate the salient features of high-quality silica deposits on Mars. The locations, types of high-silica minerals, and SiO_2 content will be studied using published data obtained from different Mars missions. The high-silica deposits are found in excess in the southern polar region of Mars. Next, potential utilization of the deposits is inferred from the predicted performance of silicon (Si) solar cells on Mars in comparison with multijunction space solar cells. The possibilities and challenges in the manufacturing of Si solar cells on Mars making use of high-concentration silica minerals is then studied. The different engineering applications of high-silica deposits and some advantages of photovoltaic power over nuclear power for a Martian settlement will be also discussed.

16.2 Spatial distribution and silica content of high-quality deposits on Mars

In Fig. 16.1, we show the spatial distribution of high-silica deposits on Mars, prepared as a map by Caltech/NASA JPL, Pasadena, CA, USA, using Gamma ray

Photovoltaics for Space
ISBN 978-0-12-823300-9, https://doi.org/10.1016/B978-0-12-823300-9.00017-0

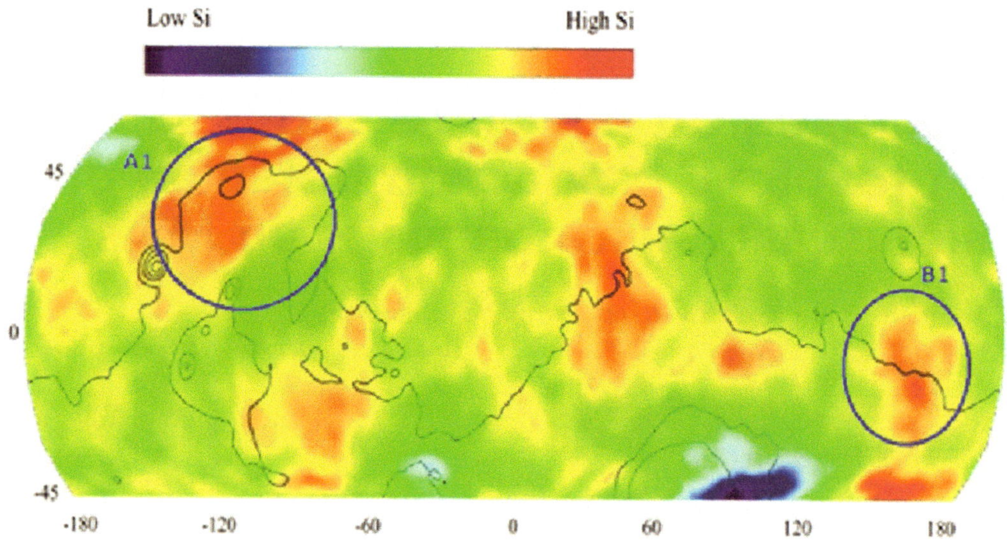

Figure 16.1 Spatial distribution of high-silica deposits on Mars. The circles indicate regions of intense volcanism in the geological past *Courtesy: NASA/JPL/University of Arizona.*

spectrometer observations from the 2001 Mars Odyssey mission [2]. From this figure, we can find high-silica regions on Mars in the northern hemisphere to 60°N and southern hemisphere to 45°S; see also Table 16.1. In a recent study it was found that some of the high-silica deposits on Mars are associated with regions of high volcanism in the geologic past [8]. Two such regions are within circles A_1 and B_1 marked in Fig. 16.1.

The thermal emission spectrometer observations from the Mars Global Surveyor orbiter mission during the years 1997–2005 also brought valuable information regarding the specific locations of high-silica deposits on Mars [3]. Relative abundance of high-silica phases is found to vary between 12 and 34 wt% (% by weight) in these locations. A recent study based on observations of sand dunes on Mars from different space missions [4,5] found that high-silica phases are observed to 75°S, almost extending to the southern polar region of Mars in the southern hemisphere of Mars. The relative abundance of high-Si phases in Martian sand dunes is inferred to vary between 13 and 38 wt%. These results are also shown in Table 16.1.

A recent study related to the chemical composition of sand dunes on Mars suggest that high-silica phases found on Mars are related to opal and silica glass [9]. Opal is related to hydrated silica found in several locations on Mars [6,7], as given in Table 16.1. Hydrated silica ($SiO_2 \cdot nH_2O$) contains mainly water and SiO_2. The average water content in this mineral on Mars is estimated to be 6.3 wt% on average [10], This suggests that hydrated silica on Mars contains >90 wt% SiO_2. Silica deposits with high SiO_2 content (>90%) have been observed or inferred on Mars by different Mars missions in locations such as Gusev crater and Gale crater [11,12]. Some studies suggest that high-quality silica

Table 16.1 Locations and relative abundance of high-silica deposits on Mars.

Mars mission/sources	Locations on Mars	Latitudinal coverage (degrees)	Relative abundance	Remarks
2001 Mars Odyssey mission JPL [2]		60°N to 45°S	Low to high	Some high-Si regions are associated with past volcanism on Mars
Mars Global Surveyor Mission: MGS-TGS data [3]	Syrtis, Tyrrhena Cimmeria lapygia, Hesperia Aonium Phrixi Mare Sirenum Meridiani, Solis, N. Acidalia	52°N to 45°S	12–34 wt%	Excess of high-silica deposits in the southern hemisphere of Mars
Mars Global Digital Dune Base (USGS) [4,5]	Sand dunes	40°N to 75°S	12–38 wt%	Excess of high-silica deposits in the southern polar region of Mars (60–75°S)
Hydrated silica on Mars: compiled from different sources/missions [6,7]	N. Syrtis Major near Antoniadi crater, Isidis Basin, Toro crater, Nili Patera, Western Hellas, Terra Sirenum, Noctis Labyrinthus, Plains near Melas Chasma, Plains near Juventae Chasma, Aeolis Mensae, Comichel crater, Ius Chasma, Nilosyrtis Mensae, Elorza crater, Nortis Labyrinthus, Iani Chaos and Jezero crater (Perseverance rover landing area)	27.82°N to 43.42°S	>90% SiO_2 content	Association with past water bodies such as hot springs; main mineral is opal; some related to past volcanism

Compiled by the authors from published sources.

deposits on Mars may be of volcanic origin. Hydrated silica is also associated with hot springs and other water bodies present on Mars during the geologic past. One such location is Jazero crater [13] where the Perseverance rover is at present involved in valuable investigations to detect biosignatures on Mars.

16.3 Expected performance of silicon solar cells on Mars

Previous Mars rovers used multijunction (MJ) III-V solar cells for power generation. Thus, we do not have direct data of Si solar cell performance on the Mars. From different published sources, we have inferred different aspects of Si solar cell performance in comparison with the performance of MJ solar cells on Mars. The radiation environment on Mars is comparable to that of the Moon. So annual radiation-related degradation of crystalline Si solar cells (covered and preirradiated cells) from the Apollo missions to the Moon during 1969–76 is taken as a reference value, and it is then extrapolated for more severe space weather conditions as well.

Performance of Apollo 14 Si solar cells was reported for sunset periods on the Moon with average temperature around 150 K [14]. This supports good low-intensity, low-temperature (LILT) performance on Mars for Si solar cells. Further, the spectral response of Si solar cells in the red region of solar spectrum is expected to be better than MJ solar cells. To infer the performance of Si solar cells under dust conditions on Mars, we have used published data on the performance of Si solar panels near desert regions of Earth. These results are summarized in Table 16.2, along with relevant information source Refs. [15–26]. It is important to note the value of the adoption of automated dust cleaning techniques [27,28] for sustained photovoltaic power generation on Mars.

16.4 Potential of silicon solar cell manufacturing on Mars

Conversion of the basic raw material (silica sands with SiO_2 content >90%) to Si wafers consists of several advanced industrial processes [29–32]. At first the silica sand is

Table 16.2 Comparison of the predicted or expected performance of Si solar cells and multijunction (MJ) solar cells on Mars.

Solar cell parameter	Silicon solar cells	MJ solar cells	References
Efficiency	12.7–26.7	29.9–33.7	[15,16]
Radiation-related degradation	1%–2% per year	1%–2% per year	[17]
LILT performance	Good	Good for special cells	[18,19]
Thermal degradation	Can be neglected on Mars	Can be neglected on Mars	—
Solar spectrum matching	Good	Satisfactory	[20]
Dust-related degradation	23%–50% per year	28%–40% per year	[21–26]

Compiled by authors from published sources.

reduced with carbon at high temperature to produce metallurgical-grade (MG) Si (SiO_2 content of 98%–99.9%). The MG Si is then further purified and converted to electronic-grade (EG) Si of very high purity. This process is complicated and there are two independent methods: (a) Siemens process and (b) metallurgical process to achieve the same. The Siemens process usually consumes double the power than the alternate technique. From EG Si we can make either polycrystalline or single-crystal silicon wafers from which Si solar cells are manufactured. Single-crystal growth of Si from a melt for which high-temperature furnaces are required consumes larger electrical power.

In Table 16.3 we have given power requirements for the above major steps in the conversion of raw silica sand to silicon wafers. From Table 16.3, we can understand that solar cell manufacturing requires high power production varying from approximately 15 to 100 kW. The prerequisites for establishing a Si solar cell manufacturing facility on Mars include the following:

(a) mining and separation of high-SiO_2 minerals from high-silica deposits on Mars;
(b) chemicals and other components for Si wafer production [33,34] to be manufactured on Mars from local resources or partially transported from Earth;
(c) provide clean rooms and automation [35] for this facility, which should be shielded from dust storms on Mars;
(d) adaptation of technical processes involved to work in low gravity on Mars;
(e) providing uninterrupted high power for the production units.

For industrial production on another planet, initially, we should have high-electric-power generators at least comparable to that at the International Space Station (ISS). The ISS uses large solar cell arrays for power generation [36]. The total area of ISS solar panels is about 2500 m^2. If we use MJ solar cells (25% efficiency) in an ISS-type large array on Mars, it will generate a maximum Beginning of Life (BOL) power of about 300 kW, and the best Si solar cells (15% efficiency) will generate in this case a peak power of about

Table 16.3 Electrical power requirements for different major processes involved in the production of silicon wafers.

Process	Electrical power requirement per unit mass of material produced (kWh/kg)	References
Silica sand to MG silicon	11–18.2	[29,30]
MG silicon to EG silicon	41–113.4	[29–32]
Polysilicon and wafer making	41	[30]
Single-crystal Si and wafer making	101.6	[30]

Compiled by authors from published sources.

200 kW. If dust-related degradation is assumed to be 50%, the power from a large MJ solar cell array is sufficient for providing power to a solar cell manufacturing facility for 1 year of operation on Mars [22]. Automated dust cleaning needs to be provided for solar cell arrays on Mars if we want to use them as a power source for a longer time. If this is successfully done, then large Si solar cell arrays can be also meet the criteria as a good high-power source on Mars at the present level of technology.

Silica deposits on Mars may also find other applications apart from photovoltaic and semiconductor devices. Silicon-containing materials, in particular, have several useful applications in different fields. Chemical-grade silicon is produced by refining MG Si through a hydrometallurgical process. It is used to manufacture silicones, sealants, lubricants, silanes, and fumed silica, a nanomaterial with many industrial applications [37]. Metallurgical silicon is used in steel smelting as well as the manufacture of cast-aluminum alloys [38].

Ferrosilicon is the most commonly used form of metallic silicon. It is produced either in a blast furnace or electric arc furnace by the reduction of quartz sand with coke in the presence of iron [39]. The melt is poured out of the furnace and solidifies in the form of a flat sheet. Ferrosilicon has a number of uses apart from steelmaking. It is a prealloy in the production of Mg ferrosilicon. It is also used to produce ductile iron, as well as during the Pidgeon process for refining high-purity magnesium. Ferrosilicon has applications in the making of heat- and corrosion-resistant ferrous silicon alloys as well as silicon steel. The latter is used in the manufacture of electromotors and transformer cores. Finally hydrated silica like that found on Mars is used in cosmetics and can be easily converted to silica gel with many scientific applications. It will be interesting to compare the physical and chemical properties of opal found on Mars and on Earth; the crystalline form of opal is often identified as a valuable gemstone [6,7,9,10].

16.5 Conclusions

Detailed investigations of quartz deposits (minerals or sands with $SiO_2 > 90\%$) on Mars are necessary for use as a valuable raw material for in situ Si solar cell manufacturing. For this purpose, dedicated sample return missions to specific locations on Mars mentioned in Table 16.1, where high-quality Si phases are found, become necessary. Dust storms occur more often in the southern hemisphere of Mars [40]. So initially, high-silica deposit locations in the northern hemisphere may be chosen as more practicable sites for Si solar manufacturing on Mars.

The possibility of manufacturing Si solar cells using abundant high-silica deposits also supports the extensive use of (Si) solar power on Mars. Space photovoltaic power is a proven technology with 60 years of advancements compared to space nuclear power, which is yet to be proven for high-power applications. Radioisotope thermal generator power is too small to use it for industrial purposes. Special nuclear power reactors for

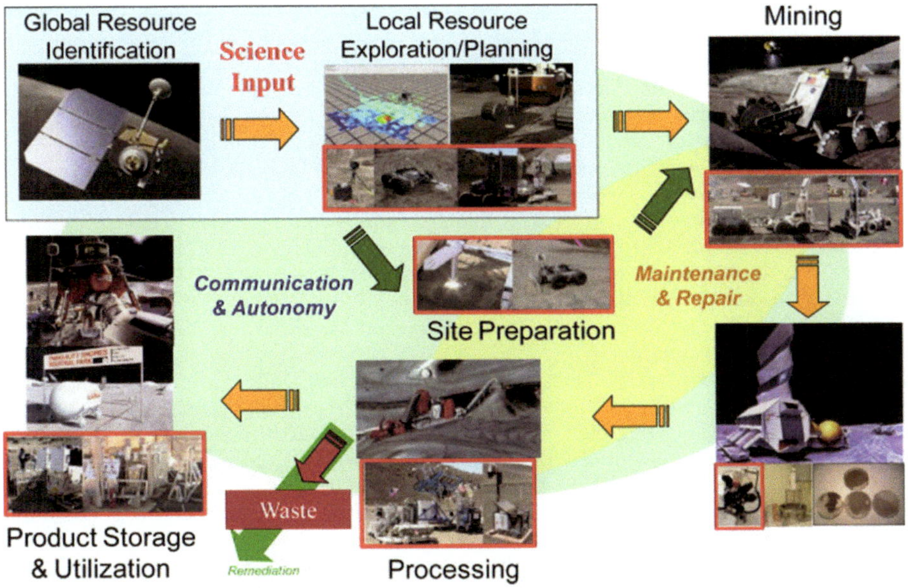

Figure 16.2 Space mining cycle that relies on energy and in situ resource utilization. *Reprinted with permission from G.B. Sanders, W.F. Larson, Integration of in-situ resource utilization into Lunar/Mars exploration through field analogs Adv. Space Res. 47 (1) (2011) 20-29, copyright (2011) Elsevier.*

space applications [41,42] that can produce high power are only at the development stage. However, their safety on Mars needs to be critically investigated. NASA has done some very preliminary planning to establish medium-power (several kW) nuclear reactors on the Moon and Mars [1,43]. These could serve to diversify energy sources for future manned missions leading to surface outposts on the Moon [44] and Mars [45] in the future; see Fig. 16.2 for an overview that summarizes key operations involving in situ resource utilization for the Moon or Mars [44].

References

[1] https://www.nasa.gov/directorates/spacetech/6_Technologies_NASA_is_Advancing_to_Send_Humans_to_Mars/. (Accessed 3 April 2021).

[2] https://photojournal.jpl.nasa.gov/catalog/PIA04256/. (Accessed 3 April 2021).

[3] A.D. Rogers, P.R. Christensen, Surface mineralogy of Martian low-albedo regions from MGS-TES data: implications for upper crustal evolution and surface alteration, J. Geophys. Res. 112 (2007), https://doi.org/10.1029/2006JE002727. E01003.

[4] L.K. Fenton, A.L. Gullikson, R.K. Hayward, H. Charles, T.N. Titus, The Mars Global Digital Dune Database (MGD3): global patterns of mineral composition and bedform stability, Icarus 330 (2019) 189–203.

[5] A.L. Gullikson, R.K. Hayward, T.N. Titus, H. Charles, L.K. Fenton, R. Hoover, N.E. Putzig, Mars Global Digital Dune Database (MGD3)–Composition, Stability, and Thermal Inertia: U.S. Geological Survey Open-File Report 2018–1164, 2018, p. 17, https://doi.org/10.3133/ofr20181164.

[6] M.R. Smith, J.L. Bandfield, E.A. Cloutis, M.S. Rice, Hydrated silica on Mars: Combined analysis with near-infrared and thermal-infrared spectroscopy, Icarus 223 (2013) 648–655.
[7] M. Pineau, L. Le Deit, B. Chauvire, J. Carter, B. Rondeau, N. Mangold, Toward the geological significance of hydrated silica detected by near infrared spectroscopy on Mars based on terrestrial reference samples, Icarus 347 (2020) 113706.
[8] T.N. Sathyan, Studies on Certain Geochemical and Biophysical Phenomena in the Terrestrial and Extraterrestrial Perspective, University of Kerala, Trivandrum, India, 2019.
[9] M.R. Smith, J.L. Bandfield, Geology of quartz and hydrated silica-bearing deposits near Antoniadi Crater: Mars, J. Geophys. Res. 117 (2012) E06007, https://doi.org/10.1029/2011JE004038.
[10] W. Rapin, B. Chauviré, T.S.J. Gabriel, A.C. Mcadam, B.L. Ehlmann, C. Hardgrove, S. Schröder, In situ analysis of opal in Gale Crater: Mars, J. Geophys. Res. Planets 123 (2018) 1955–1972, https://doi.org/10.1029/2017JE005483.
[11] S.W. Squyres, et al., Detection of silica-rich deposits on Mars, Science 320 (2008) 1063, https://doi.org/10.1126/science.1155.
[12] S. Czarnecki, C. Hardgrove, P.J. Gasda, T.S.J. Gabriel, M. Starr, M.S. Rice, et al., Identification and description of a silicic volcaniclastic layer in Gale Crater, Mars, using active neutron interrogation, J. Geophys. Res. Planets 125 (2020), https://doi.org/10.1029/2019JE006180 e2019JE006180.
[13] J.D. Tamas, et al., Orbital identification of hydrated silica in Jezero Crater, Mars, Geophys. Res. Lett. 46 (2019) 12771–12782, https://doi.org/10.1029/2019GL085584.
[14] J.R. Bates, P.H. Fang, Results of solar cell performance on lunar base derived from Apollo missions, Sol. Energy Mater. Sol. Cell. 26 (1992) 79–84.
[15] S.G. Bailey, R.P. Raffaelle, Space solar cells and arrays, in: Antonio Luque, Steven Hegedus (Eds.), Handbook of Photovoltaic Science and Engineering, second ed., John Wiley & Sons, USA, 2011, pp. 365–401, https://doi.org/10.3389/fphy.2020.631925.
[16] J. Li, A. Aierken, Y. Liu, Y. Zhuang, X. Yang, J.H. Mo, R.K. Fan, Q.Y. Chen, S.Y. Zhang, Y.M. Huang, Q. Zhang, A brief review of high efficiency III-V solar cells for space application, Front. Phys. 8 (2021) 631925, https://doi.org/10.3389/fphy.2020.631925.
[17] T.E. Girish, S. Aranya, Photovoltaic power generation on the Moon: Problems and prospects, in: V. Badescu (Ed.), Moon, Springer, Berlin, Heidelberg, 2012, pp. 367–376, https://doi.org/10.1007/978-3-642-27969-0_16.
[18] C. Baur, V. Khorenko, G. Siefer, J.C. Bourgoin, M. Casale, et al., Development status of triple junction solar cells optimized for low Intensity low temperature applications, in: 39th IEEE Photovoltaic Specialists Conference (PVSC), June 2013. TAMPA, United States. HAL Id: Hal-01057758 (2014), https://hal-onera.archives-ouvertes.fr/hal-01057758.
[19] N. Fatemi, S. Sharma, O. Buitrago, J. Crisman, P. Sharps, R. Blok, M. Kroon, C. Jalink, R. Harris, P. Stella, S. Distefano, in: Proc. Thirty-first IEEE Photovoltaic Specialists Conference, 2005, pp. 618–621.
[20] G.A. Landis, D. Hyatt, the MER Athena Science Team, The solar spectrum on the Martian surface and its effect on photovoltaic performance, in: Proc, IEEE 4th World Conference on Photovoltaic Energy Conversion, Waikoloa, 2006. https://ntrs.nasa.gov/api/citations/20110000777/downloads/20110000777.pdf.
[21] L. McMillon-Brown, T.J. Peshek, A.M. Pal, J. McNatt, Dust abrasion damage on Martian solar arrays: Experimental investigation and opportunity rover performance analysis, in: Proc. IEEE 46th Photovoltaic Specialists Conference, 2019, https://doi.org/10.1109/PVSC40753.2019.8980876C2019.
[22] J. Shahmoradi, A. Maxwell, S. Little, Q. Bradfield, S. Bakhtiyarov, M. Hassanalian, The Effects of Martian and Lunar Dust on Solar Panel Efficiency and a Proposed Solution, 2020, https://doi.org/10.2514/6.2020-1550. AIAA 2020-1550.
[23] G.A. Landis, Dust obscuration of Mars solar arrays, Acta Astronaut. 38 (1996) 888–891.
[24] A. Hussain, A. Batra, R. Pachauri, An experimental study on effect of dust on power loss in solar photovoltaic module, Renewables Wind Water Solar. 4 (2017) 9, https://doi.org/10.1186/s40807-017-0043-y.

[25] A. Ibrahim, Effect of shadow and dust on the performance of silicon solar cell, J. Basic. Appl. Sci. Res 1 (3) (2011) 222–230.
[26] M. Benghanem, A. Almohammedi, M.T. Khan, A.A. Al-Mashraq, Effect of dust accumulation on the performance of photovoltaic panels in desert countries: A case study for Madinah, Saudi Arabia, Int. J. Power Electr. Drive Syst. (IJPEDS) 9 (3) (2018) 1356–1366, https://doi.org/10.11591/ijpeds.v9n3.pp1356-1366. ISSN: 2088-8694.
[27] O.H. Attia, et al., Removal of dust from the solar panel surface using mechanical vibrator, J. Phys.: Conf. Ser. 1262 (2019), 012021.
[28] M.K. Mazumder, R. Sharma, A.S. Biris, J. Zhang, C. Calle, M. Zahn, Self-cleaning transparent dust shields for protecting solar panels and other devices, Part. Sci. Technol. 25 (1) (2007) 5–20, https://doi.org/10.1080/02726350601146341.
[29] B.S. Xakalashe, M. Tangstad, in: R.T. Jones, P. den Hoed (Eds.), Silicon Processing: From Quartz to Crystalline Silicon Solar Cells, Southern African Pyrometallurgy, 2011. Available in: https://www.pyrometallurgy.co.za/Pyro2011/Papers/083-Xakalashe.pdf (Accessed 3 April 2021).
[30] M. Fan, Z. Yu, W. Ma, et al., Life cycle assessment of crystalline silicon wafers for photovoltaic power generation, Silicon (2020), https://doi.org/10.1007/s12633-020-00670-4.
[31] A.F.B. Braga, S.P. Moreira, P.R. Zampieri, J.M.G. Bacchin, P.R. Mei, New processes for the production of solar-grade polycrystalline silicon: A review, Sol. Energy Mater. Sol. Cells 92 (2008) 418–424.
[32] G. Bye, B. Ceccaroli, Solar grade silicon: Technology status and industrial trends, Sol. Energy Mater. Sol. Cell. 130 (2014) 634–646, https://doi.org/10.1016/j.solmat.2014.06.019.
[33] E.D. Williams, R.U. Ayres, M. Heller, The 1.7 kilogram microchip: Energy and material use in the production of semiconductor devices, Environ. Sci. Technol. 36 (2002) 5504–5510, https://doi.org/10.1021/es025643o.
[34] S. Kim, et al., Chemical use in the semiconductor manufacturing industry, Int. J. Occup. Environ. Health 24 (2018) 109–118, https://doi.org/10.1080/10773525.2018.1519957.
[35] https://roboticsandautomationnews.com/2021/03/09/, (Accessed 3 April 2021); for more details: https://www.printfriendly.com/p/g/9Fzk28/. (Accessed 18 May 2021).
[36] https://www.nasa.gov/centers/glenn/about/fs06grc.html/. (Accessed 3 April 2021).
[37] Chemical Grade Silicon, Available from: https://hpqsilicon.com/silicon-applications/chemical-grade-silicon/. (Accessed 3 April 2021).
[38] Applications of MG Silicon, Available from: https://www.thoughtco.com/metal-profile-silicon-4019412/. (Accessed 3 April 2021).
[39] H.T. Ingasson, G.R. Jonsson, Control of the silicon ratio in ferrosilicon production, Control Eng. Pract. 6 (8) (August 1998) 1015–1020.
[40] M. Battalio, H. Wang, The Mars Dust Activity Database (MDAD): A Comprehensive Statistical Study of Dust Storm Sequences, vol. 354, January 2021, p. 114059.
[41] Q. Zhou, Y. Xi, G. Liu, X. Ouyang, A miniature integrated nuclear reactor design with gravity independent autonomous circulation, Nucl. Eng. Des. 340 (2018) 9–16.
[42] V. Teofilo, Space Power Systems for the 21st Century, Available from: https://arc.aiaa.org/doi/10.2514/6.2006-7288/. (Accessed 4 April 2021).
[43] A Lunar Nuclear Reactor, Available from: https://sservi.nasa.gov/articles/a-lunar-nuclear-reactor/. (Accessed 3 April 2021).
[44] G.B. Sanders, W.E. Larson, Integration of in-situ resource utilization into Lunar/Mars exploration through field analogs, Adv. Space Res. 47 (1) (2011) 20–29.
[45] S.O. Starr, A.C. Muscatello, Mars in situ resource utilization: A review, Planet. Space Sci. 182 (2020) 104824.

CHAPTER SEVENTEEN

Space nuclear power: Radioisotopes, technologies, and the future

Modeste Tchakoua Tchouaso[1], Tariq Rizvi Alam[2] and Mark Antonio Prelas[3]

[1]Department of Physics and Astronomony, Howard University, Washington, DC, United States
[2]Nuclear Environments and System Assessments, Applied Research Associates, Santa Barbara, CA, United States
[3]Electrical Engineering, University of Missouri, Columbia, MO, United States

17.1 Introduction

Space exploration has benefited from the enormous energy of the Sun to provide energy to power spacecraft and spacecraft subsystems. This solar energy is harnessed using photovoltaic (PV) cells. Solar power remains the dominant method of providing electrical power for missions in low-Earth orbit (LEO) and planets close to the Sun [1−3]. However, it becomes less efficient as one moves further away from the Sun. This is because the solar radiation flux is inversely proportional to the square of distance. The intensity of solar flux is about 25 times stronger on Earth than on Jupiter [3]. For deep-space missions, the Sun does not provide sufficient energy to power spacecraft systems for long-duration missions. Even using larger solar panels may not be enough to provide sufficient power for certain missions [2,3]. Harsh radiation fields like those around Jupiter and settling dust on the surface of solar panels are other constraints to consider when using a solar panel [3]. The operation of solar power is also affected during the long and cold lunar nights.

The first Apollo mission is a good example of where solar panels were used to provide electrical power to the experimental package, but the experiments had to shut down during the lunar nights [4]. For deep-space missions, solar power is therefore not a reliable method of generating continuous power, and an alternative method of power generation is required. Radioisotope power systems (RPSs) are the only viable alternative to solar power because of their high energy density and ability to produce continuous and reliable long-term power, independent of spacecraft attitude and distance from the Sun. RPSs also provide heat to components of spacecraft during cold lunar nights. Unlike solar power, an RPS is insensitive to harsh radiation fields. Several types of RPS have been developed over the years to provide a wide variety of power options to accommodate a range of power needs for future space missions. In this chapter, we explore different RPS concepts, their critical components, the safety of using radioisotopes in space, and the regulations surrounding space nuclear technology.

Photovoltaics for Space
ISBN 978-0-12-823300-9, https://doi.org/10.1016/B978-0-12-823300-9.00014-5

17.2 Radioisotope availability

RPSs have provided continuous, reliable, and maintenance-free power for several space missions [5]; Section 17.4.1 provides a detailed summary. The source of this energy is radioisotopes. To ensure continuous reliance on radioisotope power, the radioisotope must be available and in sufficient quantities. Feasibility studies were conducted in the 1960s to investigate possible radioisotopes for use as a heat source in thermoelectric generators [6]. The requirements of radioisotopes used in space applications and terrestrial applications are different [5,7]. The choice of radioisotope depends on the half-life, shielding requirement, power density, chemical form of radioisotope, and the availability and cost of the radioisotope [7].

A tradeoff is generally made between the various requirements. The half-life determines how long the radioisotopes can produce useful energy. Radioisotopes with half-lives greater than 10 years are generally required for space applications [5]. The radioisotope should have a large power density. The shielding required depends on the type of radioisotope under consideration. The density and weight are generally not important for terrestrial power, except in applications where there are size restrictions. However, they are critical for space applications, since the cost of space missions is related to the weight of the RPS. The chemical form of the isotope is important for safety reasons. The radioisotope of choice must be chemically and thermally stable at high temperatures. Table 17.1 shows a list of radioisotopes that are available for use in RPS; these isotopes are suitable for both terrestrial and space applications [5,7–9].

Table 17.1 Available isotopes for space radioisotope power listed in order of specific power (W/g).

Isotope	Radiation, emission	Half-life	Specific power (W/g)	Chemical form(s)
Po-210	α, γ	138 days	144	HgPo, PbPo, Po
Cm-242	α	0.45 years	120	Cm_2O_3
Ce-144	β^-, γ	285 days	25.5	CeO
Tm-170	β^-	0.35 years	13.6	Tm_2O_3
Cm-244	α	18.1 years	2.65	Cm_2O_3
Sr-90	β^-	29 years	0.935	$SrTiO_3$, SrO, $SrZrO_3$
Pu-238	α, γ	88 years	0.568	PuO_2 Pu_2C_3
Cs-137	β^-, γ	26.6 years	0.42	CsCl
Pm-147	β^-	2.62 years	0.33	Pm_2O_3
H-3	β^-	12.3 days	0.326 (Li), 0.098 (LiH)	LiH
Am-241	α, γ	432 years	0.115	AmO_2

Adapted from Ref. [8].

The preferred isotopes are often alpha radiation emitters $\left({}_{2}^{4}He\right)$ because alpha particles are highly energetic and have low penetration through matter. Alpha particles will deposit all of their energy within the transducer [7]. Polonium-210 (Po-210) is an alpha emitter that produces one gamma per 105 alpha particles. It can be obtained from the irradiation of bismuth using the reactions in Eqs. (17.1) and (17.2) [10].

$$ {}_{83}^{209}Bi + {}_{1}^{2}H \rightarrow {}_{84}^{210}Po + {}_{0}^{1}n \tag{17.1} $$

$$ {}_{83}^{209}Bi + {}_{1}^{1}H \rightarrow {}_{84}^{210}Po + \gamma \tag{17.2} $$

Po-210 has a short half-life of 138 days and a high specific energy of 144 W/g. The first radioisotope thermal generator (RTG) used Po-210 as a heat source to produce electrical power [11]. The heat source was placed between two thermocouples to create a temperature differential. The first Po-210 batteries used 57 and 146 Ci of the radioisotope and produced 1.8 and 9.4 mW of power for 57 and 146 Ci sources, respectively. The power density of this battery is several times that of a chemical battery; for example, the battery has about 10 times the specific energy capacity of an ordinary dry cell battery of the same weight [11]. Since Po-210 has a short half-life of 138 days, the performance of the battery diminished over time. For this reason, Po-210-based RTGs are limited to short-duration space applications. Po-210-based RTGs have never been used in any flight mission because of the short half-life. Curium-242 (Cm-242) and curium-244 (Ce-244) are both alpha emitters with very small gamma radiation emissions; Cm-242 is produced by irradiation of an americium-241 (Am-241) target [12].

Plutonium-238 (Pu-238) has been used for fueling most RPSs because of its long half-life of 88 years. Pu-238 has a power density of 0.54 W/g and emits a low-energy gamma and neutron radiation that can be easily shielded. It is used in the form of plutonium oxide (PuO_2). The benefits of using Pu-238 are that it is stable at high temperatures, can generate substantial heat even in small amounts, and its radiation emissions can easily be shielded. However, Pu-238 does not occur naturally. It is obtained by irradiating actinides (normally neptunium-237) in a nuclear reactor, followed by chemical purification [13]. The US production of Pu-238 was halted after the cold war when the Department of Energy (DOE) shut down its production facilities in the 1980s. During this time, the existing stockpile of Pu-238 was used to power RTGs; this is estimated to be used at a rate of 5 kg/year [14]. The current inventory of Pu-239 has diminished over time and can only support a few missions. For the United States to continue to provide power for its spacecraft and maintain its space program, it will need to restart the production of Pu-238.

There is current interest at the DOE to restart the production of Pu-238. However, reestablishing this program is expensive. Resumption of Pu-238 production can be accomplished with minimal modification of existing DOE facilities at the Idaho

National Laboratory and the Oak Ridge National Laboratory [14]. In 2012, the National Aeronautics and Space Administration (NASA) reached an agreement with the DOE to restart the production of Pu-238 to produce 300–400 g per year with Oak Ridge National Laboratory selected to lead the project [15].

Am-241 has never been used in radioisotope power generators, but it is a radioisotope being considered an alternative to Pu-238. Am-241 has a half-life of 432 years, which is longer than that of Pu-238, and therefore will power RTGs for a longer period. However, the power density of Am-241 is only 0.115 W/g, which is about one-fifth the power density of Pu-238. Am-241 also produces more penetrating radiation through the decay chain of Pu-238, which increases the shielding requirement of the radioisotope. Am-241 can be obtained as a by-product from a nuclear reactor through the fission process. The European Space Agency has conducted extensive research on using Am-241 for future space power generation with results showing that it is feasible to use Am-241 as a heat source [16,17]. There are no concrete plans for using Am-241 in future space missions.

Radioisotopes with beta emissions have also been used as heat sources. The benefit of beta sources is that they are readily available and in great quantities at a lower price than alpha sources. However, beta sources are generally associated with significant gamma radiation, and additional shielding requirements are needed to shield the gamma emissions and Bremsstrahlung X-rays. Cesium-137 (Cs-137) and cerium-144 (Ce-144) are beta emitters with strong gamma emitters but are not attractive because they require high levels of shielding. SNAP-1 used Ce-144 and ran on a Rankine power conversion system [18]. The goal of SNAP-1 was to produce 500 W of power at 10% efficiency. Pm-147 and H-3 are pure beta emitters that can be easily shielded because they have low power densities [7]. Thulium-170 (Tm-170) is a beta radiation with a very short half-life and emits both gamma- and X-ray radiation.

Radioisotopes such as strontium-90 (Sr-90, half-life of 29 years) that are attractive in terms of shielding have been used in RPSs [19]. Sr-90 is used in the form of strontium titanate because this form of Sr-90 is insoluble, noncombustible, and has a high melting point of 1910°C. Curium-242 is a beta radioisotope with a short half-life of 162 days and a high power density of 120 W/g; as discussed above, Cm-242 is produced in the reactor by irradiation of an Am-241 target [12]. SNAP-13 used Cm-242 as a heat source [20]. However, SNAP-13 was never flown in space. Due to the short half-life of Cm-242, it cannot be used for long-duration missions. Cerium-244 was studied as an alternative to Pu-238 because the source was supposed to be available from the US program to develop a breeder reactor fuel cycle. Cm-244 is attractive because of its long half-life of 18.1 years, but it produces a significant number of neutrons and gammas due to the spontaneous fission of Cm-244, which increases the shielding requirement.

17.3 Radioisotope power systems

Radioisotope Power Systems (RPSs) are compact, rugged spacecraft power systems that provide reliable and continuous power. They are best suited for long-duration missions and levels of power of 1 to 10 kW. RPSs are preferred to solar-based power in environments with limited sunlight. They can also operate within the radiation-intensive environments around Jupiter.

An RPS converts the heat emitted from the decay of radioisotope into electricity using static or dynamic conversion technologies. Dynamic technologies have moving parts that can transform heat into mechanical energy from which the desired power output is obtained. Dynamic converters include Stirling, Brayton, and Rankine converters. Dynamic systems provide efficient power with efficiencies between 25% and 38% [21,22]. The high efficiency of dynamic converters reduces the amount of radioisotope required, which has size, mass, cost, and safety implications. Static converters have no moving parts, and there is no limitation on the life of the RPS since the components are not affected by mechanical failure. However, they suffer from low conversion efficiencies. The efficiency of static converters is ~8% [7,23]. All NASA missions have used static converters; thermoelectric (TE) converters, TPV, and thermionic devices are examples of static converters.

The design of an RPS consists of two critical systems: the thermal heat source and the energy conversion system. The thermal heat source produces the decay heat from the radioisotope that is converted into electricity by the energy conversion system. The heat from the heat source is converted into electricity. Any unconverted heat is used to provide heat to spacecraft components and eventually rejected. To summarize, there are several types of energy conversion systems: general-purpose heat source technology, static energy conversion, dynamic energy conversion, and indirect conversion systems. These will be discussed in some detail throughout this section.

17.3.1 Radioisotope (general-purpose) heat source technology

The desired characteristics of a heat source include long half-life, high specific power and energy, and stable fuel form with minimum shielding requirement and high melting point. Thermal heat sources considered for use in RPSs are listed in Table 17.1; both alpha and beta radiation sources are included. Alpha particles have a short range in matter and can be stopped by a sheet of paper. Heavy nuclei that undergo alpha decay tend to produce spontaneous fission neutrons as well. The branching ratio of the spontaneous fission neutrons is small compared to alpha particles. Neutrons are significant in terms of external radiation doses. Neutrons can also be generated from the alpha interaction with low atomic number impurities like oxygen through an (α,n) reaction, which also increases the neutron dose, and thus the shielding requirement. Fast neutron shielding requires that the neutron first be moderated with a material with a low atomic number such as hydrogen or carbon. The slow neutrons are later absorbed with a material with a large thermal neutron cross-section such as boron-10, lithium-6, or cadmium. Gamma and X-ray radiation are sometimes released together with the alpha particles. The penetration of photon radiation depends on the photon energy. Photon attenuation depends on the atomic mass of the element. Materials with high atomic masses are generally used for shielding gamma radiation [24].

Few sources meet the isotope requirement in terms of power density, shielding, half-life, availability, and cost. Pu-238 is the only isotope that has been used as an RPS fuel. It offers several attractive advantages as a heat source: (1) it has a long half-life of 88 years, enabling its use for long-duration missions, (2) it has a high power density of ~0.57 W/g, (3) it has high thermal stability, (4) it exists as a stable oxide form, which is water insoluble and chemically stable, and has demonstrated good engineering properties at high temperatures, (5) It is a low alpha emitter, which reduces dose to systems and personnel, and (6) it is available and can be produced in sufficient quantity at an affordable price. The general-purpose heat source (GPHS) provides the heat in current RPSs.

The GPHS is a critical part of the RPS that provides the thermal energy from the alpha decay of Pu-238 to the thermoelectric converter (Fig. 17.1). The design of the GPHS is based on an earlier RTG design used in the LES 8/9 [25] satellite and the Voyager 1 spacecraft [26]. The design of the GPHS makes it adaptable to a wide range of

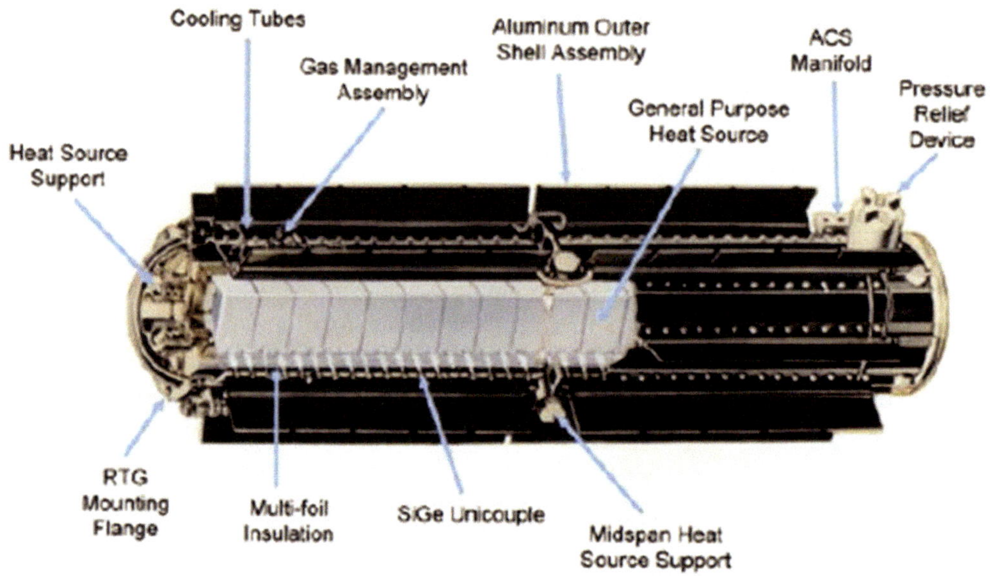

Figure 17.1 Cutaway of a general-purpose heat source radioisotope thermal generator (GPHS-RTG) system. *Courtesy: NASA.*

conversion systems and power levels. The heat source can be used with multiple statics and dynamic energy conversion systems.

The GPHS is a rectangular, parallelepiped module, each having dimensions of 93 × 97 × 53 mm and a mass of about 1.43 kg. Each GPHS-RTG contains 18 modules that constitute the heat source stack for the GPHS-RTG (Fig. 17.2). Each module consists of a fine-weave pierced fabric (FWPF) graphite composite aeroshell, two graphite impact shells (GISs), two carbon-bonded, carbon fiber (CBCF) insulating sleeves, and four fueled clads. The FWPF is a carbon-carbon composite of graphite fibers woven in three dimensions that protects the heat source from ablation, thermal shock, and thermal response entry. The aeroshell contains two GISs that provide impact protection in a potential accident. The GIS is surrounded by a CBCF cylindrical sleeve to provide additional insulation to keep the capsule from overheating during hypersonic reentry and from overcooling during subsonic descent. The thermal insulation protects against launchpad fires [27,28].

The Pu-238 is pressed into a ceramic pellet of plutonium-238 dioxide ($^{238}PuO_2$) and encapsulated in a protective case of iridium alloy, forming a fuel clad. Each of the modules contains four plutonium fuel pellets with a nominal thermal inventory of 62.5 W of thermal energy (Wt). The fuel is cylindrical with an average diameter of 27.53 mm and an average length of 27.56 mm [29]. The GPHS fuel pellets consist of approximately 150 g of plutonium oxide. The ceramic pellet fuel is chemically stable, has a high melting point and vaporization temperature, and is highly soluble. The heat source is generally enriched with 83.5% Pu-238 isotope. The specific power of the heat source is 0.4743 W/g for plutonium or 0.4181 W/g for PuO_2. The alpha decay of Pu-238 generates 0.74 cm^3 (standard temperature and pressure, STP) of helium per gram of Pu-238 in a year. The typical composition of Pu-238 is listed in Table 17.2. The Pu-238 provides 99% of the thermal power in the heat source fuel [30]. The long half-life of Pu-238 leads to a reduction of only about 0.8% of the thermal power in a year, which makes the GPHS-RTG ideal for long-duration missions. SNAP-3B and SNAP-9A systems

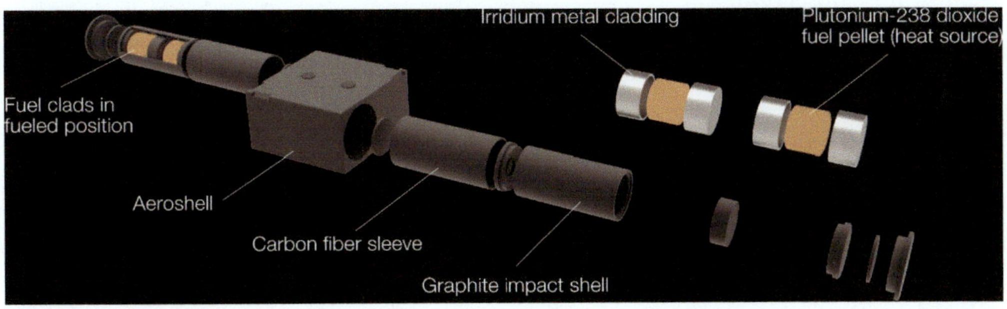

Figure 17.2 A cutaway view of general-purpose heat source (GPHS) module components and assemblies. *Courtesy: NASA.*

Table 17.2 Isotopic composition of plutonium.

Isotope	Weight %
Pu-238	83.5
Pu-239	14.01
Pu-240	1.98
Pu-241	0.37
Pu-242	0.14

Source: Public domain.

were fueled with PuO_2 molybdenum cermet, while SNAP-27 was fueled with PuO_2 microspheres [30].

The iridium alloy is chemically compatible with the plutonium oxide fuel and graphite components and has a high melting temperature. The clad iridium material protects the fuel from long-term air oxidation at elevated temperatures after Earth impact. Two of the pellets are encased in a GIS [31]. Each clad contains a sintered power fritted vent that allows helium gas generated from the decay of Pu-238 to escape to prevent pressure build-up that can crack the cladding and expose the fuel. The GPHS-RTG was used in the Ulysses [29,32], the Galileo [5,29], the Cassini-Huygens [33], and the New Horizons space missions [34]. The modules are subjected to extreme testing for a wide range of accident conditions, including simulation of multiple reentries for a single module through Earth's atmosphere, exposure to high-temperature rocket propellant fires, and impacts into solid ground; tests are conducted to show that the design meets all safety goals.

17.3.2 Static conversion technologies

Static energy converters convert the decay heat into energy without employing any movable parts. Static energy converters directly convert the heat energy into DC power. Thermoelectric conversion, thermionic conversion, and thermophotovoltaic conversion systems are three static conversion technologies that will be described in greater detail in this subsection.

17.3.2.1 Thermoelectric energy conversion: technology and recent material advances

Thermoelectric conversion is a static conversion technology that converts heat into electricity via the Seebeck effect [6,35]. Thermoelectric systems are environmentally friendly, scalable, rugged, radiation resistant, reliable, and feasible for a wide range of temperatures. However, the efficiency of such systems is low (~7%) [6]. A thermoelectric device converts the thermal energy produced due to a temperature differential induced by the heat source into electrical energy. A temperature gradient established between two points in a conductor or semiconductor creates a voltage difference. The internally generated voltage causes a current to flow through the load producing useful power. Heat is provided to the thermoelectric converter from GPHS modules. Power is generated by placing the thermoelectric element between the heat source and the heat sink. Only a portion of the heat is converted into electricity, while the unused or waste heat can be used to maintain the temperature of the electronic components.

The main component of the thermoelectric converter is the thermocouple. The thermocouple consists of a couple of metallic conductors with hot and cold end-connectors. A thermocouple is formed when junctions from a circuit formed from two dissimilar materials are connected electrically in series but thermally in parallel and the junctions are maintained at different temperatures. Heating one end of the connector with heat from the decay of a radioisotope while keeping the ends cold produces a voltage drop. The voltage drop is proportional to the combined Seebeck coefficient of the two materials and the temperature differential across the junctions. A thermoelectric generator contains several thermoelectric (TE) converters that are typically arranged in series or parallel configurations to produce the desired power characteristics. The efficiency and power output of a TE is a function of the thermocouple material and the temperature of each junction.

The conversion efficiency of thermoelectric generators depends on the thermoelectric material. The performance of a thermoelectric material depends on electrical parameters such as electrical conductivity, thermal conductivity, and the Seebeck coefficient. The characteristics of a TE material depend on the figure of merit (FOM). The nondimensional FOM (ZT), is found by multiplying the thermoelectric FOM by the absolute temperature, T. The thermoelectric efficiency, ZT, is given by Eq. (17.3):

$$ZT = \frac{\alpha^2 \sigma T}{K}, \tag{17.3}$$

where α is the Seebeck coefficient or "thermopower" (μV/K), σ is the electrical conductivity (A/μV cm), and K is the total thermal conductivity of TE material of the *p*- and *n*-legs (W/cm K). The total thermal conductivity is $k = k_e + k_{ph}$. k_e and k_{ph} represent the heat carried by the electrons and phonons.

The electrical conductivity and the Seebeck coefficient have a reciprocal relationship, which makes it difficult to optimize ZT. The Seebeck coefficient (thermopower) is a measure of the thermoelectric effect, and is the thermoelectric voltage per unit temperature difference, given by Eq. (17.4):

$$\alpha = -\frac{\Delta V}{\Delta T}. \tag{17.4}$$

The Seebeck effect can be established in a thermoelectric device by joining two dissimilar materials. Metals generally have a Seebeck coefficient of less than 10 μVK^{-1}

and are not very promising for constructing high-performance thermoelectric devices. Semiconductors have higher values of the Seebeck coefficient, with these values generally greater than 100 μVK^{-1} [6]. An ideal thermoelectric material should have a large electrical conductivity, so electrical charge can flow easily, and a low thermal conductivity to maintain a temperature gradient across the hot and cold sides. Metals have a large thermal conductivity and large electrical conductivity; this is not ideal for a thermoelectric material [6]. The electrical conductivity and the thermal conductivity are related by the power factor, determined by the product $\alpha^2 \bullet \sigma$. The thermoelectric FOM can be maximized by increasing the electrical conductivity and decreasing the thermal conductivity. However, increasing the electrical conductivity also increases the electronic thermal conductivity, which decreases the thermopower according to the Wiedemann-Franz Law. Hence, optimizing the FOM is a challeng [36].

The thermoelectric device efficiency depends on the Carnot efficiency and the material FOM. The device efficiency can be represented by Eq. (17.5). The Carnot efficiency is the maximum efficiency any heat engine can produce, but in practice, the theoretical efficiency of most engines is less than the Carnot efficiency.

$$\eta_{TE} = \eta_c \left(\frac{\sqrt{1 + ZT} - 1}{\sqrt{1 + ZT} + T_H - T_c} \right) \tag{17.5}$$

The Carnot efficiency, η_c is given by Eq. (17.6):

$$\eta_c = \frac{(T_H - T_C)}{T_H}. \tag{17.6}$$

The most widely used thermocouple materials are lead telluride (PbTe), tellurides of antimony, germanium, and silver (TAGS), lead tin telluride (PbSnTe), and silicon germanium (SiGe). Thermoelectric converters initially used tellurium-based materials such as PbTe that operate between 300 and 800 K. However, this material was subject to sublimation and required the RTG to be sealed in an inert cover gas. SiGe was later used because it operates at a much larger temperature range (800–1300 K). The RTGs that used SiGe could operate in a vacuum without the need for a cover gas because SiGe has negligible sublimation below 1300 K. Different materials are effective over different temperature ranges. Fig. 17.3A and B show the ZT values for several *n*-type and *p*-type materials, respectively, plotted over a range of temperatures [37]. The plot shows that the ZT values are temperature dependent. For the same material, the material may possess the highest ZT within a specific range while relatively low ZT at certain temperatures.

GPHS-RTGs with SiGe thermoelectric unicouples have been used in NASA's flight missions since 1976. SiGe TE has hot and cold shoe temperatures of 1273 and 573 K, respectively, with a conversion efficiency of ~5.4%. The LES 8/9 used the 312 SiGe-alloy thermoelectric elements [38]. A PbTe/TAGS thermocouple was used in the

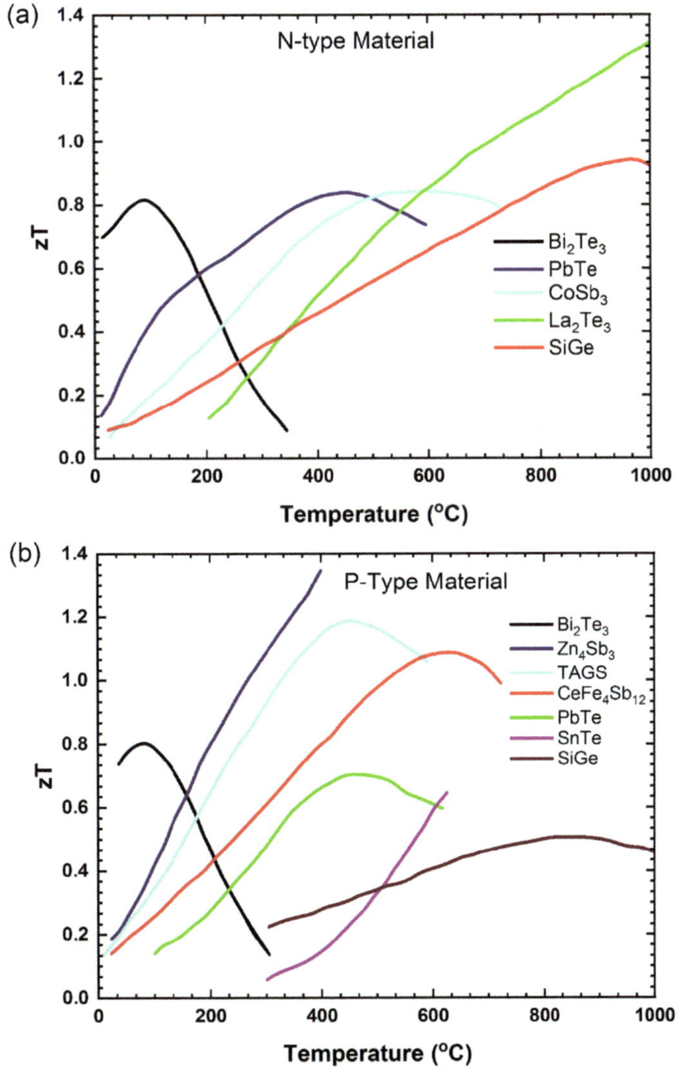

Figure 17.3 (A) The figure of merit for *n*-type thermoelectric materials; (B) the figure of merit for *p*-type thermoelectric materials. *Adapted from Ref. [37].*

multimission radioisotope thermoelectric generator (MMRTG) [39]. The list of space missions and thermoelectric materials used in these missions can be found in Table 17.3 [5,40].

There is always a need to improve on existing thermoelectric materials because high-performance materials will increase the specific power and reduce the mass of $^{238}PuO_2$ fuel in current RPSs. This has an important implication in terms of the size and cost of the RPS. The motivation for studying high-performance TE is because of the limited

Table 17.3 US spacecraft using thermoelectric materials for radioisotope power systems.

Spacecraft	Power source	No. of RPS	Thermoelectric material(s)	Mission type	Launch date	Status	Initial average RTG power (We)
Transit-4A	SNAP-3B7 RTG	1	PbTe	Navigational	06/29/1961	Successfully operated over 15 years; satellite shut down but currently in orbit	2.7
Transit-4B	SNAP-3B8 RTG	1	PbTe	Navigational	11/15/1961	Successfully operated over 9 years. Last signal in 1971; currently in orbit	2.7
Transit 5BN-1	SNAP-9A RTG	1	PbTe	Navigational	9/28/1963	Successfully operated as planned; non-RTG electrical problems caused satellite to fail after 9 months; currently in orbit	25.2
Transit 5BN-2	SNAP-9A RTG	1	PbTe	Navigational	May 12, 1963	Successfully operated after over 6 years. Currently in orbit	26.8

5BN-3	SNAP-9A RTG	1	PbTe	Navigational	4/21/1964	Spacecraft failed to achieve orbit. Reentered, burned up	25
SNAPSHOT	SNAP-10A reactor	1	N/A	Experimental satellite	March 4, 1964	Operated for 43 days, after which it shut down due to issues with voltage regulator. Currently in Earth orbit	500
Nimbus B-1	SNAP-19B2 RTG	2	N/A	Meteorological	5/18/1968	Mission aborted, power source retrieved intact, and fuel source reused for other missions	28
Nimbus III	SNAP-19B3 RTG	2	PbTe	Meteorological	4/14/1969	Successfully operated over 2.5 years; currently in orbit	28.2
Apollo 11	ALRH heater	2	N/A	Lunar	December 7, 1969	Contained two 15-W RHUs	N/A
Apollo 12	SNAP-27 RTG	1	PbTe	Lunar/ALSEP	11/14/1969	Successfully operated over 8 years; currently on lunar surface	73.6

(Continued)

Table 17.3 US spacecraft using thermoelectric materials for radioisotope power systems.—cont'd

Spacecraft	Power source	No. of RPS	Thermoelectric material(s)	Mission type	Launch date	Status	Initial average RTG power (We)
Apollo 13	SNAP-27 RTG	1	PbTe	Lunar/ALSEP	November 4, 1970	Mission aborted on way to moon. Reentered Earth's atmosphere and landed in South Pacific	73
Apollo 14	SNAP-27 RTG	1	PbTe	Lunar/ALSEP	1/31/1971	Successfully operated over 6.5 years; currently on the lunar surface	72.5
Apollo 15	SNAP-27 RTG	1	PbTe	Lunar/ALSEP	6/25/1971	Successfully operated over 6 years; currently on the lunar surface	74.7
Pioneer 10	SNAP-19 RTG	4	PbTe	Planetary/Sun escape	February 3, 1972	Successfully operated to Jupiter and beyond; spacecraft operations terminated 2003	40.7
	RHU heater	12	N/A			Contained 12 RHUs	
Apollo 16	SNAP-27 RTG	1	PbTe	Lunar/ALSEP	4/16/1972	Successfully operated over 5.5 years; currently on the lunar surface	70.9

Triad-01-1X	Transit-RTG	1	PbTe	Navigational	February 9, 1972	Currently in orbit	35.6
Apollo 17	SNAP-27 RTG	1	PbTe	Lunar/ALSEP	July 12, 1972	Successfully operated over 5 years; currently on the lunar surface	75.4
Pioneer 11	SNAP-19 RTG	4	PbTe	Planetary/Sun escape	May 4, 1973	Successfully operated to Jupiter, Saturn, and beyond. Last signal and spacecraft operation terminated 1995	39.9
	RHU heater	12	N/A			Contained 12 RHUs	
Viking 1	SNAP-19 RTG	2	PbTe	Mars lander	1975	Successfully operated for over 6 years on Mars. Operations ended in 1982	42.3
Viking 2	SNAP-19 RTG	2	PbTe	Mars lander	September 9, 1975	Successfully operated for over 4 years on Mars. Operations ended in 1982	43.1
LES 8	MHW-RTG	2,2	Si–Ge	Communication	3/14/1976	Currently in orbit	153.7
LES 9	MHW-RTG	2,2	Si–Ge	Communication	3/14/1976	Currently in orbit	154

(Continued)

Table 17.3 US spacecraft using thermoelectric materials for radioisotope power systems.—cont'd

Spacecraft	Power source	No. of RPS	Thermoelectric material(s)	Mission type	Launch date	Status	Initial average RTG power (We)
Voyager 1	MHW-RTG	3	Si–Ge	Planetary/Sun escape	May 9, 1977	Successfully Operated to Jupiter, Saturn, and beyond. Currently in interstellar space	156.7
	RHU heater	9	N/A			Contained 9 RHUs	
Voyager 2	MHW-RTG	3	Si–Ge	Planetary/Sun escape	May 9, 1977	Successfully Operated to Jupiter, Saturn, Uranus, Neptune, and beyond. Currently in interstellar space	159.2
	RHU heater	9	N/A			Contained 9 RHUs	
Galileo	GPHS-RTG	2	Si–Ge	Planetary (Jupiter)	10/18/1989	Successfully explored Venus, then orbited Jupiter; spacecraft de-orbited into Jupiter in 2003	288.4
	RHU heater	120	N/A			Galileo orbiter included 103	

					strategically placed RHUs; its atmospheric probe carried 17 of the units. Each RHU produced about 1 W of thermal power	
Ulysses	GPHS-RTG 1	Si–Ge	Solar and space physics	June 10, 1990	Successfully explored and entered Jupiter; spacecraft operations ended in 2009	283
Mars Rover Pathfinder	RHU heater 3	N/A	Planetary rover (Mars)	April 12, 1996	Landed on Martian surface and successfully deployed the first rover to study the planet's soil	N/A
Cassini	GPHS-RTG 3	Si–Ge	Planetary (Saturn)	10/15/1997	Successfully explored Venus and Jupiter; currently orbiting Saturn	295.7
	RHU heater 117	N/A			Cassini orbiter included 82 strategically placed RHUs; Huygens carried 35 units; each RHU produced ~1 Wt power	

(Continued)

Table 17.3 US spacecraft using thermoelectric materials for radioisotope power systems.—cont'd

Spacecraft	Power source	No. of RPS	Thermoelectric material(s)	Mission type	Launch date	Status	Initial average RTG power (We)
Mars MER Spirit	RHU heater		PbTe	Planetary rover (Mars)	2003	Spirit landed successfully on January 4, 2004. Mission was ended in 2011	140
Mars MER Opportunity	RHU heater		PbTe	Planetary rover (Mars)	March 7, 2003	Opportunity landed successfully in 2004. Mission ended in 2018	140
New Horizons	GPHS-RTG	1		Planetary/Sun escape	1/19/2006	Explored Jupiter. Pluto flyby July 2015. Additional exploration of Kuiper Belt and beyond underway	249.6
Mars MER Curiosity	MMRTG	8 GPHS modules	PbTe/ Germanium telluride/Silver antimony telluride (TAGS)	Planetary rover (Mars)	11/26/2011	Successfully landed on August 6, 2012; currently exploring Martian surface	113
Mars MER Perseverance and Ingenuity	MMRTG	8 GPHS modules	PbTe/germanium telluride/silver antimony telluride (TAGS)	Planetary rover and flyer (Mars)	7/30/2020	Landed February 28, 2021; both missions are underway	110

Adapted from information in Refs. [5,6,14,40].

stockpile of Pu-238 and the regulatory burden associated with handling large amounts of radioisotopes.

Skutterudite-based thermoelectric unicouples are among materials that have been studied for use in advanced radioisotope power systems (ARPSs) [5,40]. Segmented and nonsegmented skutterudite-based unicouples and cascaded SiGe unicouples with skutterudite-based segmented and nonsegmented unicouples are skutterudite-based thermoelectric materials being considered for advanced thermoelectric technology [5,6,40,41]. Results from segmented unicouples fabricated using *p*-type $CeFe_4CoSb_{12}$- and Bi_2Te_3-based alloys and *n*-type $CoSb_3$- and Bi_2Te_3-based alloys tested at cold and hot shoe temperatures of 300 and 973 K, respectively, demonstrated conversion efficiencies of ~10%. It was shown that these segmented unicouples are capable of achieving efficiency of ~14%—15% [42]. It was found that the efficiency of cascading SiGe operating at 573 K is 85% higher and uses about 46% less $^{238}PuO_2$ to produce the same amount of energy as current RTGs. Results from ($Si_{0.8}Ge_{0.2}$) and skutterudite segmented thermoelectric unicouples (STUs) with a hot side temperature of 973 K and cold side temperatures of 300, 573, and 673 K with similar total length and cross-sectional dimensions as the *p*-leg when compared showed that the STU uses half the $^{238}PuO_2$ fuel mass and radiator area and operated at higher electrical power density (>7 We/kg) than SiGe used in current RTGs (~5.5 We/kg) [43].

The challenge of using skutterudite-based unicouples is the sublimation of antimony from the legs near the hot junction, which can change the thermoelectric property of the material and degrade its performance over time. However, a sublimation suppression metallic coating (*p*-$CeFe_{3.5}Co_{0.5}Sb_{12}$ and *n*—$CoSb_3$) that is compatible with the legs of SKUs was developed by the Jet Propulsion Laboratory to address the sublimation of antimony. The suppression metallic coating significantly reduced the loss of antimony even at temperatures up to 973 K [40—44].

Another active field of research is using nanostructured or "low-dimensional" materials through the introduction of different nanostructures to tune the transport of phonons and electrons. TE nanocomposites usually have higher FOM and better performance than traditional bulk materials. Using nanotechnologies to improve energy conversion efficiency is very promising. It has been shown that the peak ZT of Bi_2Te_3-based nanomaterial improved from 1 to 1.3—1.47 [45—47]. A PbTe-based nanocomposite ($AgPb_mSbTe_{2+m}$) based on nanocomposite materials produced a ZT of ~2.2 at 800 K; the material is expected to outperform all reported bulk TE at this temperature range [48]. A SiGe/SiC nanocomposite produced a high FOM of ~1.7 at 900°C, which is twice the value for bulk SiGe TE [49]. The current challenge of using nanotechnologies is that of scalability and a methodology of integrating the technology into the current RPS technology.

There is also ongoing work to investigate materials that can use quantum well (QW) technology. The technique is expected to increase thermoelectric conversion efficiencies by more than 25%. The most interesting result from QW technology is the high values of thermoelectric FOM, ranging from ZT = 1.6 at 300 K to ZT = 3 at 550 K, for Bi-doped *n*-type PbSeTe/PbTe quantum-dot superlattice samples grown by molecular beam epitaxy [50].

The RTG is based on thermoelectric conversion. RTG has been employed for Earth-based and space-based missions. The RTG was used by the former Soviet Union to power remote lighthouses and navigation beacons [51]. The RTG has also been used extensively for space-based missions. All RTGs that have been flown by the United States comprised seven basic designs: SNAP-3/3B, SNAP-9A, SNAP-19/19B, SNAP-27, TRANSIT-RTG, MHW-RTG, and GPHS-RTG [52]. All RTGS that have flown have met or exceeded all design expectations, except for three missions that were aborted due to launch vehicle or spacecraft failures. All flight models used Pu-238 as fuel. SNAP-3B was the first RTG to be flown in space and was aboard the Navy Transit 4A spacecraft [53].

17.3.2.2 Radioisotope thermophotovoltaics

The concept of harvesting energy using radioisotope thermophotovoltaic(s) (RTPV) is similar to PV power, except that the energy is harvested from the infrared spectrum emitted by a hot surface instead of from the Sun. The benefit of using RTPV over PV power is that it uses radioisotopes that have a long shelf life and high energy density and uses PV cells that have a quicker approval process than conventional RTGs.

Thermophotovoltaic(s) (TPV) devicess are derived from and similar to PV solar systems. PV systems convert solar radiation into electricity. TPV converts infrared radiation from a heat source into electricity. RTPV are TPV that use a radioisotope as a heat source. The difference between a PV and a TPV is that a PV receives radiation from the Sun at a temperature of about 6000 K at about 150×10^6 km, whereas TPV receives radiation from a hot surface at temperatures of 1300–1800 K with a distance of a few centimeters. As a result, power delivered to the solar cell is ~ 0.1 Wcm^{-2}, while the power delivered to the TPV is between 5 and 30 Wcm^{-2}. Hence, the power density of a TPV is significantly higher than that for a PV solar cell [54]. For radioisotope heat sources, the radiation temperature is ~ 1200 K. The relatively low temperature means the power density is lower.

The concept of combining radioisotopes with PV cells is not new [55,56]. However, this system was hindered by low conversion efficiencies because of the absence of efficient conversion materials. Recently, single-junction GaAs PV devices have demonstrated efficiencies of 35.5% [57]; the methods developed to improve the efficiency of PV cells is being applied to RTPV technology [58]. The recent availability of competing materials has also sparked renewed interest in RTPV Since infrared spectrum peaks at a different wavelength (at reasonable temperatures) than solar energy, materials with different bandgap energies must be used in RTPV. To achieve high efficiency, TPV must be optimized for the infrared spectrum at practical temperatures. TPV systems have been shown to offer efficiencies greater than 15% and up to 40% [59]; see Datas and Chubb, in another chapter in this book, for a materials perspective [60].

A TPV device consists of a high-temperature radiator, a cavity that includes a filter to control the infrared spectrum, and a PV cell to convert the thermal energy into electricity. The heat source is used to heat the radiator or the emitter to typical temperatures of 1300–2000 K. The major part of this radiation is in the infrared spectral range. The close arrangement between the radiator and the PV cells eliminates any radiation loss. Hence, the operation of TPV is steady in terms of intensity, spectrum, and angle of radiation. Fig. 17.4 shows a generic TPV device structure.

The total system efficiency of an RTPV depends on the thermal efficiency and the PV cell efficiency. The system efficiency is the ratio of the electrical power output to the decay heat of the heat source. The cell efficiency is the ratio of the electrical power output of the PV cell to the thermal power emitted by the hot surface, while the thermal efficiency is the ratio of the power emitted by the hot surface to the power produced from the decay of the fuel. Spectral control of RTPVs can help increase cell efficiency. High-temperature filters can modify the emission spectrum to match the bandgap of the cells, so the emitter has high emission for photons above the cell bandgap to generate more energy and low emission from low energy photons below to reduce waste heat.

One critical aspect of RTPVs is finding an appropriate emitter material that can efficiently radiate the thermal energy generated by the heat source and a suitable energy converter that can convert infrared radiation into electrical energy. To increase efficiency, PV devices use advanced materials to utilize a broader, spectrally "tuned" range of wavelengths that can convert the infrared wavelength radiated by the emitter into electrical energy. Spectral control is achieved by using selective emitters, TPV modules, and filters [61]. The efficiency is improved by minimizing wavelengths that cannot be efficiently converted by PV cells. This can be done either using spectrally selective emitters or spectrally selective filters. Spectrally selective emitters ensure that undesired wavelengths are reduced, while a spectrally selective filter preferentially transmits those

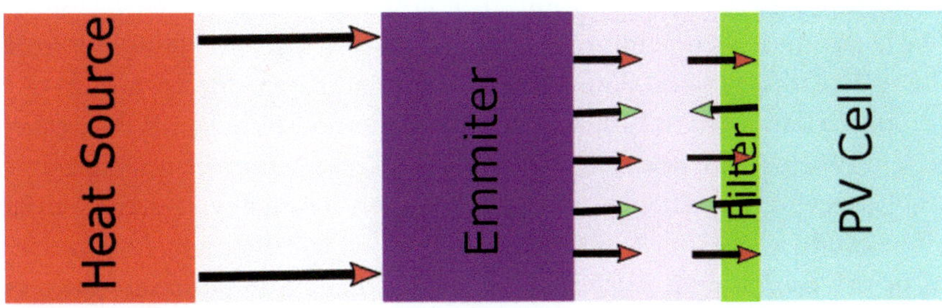

Figure 17.4 A generic TPV device with typical system components. *Source: authors.*

wavelengths that can efficiently be converted by the PV cells. The technologic challenge with selective emitters is concern about thermal stability at higher temperatures.

PV cells are relatively inefficient in harvesting solar energy because of the broad energy spectrum of the sun [54,58,59]. PV cells made from silicon and gallium arsenide will also be inefficient in TPV systems because of the significant difference in bandgap energy [58–61]. To achieve high conversion efficiency, the bandgap of the PV cell must be well matched with the infrared spectrum. One suitable material that was investigated for use in PV cells is gallium antimonide (GaSb). GaSb has several properties that makes it attractive for energy harvesting in the infrared range. The bandgap of GaSb is 0.73 eV. The efficiency of GaSb at wavelengths of between 1.5 and 1.6 μm is ~35%. When GaSb was placed at the back of GaAs to produce tandem solar cells, experimental results showed efficiencies of ~35%, which are significantly higher than that of a standalone GaAs cell [62]. The result was an indication that the efficiency of a TPV cell can be maximized by interposing a spectrally selected filter between the hot surface and the GaSb cell with the filter designed so it only transmits the wavelength that can be efficiently converted into electricity and reflects other wavelengths.

Fairchild Space and Defense Corporation, under contract with the DOE, investigated various sources of energy for the Pluto Fast Flyby (PPF). PPF involves an 8- to 9-year direct (no-gravitational assist) flight to Pluto followed by comprehensive mapping, surface composition, and atmospheric structure measurements during a brief flyby of the planet and its moon Charon. Among all systems studied (RTG, MMRTG, and Stirling engines), only RTPVs achieved NASA's desired mass reduction goal. The result of the study showed that the RTPV radiation mass represents only 40% compared to the RTG, and the RTPV requires 60% fewer GPHS modules than the RTG. This represents a significant cost savings when RTPVs are employed in missions compared to RTGs. It was also shown that replacing RTGs with RTPVs will double the generator's efficiency and specific power [63].

Creare and Edtek were awarded contracts by the US Nuclear Regulatory Commission (NRC) to develop an advanced RTPV generator using a selective emitter-based TPV power generator. Creare designed an RTPV with a 4 × 4 InGaAs array, a tandem plasma/dielectric filter for spectra control to maximize conversion efficiency, and tungsten emitters. The prototype produced a conversion efficiency of 19% and more than 50 We at an emitter temperature of 1350 K and heat rejection temperature of 300 K during phase I of the project. The project was terminated due to a lack of funding [23,64]. A contract awarded to Edtek by the NRC demonstrated an RPTV efficiency of 20% with a TPV cell and prism array. The design used a novel integrated cell front contact and concentrator scheme, a uniquely tuned frequency selective surface filter array, an improved GaSb PV cell, and a unique thermal management system that led to a reduction of the total system mass by a factor of three. The contract was terminated during phase II due to NASA funding cuts [23,65]. An improved design of this system reported efficiency of 30.1% with a specific power of 25 W/kg and a sixfold reduction in thermal size over the baseline design [66]. Studies conducted by Boeing using GaSb with Pu-238 heat source operating at 1000°C developed 573-W systems that could operate at 14.4% efficiency and exhibit a system level-specific power of approximately 9.2%. When radioisotopes with high specific power such as Cm-144 are used, it is expected to produce an even larger efficiency (20%), and specific power exceeding 60 W/kg [62]. This type of power output is desired in applications such as microsatellites and naval buoys. Table 17.4 shows the comparison of RTPV operating characteristics with different radionuclide source materials.

The challenge of using TPV is that its performance can be reduced over time due to radiation damage. However, the RTPV heat source produces alpha radiation, which has a short range in matter and will be stopped by the canister. The amount of gamma rays and neutrons generated from Pu-238 will degrade the performance of the cell over time. Neutron degradation of 0.6% per year of InGaAs device has been predicted and measured for the Pu-238 radiation heat source [67]. However, the effect of neutron irradiation will be small during the duration of the mission. The RTPV has not been tested in space, and although the concept is promising as a source of energy for future space missions, it is yet to be proven in terms of long-term reliability.

17.3.2.3 Thermionic energy conversion

Thermionic energy conversion is an energy conversion mechanism that involves the conversion of thermal energy directly into electrical energy via the emission of electrons from the hot heated surface and subsequent collection on a cooler surface and the return of the electrons to the emitting surface through an external load. A thermionic converter uses heat as its source of energy. It provides silent, maintenance-free operation, it is lightweight, and offers high specific power. Thermionic conversion systems are expected to have the highest specific power of any direct energy conversion system because their

Table 17.4 Comparison of RTPV operating characteristics with different radionuclide source materials.

Isotope configuration	Operating temperature (K)	Percent efficiency	Output power (Watts)	Mass (kg)	Specific power (W/kg)
GPHS (1 × 15)	1100	12.4	458	58	7.9
GPHS (2 × 2 × 15)	1100	14.4	573	61.9	9.25
Sr-90	1200	17.0	487	41.2	11.8
Sr-90	1200	17.0	49	3.85	12.7
Cm-244	1200	17.0	413	7.56	54.6
Cm-244	1200	19.9	6	0.10	60.0
RTG	1000	7.42	278	58.6	4.74

Adapted from Ref. [62].

operating temperatures are much higher. To make thermionic conversion attractive as a means of generating electrical energy, the material used in both the heat source and the converter must operate at relatively high temperatures (1200–1800°C). However, not all materials are stable at this temperature, which is a significant issue for thermionic systems.

A thermionic energy conversion system consists of two electrodes: the hot cathode (emitter) and the cold anode (collector), which are separated by a gap. The electrodes are heated to sufficiently high temperatures that the electrons are "boiled" off. The electrons travel to the cooler collector where they condense. The thermionic mechanism produces a potential difference between the electrodes. When a load is connected, the potential difference drives a current. The difference between the collector and the emitter work function is related to the output voltage. The gap between the two electrodes is filled with extremely low-pressure gas or vapor to reduce space charge effects. Cesium is the most widely used gas because it has the lowest ionization potential of any element. Cesium surrounds both emitter and collector, leading to high conversion efficiency. However, cesium tends to introduce other effects that tend to reduce performance. Fig. 17.5 shows the thermionic conversion principle.

The thermionic efficiency is the ratio of the useful power delivered to the external load to the power input to the emitter. The efficiency is given by Eq. (17.7):

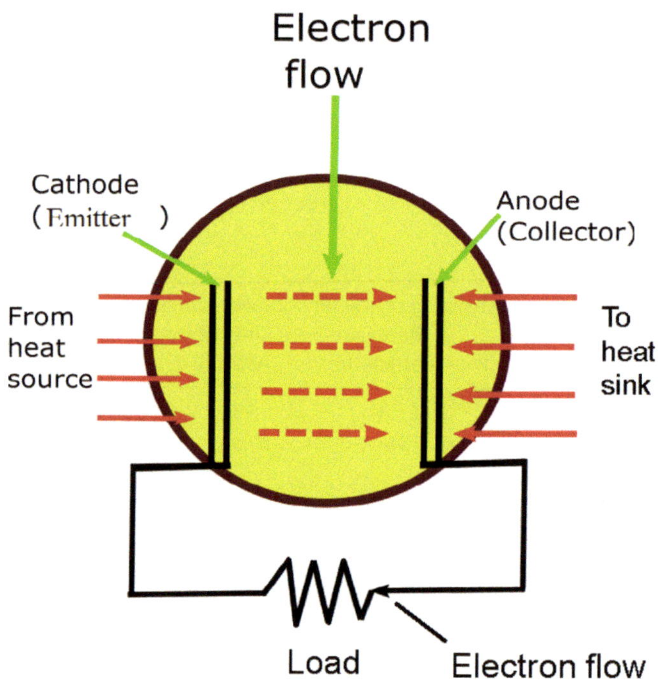

Figure 17.5 Diagram of thermionic device operation. *Source: authors.*

$$\eta = \frac{J\,V}{Q_e + Q_r + Q_c} \tag{17.7}$$

This ideal equation ignores transport losses and space charge effects [68]; J is the output current density, V is the corresponding output voltage, Q_e is the electron cooling of the emitter, Q_r is the radiant heat lost by the system, and Q_c is the conductive heat losses through lead wires and support.

SNAP-13 was based on radioisotope thermionic conversion. It used cerium-242 radioisotope as the heat source. It also used a cesium vapor as a thermionic converter operating at emitter temperature between 1600 and 1700 K and produced 12.2 W at an efficiency of 8%. The reactor was expected to operate for 120 days. SNAP-13 used planar diode converters with the capsule insulated with multifoil insulators. The device was fueled and tested but never flown into orbit [69]. The TOPAZ reactor is based on thermionic energy conversion [70]. However, thermionic conversion is not the desired method of power production because it is difficult to achieve long-term reliability.

17.3.3 Dynamic energy conversion systems

The development of dynamic conversion systems is driven by the desire to improve the efficiency of the current RPS in terms of cost, size, availability, and safety. Dynamic converters offer advantages of improved efficiency, long lifetime, and attractive scaling. NASA has invested enormous resources into developing dynamic conversion systems because they are projected to produce higher conversion efficiencies than static conversion systems. The current dynamic systems under consideration are the Brayton, Rankine, and Stirling converters. The current challenge of using dynamic converters is that the reliability of these systems has not been established since they have never been used in long-duration missions.

17.3.3.1 Stirling power converter

The Stirling engine is a reversible thermodynamic cycle that uses reciprocating pistons and a single-phase gaseous working fluid. Most of NASA's efforts have focused on the development of Stirling engines rather than Brayton, Rankine, and RTPV. The Stirling converter is attractive because of its high conversion efficiency of 23% compared to 6% −8% efficiency for thermoelectric technology [71]. It also uses only 25% of the Pu-238 required to produce similar power as an RTG. A Stirling RPS could be used in missions such as Lunar Geophysical Network, Europa (although, it was recently baselined with solar power), Titan Saturn System Mission, Saturn Probe, Uranus Orbiter and Probe, Trojan Tour, Enceladus Orbiter, and Io Orbiter [72].

The Advanced Stirling Radioisotope Generator (ASRG) was under development, starting in 2000, jointly sponsored by the DOE and NASA under contract with Lockheed Martin with key contributions from NASA Glenn Research Center

(Cleveland, OH), Lockheed Martin (Valley Forge, Pa), and SunPower and was based on Stirling conversion technology. The ASRG uses two GPHS modules, each providing a total power input of 500 Wt and generating 130 W of electrical energy (We) [73]. The remaining 75% of the heat was used to keep systems and instruments at their operating temperature. The ASRG has a modular design to meet a variety of missions. A free-piston Stirling engine operates by a thermodynamic cycle. It has two reciprocating components: the displacer and the piston. The ASRG generates electrical energy by first converting the heat from a GPHS into the high-speed kinetic motion of a small piston and a displacer. The piston is sealed in a closed cylinder suspended in helium gas. The movement of the displacer and the piston is synchronized, with the piston rapidly forcing the helium gas to move rapidly back and forth from the hot side (heated by the GPHS) and the passive cooler end through a series of heat exchangers [21]. A linear alternator connected to the piston converts mechanical energy to electrical power. The helium gas also functions as a "hydrostatic" bearing, keeping the displacer and the piston from rubbing the walls of the cylinder and eliminating any physical wear, so the ASRG can operate for its intended life of 14 years. The ASRG uses heat from the decay of Pu-238. Each ASRG contains two sets of pistons and a displacer known as the Advanced Stirling Convertor (ASC). The two connectors are arranged end to end in the middle of the GPHS, which helps to cancel out the small linear vibrations produced by the pistons when their motion is synchronized. The far end of the converter is connected to the GPHS. The ASRG contract with Lockheed Martin was terminated in 2012 due to budget constraints.

Table 17.5 shows the key difference between past GPHS-RTG and ASRG. There are several advantages of using the ASRG compared to GPHS-RTG. The ASRG (two units) uses fewer GPHS units than the RTG (18 units) and has a larger specific power. It uses less Pu-238 than GPHS-RTG and has a projected efficiency of 30%, which is more than four times larger than GPHS-RTG. These characteristics made ASRG attractive for space missions. However, the ASRG was never flown into space. Results from experiments performed using ASRG showed power variations within 175 h of operation.

Table 17.5 The differences between GPHS-RTG, MMRTG, and ASRG.

	GPHS-RTG	MMRTG	ASRG
Electrical output, BOM, We	285	125	~150
Heat input, BOM, We	4500	2000	500
RPS system efficiency, BOM, %	6.3	6.3	30
Total system weight, kg	56	44.2	~22
Specific power, We/kg	5.1	2.8	8
Number of GPHS modules	18	8	2
GPHS module weight, kg	25.7	12.9	3.2
Pu-238 weight, kg	7.6	3.5	0.88

Adapted from Ref. *[14]*.

NASA under contract with Sunpower has been supporting research on the development of ASC. The ASC is expected to increase the power density from 3.5 to7 We/kg because of its reduced envelope and lighter mass compared to previous versions of the Stirling converter [21]. Fig. 17.6 is an image of an ASC.

17.3.3.2 Brayton power converter

The Closed Brayton Cycle (CBC) was demonstrated by NASA using the Brayton Rotating Unit (BRU) [74]. A significant advantage of the closed Brayton conversion is that it can be used with a wide range of heat sources for space application. It also has a power scaling advantage in that it can be used to generate power from as low as a kilowatt to as large as multimegawatt. Applications of CBC include global communication satellites, power for interplanetary explorations using electric propulsion (10−100 kWe), and large lunar/Mars outpost (100−1000 kWe). The heat source used in CBC could be obtained from solar concentrators, radioisotope power sources, or a nuclear reactor. The development of CBC will reduce the Pu-238 used and conserve the limited amount of Pu-238.

The Brayton power converter consists of a heater, a turbine, a compressor, an alternator, and a separate heat exchanger for the heat source and waste heat rejection.

Figure 17.6 Superpower Advanced Stirling Converter (ASC-1A) including an all Mar-M-247 heater head allowing 850°C operation. *Courtesy: NASA.*

The heater configuration depends on the selected heat source, while the cooler depends on the mission configuration consideration. The heat exchanger, also called the recuperator, is used to transfer heat within the cycle and improve efficiency. An inert gas, often a mixture of helium and xenon, is used as a working fluid. The working fluid undergoes a thermodynamic cycle. Since the working fluid is inert, there is no concern about corrosion that can limit the life of the engine. Helium is used because of its superior heat transfer properties in the heater, cooler, and recuperator, while Xe adjusts the molecular weight to maximize turbine and compressor efficiency. The working fluid is heated in the heat exchanger, expanded through the turbine, passed through the gas cooler, and pressurized by the compressor before reentering the heat source heat exchanger to be heated again to complete the thermodynamic cycle. As the working fluid moves through the recuperator, energy is extracted, while waste heat is rejected in the cooler [75]. A rotary alternator attached to the turbine shaft produces alternating current electrical output.

The closed Brayton power and the Rankine conversion cycles were assessed for integration with the Dynamic Isotope System (DIPS) program sponsored by DOE as a technology that can provide reliable power for a wide range of energy for space missions and a mission duration of 7–10 years [76]. The DIPS concept was demonstrated at a power level of 1.3 kWe [77]. Assessment of the Stirling engine and CBC approaches shows that for low-power radioisotope systems, Stirling is the preferred choice from a mass perspective. The mass cross-over point in which the Brayton will offer a smaller specific mass is between 10 and 50 kW, which is beyond the limit of radioisotope thermal power. High-power spacecraft reactor systems results indicate that the Brayton is the clear favorite [78]. Studies were also conducted to incorporate the CBC with the SP-100 space reactor power system. SP-100 is a fast spectrum, liquid-cooled lithium reactor that uses a SiGe doped with gallium phosphide (GaP) thermoelectric module for heat conversion. The CBC/SP-100 reactor concept involves using a direct-cooled SP-100 reactor. The studies show that SP-100 and CBC technologies provide an attractive combination of advanced reactor technology and represent a highly efficient power conversion technology [79].

17.3.3.3 Rankine converter

The Rankine cycle, developed in 1859 by Scottish engineer W.J.M. Rankine, is the most widely used stationary source of power. It is the first class of dynamic engines to be employed for space nuclear power (SNP) [76,77]. In a Rankine cycle, a liquid is pumped under pressure into a boiler where heat is added. The vapor then drives a turbine. The turbine expands the working fluid producing mechanical energy that is converted to electrical energy by the generators. The low-pressure liquid that emerges from the expansion process is then condensed into liquid and pushed back to a heat source where the process begins again.

Organic working fluids are attractive for small systems working at low temperatures. An organic Rankine cycle (ORC) power system is a refinement of the water Rankine cycle using an organic working fluid instead of water. The organic liquid has a higher vapor pressure and a lower boiling point than water. Using organic working fluids prevents corrosion of the blades, leading to longer system life and improved reliability. The ORC is different from the CBC because it uses a phase change of the working fluid. The power conversion efficiency is strongly dependent on the heat rejection temperature. The lower the rejection temperature is, the higher is the efficiency. On the other hand, the lower heat rejection temperature requires a larger radiation area. An ORC system requires a larger radiation area for the same cycle efficiency as the CBC.

The first class of dynamic heat engines to be applied to SNP systems was the Rankine engine. The first demonstration was with SNAP-1 in the 1950s. SNAP-1 was based on a mercury Rankine cycle heat engine. SNAP-1 used Ce-144 radioisotope and successfully operated for 2500 h. SNAP-1 was never deployed in space [18]. SNAP-2 also used a mercury Rankine cycle. SNAP 2 was an experimental reactor (SER). It generated a total energy of 224,650 kW-h during the life of the reactor [80,81]. SNAP-8 also used a mercury Rankine cycle [82,83]. SNAP-50/SPUR used a potassium working fluid [84,85]. The kilowatt isotope power system (KIPS) was also based on the Rankine cycle [86].

17.3.4 Photon intermediate direct energy conversion

Photon intermediate direct energy conversion (PIDEC) can be used with a variety of nuclear sources: fission [87,88], fusion [89], and radioisotopes [90,91]. The main advantage of PIDEC is through the use of waveguides to transport photons through shielding materials into PV cells to protect the cells from radiation damage. A long-term test demonstrated long-term operation of a silicon PV cell using both alpha sources and energetic beta particles [92–95]. Since PIDEC uses high-grade photons as its source, it can serve as a topping cycle. The residual high-grade heat is then useable for other energy conversion cycles [91].

RTGs use a highly robust but relatively inefficient (6%–7%) energy transducer, utilizing the Seebeck effect, for energy conversion. A simple means of extending the lifetime of the isotope stockpile is to increase the energy conversion efficiency of the transducer. There has been work done on improving the efficiency of a single-cycle system such as certifying Brayton cycles and Stirling engines for space missions. Increasing energy conversion efficiency will extend the lifetime of the isotope stockpiles. A multicycle approach can also extend the lifetime of isotope stockpiles by increasing efficiency as well as being a conduit for adding new isotopes to the discussions. The key to implementing a multicycle approach is to use a topping cycle that has an operating temperature sufficiently high to provide the input temperature for lower stage energy converters (Fig. 17.7).

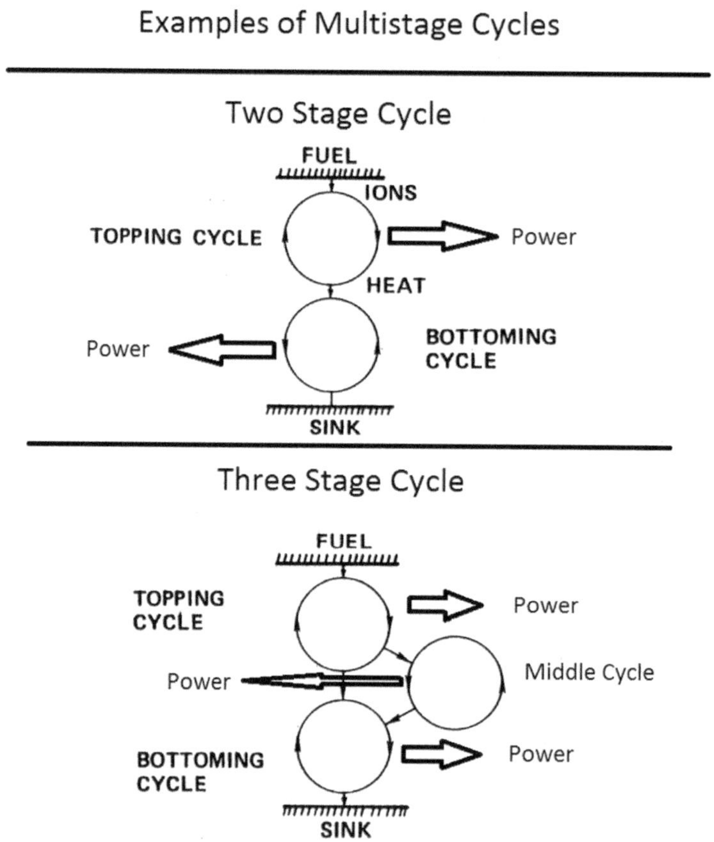

Figure 17.7 An illustration of a two-stage and three-stage cycle approach. In the two-stage approach, the ions from the fuel (e.g., radioisotope) provide the energy for the top cycle. The operating temperature of the top cycle is sufficiently high to be used as the high-temperature input of a bottom cycle. In the three-stage cycle, the operating temperature of the second stage cycle would have to be sufficient to provide the high-temperature leg for the third cycle. *Source: authors.*

Through a combination of a top cycle capable of operating at a reasonably high temperature (between 800 and 1050 K) and a bottom cycle that has a reasonably high energy conversion efficiency in this temperature range, it is possible to maximize the efficiency of a dual-cycle conversion system (DCCS) for converting the power being generated from a nuclear source into electrical power. The nuclear source includes fission, fusion, and radioisotopes. A study using Kr-85 was used to demonstrate the utility of a DCCS [91]. The DCCS using the Kr-85 isotope does possess the capability of having a primary cycle with a high operating temperature and a bottom cycle that can efficiently convert the residual thermal energy into electrical power. In this scenario, the PIDEC system is used as the top cycle along with a bottom cycle such as the Brayton cycle or the Stirling engine.

The Kr-85-based DCCS essentially requires that the radioisotope be used as a volume source [7]. The PIDEC cycle is an indirect energy conversion system, so there is a primary and a secondary transducer. The Kr-85 served a dual purpose; it was both the source of ionizing radiation (e.g., beta particles) as well as being a fluorescence-producing transducer (Fig. 17.8) [90]. The ionization and excitation created by the interaction of the beta particles with the primary transducer is used to create KrCl* excited dimer (excimer) fluorescence. The KrCl* excimer is chosen for the base case studied in this paper because of its high fluorescence efficiency and good spectral match with the secondary transducer (31%, i.e., conversion of ionization and excitation caused by ionizing radiation to narrow band fluorescence emission at 222 ± 10 nm) [88]. The simplest and most efficient geometric configuration for the fluorescer gas is a sphere, as shown in Fig. 17.8.

Excimer fluorescers are optically thin (meaning virtually no self-absorption of the excimer photons) because they have an unbound ground state. Thus, the scale of the fluorescer cell is not an issue that negatively impacts the photon transport efficiency from the fluorescer to the secondary transducer (i.e., PV cells). The sphere is a nonreentrant geometry, which means that regardless of the emission angle, the excimer photons will travel unimpeded through the optically thin fluorescence medium and intersect with the PV cells on the walls of the pressure vessel where the photons are absorbed [96]. The bank of PV cells is spectrally matched to the emission of the excimer fluorescence (i.e., matching the bandgap of the semiconductor, E_g, to the average energy of the photons from the fluorescer, $h\nu$ [97]). KrCl* excimer photons have a 96.4% spectral match with diamond PV cells; see Table 17.6. Due to the 5.47-eV bandgap, diamond PV cells will have a low dark current even at elevated temperatures (~800−1050 K). The optical

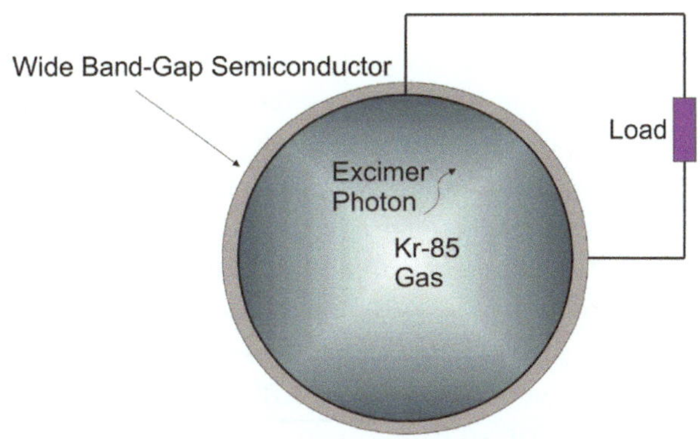

Figure 17.8 Illustration of a Kr-85-powered PIDEC cycle. The Kr-85 gas is used as both the source of ionizing radiation and as a fluorescence source (e.g., the primary transducer). It produces excimer fluorescence that then interact with the diamond PV cells (the secondary transducer) surrounding the fluorescer. *Source: authors.*

Table 17.6 The theoretical maximum spectral matching efficiency (η_{PV}) and the efficiency for the production of ion-to-electrical current (η_{IE}) are shown for selected rare-gas, rare-gas halide, and alkali metal excimer fluorescers. The fluorescence efficiency (η_f), the energy of the photon (E_λ), and the bandgap of the photovoltaic cell (E_g) are also shown.

Excimer	η_f	E_λ (eV)	Photovoltaic material	Bandgap E_g (eV)	$\eta_{PV} = E_g/E_\lambda$	$\eta_{IE} = \eta_{PV} \cdot \eta_f$
Ar_2*	0.5	9.6	AlN	6.2	0.65	0.32
Kr_2*	0.47	8.4	AlN	6.2	0.74	0.35
	0.47	8.4	Diamond	5.47	0.65	0.31
F_2*	0.44	7.8	AlN	6.2	0.79	0.35
	0.44	7.8	Diamond	5.47	0.70	0.31
Xe_2*	0.48	7.2	AlN	6.2	0.86	0.41
	0.48	7.2	Diamond	5.47	0.76	0.36
ArF*	0.35	6.4	AlN	6.2	0.97	0.34
	0.35	6.4	Diamond	5.47	0.85	0.30
KrBr*	0.33	6	Diamond	5.47	0.91	0.30
KrCl*	0.31	5.6	Diamond	5.47	0.98	0.30
Na_2*	0.46	2.84	ZnSe	2.7	0.95	0.44
	0.46	2.84	SiC (3C)	2.3	0.81	0.37
Li_2*	0.42	2.7	$CuAlSe_2$	2.6	0.96	0.40
	0.42	2.7	SiC (3C)	2.3	0.85	0.36
Hg_2*	0.21	2.58	GaS	2.5	0.97	0.20
	0.21	2.58	SiC	2.4	0.93	0.20
ArO*	0.11	2.27	GaP	2.2	0.97	0.11
	0.11	2.27	GaAlAs	2.2	0.97	0.11
KrO*	0.13	2.27	GaP	2.2	0.97	0.13
	0.13	2.27	GaAlAs	2.2	0.97	0.13
XeO*	0.15	2.27	GaP	2.2	0.97	0.15
	0.15	2.27	GaAlAs	2.2	0.97	0.15

* Designation of an excimer.
Source: authors.

properties of diamond decrease at temperatures above 1500°C because of graphitization [98], and in the presence of oxygen, diamond will oxidize at temperatures higher than 600°C [99,100]. Otherwise, diamond is chemically inert. The diamond PV cells in the proposed DCCS operate well below the graphitization temperature and are not exposed to oxygen. So, diamond PV cells are capable of operating at temperatures between 800 and 1000 K. Diamond is one of most radiation-hardened materials known. It has been reported to be resistant to gamma radiation. For example, in one study, no changes in sensitivity or response times were observed after diamond was exposed to ^{60}Co at dose levels as high as 250 MRad [101,102].

Additionally, in studies where diamond was exposed to high levels of gamma rays, fast neutrons, thermal neutrons, and alpha particles produced by the B-10(n, α)Li-7 reaction in a high-flux nuclear reactor core (thermal and fast flux on the order of 10^{14} neutrons/cm^2-s) for 1 month, there was little damage to the crystal after annealing at 600°C [103]. This study is significant because the total dose to the diamond in an operating high-flux nuclear reactor over 1 month is far greater than the dose expected during a 20 or so year lifetime of a DCCS powered by Kr-85. The gamma rays from Kr-85 are not particularly hard (514 keV). The diamond substrate is thin, so the number of interactions that the gamma rays will have in the PV cell volume is small. The main interaction will be Compton scattering. Thus, for the population density of electrons with energies greater than 200 keV, the energy cut-off for the creation of primary knock-on atom will not be high [7]. Furthermore, the elevated diamond cell temperature will enhance the annealing repair of Frankel defects, which will further inhibit radiation damage effects. Thus, the performance degradation over time from gamma radiation should be minimal. The high operating temperature of the PIDEC top cycle is a critical feature for the dual-cycle approach as previously discussed.

A top cycle operating temperature between 873 and 900 K was selected to be conservative on the choice of high-temperature pressure vessel materials. This temperature range is sufficient to power the NASA-tested free-piston Stirling engine (FPSE) as the bottom cycle (Fig. 17.9) [104]. It should be noted that even though a temperature range of 873—900 K was chosen for the base design, it is feasible to push the design criteria, using advanced pressure vessel materials, toward higher operating temperatures

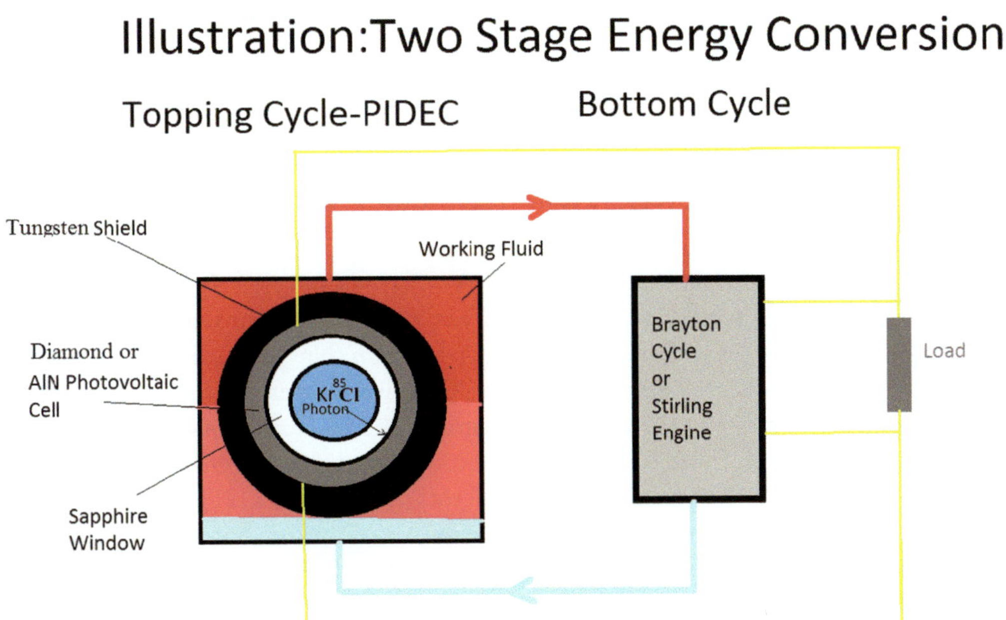

Figure 17.9 An illustration of a two-stage cycle based on a PIDEC top cycle using a KrCl* excimer fluorescer and a bottom cycle using either a Mini-BRU or a free-piston Stirling engine. *Source: authors.*

(>1000 K), which would be sufficient for the NASA-tested Mini-BRU (which has a slightly higher efficiency than the FPSE).

The study demonstrated that a DCCS device using Kr-85 and a FPSE with a thermal power of 1322 W thermal power produced electricity at an estimated efficiency of 45%. Thus, the electrical power was estimated to be 595 W. The specific power for a device without a shield (shielding strategy 1) is then 6.36 W/kg; the mass of a 4.7-cm or a 6-cm shadow shield reduces the specific power to 2.65 W/kg (shielding strategy 2) or 2.28 W/kg (shielding strategy 3), respectively.

In contrast, a Mini-BRU designed for an operating temperature of 873 K is more efficient than a FPSE. In this case, a 30% efficient Mini-BRU used with the 1322-W thermal power DCCS produces electricity at an estimated efficiency of 51%. Thus, the electrical power produced by this configuration is estimated to be ~674 W. So, the estimated specific power without a shield (shielding strategy 1) would be 7.25 W/kg. If a 4.7-cm shadow shield is used, that would lead to a DCCS with a specific power of 3.02 W/kg, and a 6-cm shadow shield would lead to a DCCS with a specific power of 2.58 W/kg.

17.4 Practical aspects of space nuclear power systems

The development of self-sustaining nuclear-based reactor power will meet the needs of missions that are not considered because of the low power output of RTGs. High power requirements will be needed to support activities such as scientific experimentation, in situ mining and processing, astronomical observations, and surface explorations. This section of the chapter addresses practical aspects of space nuclear power (SNP) systems including systems used in space, safety aspects, and regulations.

17.4.1 Space reactor power system technologies

The first and only US reactor to be flown into space was SNAP-10A. The SNAP-10A design was based on SNAP-10 [105] and SNAP-2 [81] concepts. It was designed to produce 500 We of reactor power. SNAP-10A was a thermal reactor that used uranium-zirconium hydride (U-ZrH_x) fuel and a sodium-potassium (Na—K) liquid-metal primary coolant. It was launched into space on April 3, 1965, and it operated for 43 days until it shut down due to a failure in the voltage regulator in the spacecraft. SNAP-10A reactor demonstrated that it was feasible to operate a liquid-cooled reactor remotely in space.

SNAP-2 and SNAP-8 were also thermal reactors that used U-ZrH_x fuel and a Na—K coolant but were coupled to a Rankine-cycle power conversion system. These two reactors were tested but never flown into space. The United States developed the SP-100 system between 1978 and 1995. SP-100 was designed to be able to couple with various static and dynamic power systems. It had a modular design, so the reactor power level could be varied by changing the number of fuel pins and the number of thermoelectric converts. The SP-100 reactor could also be configured in several arrangements for lunar and Mars applications. During development, several systems designs evolved to incorporate features that can enhance survivability during launch [106,107].

The SP-100 was based on using a fast spectrum, lithium-cooled reactor as a heat source, and SiGe doped with GaP thermoelectric modules for power conversion. The fuel pins were made from uranium nitride (UN) enclosed in Nb—Zr fuel cladding. Lithium was used as a coolant and flowed into the reactor vessel through heat pipes. The reactor heat was removed by circulating lithium, which was heated to about 1350 K. In the SP-100 SRPS, the heated lithium flowed into a thermoelectric electromagnetic pump and onto the SiGe/GaP thermoelectric power converters.

Reactor control was achieved using beryllium oxide (beryllia or BeO) reflectors. The in-core safety rods used B_4C material to achieve shutdown and reactor subcriticality during accident conditions. SP-100 was designed to prevent inadvertent criticality during handling or in accident situations. It has two independent control elements that are locked in their shut-down positions during transport, handling, launch, ascent, and final orbit acquisition. The SP-100 study shows that it could be integrated with a Stirling engine with a predicted eightfold increase in thermal to electric efficiency [108]. Some studies used SP-100 with a CBC and reported improvements of 16%—18% compared to 4%—5% when TE was used for a 10—100 kWe power range [79]. The program was dismantled before the reactor was ever tested.

Most of the USSR reactors are coupled with thermoelectric generators and used for short-term LEO military satellites. USSR developed the Romashka, a reactor with a thermoelectric converter similar to SNAP-10 and TOPAZ thermionic reactors. The TOPAZ power system uses thermionic energy conversion to generate electrical energy. TOPAZ was the first thermionic nuclear reactor in the world. TOPAZ I used a multicell construction and was developed in Moscow by Red Star. Red Star also developed the "Bouk reactor" that was flown on 33 Radar Ocean Reconnaissance Satellite missions [109]. The USSR flew at least two TOPAZ 1 systems in space: Cosmos 1818 operated for 6 months, and Cosmos 1867 operated for a year. The Cosmos satellites were powered by 10 KWe based on TOPAZ design [109]. The difference between Cosmos 1818 and Cosmos 1867 systems is that the former used thermionic fuel elements (TFE) emitters without tungsten coatings, while the later used TFE tungsten coated with W-184, which limited its operation because it relied on twice as much cesium to produce the same power output as the first system. Its operation was limited because it ran out of cesium [70]. It also used urania fuel, cathodes from tungsten alloy, and anodes from niobium alloy VN-2, beryllia insulators, and cesium vapor in the interelectrode gap [109].

TOPAZ II used a single-cell construction of the TFEs and had an average conversion efficiency of 5%. The single-cell TFE in TOPAZ II enables nonnuclear testing of the unit in which the tungsten electric heater could be replaced with nuclear fuel. TOPAZ II could generate between 4.5 and 5.5 kWe for 3 years of continuous and autonomous operation in space. The reactor was fueled with UO_2 fuel pellets enriched with 96%U-235. The reactor contains 27 kg of U-235 and was moderated with ZrH1.85r and used NaK coolant. The reactor core had a dimension of 36.5 cm high and 26 cm in diameter. TOPAZ II was never flight demonstrated, but the system was extensively tested [110]. TOPAZ 3 reactor provides greater power levels, while TOPAZ 4 had an improved multicell thermionic fuel element [111]. The USSR may have also worked on several other reactor concepts, but most of this information is classified.

The Kilowatt Reactor Using Stirling Technology (KRUSTY) is a prototypic nuclear power reactor currently under development in the United States. It is a compact, low-cost, scalable fission power system that is currently being developed for science and exploration. KRUSTY is a prototype nuclear reactor test of a 5 kW kilopower space reactor. The reactor comprises the core, the heat pipes, a reflector, absorber rod, and shield. It uses HEU U7.5Mo fuel and a BeO reflector. The heat pipe is made from Haynes 230 and uses sodium as the working fluid. It contains a molybdenum material diffusion barrier placed between the Haynes and the UMO. It uses B4C neutron shield and SS316 gamma shield. The reactor is solid state, with the control rod being the only moving part. Experimental results with the KRUSTY nuclear system conducted at various temperatures and power levels for 28 conservative hours show that the system operated as expected, and the reactor is highly tolerant to possible failure conditions and transients. Each of the Stirling converters produced ~90 We and component efficiency of ~35% and overall system efficiency of ~25% when the KRUSTY was testing at thermal energies between 1.5 and 5 kWt with a fuel temperature of 880°C [112]. Potential missions that could use the KRUSTY include surface missions such as missions to the Moon and Mars and deep-space missions such as Pluto orbiter and survey of Titan-Saturn system. Table 17.7 shows the principal reactors that were developed in the United States [5,112].

17.4.2 Safety of nuclear space power

The presence of nuclear sources in space and the consequences of potential harm to people and the environment from a nuclear incident require that safety should be incorporated into the design of RPSs. All RTGs that have flown in space have used Pu-238, which is a hazardous material. Pu-238 is an alpha emitter, and the primary hazard from alpha producing radioisotopes is the inhalation of fine powders. Plutonium possesses a serious health risk when inhaled, accumulating in the lungs, liver, and bones, inducing cancer [113,114].

Three US spacecraft accidents have occurred with RPSs. In all three cases, the RPS performed as designed. In 1964, the Transit-5-BN-3 mission was aborted because of a launch vehicle failure resulting in the burn-up during reentry. Some amount of the plutonium fuel was released into the upper atmosphere. In May 1968, another accident

Table 17.7 Principal US space nuclear reactors.

Power plant	Purpose	Power level	Operating temp. (K)	Period	Type reactor	Fuel	Converter	Development level
Rover (inside NERVA)	Propulsion	365–500 MWt	2450	1955–73	Epithermal	UC	–	Twenty reactors tested. Demonstrated all components of flight engine >2 h. Ready for flight engine development
Fluidized bed reactor	Propulsion	1000 MWt	3000	1958–73	Thermal	UC-ZrC	–	Cold flow, bed dynamics experiments successful
Gaseous core reactors	Propulsion electricity	4600 MWt	1000–1500	1959–78	Fast	Uranium plasma UF_6	Brayton	Successful critical assembly of UF_6
SNAP-2	Electricity	3 kWe	820	1957–63	Thermal	Uranium zirconium hydride	Mercury Rankine	Precursor for SNAP-10A
SNAP-10A	Electricity	0.5 kWe	810	1960–66	Thermal	Uranium zirconium hydride	Thermoelectric	Flight tested reactor 43 days
SNAP-8	Electricity	30–60 kWe	975	1960–70	Thermal	Uranium zirconium hydride	Mercury Rankine	Tested two reactors. Demonstrated 1 year operation
Advanced hydride reactors	Electricity	5 kWe	920	1970–73	Thermal	Uranium zirconium hydride	Thermoelectric and Brayton	PbTe thermoelectric test for 42,000 h
SNAP-50	Electricity	300–1200 kWe	1365	1962–65	Fast	UN, UC	Potassium Rankine	Fuels tested to 6000 h
Advanced metal cooled reactor	Electricity	300 kWe	1480	1965–73	Fast or thermal driver	UO_2 U–ZrC	In-core thermionics	Thermionic demonstrated >1 year operation
Nuclear electric propulsion	Electricity	400 KWe	1675	1974–81	Fast	UO_2	Out of core thermionics	Limited testing on thermionic elements
SPAR/SP-100	Electricity	100 kWe	1500	1979 to present	Fast	UO_2	Thermoelectric	Limited testing on core heat pipes and advanced thermoelectric materials
KRUSTY	Electrical	1 and 10 kWe	1123	2015 to present	Thermal	U–8Mo	Stirling power conversion	Prototypic nuclear-powered test of a 5-kW_t space reactor

Adapted from Refs. [5,14].

occurred during the launching of the Nimbus B-1. The heat source was recovered about 90 m under water off the coast of California. No plutonium was released. The fuel was retrieved, repurposed, and used in later missions. In April 1970, the Apollo 13 mission to the Moon was aborted as a result of an oxygen tank explosion in the spacecraft service module. Upon reentry into the Earth's atmosphere, the lunar excursion module with a SNAP-27 RTG exploded into the South Pacific Ocean and the heat source fell into the ocean. However, there was no release of nuclear fuel [52]. Following these accidents, NASA reevaluated its potential mission accident scenarios and a series of safety test programs were initiated to ensure that any technology that is used in space demonstrated that it is safe during any potential accident scenario. Safety was incorporated into every phase of the design, testing, manufacturing, and operation of space nuclear systems.

Plutonium dioxide, the heat source used in RPS, is protected by housing it in the GPHS, designed with safety in mind to keep the fuel contained or immobilized to prevent inhalation or ingestion. Multiple layers of protection exist to protect the fuel. The GPHS consists of the fuel, the fuel cladding, the GIS, the CBCF insulation, and the FWPF aeroshell. Each GPHS module contains four fuel pellets made of high-temperature PuO_2 ceramic. Plutonium dioxide is used in a ceramic form so that it would only break primarily into large, noninhalable, and nonsoluble pieces, rather than fine particles that could be harmful to people or damage the environment. The ceramic form of the fuel is heat resistant and has low solubility in water.

The GPHS source is subjected to various tests to assess its performance under a wide range of accidental conditions such as launched pad explosions, projectile impacts, propellant fires, impacts, and atmospheric reentry. Two fuel clads are encased in a cylindrical GIS made of FWPF, a carbon-carbon composite material. The GIS is designed to protect the fuel clads from postulated impacts. A CBCF insulator surrounds each GIS and limits the peak temperature of the fuel clad during inadvertent reentry. The aeroshell is designed to contain the two GIS assemblies under a wide range of postulated reentry scenarios. The aeroshell also protects the fuel clads from postulated launched vehicle explosion overpressures and fragment impacts and protects it in the event of a propellant fire. The RPS designed by the United States is generally considered to be very safe and reliable. Every RPS that has been flown by the United States has worked as designed and has exceeded its operational design life.

17.4.3 Regulation of space nuclear power

There are stringent regulations in place for SNP. Each nation is responsible for the safety of its source to prevent accidental release of radiation. According to the United Nations resolutions on the use of nuclear power sources in outer space, sources should only be used based on a thorough safety assessment, including probabilistic risk analysis, with a particular interest in the risks of accidental exposure of the public to harmful radiation or radioactive material. The United Nations resolution requires that "States launching in space onboard ensure with a high degree of confidence that the hazards, in foreseeable operational or accidental circumstances, are kept below acceptable levels. States are required to bear the international responsibility for national activities involving the use of nuclear power sources in outer space, whether such activities are carried on by governmental agencies or by non-governmental entities, and for assuring that such national activities are carried out in conformity with that Treaty and the recommendations contained in these Principles" [115,116]. The United States and Russia are members of the UN Security Council and are bound by this international law. The International Atomic Energy Agency has a safety framework that provides high-level guidance that addresses unique nuclear safety considerations for relevant launch, operation, and end-of-service mission phases of SNP applications [117].

In the United States, the DOE and the NRC are tasked with the possession, use, and production of nuclear materials and facilities. DOE is subject to the National Environmental Policy Act (NEPA) that requires that agencies take into account the impact on the environment from the nuclear material operation, transportation, and storage [118]. The DOE provides the environmental impact statement that NASA uses to show compliance with NEPA. Only the DOE is authorized to own SNP systems. As a result, NASA must work with the DOE to manufacture, launch, and operate RPSs in space. The DOE is responsible for research, technologic development, design, production, testing, evaluation, delivery, space vehicle integration, launch, and operation of the RPS. The DOE has established rules, specifications, development, testing, transport, and handling of the RPS. The safety approach used in the United States for nuclear facilities can be found on a DOE website [119].

The United States flight safety review and launch approval process for space power were established by the Presidential Directive/National Security Council Memorandum 25 (PD/NSC-25, 1977) [120]. The regulations specify that safety should be engineered into systems during their design and development, and systems and processes should be designed and implemented to reduce radiation exposures to as low as reasonably achievable. As part of this process, the DOE prepares a series of detailed safety analysis reports (SARs) that characterize the radiological risk for each mission. Each space mission involving the launch of a nuclear system shall report its SAR to the DOE. These SARs shall include a Preliminary Safety Analysis Report, an Updated Safety Analysis Report, and a Final Safety Analysis Report. The SAR is very extensive and includes all aspects of mission safety. Each safety analysis report includes the following: (1) A reference document report that contains a description of the mission and flight summary, power conversion subsystem, ground support equipment, mission profile, launch vehicle, reference trajectory and launch site, range, and radiological safety operations, and safety-related to systems and subsystems; (2) an accident model document that contains accident and radiological models and data, vehicle and reactor failure mode analysis, nuclear response to accident environment and mission failure evaluation; and (3) a nuclear risk analysis document that includes the prelaunch, launch, and ascent phases and orbit and/or flight trajectory phases. The entire approval process takes about 3 years, although it could take as long as 8 years.

17.5 Conclusions: Future of nuclear power technologies

In this chapter, we have discussed the available radioactive sources for space applications, critical components of SNP systems, future SNP systems, and safety as well as regulations regarding use of SNP. Space nuclear power systems are desirable because they offer high power densities and provide continuous, reliable, maintenance-free, and long-lasting power independent of the distance from the Sun and the radiation environment. Thermoelectric conversion using RTG is the dominant method of SNP. The RTG has been used to provide safe and reliable power for several space missions and has demonstrated long-term reliability. There is ongoing research to develop better thermoelectric material to improve the conversion efficiency of thermoelectric conversion, i.e., RTG, which is driven by the need to reduce the amount of Pu-238 used in current missions.

Different RPSs are also considered because of the need to provide a range of power options for different space missions. Some of these technologies, although promising, have not established long-term reliability. The development of self-sustaining nuclear-based reactor power will meet the need of missions that are not considered because of the low power output of RTGs. High power requirements will be needed to support activities such as scientific experimentation, in situ mining and processing, astronomical observations, and surface exploration. The PIDEC is another technology that may find use in space PV power systems. Photon intermediate direct energy conversion separates PV device materials from ionizing radiation with shielding, thus eliminating concerns over radiation damage; the device is also capable of operating as a topping cycle, thus improving the overall efficiency of an energy conversion system based on multicycles.

References

[1] S.G. Bailey, D.J. Flood, Space photovoltaics, Prog. Photovolt. Res. Appl. 6 (1) (1998) 1–14.

[2] S.G. Bailey, R.P. Raffaelle, Chapter II-4-B – Operation of solar cells in a space environment, in: S.A. Kalogirou (Ed.), McEvoy's Handbook of Photovoltaics Fundamentals and Applications, third ed., Academic Press, New York, NY, USA, 2012, pp. 987–1003.

[3] A. Bermudez-Garcia, P. Voarino, O. Raccurt, Environments, needs and opportunities for future space photovoltaic power generation: A review, Appl. Energy 290 (2021), 116757.

[4] S. Ulamec, J. Piele, E. Trollope, How to survive a Lunar night, Planet. Space Sci. 58 (14–15) (2010) 1985–1995.

[5] G.L. Bennett, Mission interplanetary: Using radioisotope power to explore the solar system, Energy Convers. Manag. 49 (3) (March 2008) 382–392.

[6] D. Beretta, N. Neophytou, J.M. Hodges, M.G. Kanatzidis, D. Narducci, M. Martin- Gonzalez, M. Beekman, et al., Thermoelectrics: from history, a window to the future, Mater. Sci. Eng. R Rep. 138 (October 2019), 100501.

[7] M. Prelas, M. Boraas, F. De La Torre Aguilar, J.D. Seelig, M.T. Tchouaso, D. Wisniewski, Nuclear Batteries and Radioisotopes, Springer International Publishing, 2016.
[8] X. Wang, R. Liang, P. Fisher, W. Chan, J. Xu, Critical design features of thermal-based radioisotope generators: A review of the power solution for polar regions and space, Renew. Sustain. Energy Rev. 119 (March 2020), 109572.
[9] E.D. Arnold, Handbook of Shielding Requirements and Radiation Characteristics of Isotopic Power Sources for Terrestrial, Marine, and Space Applications, Report ORNL-3576,, April 1, 1964, https://doi.org/10.2172/4028157. DOE Contract Number W-7405-ENG-26.
[10] A. Younes, G. Montavon, C. Alliot, M. Mokili, F. Haddad, D. Deniaud, J. Champion, A route for polonium 210 production from alpha-particle irradiated bismuth-209 target, Radiochim. Acta 102 (8) (2014) 681–689.
[11] K.C. Jordan, J.R. Birden, Thermal Batteries Using Polonium-210, Report MLM-984, June 2, 1954, https://doi.org/10.2172/4406897.
[12] N. Shinohara, Y. Hatsukawa, K. Hata, N. Kohno, Measurement of fission product yields from neutron-induced fission of americium-241, J. Nucl. Sci. Technol. 36 (3) (March 1999) 232–241.
[13] D.T. Rankin, W.R. Kanne Jr., M. Louthan Jr., D.F. Bickford, J.W. Congdon, Production of Pu-238 Oxide Fuel for Space Exploration, 2000. Available at: https://cntaware.org/wp-content/uploads/2019/12/179rankin.pdf.
[14] W.W. Hoover, R.L. McNutt Jr., Radioisotope Power Systems: An Imperative for Maintaining U.S. Leadership in Space Exploration, National Research Council, National Academies Press, 2009, p. 74. Available at: http://www.nap.edu/catalog/12653.html.
[15] S.P. Melnikov, A.A. Sinyanskii, A.N. Sizov, G.H. Miley, Lasers with Nuclear Pumping, Springer, 2015.
[16] L. Summerer, K. Stephenson, Nuclear power sources: A key enabling technology for planetary exploration (2), in: Proc. Inst. Mech. Eng. Part G: J. Aerospace Eng., vol 225, 2011, pp. 129–143, https://doi.org/10.1243/09544100JAERO766. Available online at:.
[17] K. Stephenson, M. Landgraf, Hybrid radioisotope-solar power systems as a key to sustained lunar exploration, 2019, in: European Space Power Conference (ESPC), 2019, pp. 1–4, https://doi.org/10.1109/ESPC.2019.8932065. Available at: https://ieeexplore.ieee.org/document/8932065.
[18] H.D. Raleigh, Systems for Nuclear Auxiliary Power (SNAP): A Literature Search, U.S. Atomic Energy Commission, Division of Technical Information, Oak Ridge, TN, June 1964. Available at: https://www.osti.gov/servlets/purl/4034092.
[19] W.R. Corliss, D.G. Harvey, Radioisotopic Power Generation, Prentice-Hall, Inc., Englewood Cliffs, NJ, USA, 1964, p. 304.
[20] E.S. Pedersen, Heat-Sterilizable Power Source Study for Advanced Mariner Missions, Technical Memorandum 33-180, Caltech Jet Propulsion Laboratory, July 1, 1964. Pasadena, CA, Available at: https://ntrs.nasa.gov/citations/19660014188.
[21] J. Chan, J.G. Wood, J.G. Scheiber, Development of Advanced Stirling Radioisotope Generator for Space Exploration, NASA TM-2007-214806, May 2007. Available at: https://ntrs.nasa.gov/api/citations/20070021756/downloads/20070021756.pdf.
[22] W.A. Wong, J.G. Wood, K. Wilson, Advanced Stirling Convertor (ASC)—From Technology Development to Future Flight Product, NASA TM-2008-215282, December 2008. Available at: https://ntrs.nasa.gov/citations/20090007956.
[23] D.J. Anderson, W.A. Wong, K.L. Tuttle, An Overview and Status of NASA's Radioisotope Power Conversion Technology, NASA TM-2005-213980, AIAA–2005–5713, November 2005. Available at: https://permanent.fdlp.gov/lps123484/TM-2005-213980.pdf.
[24] G.F. Knoll, Radiation Detection and Measurement, fourth ed., John Wiley & Sons, Hoboken, NJ, USA, 2010.
[25] W.W. Ward, Developing, Testing, and Operating Lincoln Experimental Satellites 8 and 9 (LES-8/9), M.I.T. Lincoln Laboratory Technical Note, January 16, 1979, 1979-3.
[26] R.L. Heacock, The Voyager spacecraft, Proc. Inst. Mech. Eng. 194 (1) (1980) 211–224.

[27] A. Schock, Design Evolution and Verification of the General-Purpose Heat Source, Report No. DOE/ET/33021-T1, Fairchild Space and Electronics Co., Germantown, MD, USA, January 1, 1980, https://doi.org/10.2172/5355852.

[28] T.A. Cull, General-Purpose Heat Source Development: Extended Series Test Program Large Fragment Tests, Report LA-11597-MS, August 1, 1989, https://doi.org/10.2172/5857636.

[29] G.L. Bennett, R.J. Hemler, A. Schock, Development and Use of the Galileo and Ulysses Power Sources, Fairchild Space Company; National Aeronautics and Space Administration, Jerusalem, Israel, October 9–14, 1994, https://doi.org/10.2172/1033366. IAF-94-R.1.362.

[30] G.H. Rinehart, Design characteristics and fabrication of radioisotope heat sources for space missions, Prog. Nucl. Energy 39 (3–4) (2001) 305–319.

[31] G. Bennett, J. Lombardo, R. Hemler, G. Silverman, C. Whitmore, et al., Mission of daring: The general-purpose heat source radioisotope thermoelectric generator, in: AIAA 2006-4096, 4th International Energy Conversion Engineering Conference and Exhibit (IECEC), June 26–29, 2006, https://doi.org/10.2514/6.2006-4096. San Diego, CA.

[32] K.-P. Wenzel, R. Marsden, D. Page, E. Smith, Ulysses: The first high-latitude heliospheric mission, Adv. Space Res. 9 (4) (1989) 25–29.

[33] D.L. Matson, L.J. Spilker, J.-P. Lebreton, The Cassini/Huygens mission to the Saturnian system, in: The Cassini-Huygens Mission, 2003, pp. 1–58.

[34] Y. Guo, R.W. Farquhar, New Horizons mission design, Space Sci. Rev. 140 (1–4) (2008) 49–74.

[35] E. Velmre, Thomas Johann Seebeck (1770–1831), Est. J. Eng. 13 (4) (2007) 276–282.

[36] M.H. Elsheikh, D.A. Shnawah, M.F.M. Sabri, S.B.M. Said, M.H. Hassan, M.B.A. Bashir, M. Mohamad, A review on thermoelectric renewable energy: Principle parameters that affect their performance, Renew. Sustain. Energy Rev. 30 (2014) 337–355.

[37] A. Mazzetti, M.G. Pret, G. Pinarello, L. Celotti, M. Piskacev, A. Cowley, Heat to electricity conversion systems for moon exploration scenarios: A review of space and ground technologies, Acta Astronaut. 156 (March 2019) 162–186.

[38] C. Kelly, MHW RTG performance during LES 8/9 and Voyager missions, in: Transactions of the Fourth Symposium on Space Nuclear Power Systems (SEE N88-24254 17-73), New Mexico Univ., 1987, pp. 197–200.

[39] F. Ritz, C.E. Peterson, Multi-mission radioisotope thermoelectric generator (MMRTG) program overview, in: 2004 IEEE Aerospace Conference Proceedings, vol. 5, 2004, pp. 2950–2957, https://doi.org/10.1109/AERO.2004.1368101.

[40] J. Yang, T. Caillat, Thermoelectric materials for space and automotive power generation, MRS Bull. 31 (3) (March 2006) 224–229.

[41] M.S. El-Genk, Energy conversion technologies for advanced radioisotope and nuclear reactor power systems for future planetary exploration, in: Twenty-First International Conference on Thermoelectrics (Proceedings ICT '02), 2002, pp. 375–380, https://doi.org/10.1109/ICT.2002.1190342.

[42] T. Caillat, J.-P. Fleurial, G. Snyder, A. Borshchevsky, Development of high efficiency segmented thermoelectric unicouples, in: Paper Presented at the Proceedings ICT2001. 20 International Conference on Thermoelectrics (Cat. No. 01TH8589), 2001.

[43] M.S. El-Genk, H.H. Saber, J. Sakamoto, T. Caillat, Life tests of a skutterudites thermoelectric unicouple (MAR-03), in: Proceedings 22nd International Conference on Thermoelectrics (ICT'03), 2003, pp. 417–420, https://doi.org/10.1109/ICT.2003.1287537.

[44] M.S. El-Genk, H.H. Saber, T. Caillat, Efficient segmented thermoelectric unicouples for space power applications, Energy Convers. Manage. 44 (11) (2003) 1755–1772.

[45] Y. Cao, X. Zhao, T. Zhu, X. Zhang, J. Tu, Syntheses and thermoelectric properties of Bi_2Te_3/Sb_2Te_3 bulk nanocomposites with laminated nanostructure, Appl. Phys. Lett. 92 (14) (2008), 143106.

[46] Y. Ma, Q. Hao, B. Poudel, Y. Lan, B. Yu, D. Wang, G. Chen, Z. Ren, Enhanced thermoelectric figure-of-merit in *p*-type nanostructured bismuth antimony tellurium alloys made from elemental chunks, Nano Lett. 8 (8) (2008) 2580–2584.

[47] B. Poudel, Q. Hao, Y. Ma, Y. Lan, Y.A. Minnich, et al., High-thermoelectric performance of nanostructured bismuth antimony telluride bulk alloys, Science 320 (5876) (2008) 634–638.

[48] K.F. Hsu, S. Loo, F. Guo, W. Chen, J.S. Dyck, C. Uher, et al., Cubic $AgPb_mSbTe_{2+m}$: bulk thermoelectric materials with high figure of merit, Science 303 (5659) (2004) 818–821.
[49] S. Bathula, M. Jayasimhadri, B. Gahtori, A. Kumar, A. Srivastava, A. Dhar, Enhancement in thermoelectric performance of SiGe nanoalloys dispersed with SiC nanoparticles, Phys. Chem. Chem. Phys. 19 (36) (2017) 25180–25185.
[50] T. Harman, M. Walsh, G. Turner, Nanostructured thermoelectric materials, J. Electron. Mater. 34 (5) (2005) L19–L22.
[51] W. Standring, M. Dowdall, M. Sneve, Ø.G. Selnæs, I. Amundsen, Environmental, health and safety assessment of decommissioning radioisotope thermoelectric generators (RTGs) in northwest Russia, J. Radiol. Prot. 27 (3) (2007) 321.
[52] G.R. Schmidt, T.J. Sutliff, L.A. Dudzinski, Radioisotope power: A key technology for deep space exploration, in: B.N. Singh (Ed.), Radioisotopes – Applications in Physical Sciences, Intech Open, 2011, pp. 419–456, https://doi.org/10.5772/22041.
[53] R.L. Cataldo, G.L. Bennett, US space radioisotope power systems and applications: Past, present and future, in: B.N. Singh (Ed.), Radioisotopes - Applications in Physical Sciences, Intech Open, 2011, pp. 473–496, https://doi.org/10.5772/23914.
[54] T. Coutts, A review of progress in thermophotovoltaic generation of electricity, Renew. Sustain. Energy Rev. 3 (2) (1999) 77–184.
[55] D.G. Gritton, R.C. Bourke, Radioisotope-photovoltaic energy conversion system, Adv. Energy Convers. 5 (2) (July 1965) 119–145.
[56] J.K. Tong, W.-C. Hsu, Y. Huang, S.V. Boriskina, G. Chen, Thin-film 'thermal well' emitters and absorbers for high-efficiency thermophotovoltaics, Sci. Rep. 5 (2015) 10661, https://doi.org/10.1038/srep10661.
[57] Best Research-Cell Efficiency Chart. https://www.nrel.gov/pv/cell-efficiency.html/. (Accessed 10 January 2022).
[58] L.M. Fraas, J.E. Avery, J. Martin, V.S. Sundaram, G. Girard, et al., Over 35-percent efficient GaAs/GaSb tandem solar cells, IEEE Trans. Electron. Dev. 37 (2) (February 1990) 443–449, https://doi.org/10.1109/16.46381.
[59] A. Datas, Optimum semiconductor bandgaps in single junction and multijunction thermophotovoltaic converters, Sol. Energy Mater. Sol. Cell. 134 (2015) 275–290.
[60] A. Datas, D.L. Chubb, Thermophotovoltaic energy conversion in space, in: S.G. Bailey, A.F. Hepp, D.C. Ferguson, R.P. Raffaelle, S.M. Durbin (Eds.), Space Photovoltaics: Materials, Missions and Alternative Technologies, Elsevier, Cambridge, MA, USA, 2022, pp. 197–214.
[61] Z. Omair, G. Scranton, L.M. Pazos-Outón, T.P. Xiao, M.A. Steiner, et al., Ultraefficient thermophotovoltaic power conversion by band-edge spectral filtering, Proc. Nat. Acad. Sci. 116 (31) (July 30, 2019) 15356–15361, https://doi.org/10.1073/pnas.1903001116.
[62] A.C. Day, W.E. Horne, M.D. Morgan, Application of the GaSb solar cell in isotope-heated power systems, in: IEEE Conference on Photovoltaic Specialists, vol. 2, 1990, pp. 1320–1325, https://doi.org/10.1109/PVSC.1990.111826.
[63] A. Schock, M. Mukunda, C.T. Or, V. Kumar, G. Summers, Radioisotope Thermophotovoltaic (RTPV) Generator and its Application to the Pluto Fast Flyby, Report FSC-ESD-217-93-519, January 16, 1994, https://doi.org/10.2172/1033373.
[64] D. Wilt, D. Chubb, D. Wolford, P. Magari, C. Crowley, Thermophotovoltaics for space power applications, in: AIP Conf. Proc, vol. 890, 2007, p. 335, https://doi.org/10.1063/1.2711751.
[65] A. Schock, C.T. Or, V. Kumar, Design and integration of small RTPV generators with new millennium spacecraft for outer solar system, Acta Astronaut. 41 (12) (1997) 801–816.
[66] W.E. Horne, M.D. Morgan, S.B. Saban, Performance tuned radioisotope thermophotovoltaic space power system, in: AIP Conf. Proc, vol 420, 1998, p. 1385, https://doi.org/10.1063/1.54764.
[67] P. Bermel, S.V. Boriskina, Z. Yu, K. Joulain, Control of radiative processes for energy conversion and harvesting, Opt. Express 23 (24) (2015) A1533–A1540, https://doi.org/10.1364/OE.23.0A1533.
[68] R. Allen, C. Skeen, Materials Requirements for Thermionic Energy Conversion, Report Number AED-Conf-63-036-3, American Societies for Metals, Western Metal Congress, Los Angeles, CA, USA, 18-21 March 1963. https://www.osti.gov/servlets/purl/4876316.

[69] G. Miskolczy, D.P. Lieb, Radioisotope thermionic converters for space applications, in: Proceedings of the 25th Intersociety Energy Conversion Engineering Conference, 1990, pp. 222–226, https://doi.org/10.1109/IECEC.1990.716874.
[70] S.S. Voss, in: TOPAZ II design evolution, in: space nuclear power and propulsion: Eleventh Symp., AIP Conf. Proc., vol. 301, 1994, pp. 791–802, https://doi.org/10.1063/1.2950271.
[71] NASA Radioisotope Power Systems. https://rps.nasa.gov/. (Accessed 10 January 2022).
[72] NASA Science Solar System Exploration. https://solarsystem.nasa.gov/. (Accessed 10 January 2022).
[73] L.S. Mason, Dynamic energy conversion: Vital technology for space nuclear power, J. Aero. Eng. 26 (2) (2013) 352–360.
[74] J.E. Davis, Design and Fabrication of the Brayton Rotating Unit, Contractor Report 1870, NASA HQ, Washington, D.C., March 1, 1972. https://ntrs.nasa.gov/citations/19720012386.
[75] R. Harty, W. Otting, C. Kudija, Applications of Brayton cycle technology to space power, IEEE Aero. Electron. Syst. Mag. 9 (1) (1994) 28–32.
[76] G.L. Bennett, J.A. Janzen, J.J. Lombardo, A.S. Mehner, The dynamic isotope power system technology program, in: Proc. Of the 23rd Intersociety Energy Conversion Engineering Conference, vol. 3, 1989, pp. 131–136, 31 July 1988 through 5 August 1988.
[77] G.L. Bennett, J.J. Lombardo, Technology development of dynamic isotope power systems for space applications, in: 22nd Intersociety Energy Conversion Engineering Conference, AIAA 1987-9049, August 10–14, 1987, https://doi.org/10.2514/6.1987-9049.
[78] L.S. Mason, A comparison of Brayton and Stirling space nuclear power systems for power levels from 1 kilowatt to 10 megawatts, NASA/TM 2001-210593, in: Space Technology and Applications International Forum—2001, AIP Conference Proceedings, vol 552, 2001, pp. 1017–1022, https://doi.org/10.1063/1.1358045.
[79] D.F. Owen, SP-100/Brayton power system concepts, in: Proceedings of the 24th Intersociety Energy Conversion Engineering Conference, vol. 2, 1989, pp. 1257–1262, https://doi.org/10.1109/IECEC.1989.74630.
[80] M. Hulin, J. Beall, Preliminary Results of the SNAP 2 Experimental Reactor, Report No. NAA-SR-5991, North American Aviation, Inc., Canoga Park, CA, USA, April 1, 1961, https://doi.org/10.2172/4048774.
[81] A.A. Jarrett, SNAP 2: Summary Report, Report No. AI-AEC-13068, North American Aviation, Inc., Canoga Park, CA, USA, July 27, 1973, https://doi.org/10.2172/4430852.
[82] A. Nice, S. Bradley, SNAP-8 electrical system, IEEE Trans. Aerospace 2 (2) (1964) 867–873.
[83] H.W. Savage, E.L. Compere, B. Fleischer, W.R. Huntley, R.E. MacPherson, A. Taboada, SNAP-8 Corrosion Program Summary Report, NASA CR-69822 (ORNL-3898), Oak Ridge National Lab., TN, December 1965.
[84] E.M. Douthett, SNAP-50/SPUR program, Prog. in Astronauts. Rocketry 16 (1966) 541–560.
[85] A.E. King, J.L. McCabria, C.C. Kouba, J.G. Hoop, SNAP 50/SPUR 467 kVA Experimental Generator Design Summary, Westinghouse Electric Corp., Lima, OH, USA, May 1, 1967, pp. 385, Aerospace Electrical Div., USAF Technical Report AFAPL TR-67-7.
[86] R. Brouns, E. Krueger, R. Niggemann, G. Sorenson, F. Russo, KIPS: Kilowatt isotope power system, Proc. Of the 12th Intersoc. Energy Conversion Engineering Conference, 1357–1361.
[87] M.A. Prelas, S.K. Loyalka, A review of the utilization of energetic ions for the production of excited atomic and molecular states and chemical synthesis, Prog. Nucl. Energy 8 (1) (1981) 35–52.
[88] M.A. Prelas, F.P. Boody, G.H. Miley, J.F. Kunze, Nuclear driven flashlamps, Laser Part. Beams 6 (01) (1988) 25–62, https://doi.org/10.1017/S0263034600003803.
[89] M. Prelas, J. Romero, E. Pearson, A critical review of fusion systems for radiolytic conversion of inorganics to gaseous fuels, Nucl. Technol. Fusion 2 (2) (1982) 143.
[90] M. Prelas, E. Charlson, E. Charlson, J. Meese, G. Popovici, T. Stacy, Diamond photovoltaic energy conversion, in: M. Yoshikawa, M. Murakawa, Y. Tzeng, W.A. Yarbrough (Eds.), Second International Conference on the Application of Diamond Films and Related Materials, 1993, pp. 329–334. MY – Tokyo, Japan.
[91] M.A. Prelas, M.T. Tchouaso, High efficiency dual-cycle conversion system using Kr-85, Appl. Radiat. Isot. 139 (2018) 70–80.

[92] R.J. Schott, Photon Intermediate Direct Energy Conversion Using a Sr-90 Beta Source, Ph.D. thesis, University of Missouri, 2012. Retrieved from: http://hdl.handle.net/10355/33103.
[93] C.L. Weaver, PIDECα: Photon Intermediate Direct Energy Conversion Using the Alpha Emitter Polonium-210, Ph.D. thesis, University of Missouri, 2012. Retrieved from: http://hdl.handle.net/10355/15908.
[94] R.J. Schott, C.L. Weaver, M.A. Prelas, K. Oh, J.B. Rothenberger, R.V. Tompson, D.A. Wisniewski, Photon intermediate direct energy conversion using a ^{90}Sr beta Source, Nucl. Technol. 181 (2) (2013) 5.
[95] C.L. Weaver, R.J. Schott, M.A. Prelas, D.A. Wisniewski, J.B. Rothenberger, E.D. Lukosi, K. Oh, Radiation resistant PIDEC alpha cell using photon intermediate direct energy conversion and a Po-210 source, Appl. Radiat. Isot. 132 (2018) 110–115, https://doi.org/10.1016/j.apradiso.2017.11.026.
[96] J.B. Javedani, M.A. Prelas, Concentrating properties of simple two-dimensional geometries for isotropic light, in: H. Hora, G.H. Miley (Eds.), Laser Interaction and Related Plasma Phenomena, Springer, 1986, pp. 155–165.
[97] M. Prelas, E. Charlson, F. Boody, G. Miley, Advanced nuclear energy conversion using a two step photon intermediate technique, Prog. Nucl. Energy 23 (3) (1990) 223–240.
[98] X. Yan, J. Wei, J. Guo, C. Hua, J. Liu, et al., Mechanism of graphitization and optical degradation of CVD diamond films by rapid heating treatment, Diam. Relat. Mater. 73 (2017) 39–46.
[99] M.A. Prelas, G. Popovici, L.K. Bigelow, Handbook of Industrial Diamonds and Diamond Films, Taylor & Francis, 1997.
[100] J.-C. Pu, S.-F. Wang J.C, J.C. Sung, High-temperature oxidation behavior of nanocrystalline diamond films, J. Alloys Compd. 489 (2) (January 2010) 638–644.
[101] C. Bauer, I. Baumann, C. Colledani, J. Conway, P. Delpierre, et al., Radiation hardness studies of CVD diamond detectors, Nucl. Instrum. Methods Phys. Res. Sect. A Accel. Spectrom. Detect. Assoc. Equip. 367 (1–3) (1995) 207–211.
[102] A. Brambilla, P. Chambaud, D. Tromson, P. Bergonzo, F. Foulon, Thin CVD diamond detectors with high charge collection efficiency, IEEE Trans. Nucl. Sci. 49 (1) (Feb. 2002) 277–280, https://doi.org/10.1109/TNS.2002.998653.
[103] S. Khasawinah, G. Popovici, J. Farmer, T. Sung, M. Prelas, J. Chamberlain, H. White, Neutron irradiation and annealing of 10 B doped chemical vapor deposited diamond films, J. Mater. Res. 10 (10) (1995) 2523–2530.
[104] L.S. Mason, J.G. Schreiber, A Historical Review of Brayton and Stirling Power Conversion Technologies for Space Applications, NASA/TM—2007-214976, November 2007. Retrieved from: http://gltrs.grc.nasa.gov.
[105] P. Merrill, D. Forejt, P. Pityk, G. Spira, P. Kueser, SNAP-10 experimental thermoelectric generator, Adv. Energy Convers. 2 (1962) 281–285.
[106] N.A. Deane, S.L. Stewart, T.F. Marcille, D.W. Newkirk, SP-100 reactor and shield design update, in: Proceedings of the Ninth Symposium on Space Nuclear Power Systems, AIP Conference Proceedings, vol 246, 1992, pp. 97–106, https://doi.org/10.1063/1.41857.
[107] S.F. Demuth, SP100 space reactor design, Prog. Nucl. Energy 42 (3) (April 2003) 323–359.
[108] H.W. Brandhorst Jr., P.A. Chapman Jr., New 5 kW free-piston Stirling space convertor developments, Acta Astronaut. 63 (1–4) (July–August 2008) 342–347.
[109] G.L. Bennett, Space nuclear power: Opening the final frontier, in: AIAA 2006-4191, 4th International Energy Conversion Engineering Conference and Exhibit (IECEC), June 26–29, 2006, https://doi.org/10.2514/6.2006-4191. San Diego, CA.
[110] S.S. Voss, The TOPAZ II space reactor response under accident conditions, LA-UR-93-4409 (CONF-940402-2), in: International Meeting Advanced Reactors Safety, Los Alamos National Lab., NM, USA, April 17–21, 1994, p. 9. Pittsburgh, PA, USA.
[111] V.P. Nickitin, B.G. Ogloblin, A.N. Luppov, N.N. Ponomarev-Stepnoi, V.A. Usov, Y.V. Nicolaev, J.R. Wetch, "TOPAZ-2" thermionic space nuclear power system and the perspectives of its development (2), in: Proceedings of the Eighth Symposium on Space Nuclear Power Systems, AIP Conf. Proc vol. 217, 1991, pp. 631–635, https://doi.org/10.1063/1.40116. Available at:.

[112] D.I. Poston, M.A. Gibson, R.G. Sanchez, P.R. McClure, Results of the KRUSTY nuclear system test, Nucl. Technol. 206 (Suppl. 1) (2020) S89−S117.
[113] G.L. Voelz, J. Lawrence, E.R. Johnson, Fifty years of plutonium exposure to the Manhattan Project plutonium workers: An update, Health Phys. 73 (4) (1997) 611−619.
[114] F.D.L.T. Aguilar, N.E. White, M.A. Prelas, R.V. Tompson Jr., S.K. Loyalka, Space nuclear power system accidents: Doses from Pu-238 and Am-241 inhalation, Prog. Nucl. Energy (2017) 171−182.
[115] U.N. General Assembly, Principles Relevant to the Use of Nuclear Power Sources in Outer Space, Res. 47/68, Dec. 14, 1992 available at: https://nuke.fas.org/space/principles.pdf.
[116] https://www.unoosa.org/oosa/en/ourwork/copuos/index.html. (Accessed 20 January 2022).
[117] International Atomic Energy Agency, Safety Framework for Nuclear Power Source Applications in Space, International Atomic Energy Agency, Vienna, Austria, 2009, pp. 17, available at: https://www.iaea.org/sites/default/files/safetyframework1009.pdf.
[118] National Environmental Policy Act. https://www.epa.gov/nepa/. (Accessed 20 January 2022).
[119] https://www.energy.gov/ehss/facility-safety/. (Accessed 20 January 2022).
[120] Presidential Directive NSC-25, Scientific or Technological Experiments with Possible Large-Scale Adverse Environmental Effects and Launch of Nuclear Systems into Space, The White House, Washington, D.C, December 14, 1977. Available at: https://irp.fas.org/offdocs/pd/pd25.pdf.

Index

Note: 'Page numbers followed by "*f*" indicate figures and "*t*" indicate tables.'

B

C

G

H

N

T

U

V

W

X

Y

Z